Robust Control Engineering

Robust Control Engineering
Practical QFT Solutions

Mario García-Sanz

CRC Press
Taylor & Francis Group
Boca Raton London New York

CRC Press is an imprint of the
Taylor & Francis Group, an **informa** business

CRC Press
Taylor & Francis Group
6000 Broken Sound Parkway NW, Suite 300
Boca Raton, FL 33487-2742

First issued in paperback 2021

© 2017 by Taylor & Francis Group, LLC
CRC Press is an imprint of Taylor & Francis Group, an Informa business

No claim to original U.S. Government works

ISBN 13: 978-1-03-209674-2 (pbk)
ISBN 13: 978-1-138-03207-1 (hbk)

Visit the Taylor & Francis Web site at
http://www.taylorandfrancis.com

and the CRC Press Web site at
http://www.crcpress.com

"If you want to find the secrets of the universe, think in terms of energy, **frequency** and vibration". Nikola Tesla (1856–1943).

"If you believe in the **frequency domain**, you believe in QFT". Isaac Horowitz to Mario García-Sanz (Pamplona, Spain), 2000.

"QFT provides one of the most practical robust control design methods available... It is now becoming a much more accessible and valuable method, mainly because the availability of good design software... Prof. Garcia-Sanz has developed a very easy to use design package". Michael Grimble, founder of the industrial control centre (Glasgow, UK), E-news August 2016.

"The most valuable companies in the future won't ask what problems can be solved with computers alone. Instead, they'll ask: how can computers help humans solve hard problems". Peter Thiel, co-fouder of PayPal and Palantir, in his book *Zero to One*, 2014.

"Thanks for making this amazing toolbox!!". "The QFT Control Tollbox proved to be an excellent tool for robust controller design!!". "One of the best courses I've taken!!". CWRU students who attended Prof. Garcia-Sanz QFT control courses.

"The investment in training will pay off in the long term. People can't deliver on what they do not know how to do... you have to upgrade capabilities... A well-trained team is often a company's most significant advantage—and an incompetent one its biggest liability". David Hoffeld, CEO Hoffeld Group, in his book *The Science of Selling*, 2016.

To Marta, María, Pablo, and Sofía

Contents

Preface

Control engineering is the discipline that applies mathematics, physics, and technology to the design of smart machines that control automatically physical systems. These smart machines, also known as *control systems*, use feedback loops with sensors that collect information of the plant being controlled, computers or circuits that based on this information calculate the actions to be applied, and actuators that modify the system in real time according to these actions. The system to be controlled can be a mechanism, an electrical network, a thermal process, a hydraulic or aerodynamic system, a chemical reactor, a spacecraft, or even a financial or biomedical system.

In the real world, the mathematical description of a system is not accurate and includes typically model uncertainty or even variation with time. *Robust control engineering* is the area of control engineering that takes into account this model uncertainty and designs reliable solutions to automatically control these physical systems.

Control engineering plays a primary role in the design of many engineering systems. I have worked in this field for over three decades. I started my journey in the 1980s, applying adaptive control solutions to industry with Professor Julian Florez at CEIT (Spain). During the 1990s, I had the privilege to work with Professor Peter Wellstead at UMIST (UK). Peter was one of the best control engineers and professors I ever met. He showed me the beauty of the frequency domain. Sadly, Peter passed away some months ago after an extraordinary and generous life dedicated to his students and to the most challenging control projects. Also in the 1990s, I had the honor to work with Professor Isaac Horowitz and Professor Constantine Houpis (AFIT, U.S. Air Force Institute of Technology). Isaac Horowitz was the pioneer of the robust control techniques, the father of the *quantitative feedback theory* (QFT), and one of the most influential control engineers in the history of control. Constantine (Dino) Houpis was the "teacher" and the driving force of QFT for many years. With these mentors and friends, I had the opportunity to develop new QFT control theory for multi-input multi-output plants, distributed parameter systems, time-delay processes, nonlinear switching control systems, unstable systems, and feedforward control. In addition, I have applied all these new ideas to many commercial control solutions for industry and space agencies, including satellites, wind turbines, water treatment plants, power systems, and radio telescopes with NASA-JPL, ESA-ESTEC, US-AFIT, NRAO-GBT, GMRT, Gamesa, Acciona, MTorres, IngeTeam, CENER, Eaton Corporation, Enercon, Siemens, Iberdrola, REE, Sener, EEQ, and many others.

This book, *Robust Control Engineering: Practical QFT Solutions*, summarizes my experience in the field. It presents the fundamentals of the QFT robust control methodology and its application to a large collection of real-world cases, all in 10 chapters, 10 appendices and over 50 detailed examples, case studies, projects, and problems. In particular, the book includes:

1. The fundamentals of the QFT robust control technique as well as advanced practical methodologies for a number of real-world control problems, including: (a) unstable systems, (b) time-delay and non-minimum phase systems, (c) distributed parameter systems, (d) plants with large model uncertainty, (e) high-performance

systems, (f) nonlinear systems, (g) multi-input multi-output plants, (h) systems with a different number of sensors and actuators, and (i) analog and digital controller implementation solutions.

2. A large collection of over 50 detailed real-world case studies, examples, and problems, including satellites with flexible appendages, multi-megawatt variable-speed wind turbines, wastewater treatment plants, large radio telescopes, active suspension systems, DVD controllers, central heating systems, hydraulic systems, DC motors, pasteurization processes with heat exchangers, interconnected electrical micro-grids, and distillation columns.

3. The original MATLAB® code of the algorithms and calculations used in the resolution of the examples presented in each chapter, including cases of stable and unstable systems, time-delay systems, distributed parameter systems, plants with large model uncertainty, switching control systems, nonlinear dynamic controllers, and multi-input multi-output plants.

4. The use of the *QFT control toolbox* (QFTCT) for MATLAB. This is the toolbox we developed and continuously improve to easily apply the QFT control theory to real-world projects with industry. Over the years, the toolbox has been widely used by industry, space agencies, research centers, and universities to design control solutions and servo systems, including the European Space Agency ESA-ESTEC, NASA-JPL, wind energy companies, large radio telescopes, power systems, and water treatment plant companies. The toolbox (1) deals with plants with model uncertainty, (2) is able to work with multi-objective performance specifications, (3) keeps the engineering understanding of the design in the frequency domain, and (4) gives solutions from simple PID regulators to more advanced control strategies when necessary. The user's guide for the QFTCT is in Appendix 2. Project files for some of the problems presented in this book and a demo version of the toolbox can be found at http://cesc.case.edu. The student and standard versions of the QFTCT is at http://codypower.com. For additional information, see http://crcpress.com.

5. The material presented in the book has been extensively classroom tested in undergraduate and graduate courses at several universities worldwide, as well as in special control courses for industry. Students have always evaluated these courses as excellent, receiving many teaching awards over the years.

Robust Control Engineering: Practical QFT Solutions is an ideal textbook for undergraduate and graduate students and control engineers. It presents practical methodologies to design reliable control systems, bridging the gap between successfully tested theory and real-world control system implementation. The book is indispensable for engineers and researchers designing reliable control solutions for industrial, energy, environmental, biomedical, chemical, electrical, mechanical, and aerospace applications. Unlike most books on the subject, this book presents a large collection of successful real-world cases and projects. The book can be used as a self-study reference by the engineer in practice and in undergraduate and graduate-level courses.

The book starts where a typical undergraduate control course ends. This means that it is assumed that the reader is familiar with differential equations, basic linear algebra, time-domain and frequency-domain control analysis and design, single-input single-output systems, and elementary state-space theory. This corresponds to the material in, for example, Houpis and Sheldon (2013), *Linear Control System Analysis and Design with MATLAB*, CRC Press; or Dorf and Bishop (2016), *Modern Control Systems*, Pearson, or any equivalent book.

Robust Control Engineering is both, a consolidated *Science* and a beautiful *Art* at the same time. Every control-engineering project combines science and art components. QFT bridges the gap between the control theory and the design of real-world solutions. I hope you enjoy this journey as much as I have enjoyed it.

Mario García-Sanz
Cleveland, Ohio

MATLAB® is a registered trademark of The MathWorks, Inc. For product information, please contact:

The MathWorks, Inc.
3 Apple Hill Drive
Natick, MA 01760-2098, USA
Tel: 508 647 7000
Fax: 508-647-7001
E-mail: info@mathworks.com
Web: www.mathworks.com

Acknowledgments

I would like to express special thanks to my former graduate PhD and Master students in advancing the state of the art of the QFT control field over the last 25 years, especially to Dr. Xabier Ostolaza, Dr. Juan Carlos Guillén, Dr. Montserrat Gil, Dr. Arturo Esnoz, Dr. Igor Egaña, Dr. Marta Barreras, Dr. Juan Jose Martín, Dr. Irene Eguinoa, Dr. Alejandro Asenjo, Dr. Jorge Elso, Carlos Molins, Augusto Mauch, Pablo Vital, Javier Villanueva, Manu Motilva, Juan Antonio Osés, Javier Castillejo, Mikel Iribas, Ana Huarte, Maria Brugarolas, Xabier Montón, Asier Oiz, and Daniel Casajus at the Public University of Navarra (Spain), and to Dr. Ion Irizar, Dr. Maria Jose Mercado, Dr. Timothy Franke, Dr. Trupti Ranka, Dr. Fa Wang, Dr. Sameer Alsharif, Tipakorn Greigarn, Nicholas White, Nicholas Tierno, Katherine Faley, Gerasimos Houpis, Erica Pettit, William Lounsbury, Harry Labrie, Bowen Weng, Tony Joy, John O'Brian, Manuel Casado, Laura Wheeler, Julio Cesar Cavalcanti dos Santos, Inigo Jimenez, Inigo Ostiz, Asier Diez de Ulzurrun, Yingkang Du, Chao Lin, Shiyu Sun, and Dr. Amir Sajadi at Case Western Reserve University (USA).

In addition, I want to acknowledge many colleagues and friends at Case Western Reserve University (Ohio), the Public University of Navarra (Spain), Tecnun/CEIT (Centre of Studies and Technical Research of Gipuzkoa) and MTorres (Spain), UMIST (University of Manchester Institute of Science and Technology) and Oxford Universities (UK), NASA-JPL (NASA's Jet Propulsion Laboratory) (California), ESA-ESTEC (The European Space Agency- European Space Research and Technology Centre) (The Netherlands), the U.S. Air Force Institute of Technology (Ohio), and the Green Bank Observatory (West Virginia), and especially to Professor Constantine Houpis, Professor Isaac Horowitz, Professor Peter Wellstead, Professor John Edmunds, Professor Neil Munro, Professor Olaf Wolkenhauer, Professor Ron Daniel, Professor Joaquin Casellas, Professor Julian Florez, Professor Pedro Dieguez, Professor Juan Sandoval, Dr. Fred Hadaegh, Dr. Boris Lurie, Dr. Paul Brugarolas, Dr. Christian Philippe, Dr. Samir Bennani, Dr. Robert Ewing, Jose Angel Fernandez, Carlos Aguerri, Ignacio Eizaguirre, Manuel Torres, Tim Weadon, John Ford, Dr. Richard Prestage, Professor Per Olof Gutman, Professor Eduard Eitelberg, Professor Edward Boje, Professor Paluri Nataraj, Professor Suhada Jayasuriya, Professor Oded Yaniv, and Professor Michael Grimble for their help and cooperation over the years. Finally, I also would like to thank Harry Morton and Tim Weadon (National Radio Astronomy Observatory-Green Bank Telescope [NRAO-GBT]) for the excellent picture of the Green Bank Telescope they provided for the cover.

Author

Professor Mario García-Sanz is a pioneer in the QFT robust control arena. Over the last 25 years, he has developed new QFT control theory for multi-input multi-output plants, distributed parameter systems, time-delay processes, nonlinear switching and feedforward control, including also methods to apply the Nyquist stability criterion in the Nichols chart, and to calculate QFT templates and bounds. In addition, he has designed many commercial control solutions for industry and space agencies. Customers include NASA-JPL, ESA-ESTEC, US-AFIT, NRAO-GBT, GMRT, Gamesa, Acciona, MTorres, IngeTeam, CENER, Eaton Corporation, Enercon, Siemens, Iberdrola, REE, Sener, EEQ, and others. With over 20 industrial patents and 200 research papers, Dr. García-Sanz is one of the inventors of the TWT direct-drive variable-speed pitch-control multi-megawatt wind turbine, the EAGLE airborne wind energy system, the TWT variable-speed hydro-wind turbine, the DeltaGrids optimal planning algorithms for electrical distribution networks, and of numerous advanced industrial controllers. In addition, he has been the principal investigator of over 50 funded research projects for industry, and worked as an international expert on wind turbine design and control in patent litigation at the British Court in London. As a full professor at the Public University of Navarra (Spain) and senior advisor for European wind energy companies, he played a central role in the design and field experimentation of multi-megawatt wind turbines for industry, including the advice of many PhD students and engineers in the field. Dr. García-Sanz is currently a professor and founding director of the *Control and Energy Systems Center*, and the inaugural Milton and Tamar Maltz Endowed Chair in Energy Innovation at Case Western Reserve University (http://cesc.case.edu). He also has been NATO/RTO lecture series director for Advanced Controls, visiting professor at the Control Systems Centre, UMIST (UK); at Oxford University (UK); at the Jet Propulsion Laboratory NASA-JPL (California); and at the European Space Agency ESA-ESTEC (The Netherlands), and has given invited seminars in over 20 countries. He founded CoDyPower LLC, a consulting firm specialized in control systems, energy innovation, and optimum planning of electrical distribution networks (http://codypower.com). Professor García-Sanz's three CRC Press books *Quantitative Feedback Theory: Theory and Applications* (2006), *Wind Energy Systems: Control Engineering Design* (2012), and *Robust Control Engineering: Practical QFT Solutions* (2017) are among the reference books in QFT robust control and wind turbine control. His *QFT Control Toolbox for MATLAB* is considered as the top tool for designing QFT robust control systems. Dr. García-Sanz is subject editor of the *International Journal of Robust and Nonlinear Control* and was awarded the IEE Heaviside Prize (UK) in 1995, the BBVA research award (Spain) in 2001, and the CWRU Diekhoff Teaching Award (USA) in 2012 among other prizes.

1

Introduction

1.1 The Control Engineer's Leadership

Control engineering plays a primary role in the design of technology. Control concepts are often the key factor to achieve transformational ideas in multidisciplinary systems, including aerospace missions, energy plants, chemical processes, power systems, transportation, robotics, and environmental and industrial projects.

The increasing complexity of technology has dramatically changed the way we study engineering at university. In the last few decades, as systems became more and more complex, engineering careers turned out to be much more specialized. The new engineers have a deeper knowledge of some aspects at the cost of a much narrower picture. As a consequence, we have experienced a more and more sequential way of working in the engineering departments in industry. Under a typical new project, product, or system, companies follow a consecutive approach to the problem, starting for instance with some mechanical engineering department discussions, continuing perhaps with aerodynamics or fluids departments, then electrical or electronics departments and so on, to finish up with the control engineering department. A very independent and sequential way of working that merges all the components of the design at a very late stage.

This sequential approach significantly limits the possibilities of the final design. Also, it was not the approach in the late nineteenth century, before the specialization age. Far from a too narrow image, great engineers of that period like Nikola Tesla, Oliver Heaviside, Charles Brush, and Wilbur and Orville Wright had a more complete and multidisciplinary picture of the projects. This characteristic opened them to the possibility of more optimal designs. Giving necessary credit to specialization, we need today to recover the multidisciplinary and concurrent engineering approach for industry. A person that can play a central role in this mission is the *control engineer*, as he or she is trained to apply control concepts to understand the interactions among subsystems, and coordinate the different disciplines to achieve a better system dynamics, controllability, and optimal design. Figure 1.1 shows the details of this approach.

A new engineering system can typically involve many aspects that deeply affect each other with many internal feedback loops, including physics, mathematics, economics, reliability, efficiency, certification, regulations, implementation, marketing, interaction with the environment, maintenance, etc. This is illustrated in the project's road map shown in Figure 1.1. In general, any single aspect in this big picture can affect the rest of the areas, and eventually be the critical bottleneck of the entire project.

As an example, the selection of a specific airfoil for the blades of a wind turbine affects the rotor speed, control system, and mechanical loads, which could require modifying the electrical design of the generator and the structure of the tower and foundation,

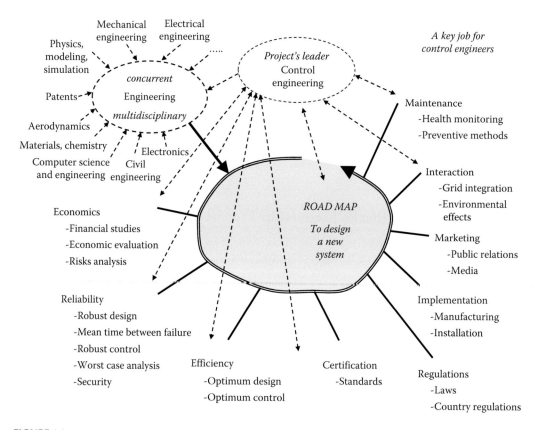

FIGURE 1.1
Road map to design new engineering systems, with a concurrent engineering approach, and the primary role of the *control engineer.*

which in turn could change the economics, reliability, efficiency, and maintenance of the project. Analogously, a marketing issue or a regulation matter could require a noise-level reduction, which eventually could result in modifying the rotor speed or blade design of such a wind turbine, which again could require modifying the control system, electrical design of the generator, tower structure and foundation, which could change the economics, reliability, efficiency, and maintenance of the project. Also, a transport limitation or an economic specification could require shorter blades of the wind turbine, which affect the rotor speed, which could require modifying the control system and electrical design of the generator, which in turn could change the mechanical loads, structure of the tower, foundation, economics, reliability, efficiency and maintenance of the entire project, etc.

This big picture definitely requires a concurrent engineering way of thinking, with the leadership of a multidisciplinary engineer like the control engineer, who naturally and simultaneously can consider all the aspects of the road map to achieve an optimal solution (see Figure 1.1).

In other words, *control engineering* is not only related to the design of the specific control strategies to automatically regulate systems, but also to the understanding of the interactions among the subsystems, to the coordination of the different disciplines to improve the system dynamics, controllability, and optimal design, and eventually to provide leadership to the project team.

FIGURE 1.2
Wright brothers' unstable airplane. December 17, 1903: the World's first heavier-than-air, powered, and controlled human flight (US Library of Congress).

Two examples of these ideas are discussed next. The first one is the world's first successful, heavier-than-air, powered airplane (see Figure 1.2). It was invented by the Wright brothers, Wilbur (1867–1912) and Orville (1871–1948), in Dayton, Ohio. The Wrights had not only extraordinary mechanical engineering skills, but also a deep understanding of practical control engineering concepts learnt while working in their bicycle shop in Dayton. For a long time before the brothers, all attempts to fly considered only stable airplanes and were unsuccessful. The Wright brothers, however, realized that the agility of the airplane to deal with the wind disturbances was a key aspect, and proposed a new revolutionary idea based on a control engineering understanding: an intrinsically unstable airplane (to speed up the response to disturbances) with a longitudinal and lateral control system (regulated by the pilot) able to stabilize the airplane.

As mentioned before, the brothers developed an integrated view of the design with a *control engineering* vision of the problem as a central role from the very beginning. They applied control concepts to understand the interactions of the subsystems, and coordinated the different disciplines to achieve better airplane dynamics, controllability, and optimal design.

The second example is the world's first successful automatically operating wind turbine for electricity generation. It was designed and erected by Charles F. Brush in Cleveland, Ohio, in 1887. The wind turbine operated for 20 years, delivering 12 kW of power to Brush's home on 37th Euclid Avenue in Cleveland. The wind turbine was a giant, the world's largest, with a rotor diameter of 17 m and 144 blades made of cedar wood. The electrical generator power output was measured at 12 kW at 500 rpm (full load) and was used to supply energy via underground conductors to 408 battery cells, 350 incandescent lights, 2 arc lights, and 3 electric motors in Brush's mansion (see Figure 1.3).

Charles Brush described the wind turbine as quote *"self-regulating, requiring no person to attend it."* The turbine had what it is probably the world's first automatic *DC voltage regulator*: a group of automatic switching control devices arranged so that the dynamo was into effective action at 330 rpm, and the DC voltage at the batteries was kept between 75 and 90 V. In addition, the wind turbine had a *yaw control system* composed of a large tail (18 × 6 m), an auxiliary vane perpendicular at one side, and a mechanism of weights and pulleys. Under normal wind conditions, the yaw control system was able to point the

FIGURE 1.3
Charles Brush's wind turbine, 1887. The world's first automatically operating wind turbine for electricity generation (Kelvin Smith Library, CWRU).

turbine rotor perpendicular to the wind direction for maximum efficiency. Eventually, under high winds the yaw control system was also able to turn the rotor out of the wind direction, reducing the mechanical loads and reducing the rotor velocity.

As with the Wright brothers, Charles Brush had also an integrated view of the design of the wind turbine, with a *control engineering* vision of the problem as a central role. He applied control concepts to understand the interactions of the electrical, mechanical, and aerodynamic subsystems, and coordinated the different disciplines to achieve better wind turbine dynamics, controllability, and optimal design.

The next section introduces the quantitative feedback theory (QFT). It is a powerful control system design tool that can be applied to understand the dynamics and interactions among subsystems discussed in this section (concurrent engineering), including model uncertainty, performance specifications and control design to improve the system dynamics, controllability, and optimization.

1.2 QFT Robust Control Engineering

Automatic control systems are everywhere. Electrical power systems, airplanes, factories, wind turbines, homes, cars, trains, water treatment plants, spacecraft, and so on need control systems to work. *Control engineering* is the discipline that applies mathematics, physics, and technology to the design of smart machines that automatically control physical systems. These smart machines, also known as *control systems*, use sensors to measure the variables of the physical system being controlled, computers to calculate the actions to be applied, and actuators to adjust the system in real time. The physical system to be controlled can be a mechanism, an electrical network, a thermal process, an aero/fluid-dynamical system, a chemical reactor, or even a financial or biomedical system.

In the real world, the mathematical description of the dynamics of a physical system is not accurate, as it often presents some model uncertainty or changes with time. This uncertainty is a consequence of unknown dynamics, unidentified high-frequency components, inaccuracies in the parameter estimation, errors in the sensors and actuators, system nonlinearities, plant aging, temperature changes, etc.

Model uncertainty makes the design of the controller much more difficult, as it has to meet the control specifications not only for a single plant with fixed parameters, but for each and every one of the plants within the uncertainty. The independent study of a control solution for each plant within the uncertainty, that is, the brute-force solution, is very time consuming or even impossible. *Robust control engineering* is the area of automatic control that takes into account these model uncertainties and designs reliable control solutions for real-world systems.

QFT gives an answer to this paradigm. It gives a control solution at once, a solution that works for each plant within the uncertainty and for all control specifications. In other words, it is a *robust control methodology*: gives just one fixed controller (opposite to adaptive control) that is able to deal with all the plants within the model uncertainty.

QFT is also a multi-objective control engineering methodology that can deal simultaneously with not just one, but many performance specifications at the same time, including stability, disturbance rejection, reference tracking, model matching, noise rejection, actuator limitations, reduction of vibrations, etc.

QFT is a frequency domain methodology that deals with frequency tools that are familiar for control engineers. Also, it is a very transparent technique that includes graphics and concepts that keep the engineering understanding at each and every step of the design. QFT quantifies the balance among the controller structure (complexity, order), cost of feedback (bandwidth, gains), performance specifications (control objectives), and plant uncertainty (models) at each frequency of interest.

QFT has been successfully applied to a wide variety of control problems, including stable and unstable plants, minimum and non-minimum phase systems, single-input single-output (SISO) and multiple-input multiple-output (MIMO) processes, with linear and nonlinear characteristics, long time delay, distributed parameter systems (DPS), time-varying plants, feedforward control topologies, and multi-loop systems. It has been also used in many real-world applications, like radio telescopes servo systems, wind turbine control, water treatment plants, spacecraft control, manufacturing, flight control, power systems, robotics, motion control, and chemical reactors. As an example, Figure 1.4 shows some industrial and commercial applications where the author has applied QFT robust control solutions.

Computer-aided design (CAD) tools facilitate the use of QFT. The *QFT Control Toolbox* (or QFTCT) for MATLAB® is used in the examples of this book (see Figure 1.5). The QFTCT is the professional, interactive, and user-friendly toolbox for MATLAB of CoDyPower LLC. It applies QFT to the design of automatic robust control systems. The software has been developed by Professor Mario García-Sanz. Over the years, the toolbox has been widely

FIGURE 1.4
Some industrial and commercial projects where the author has applied QFT robust control solutions: (a) NRAO Green Bank Telescope, (b) NASA-ESA space-telescope formation flying missions, (c) MTOI multi-megawatt wind turbines, (d) Crispijana water treatment plant, and (e) MTorres manufacturing.

applied by industry, space agencies, research centers, and universities to design control solutions and servo systems, including the European Space Agency ESA-ESTEC, NASA-JPL, wind energy companies, large radio telescopes, power systems, water treatment plants, etc.

The toolbox (1) deals with plants with model uncertainty, (2) is able to work with multi-objective performance specifications, (3) keeps the engineering understanding of the design in the frequency domain, and (4) gives solutions from simple proportional integral derivative (PID) regulators to more advanced control strategies when necessary. The user's guide for the QFTCT is in Appendix 2.

Project files for some of the problems presented in this book and a demo version of the toolbox can be found at http://cesc.case.edu. The student and standard versions of the QFTCT is at http://codypower.com. For additional information, see http://crcpress.com.

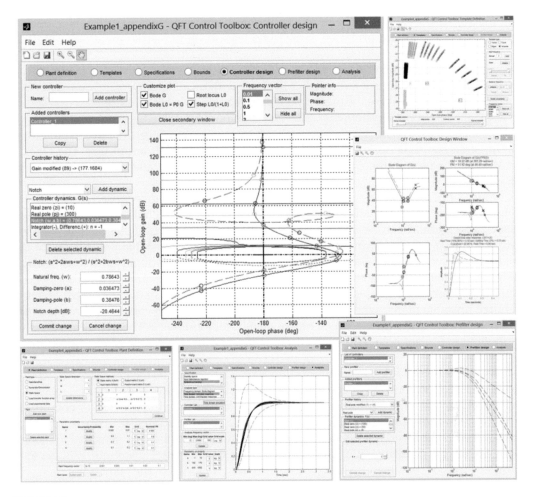

FIGURE 1.5
The QFTCT for MATLAB.[175]

1.3 Book's Outline

This book includes 10 chapters, 10 appendices, and over 50 detailed examples, case studies, projects, and problems. It introduces the fundamentals of the quantitative robust control (QFT) as well as practical advanced control solutions for real-world projects, including unstable, time-delay, non-minimum phase, or distributed parameter systems, plants with large model uncertainty, with high-performance specifications, nonlinear components, multi-input multi-output (MIMO) characteristics, or asymmetric topologies. *Robust Control Engineering: Practical QFT Solutions* is an ideal textbook for undergraduate and graduate students and control engineers. It presents practical methodologies to design reliable control systems, bridging the gap between successfully tested theory and real-world control system implementation. The book is indispensable for engineers and researchers designing reliable control solutions for industrial, energy, environmental, biomedical, chemical, electrical, mechanical, and aerospace applications. Unlike most books on the control area, this book presents a large collection of successful real-world cases and projects, including commercial wind turbines, wastewater treatment plants, satellites with flexible append-ages, spacecraft flying in formation, large radio telescopes, industrial manufacturing sys-tems, etc.

Chapter 2 presents the fundamentals of the QFT. The basic steps of the QFT methodology are introduced in Sections 2.2 through 2.10 along with an illustrative direct current (DC) motor example. To reinforce the presentation, the chapter includes an additional example of an airplane flight control system that follows the methodology step by step. Finally, Sections 2.11 through 2.14 complete the chapter with model-matching control systems, feedforward control strategies, PID design and tuning, and practical control design tips. The QFTCT for MATLAB is used in the examples. Appendix 2 presents a detailed user's guide for the QFTCT.

Chapter 3 introduces the practical and comprehensive method proposed by the author to compute the Nyquist stability criterion directly in the Nichols chart (magnitude/phase). The method can be applied to linear time invariant (LTI) closed-loop systems with minimum and non-minimum phase zeros, stable and unstable poles, poles at the origin with diverse multiplicity and systems defined by nonrational functions, such as plants with time delay. The chapter also shows how to apply the method with a large col-lection of illustrative examples. In addition, it gives guidelines to design controllers that stabilize unstable plants when dealing with frequency domain control techniques, and in particular with the QFT robust control design methodology. The algorithm for this methodology is also included in the QFTCT, and the MATLAB code of the algorithm in Appendix 3.

The presence of time-delay or non-minimum phase zeros limits the performance achiev-able by closed-loop control systems. For open-loop stable systems, the tracking problem of a time-delay system can be improved substantially by introducing time-delay compensa-tion. One of the most popular and effective time-delay control strategies in use is the Smith predictor (SP). However, it is also well known that the SP may be very sensitive to model-plant mismatch, resulting in a poor performance when model uncertainty is present. Chapter 4 introduces a method based on QFT to design SPs when the plant is not precisely known. An illustrative example shows how to effectively apply the method step by step to a plant with time-delay and model uncertainty. Appendix 4 includes the MATLAB code for the algorithms of the example. Finally, the same methodology is extended to deal with non-minimum phase systems.

A distributed parameter system (DPS) is a system in which the dependent variables are functions of both time and space. The model that describes a DPS usually involves partial differential equations (PDE). Examples for DPS include flexible robots, heat trans-fer problems, large mechanical structures, spacecraft with flexible appendages, electri-cal transmission lines, etc. Chapter 5 presents a practical methodology based on QFT to design control solutions for DPS. The method includes a spatial distribution of the location of the relevant points, allowing the designer to place the actuator, disturbances, sensor and control objectives in any spatial location of interest, including location uncer-tainty. From this topology, new stability and performance specifications, transfer func-tions, and quadratic inequalities are introduced. The simplicity and excellent practical results of the methodology are also illustrated with a well-known PDE example. Finally, Appendix 5 includes the MATLAB code for the design of the QFT robust control solution of the example.

Chapter 6 presents a practical QFT method to design robust controllers that work under a switching mechanism. The method is capable of optimizing performance and stability simultaneously, going beyond the classical linear limitations and giving a solution for the well-known robustness-performance trade-off. Based on the frequency domain approach, the method combines a graphical stability criterion for switching linear systems and the robust QFT technique. The formulation is applied to four illustrative examples. Appendix 6 includes the MATLAB code utilized in the examples.

Chapter 7 introduces a nonlinear dynamic control (NDC) methodology. It is a practical control design method based on QFT that can deal with one or several nonlinearities in the system, either in the plant, and/or deliberately introduced in the controller to go beyond the linear limitations and achieve a high performance. The chapter starts with the analysis of the well-known circle stability criterion to study systems with one, bounded and static or even time-variant nonlinearity. Based on this criterion, Section 7.3 introduces a practical engineering methodology to design controllers for systems with one nonlinearity in the plant (e.g., saturation or dead zone in the actuator), and the same nonlinearity deliberately introduced in the controller. The method, known as *NDC—one nonlinearity*, is applied to the design of an anti-windup strategy for a classical PID controller in Section 7.4. Then, Section 7.5 expands the methodology to systems with more than one nonlinearity: *NDC – several nonlinearities*. In this case, the nonlinear blocks can be either in the plant (such as saturation, dead zone, relay, hysteresis, or friction), and/or in the controller (to go beyond the classical linear limitations). Example 7.2 applies this new methodology to an illustrative example of a DC motor with saturation and a PID with nonlinear dynamics, and Appendix 7 includes the MATLAB code utilized in the example.

Chapter 8 analyzes the main characteristics of multi-input multi-output (MIMO) plants and introduces two methodologies (methods 1 and 2) to design full-matrix QFT controllers for MIMO systems with model uncertainty. Sections 8.1 through 8.4 analyze the special properties that define MIMO systems and propose some tools to understand their dynamics and design appropriate control systems. Sections 8.5 and 8.8 present the non-diagonal MIMO QFT Method 1, including both theory and a detailed 2×2 heat exchanger example. Sections 8.6 and 8.9 present the non-diagonal MIMO QFT Method 2, including both theory and the detailed 2×2 heat exchanger example. The MATLAB code that develops the examples and cases presented in this chapter are also included in Appendix 8.

Chapter 9 proposes control topology solutions for nine common problems that have a different number of sensors and actuators in the system. It includes design criteria, block diagram topologies, and algorithm tuning for cascade control, feedforward control, override control, ratio control, mid-range control, load-sharing control, split-range control, inferential control, and auctioneering control.

Chapter 10 studies both the analog implementation of controllers with active resistor–capacitor (RC) electrical circuits and the digital implementation with micro-controllers and algorithms. The chapter also studies the resiliency of the implemented controllers with a fragility analysis method based on QFT.

The book also contains five detailed and real-world QFT control design case studies, including a variety of problems from SISO plants to MIMO systems, advanced topologies, and even NDC solutions.

Case study CS1 designs a robust QFT attitude control system (ACS) for a satellite with flexible solar panels. The design considers the model uncertainty introduced by the fuel consumption and the imprecise knowledge of the damping coefficients. It accomplishes four simultaneous control objectives: (1) stability, (2) reference tracking and regulation of the satellite angle, (3) rejection of unpredictable disturbances, and (4) minimization of the solar panel vibrations.

Case study CS2 designs a robust QFT control system for a variable-speed pitch-controlled gearless wind turbine. The design considers the model nonlinearities and uncertainty introduced by the aerodynamics. It accomplishes five simultaneous control objectives: (1) stability, (2) tracking and regulation of the rotor speed by pitching the angle of the blades, (3) rejection of unpredictable wind disturbances, (4) minimization of the tower fore-aft oscillations, and (5) attenuation of the blades' flap-wise vibrations.

Case study CS3 designs a robust 2×2 MIMO QFT control system for a wastewater treatment plant with an activated sludge process that simultaneously reduces the concentration of ammonia and nitrates in the plant effluent. The design considers the model uncertainty and MIMO loop interaction. It accomplishes four simultaneous control objectives: (1) stability, (2) reference tracking, (3) rejection of unpredictable disturbances, and (4) minimization of loop interaction.

Case study CS4 designs two robust QFT control solutions for the velocity and position loops of a radio telescope servo system. The first solution is based on the classical QFT methodology and the second one proposes an NDC strategy. Both designs accomplish five simultaneous control objectives: (1) stability, (2) tracking of the azimuth axis telescope position, (3) regulation of the azimuth axis telescope velocity, (4) rejection of unpredictable wind disturbances, and (5) reduction of dish and feed-arm vibrations.

Case study CS5 demonstrates the feasibility of the sequential non-diagonal MIMO robust QFT control strategies to regulate simultaneously the position and attitude of a 6×6 spacecraft telescope with large flexible appendages. The spacecraft is part of a multiple formation flying constellation of a European Space Agency cornerstone mission. The controller satisfactory meets all the astronomical, engineering, and control requirements.

The book also proposes 8 detailed QFT control design projects and 12 quick problems in Appendix 1, including SISO plants, unstable systems, advanced topology cases, MIMO processes, and time-delay systems. Project P1 offers a vehicle active suspension system, Project P2 a DVD-disk head control system, Project P3 an inverted pendulum, Project P4 two interconnected multi-generator micro-grids, with frequency and active power control requirements, Project P5 a 2×2 distillation column, Project P6 a central heating system with long time delays, Project P7 a feedforward and cascade control of a multi-tank, and Project 8 an attitude control system for a satellite with fuel tanks partially filled. The collection of quick problems, Q1–Q12, offers the reader an opportunity for practical training.

Finally, a long reference section compiles the main references used in the book. This section is arranged according to subject, and chronologically within each subject.

The topics and references are: Books of QFT and frequency-domain methods[1-7]; Special issues about QFT[8-12]; International QFT symposia[13-21]; Tutorials about QFT[22-26]; History of QFT[27-31]; First QFT papers[32-34]; QFT templates[35-53]; QFT bounds[54-75]; QFT loop-shaping[76-88]; Existence conditions for QFT controllers[89-91]; MIMO QFT[92-121]; Time-delay systems[122]; Unstable systems[123-125]; Digital QFT[126-129]; Distributed parameter systems[130-137]; Feedforward[138,121]; Non-minimum phase systems[139-144]; Multi-loop systems[145-148]; Nonlinear systems[149-159]; Linear-time-variant LTV systems[160-163]; Stability analysis and controller design in the Nichols chart[164-165]; CAD tools for QFT controller design[166-175]; Real-world applications with QFT[176-222]; NATO/RTO Lecture Series about QFT[223-227]; Miscellaneous of QFT[228]; Books of control engineering[229-263]; General MIMO systems[264-314]; Papers of Fragility[315-325]; Papers of hybrid/switching control systems[326-336]; Miscellaneous control[337-357]; Miscellaneous[358-367].

1.4 Courses and Modules

The prerequisites needed to understand and follow the topics introduced in this book are just the *fundamentals of automatic control*. Examples of courses about fundamentals of control are in References 259, 262, and 263.

The material presented in the book can be used as a self-study reference and in undergraduate and graduate-level courses. The topics of the book can be organized in 10 modules, M1–M10. Thinking on a classical course of three credit-hours in 15 weeks, these modules can be used to develop courses on *robust control engineering* and *applied control*, as shown in Figure 1.6.

Modules

M1. *Fundamentals of QFT robust control.* [4 weeks]
- Chapter 2. QFT robust control.
- Example 2.1. Sections 2.3 through 2.10. Armature-controlled DC motor.
- Example 2.2. Airplane flight control system.
- Case study CS1. Satellite with flexible appendages.
- Case study CS2. Wind turbine pitch control.
- Project P1. Appendix 1. Vehicle active suspension system.
- Project P2. Appendix 1. DVD-disk head control.
- Project P8. Appendix 1. Satellite with fuel tanks partially filled.
- Problem Q2. Appendix 1. Control of first-order system with uncertainty.
- Problem Q3. Appendix 1. Control of third-order state space system.
- Problem Q4. Appendix 1. Field-controlled DC motor.
- Problem Q5. Appendix 1. Formation flying spacecraft control. Deep space.
- Problem Q6. Appendix 1. Helicopter control.
- Problem Q7. Appendix 1. Two cart problem.

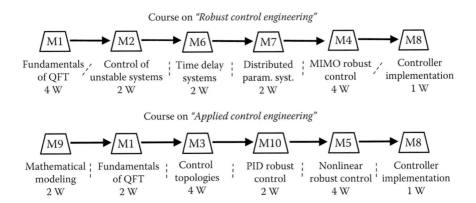

FIGURE 1.6
Example of two possible courses built with the modules of this book.

M2. *Control solutions for unstable systems.* [2 weeks]
- Chapter 3. Unstable systems and control solutions.
- Examples 3.1 through 3.9, Section 3.5. Analysis of stability.
- Examples 3.10 through 3.15, Section 3.6. Design of control solutions to stabilize plants.
- Example P0, Section 3.7. Design of multiple control solutions.
- Project P3. Appendix 1. Inverted pendulum.

M3. *Control topologies with different number of sensors and actuators.* [4 weeks]
- Chapter 9. Control topologies.
- Section 9.2. Cascade control.
- Section 9.3a. Feedforward control for disturbance rejection.
- Section 9.3b. Feedforward control for model matching (reference tracking).
- Section 9.4. Override control.
- Section 9.5. Ratio control.
- Section 9.6. Mid-range control.
- Section 9.7. Load-sharing control.
- Section 9.8. Split-range control.
- Section 9.9. Inferential control.
- Section 9.10. Auctioneering control.
- Example 2.2. Section 2.12. Feedforward/feedback for a flight control system.
- Case study CS4. Cascade control for a radio telescope servo system.
- Project P3. Appendix 1. Override control for inverted pendulum.
- Project P4. Appendix 1. Load-sharing control for interconnected micro-grids.
- Project P7. Appendix 1. Feedforward/cascade control of a multi-tank system.

M4. *MIMO robust control.* [4 weeks]
- Chapter 8. Multi-input multi-output control. Methods 1 and 2.
- Example 8.1. SISO control of a 2×2 heat exchanger system.
- Example 8.1, Section 8.8. MIMO QFT control (Method 1), 2×2 heat exchanger.
- Example 8.1, Section 8.9. MIMO QFT control (Method 2), 2×2 heat exchanger.
- Case study CS3. 2×2 Waste water treatment plant.
- Case study CS5. 6×6 spacecraft telescope control.
- Project P5. Appendix 1. 2×2 distillation column.
- Problem Q8. Appendix 1. Two flow problem.
- Problem Q9. Appendix 1. 2×2 MIMO system.
- Problem Q10. Appendix 1. 2×2 MIMO system.
- Problem Q11. Appendix 1. Spacecraft flying in formation. Low Earth orbit.
- Problem Q12. Appendix 1. 3×3 MIMO system.

M5. *Nonlinear robust control.* [4 weeks]
- Chapter 6. Gain scheduling/switching control solutions.

- Examples 6.1 and 6.2, Section 6.4. Systems with gain, zero, and pole variations.
- Example 6.3, Section 6.4. Two-mass-spring benchmark system.
- Chapter 7. Nonlinear dynamic control: one or several nonlinearities.
- Example 7.1, Section 7.4. System with actuator saturation. PID with anti-windup.
- Example 7.2. DC motor with saturation. PID with nonlinear dynamics.
- Case study CS4. Radio telescope servo system.

M6. *Time-delay systems and non-minimum phase systems.* [2 weeks]
- Chapter 4. Time-delay and non-minimum phase systems.
- Example 4.1. Sections 4.2 and 4.4. First-order system with delay and SP.
- Examples 4.2 and 4.3. Section 4.5. Non-minimum and minimum phase systems.
- Project P6. Appendix 1. Central heating system with long-time delays.

M7. *Distributed parameter systems.* [2 weeks]
- Chapter 5. Distributed parameter systems. Analysis and robust control.
- Example 5.1, Section 5.5. Heat transmission system with distributed temperature.

M8. *Analog and digital implementation of controllers.* [2 weeks]
- Chapter 10. Controller implementation.
- Section 10.2. Analog implementation.
- Section 10.3. Digital implementation.
- Section 10.4. Fragility analysis with QFT.
- Example 10.1, Section 10.3.2. Digital Proportional Integral (PI) algorithm.
- Example 10.2, Section 10.3.5. Digital Pulse Width Modulation (PWM)-PI algorithm.
- Example 10.3, Section 10.4. Fragility analysis of a control system.

M9. *Mathematical modeling of dynamic systems.* [2 weeks]
- Example 2.1, Section 2.2. Armature-controlled DC motor.
- Example 5.1, Section 5.5. Heat conduction distributed temperature system.
- Section 10.2. Active RC circuits with operational amplifiers.
- Case study CS1. Satellite with flexible appendages.
- Case study CS2. Wind turbine.
- Case study CS4. Radio telescope.
- Project P1. Appendix 1. Vehicle active suspension system.
- Project P3. Appendix 1. Inverted pendulum.
- Project P8. Appendix 1. Satellite with fuel tanks partially filled.

M10. *Design of PID controllers.* [2 weeks]
- Section 2.13. PID control: design and tuning with QFT.
- Example 2.1. Sections 2.3 through 2.10. Armature-controlled DC motor.

- Example 7.1, Section 7.4. System with actuator saturation. PID with anti-windup.
- Example 7.2. DC motor with saturation. PID with nonlinear dynamics.
- Section 10.2. Analog implementation of PID.
- Section 10.3.3. Position and velocity digital PID algorithms.
- Example 10.1, Section 10.3.2. Digital PI algorithm.
- Example 10.2, Section 10.3.5. Digital PWM-PI algorithm.
- Example 4.1. Sections 4.2, and 4.4. First-order system with delay and SP.
- Example 5.1, Section 5.5. Heat conduction distributed temperature system.
- Example 8.1. SISO control of a 2×2 heat exchanger system.
- Example 8.1, Section 8.8. MIMO QFT control (Method 1), 2×2 heat exchanger.
- Example 8.1, Section 8.9. MIMO QFT control (Method 2), 2×2 heat exchanger.
- Case study CS2. Wind turbine pitch control.
- Case study CS4. Radio telescope servo system.

Case studies
- Satellite with flexible solar panels (Case study CS1).
- Wind turbine pitch control (Case study CS2).
- Waste water treatment plant (Case study CS3).
- Radio telescope servo system (Case study CS4).
- Spacecraft telescope MIMO control (Case study CS5).

Projects. Appendix 1
- Vehicle active suspension system (Project P1).
- DVD-disk head control (Project P2).
- Inverted pendulum (Project P3).
- Interconnected micro-grids, frequency, and active power control (Project P4).
- 2×2 distillation column (Project P5).
- Central heating system with long-time delays (Project P6).
- Multi-tank hydraulic system (Project P7).
- Satellite with fuel tanks partially filled (Project P8).

Quick Problems. Appendix 1
- Definition of uncertainty (Problem Q1).
- Control of first-order system with uncertainty (Problem Q2).
- Control of third-order state space system with uncertainty (Problem Q3).
- Field-controlled DC motor (Problem Q4).
- Formation flying spacecraft control. Deep space (Problem Q5).
- Helicopter control (Problem Q6).
- Two cart problem (Problem Q7).
- Two flow problem (Problem Q8).
- 2×2 MIMO system (Problem Q9).

- 2×2 MIMO system (Problem Q10).
- Spacecraft flying in formation in low Earth orbit (Problem Q11).
- 3×3 MIMO system (Problem Q12).

MATLAB code included in the book
- Armature-controlled DC motor (Example 2.1).
- Feedforward/feedback airplane flight control (Example 2.2).
- Unstable systems, mp and nmp (Examples 3.1 through 3.15).
- First-order system with time delay (Example 4.1).
- Heat conduction distributed temperature system (Example 5.1).
- Switching system varying poles and zeros (Example 6.1).
- Switching system, common quadratic Lyapunov function (Example 6.2).
- System with actuator saturation. PID with anti-windup (Example 7.1).
- DC motor with saturation. PID with nonlinear dynamics (Example 7.2).
- Heat exchanger MIMO system (Example 8.1).
- Appendices A3 through A8.

2

QFT Robust Control

2.1 Introduction

Designing reliable and high-performance control systems is an essential priority of control engineering projects. In practical circumstances, the presence of model uncertainty challenges the design. One robust control approach for these cases is the quantitative feedback theory (QFT). Deeply rooted in the classical frequency domain, QFT provides control solutions that guarantee the achievement of a multi-objective set of performance specifications for every plant within the model uncertainty. It balances the trade-off between the simplicity of the controller structure and the minimization of its activity at each frequency of interest.

QFT is a robust control engineering design methodology that uses the feedback to simultaneously and quantitatively (1) reduce the effects of plant uncertainty and (2) satisfy performance control specifications. The method searches for a controller that guarantees the satisfaction of the required performance specifications for every plant within the model uncertainty (robust control).

Hendrik Bode introduced many of the frequency-domain fundamentals in his seminal book *Network Analysis and Feedback Amplifier Design*, published in 1945.[1] The book strongly influenced the understanding of automatic control theory for many years, especially where system sensitivity and feedback constraints are concerned.

Almost 20 years later, in 1963, a new influential book entitled *Synthesis of Feedback Systems*,[2] written by Isaac Horowitz, proposed a formal combination of the frequency-domain methodology with plant model uncertainty (robust control) under a quantitative analysis. The new book addressed an extensive set of sensitivity problems in feedback control. It was the first work in which a control problem was treated quantitatively in a systematic way. The book laid the foundation for a new control design methodology that had been introduced briefly, also by Horowitz, in a previous paper (1959),[32] the one that became known as QFT in the early 70s.

QFT has been successfully applied to a wide variety of control problems, including stable and unstable plants, minimum and non-minimum phase systems, single-input single-output (SISO) and multiple-input multiple-output processes (MIMO), with linear and nonlinear characteristics, long-time delay, distributed parameter systems, time-varying plants; and has been combined with feedforward control topologies, multi-loop systems, etc. Also it has been used in many real-world applications, such as flight control, wind energy, water treatment plants, spacecraft, power systems, mechanical systems, motion control, and chemical reactors—see References 176–222.

Computer-aided design (CAD) tools facilitate the use of QFT. The *QFT Control Toolbox* (or QFTCT) for MATLAB is used in the examples through this book. It is the interactive object-oriented CAD tool developed by the author to design QFT control solutions. It has been applied to many real-world projects, including the European Space Agency ESA-ESTEC, NASA-JPL, wind energy companies, large radio telescopes, power systems, water treatment plants, etc. Appendix 2 presents a detailed user's guide for the QFTCT. A demo version of the toolbox can be found at http://cesc.case.edu. The student and standard versions of the QFTCT is at http://codypower.com. For additional information, see http://crcpress.com

QFT is also a multi-criteria and transparent control engineering methodology. It quantifies the balance among the controller structure, cost of feedback, performance specifications, and model plant uncertainty at each frequency of interest. The basic steps of the QFT methodology are summarized in Figure 2.1 and presented in Sections 2.2 through 2.10 along with an illustrative example of a DC motor (Example 2.1). Afterward, an additional case of an airplane flight control system (Example 2.2) shows more details of the methodology. Then, Sections 2.11 through 2.13 complete the chapter applying QFT to model-matching control systems, feedforward control strategies, and PID design and tuning. Finally, Section 2.14 discusses some practical QFT control design tips. For a better understanding, we recommend the study of each section and the proposed examples

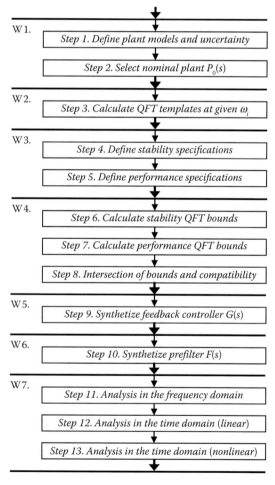

FIGURE 2.1
QFT-controller design methodology (Windows QFTCT: W1–W7).

along with the *QFTCT*—see also Appendix 2. The steps of the QFT methodology are included as windows of the toolbox, as shown in Figure 2.1, W1–W7.

2.2 Plant Modeling: Step 1

See also *Plant definition* window of the QFTCT and W1 in Figure 2.1.

The design of a reliable control system needs a good model of the plant dynamics. The description of the physics of the plant is key to understand the possibilities and limitations of the control system, and to design an appropriate controller to meet the specifications. We can never emphasize too much the importance of modeling. The reliability and performance of a control system depends primarily on the understanding of the plant model.

A model that captures the dynamics of the plant is usually composed by input/output transfer functions in the Laplace domain $P(s)$—see Equation 2.1, or by state-space matrices in the time domain (A, B, C, and D)—see Equation 2.2.

$$P(s) = \frac{y(s)}{u(s)} = \frac{b_m s^m + b_{m-1} s^{m-1} + \cdots + b_1 s + b_0}{s^n + a_{n-1} s^{n-1} + \cdots + a_1 s + a_0}, \quad n \geq m \tag{2.1}$$

$$\begin{aligned} \dot{x}(t) &= Ax(t) + Bu(t) \\ y(t) &= Cx(t) + Du(t) \end{aligned} \tag{2.2}$$

where $u(s)$ and $y(s)$ are the input and output of the plant respectively, x is the array with state variables, and $P(s) = C(sI - A)^{-1}B + D$.

In general, all real-world systems present some degree of model uncertainty. This means that Equations 2.1 and 2.2 are not fixed or exact expressions, but equations with parameters that belong to an interval. The uncertainty affects either the parameters (parametric)—see Equations 2.1 and 2.3, or an additional block (nonparametric)—see Equation 2.4.

$$a_k \in [a_{k\min}, a_{k\max}], b_r \in [b_{r\min}, b_{r\max}], k = 1, \ldots, n, r = 1, \ldots, m \tag{2.3}$$

$$P(s) = P_0(s)(1 + \Delta(s)), 0 \leq \Delta(s) \leq 1 \text{ at each frequency} \tag{2.4}$$

This model uncertainty is a consequence of many factors, including unknown dynamics, unknown high-frequency components, inaccuracies measuring the parameters, sensor and actuator errors, system nonlinearities, plant aging, temperature changes, etc.

Model uncertainty makes the controller design much more difficult. With uncertainty, the controller has to meet the specifications not only for a single plant with fixed parameters, but also for each and every one of the plants within the uncertainty. The design of a controller for each plant within the uncertainty, that is, the brute-force solution, is not reasonable. It is very time consuming, or almost impossible, to analyze every plant within the uncertainty.

In this context, QFT is a powerful technique for these cases with uncertainty. It gives a control solution at once, a solution that works for each plant within the uncertainty and for all specifications. In other words, it is a *robust control methodology*: gives just one fixed controller (opposite to adaptive control) that is able to deal with all the plants within the model uncertainty.

The plant model can be obtained using analytical and experimental techniques. See the References 236, 360, 361, and 362 for additional information. The next example illustrates how to model an electromechanical system, including two cases of the parametric uncertainty, either with independent or interrelated parameters.

EXAMPLE 2.1: ARMATURE-CONTROLLED DC MOTOR

Direct current (DC) electrical motors with constant field are a very popular component of modern servomechanisms. They are also known as armature-controlled DC motors and can directly control the torque that causes the armature to rotate. The model derived in this example is also valid for other type of motors that, using some additional power electronics, can control the torque similarly.

A DC motor consists of a rotating part or armature that contains wire conductors wrapped around an iron core. These conductors have an inductance L_a and a resistance R_a—see Figure 2.2. A commutator composed of a split ring with metal contact segments in the mobile part and brushes in the fixed part connects the wires of the rotating armature to the fixed part of the motor. The armature is inside a constant magnetic field B_m produced by magnets or electromagnets in the stator. When an external voltage $v(s)$ is applied to the conductors of the armature, a current $i_a(s)$ is produced. The constant magnetic field B_m reacts with the armature current $i_a(s)$ creating a torque $T_m(s)$ that causes the armature to rotate.

The voltage $v(s)$ can be used to control the torque $T_m(s)$ and then the rotational velocity $\omega_m(s)$ of the motor. In this case, the DC motor is said to be *armature-controlled*.

The model that describes the dynamics of the armature-controlled DC motor has the diagram shown in Figure 2.2 and the electromechanical expressions presented in Equations 2.5 through 2.8. The torque $T_m(s)$ produced by the motor is proportional to the armature current $i_a(s)$, and the voltage $v_e(s)$ produced in the conductors of the armature and opposed to the current is proportional to the rotational velocity $\omega_m(s)$, so that

$$T_m(t) = K_T\, i_a(t) \tag{2.5}$$

$$v_e(t) = K_e \omega_m(t) \tag{2.6}$$

where K_T is the torque constant of the motor and K_e the voltage constant. Notice that $K_T = K_e$. Now, applying Kirchhoff's voltage law to the armature circuit—see Figure 2.2, we have

$$v(t) = R_a\, i_a(t) + L_a \frac{d}{dt} i_a(t) + v_e(t) \tag{2.7}$$

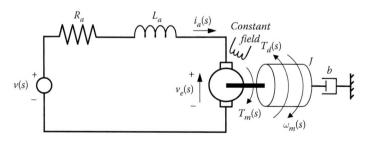

FIGURE 2.2
Armature-controlled DC motor.

And applying Newton's second law to the mechanical part of the motor,

$$J \frac{d}{dt} \omega_m(t) = T_m(t) - b\omega_m(t) - T_d(t) \tag{2.8}$$

where J is the inertia of all the rotating parts of the motor, b the damping coefficient for the losses that are proportional to the motor velocity $\omega_m(t)$, and $T_d(t)$ are the external torques or loads applied to the motor. Combining Equations 2.5 through 2.8, we have

$$v(t) = R_a i_a(t) + L_a \frac{d}{dt} i_a(t) + K_e \omega_m(t) \tag{2.9}$$

$$J \frac{d}{dt} \omega_m(t) = K_T i_a(t) - b\omega_m(t) - T_d(t) \tag{2.10}$$

Now, rearranging both equations to have the first order derivatives on the left-hand side, we get

$$\frac{d}{dt} i_a(t) = \frac{1}{L_a} [v(t) - R_a i_a(t) - K_e \omega_m(t)] \tag{2.11}$$

$$\frac{d}{dt} \omega_m(t) = \frac{1}{J} [K_T i_a(t) - b\omega_m(t) - T_d(t)] \tag{2.12}$$

and choosing the state variables as $x_1 = i_a$ and $x_2 = \omega_m$, the state-space model for the DC motor is

$$\dot{x} = \begin{bmatrix} \dot{x}_1 \\ \dot{x}_2 \end{bmatrix} = \begin{bmatrix} \dfrac{-R_a}{L_a} & \dfrac{-K_e}{L_a} \\ \dfrac{K_T}{J} & \dfrac{-b}{J} \end{bmatrix} \begin{bmatrix} x_1 \\ x_2 \end{bmatrix} + \begin{bmatrix} \dfrac{1}{L_a} & 0 \\ 0 & \dfrac{-1}{J} \end{bmatrix} \begin{bmatrix} v \\ T_d \end{bmatrix} = Ax + Bu$$

$$y = \begin{bmatrix} 1 & 0 \\ 0 & 1 \end{bmatrix} \begin{bmatrix} x_1 \\ x_2 \end{bmatrix} = Cx \tag{2.13}$$

Applying $P(s) = C(sI - A)^{-1}B$, the Laplace "s" transfer functions are

$$\begin{bmatrix} i_a(s) \\ \omega_m(s) \end{bmatrix} = \begin{bmatrix} p_{11}(s) & p_{12}(s) \\ p_{21}(s) & p_{22}(s) \end{bmatrix} \begin{bmatrix} v(s) \\ T_d(s) \end{bmatrix} \tag{2.14}$$

with

$$p_{11}(s) = \frac{(Js + b)}{(Js + b)(L_a s + R_a) + K_T K_e} \tag{2.15}$$

$$p_{12}(s) = \frac{K_e}{(Js + b)(L_a s + R_a) + K_T K_e} \tag{2.16}$$

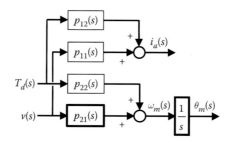

FIGURE 2.3
Armature-controlled DC motor model.

$$p_{21}(s) = \frac{K_T}{(Js + b)(L_a s + R_a) + K_T K_e} \qquad (2.17)$$

$$p_{22}(s) = \frac{-(L_a s + R_a)}{(Js + b)(L_a s + R_a) + K_1 K_e} \qquad (2.18)$$

Also, the motor angular position $\theta_m(s)$ is easily calculated by taking the integral of the rotational velocity—see Equation 2.19.

$$\theta_m(s) = \frac{1}{s}[p_{21}(s)v(s) + p_{22}(s)T_d(s)] \qquad (2.19)$$

Figure 2.3 and Equations 2.15 through 2.18 show the complete 2×2 model for the armature-controlled DC motor. A classical control system for this motor is shown in Figure 2.4. It is composed of two cascade control loops. The inner loop controls the velocity $\omega_m(s)$, and the outer loop the angular position $\theta_m(s)$. The control signal is the voltage $v(s)$ applied to the armature. The external disturbances or loads are characterized by the input $T_d(s)$. Also, the model gives a second additional output: the armature current $i_a(s)$. $r_\theta(s)$ is the tracking reference for the angular position $\theta_m(s)$. $n_\omega(s)$ and $n_\theta(s)$ are the sensor noise for the inner and outer loops, respectively, and $H_\omega(s)$ and $H_\theta(s)$ the feedback dynamics, if any. The control system is composed of the feedback controllers $G_\omega(s)$ and

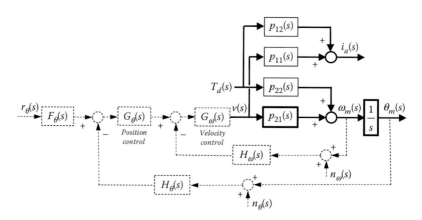

FIGURE 2.4
DC motor. Position and velocity control systems.

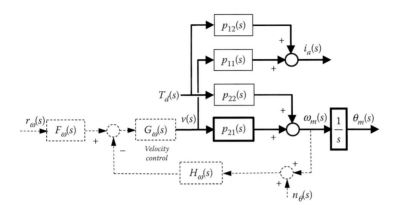

FIGURE 2.5
DC motor. Velocity control system.

$G_\theta(s)$ and the prefilter $F_\theta(s)$. If we only need to control the motor velocity, then the control system is simpler, as shown in Figure 2.5. In this case, $r_\omega(s)$ is the tracking reference for the velocity, the $G_\omega(s)$ feedback controller, and $F_\omega(s)$ the prefilter.

Parametric Uncertainty

A correct definition of the model uncertainty is always important. Every real system has some degree of uncertainty. The reliability of the final controller depends on the appropriate description of the uncertainty. If the uncertainty covers all the possibilities of the plant, a controller that meets all the QFT bounds (see next sections) will assure a good performance for every possible plant within the uncertainty. However, an over-sized uncertainty would make it more difficult or even impossible to find an appropriate controller for the problem. In fact, there is *a trade-off between the size of the uncertainty and the achievable performance of the controller*. This stresses the importance of the precise definition of the uncertainty, covering all the possibilities of the real plant—for robustness, but not more—for performance.

To illustrate these ideas, we present below two cases of parametric uncertainty for the DC motor. The first one (*Case a*) proposes a situation where there are five physical parameters with independent uncertainty. The second one (*Case b*) proposes a problem where four of the previous physical parameters are interrelated with each other, as a function of the temperature T_{emp}. As we will see in Section 2.4, although the minimum and maximum values of each parameter of *Case a* are the same as in *Case b*, the uncertainty in the first case is much larger than in the second one.

Case A

The inertia J, the damping coefficient b, the inductance L_a, the resistance R_a, and the torque coefficient K_T are independent. The transfer function of the plant to be controlled is

$$p_{21}(s) = \frac{\omega_m(s)}{v(s)} = \frac{K_T}{J L_a s^2 + (J R_a + b L_a)s + (b R_a + K_T K_e)} \quad (2.20)$$

with five parameters with independent uncertainty—see also Table 2.1,

$$\begin{aligned} &J \in [J_{min}, J_{max}], \quad b \in [b_{min}, b_{max}] \\ &L_a \in [L_{a\,min}, L_{a\,max}], \quad R_a \in [R_{a\,min}, R_{a\,max}] \\ &K_T \in [K_{T\,min}, K_{T\,max}], \quad K_e = K_T \end{aligned} \quad (2.21)$$

TABLE 2.1

Case a. Parametric Uncertainty for DC Motor Example 2.1

		Minimum	Maximum	Nominal	Grid
J	N m/s^2	4×10^{-5}	8×10^{-5}	6×10^{-5}	5
b	N m/(rad/s)	6×10^{-5}	9×10^{-5}	6×10^{-5}	3
L_a	H	1×10^{-3}	3×10^{-3}	1×10^{-3}	5
R_a	Ω	0.5	0.7	0.7	4
K_T	N m/A	0.04	0.06	0.04	3
K_e	V/(rad/s)		$K_e = K_T$		

In this case, and for the future calculation of the QFT templates—see Section 2.4, we define a discrete number of plants by gridding each parameter with uncertainty between the minimum and the maximum value—see Table 2.1. The column *"grid"* defines the total number of values for each parameter, including both extremes. In this way, the total number of plants is: $5 \times 3 \times 5 \times 4 \times 3 = 900$ plants. The distribution between the minimum and the maximum, with a *"grid"* number of points, can be typically selected as either a linear or a logarithmic distribution. For instance, for the parameter R_a, $grid = 4$, the linear or logarithmic distributions are, respectively:

1. *linspace*(0.5,0.7,4) = [0.5000   0.5667   0.6333   0.7000].
2. *logspace*(log10(0.5),log10(0.7),4) = [0.5000   0.5593   0.6257   0.7000].

The calculation of the 900 plants needs five nested for-loops. It can be done using the QFTCT—see Figure 2.6 and Appendix 2, or just the following MATLAB code.

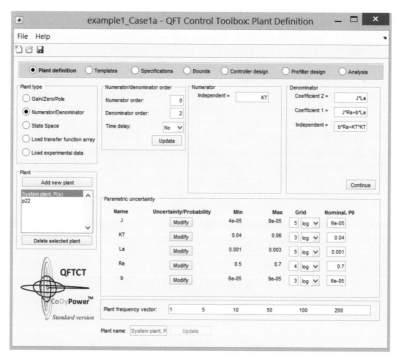

FIGURE 2.6
Plant $p_{21}(s)$ with parametric uncertainty—*Case a.* QFTCT.

```
% -- Parameters: minimum "m", maximum "M", and grid
Jm = 4e-5; JM = 8e-5; i1m = 5;
bm = 6e-5; bM = 9e-5; i2m = 3;
Lam = 1e-3; LaM = 3e-3; i3m = 5;
Ram = 0.5; RaM = 0.7; i4m = 4;
KTm = 0.04; KTM = 0.06; i5m = 3;

% -- Gridding
Jv = logspace(log10(Jm),log10(JM),i1m);
bv = logspace(log10(bm),log10(bM),i2m);
Lav = logspace(log10(Lam),log10(LaM),i3m);
Rav = logspace(log10(Ram),log10(RaM),i4m);
KTv = logspace(log10(KTm),log10(KTM),i5m);

% -- Plants
c = 0;
for i1=1:i1m
  J = Jv(i1);
  for i2=1:i2m
    b = bv(i2);
    for i3=1:i3m
      La = Lav(i3);
      for i4=1:i4m
        Ra = Rav(i4);
        for i5=1:i5m
          KT = KTv(i5);
          c = c + 1;
          P(1,1,c) = tf(KT,[J*La (J*Ra+b*La) (b*Ra+KT*KT)]);
        end
      end
    end
  end
end
```

Case B

In this second option, the damping coefficient b, the inductance L_a, the resistance R_a, and the torque coefficient K_T are interrelated as functions of the temperature T_{emp}, so that

$$b = \alpha_2 T_{emp} + \alpha_1$$
$$L_a = \alpha_4 T_{emp} + \alpha_3$$
$$R_a = \alpha_6 T_{emp} + \alpha_5 \qquad (2.22)$$
$$K_T = \alpha_8 T_{emp} + \alpha_7$$
$$K_e = K_T$$

and being α_1 to α_8 known and fixed coefficients:

$$\alpha_1 = 1.6767 \times 10^{-4}; \quad \alpha_2 = -3.3333 \times 10^{-7}; \quad \alpha_3 = 0.0082; \quad \alpha_4 = -2.2222 \times 10^{-5}$$
$$\alpha_5 = -0.0178; \quad \alpha_6 = 0.0022; \quad \alpha_7 = 0.1118; \quad \alpha_8 = -2.2222 \times 10^{-4}$$

Then the transfer function $p_{21}(s)$ becomes

$$p_{21}(s) = \frac{\omega_m(s)}{v(s)} = \frac{\beta_3}{\beta_2 s^2 + \beta_1 s + \beta_0} \qquad (2.23)$$

TABLE 2.2

Case b. Parametric Uncertainty for DC Motor. Example 2.1

		Minimum	Maximum	Nominal	Grid
J	N m/s^2	4×10^{-5}	8×10^{-5}	6×10^{-5}	10
T_{emp}	Kelvin	233	323	323	20

with

$$
\begin{aligned}
\beta_3 &= (\alpha_8 T_{emp} + \alpha_7) \\
\beta_2 &= J(\alpha_4 T_{emp} + \alpha_3) \\
\beta_1 &= J(\alpha_6 T_{emp} + \alpha_5) + (\alpha_2 T_{emp} + \alpha_1)(\alpha_4 T_{emp} + \alpha_3) \\
\beta_0 &= (\alpha_2 T_{emp} + \alpha_1)(\alpha_6 T_{emp} + \alpha_5) + (\alpha_8 T_{emp} + \alpha_7)^2
\end{aligned}
\tag{2.24}
$$

and with only two parameters with uncertainty, the inertia J and the temperature T_{emp}—see also Table 2.2,

$$
J \in [J_{\min}, J_{\max}], \ T_{emp} \in [T_{emp\min}, T_{emp\max}]
\tag{2.25}
$$

In this case, the calculation of the plants needs only two nested for-loops. It can be easily done by using the QFTCT—see Figure 2.7 and Appendix 2, or the following MATLAB code.

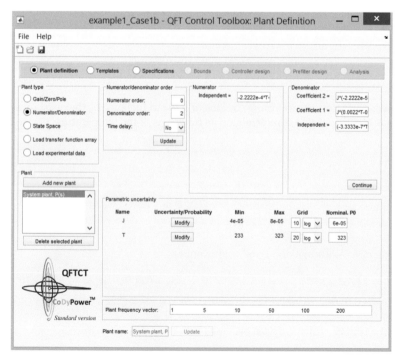

FIGURE 2.7
Plant $p_{21}(s)$ with parametric uncertainty—*Case b.* QFTCT.

```
% -- Parameters: minimum "m", maximum "M", and grid
Jm = 4e-5; JM = 8e-5; i1m = 10;
Tempm = 233; TempM = 323; i2m = 20;
a1 = 1.6767e-4; a2 = -3.3333e-7; a3 = 0.0082; a4 = -2.2222e-5;
a5 = -0.0178; a6 = 0.0022; a7 = 0.1118; a8 = -2.2222e-4;

% -- Gridding
Jv = logspace(log10(Jm),log10(JM),i1m);
Tempv = logspace(log10(Tempm),log10(TempM),i2m);

% -- Plants
c = 0;
for i1=1:i1m
  J = Jv(i1);
  for i2=1:i2m
    Temp = Tempv(i2);
    b3 = (a8*Temp+a7);
    b2 = J*(a4*Temp+a3);
    b1 = J*(a6*Temp+a5)+(a2*Temp+a1)*(a4*Temp+a3);
    b0 = (a2*Temp+a1)^(a6*Temp+a5)+(a8*Temp+a7)^2;
    c = c + 1;
    P(1,1,c) = tf(b3,[b2 b1 b0]);
  end
end
```

Frequencies

As a frequency-domain methodology, QFT works with an array of frequencies of interest—second gridding. The array of frequencies can be selected by inspection of the Bode diagram of a number of plants within the uncertainty.

Figure 2.8 shows the Bode diagram of the plant $p_{21}(s)$ and the uncertainty according to *Case a*. In this case, we select an array of frequencies starting at 1 rad/s, ending at

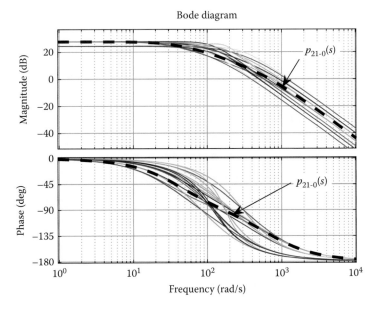

FIGURE 2.8
Bode diagram. Example 2.1, DC motor. Plant $p_{21}(s)$ with uncertainty as *Case a*. See also the nominal plant $p_{210}(s)$.

10,000 rad/s, and adding some more frequencies to properly represent the slopes. The selected array, also included in Figure 2.6—QFTCT, is:

$$\omega = \begin{bmatrix} 1\ 5\ 10\ 50\ 100\ 200\ 300\ 500\ 700\ 1000\ 2000\ 5000\ 10,000 \end{bmatrix} \text{rad/s} \qquad (2.26)$$

2.3 The Nominal Plant: Step 2

See also *Plant definition* window of the QFTCT and W1 in Figure 2.1.

The nominal plant $P_0(s)$ is a designated plant within the model uncertainty. Any plant can be selected as the nominal plant. The final controller designed by QFT will be the same, no matter what nominal plant is chosen. As we will see in the next sections, the nominal plant affects the QFT bounds—see Section 2.7, and the open loop transfer function $L_0(s) = P_0(s)G(s)$ in the same way. As a result, as the controller $G(s)$ gives the relative distance between the bounds and $L_0(s)$, the selection of the nominal plant has no effect on the controller $G(s)$.

Going back to Example 2.1 (*armature-controlled DC motor*), the designated nominal plant is the one at $T_{emp0} = 323$ K and $J_0 = 6 \times 10^{-5}$ N m/s^2 (in *Case b*)—see Table 2.2, or the one with $b_0 = 6 \times 10^{-5}$ N m/(rad/s), $L_{a0} = 1 \times 10^{-3}$ H, $R_{a0} = 0.7\ \Omega$, $K_{T0} = 0.04$ N m/A, $K_{e0} = 0.04$ V/(rad/s), and $J_0 = 6 \times 10^{-5}$ N m/s^2 (in *Case a*)—see Table 2.1. Notice that to facilitate further analysis, the selected nominal plant is the same in both cases. Then, the transfer function $p_{12-0}(s)$ is—see also Figure 2.8, dashed line,

$$\left. \frac{\omega_m(s)}{v(s)} \right|_{\text{nominal}} = p_{21-0}(s) = \frac{K_{T0}}{(J_0 s + b_0)(L_{a0} s + R_{a0}) + K_{T0} K_{e0}} =$$

$$= \frac{0.04}{6 \times 10^{-8} s^2 + 4.206 \times 10^{-5} s + 0.001642} \qquad (2.27)$$

2.4 QFT Templates: Step 3

See also *Templates* window of the QFTCT and W2 in Figure 2.1.

The QFT templates are the projection of the transfer function $P(j\omega)$ onto the Nichols chart (NC), considering each parameter within the uncertainty, and at each frequency of interest ω—see Section 2.2.

Figure 2.9 shows the QFT templates for the Example 2.1, *Case a*. They are calculated by using the uncertainty defined in Equations 2.20 and 2.21 and Table 2.1, and the array of frequencies of interest defined in Equation 2.26. Notice that instead just one line—one point at each frequency, we have a set of points at each frequency (the template) due to the uncertainty. Also, notice that the templates are just vertical lines at low frequencies, become wider at middle frequencies, and finally collapse again into vertical lines at high frequencies. This is a common property that can be used to pick the interval of frequencies of interest. The template for $\omega = 500$ rad/s, which is in the middle of Figure 2.9, is also shown in Figure 2.10a with more detail.

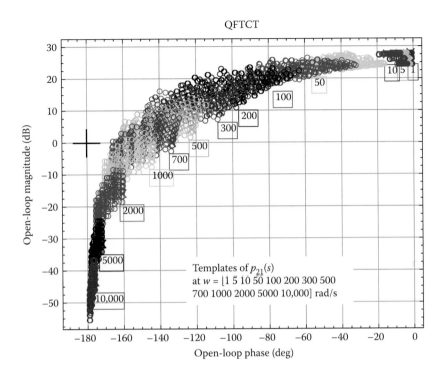

FIGURE 2.9
Nichols chart. Example 2.1, DC motor. Templates. Plant $p_{21}(s)$. *Case a.*

Additionally, Figure 2.10b shows the QFT template of *Case b* at the same frequency $\omega = 500$ rad/s. As we said in Section 2.2, the uncertainty in *Case a* (independent parameters) is much larger than in *Case b* (interdependent parameters). Comparing both templates, we see that the template in the first case (Figure 2.10a) is larger than the one in the second case (Figure 2.10b). In general, the second case allows the designer to achieve a higher

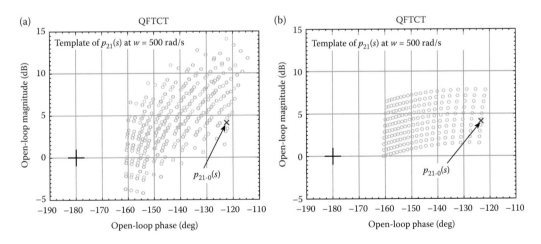

FIGURE 2.10
Nichols chart. Example 2.1, DC motor. Template of plant $p_{21}(s)$ and nominal plant $p_{210}(s)$, $\omega = 500$ rad/s. (a) *Case a* and (b) *Case b*.

control performance. As we saw, there is *a trade-off between the size of the uncertainty and the achievable performance of the controller.* The nominal plant at $\omega = 500$ rad/s, $p_{21\text{-}0}(\omega = 500)$, is also represented in both cases.

2.5 Stability Specifications: Step 4

See also *Specifications* window of the QFTCT and W3 in Figure 2.1.

The stability margins, *gain margin GM* and *phase margin PM*, are commonly used to measure the degree of stability of a closed-loop system. Figure 2.11 shows an example of a stable closed-loop system, with the stability margins represented in the Bode diagram and in the Nichols chart.

Another way to measure the stability is with M_c circles, which represent the locus of constant magnitude W_s of the closed-loop transfer function in the Nichols chart—see Figure 2.12. That is to say: $|P(s)G(s)/[1 + P(s)G(s)]| = W_s = $ constant. As the M_c circles enclose the (0 db, −180°) point, they are also related with the stability margins. Equation 2.28 defines the M_c circle for a constant magnitude W_s, Equation 2.29 describes the M_c circle (W_s) as a function of *PM*, and Equations 2.30 and 2.31 show *PM* and *GM* as a function of the M_c circle (W_s), respectively.

$$\left| \frac{P(j\omega)G(j\omega)}{1 + P(j\omega)G(j\omega)} \right| \leq W_s \tag{2.28}$$

$$W_s = \frac{0.5}{\cos((\pi/180)((180 - PM)/2))}, \text{in magnitude} \tag{2.29}$$

$$PM = 180 - 2\left(\frac{180}{\pi}\right)\text{acos}\left(\frac{0.5}{W_s}\right), \text{in deg} \tag{2.30}$$

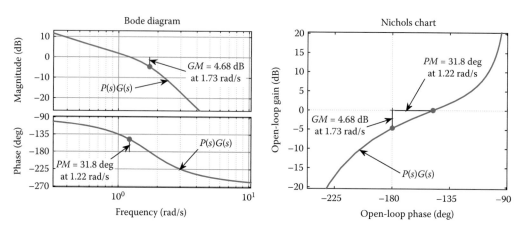

FIGURE 2.11

GM and *PM* stability specs in the Bode diagram and the Nichols chart for $L_1(s) = P_1(s)G_1(s) = 3.5/[s\,(s^2 + 2\,s + 3)]$.

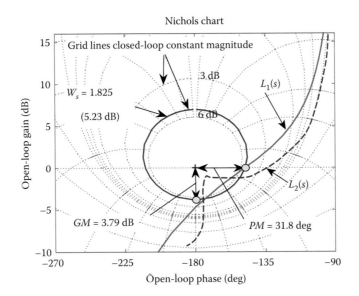

FIGURE 2.12
Stability specification in the Nichols chart for $W_s = 1.825$ (5.23 dB).

$$GM = 20\log_{10}\left(1+\frac{1}{W_s}\right), \text{in dB} \tag{2.31}$$

Using the same phase margin of the example shown in Figure 2.11, $PM = 31.8°$ for $L_1(s)$, the equivalent closed-loop constant magnitude is $W_s = 1.825$ (5.23 dB) according to Equation 2.29. Also, using Equation 2.31, the gain margin is $GM = 3.79$ dB. Figure 2.12 shows the details.

As a result, we can say that it is possible to work either directly with the stability margins GM and PM or with the M_c circle W_s. The second option (W_s) is more reliable than the stability margins GM/PM option, as it prevents cases such as $L_2(s) = P_2(s)G_2(s)$, showed in Figure 2.12, which is correct from the GM and PM point of view, but still very close to be unstable—it is quite close to encircle the point (0 db, –180°). For this reason, QFT uses M_c circles to define the stability of the closed-loop system. Chapter 3 completes this discussion and develops a comprehensive study of stability in the frequency domain.

2.6 Performance Specifications: Step 5

See also *Specifications* window of the QFTCT and W3 in Figure 2.1.

The classical two-degree of freedom (2DOF) control system includes a loop compensator $G(s)$ and a prefilter $F(s)$ in series with the feedback loop—see Figure 2.13. Based on this structure, the following well-known equations for the plant output $y(s)$, control signal $u(s)$, and signal error $e(s)$ are obtained:

$$y = \frac{PG}{1+PGH}Fr + \frac{1}{1+PGH}(PGd_e + Pd_i + d_o + Md) - \frac{PGH}{1+PGH}n \tag{2.32}$$

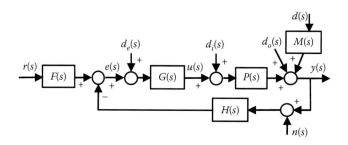

FIGURE 2.13
Classical 2DOF control system.

$$u = \frac{G}{1+PGH}Fr + \frac{G}{1+PGH}(d_e - HPd_i - Hd_o - HMd) - \frac{GH}{1+PGH}n \tag{2.33}$$

$$e = \frac{1}{1+PGH}Fr - \frac{H}{1+PGH}(PGd_e - Pd_i - d_o - Md) - \frac{H}{1+PGH}n \tag{2.34}$$

where $P(s)$ denotes the plant with uncertainty, $H(s)$ the dynamics of the sensor, filters, and feedback path, $M(s)$ the dynamics of generic disturbances $d(s)$ over the plant output, and $r(s)$, $n(s)$, $d_e(s)$, $d_i(s)$, $d_o(s)$ represent the reference signal for tracking, the sensor noise, and the external disturbances at the error signal, plant input and plant output, respectively. The Laplace variable s is omitted in the equations for simplicity.

As developed originally by Horowitz,[33] QFT uses the transfer functions between the inputs $[r(s), n(s), d_e(s), d_i(s), d_o(s), d(s)]$ and the outputs $[y(s), u(s), e(s)]$, see Equations 2.32 through 2.34, to define the stability and performance specifications in terms of inequalities in the frequency domain. In particular, we express each specification as $\{|T_k(j\omega_i)| \le \delta_k(\omega_i),$ $\omega_i \in \Omega_k, k = 1, 2, \ldots; i = 1, 2, \ldots\}$, where $T_k(j\omega_i)$ are the transfer functions defined in Equations 2.32 through 2.34, and $\delta_k(\omega_i)$ the magnitude of each specification in the frequency domain. Making $H(s) = 1$ without loss of generality, the resulting inequalities are:

- *Type 1: Stability specification*

$$|T_1(j\omega)| = \left|\frac{P(j\omega)G(j\omega)}{1+P(j\omega)G(j\omega)}\right| \le \delta_1(\omega) = W_s, \omega \in \Omega_1$$

$$\text{where } |T_1(j\omega)| = \left|\frac{y(j\omega)}{F(j\omega)r(j\omega)}\right| \tag{2.35}$$

- *Type 2: Complementary sensitivity specification*

$$|T_2(j\omega)| = \left|\frac{P(j\omega)G(j\omega)}{1+P(j\omega)G(j\omega)}\right| \le \delta_2(\omega), \omega \in \Omega_2$$

$$\text{where } |T_2(j\omega)| = \left|\frac{y(j\omega)}{F(j\omega)r(j\omega)}\right| = \left|\frac{y(j\omega)}{n(j\omega)}\right| = \left|\frac{u(j\omega)}{d_i(j\omega)}\right| = \left|\frac{y(j\omega)}{d_e(j\omega)}\right| = \left|\frac{e(j\omega)}{d_e(j\omega)}\right| \tag{2.36}$$

- *Type 3: Sensitivity or disturbances at plant output specification*

$$|T_3(j\omega)| = \left|\frac{1}{1 + P(j\omega)G(j\omega)}\right| \le \delta_3(\omega), \omega \in \Omega_3$$ (2.37)

$$\text{where } |T_3(j\omega)| = \left|\frac{y(j\omega)}{d_o(j\omega)}\right| = \left|\frac{e(j\omega)}{d_o(j\omega)}\right| = \left|\frac{e(j\omega)}{F(j\omega)r(j\omega)}\right| = \left|\frac{e(j\omega)}{n(j\omega)}\right|$$

- *Type 4: Disturbances at plant input specification*

$$|T_4(j\omega)| = \left|\frac{P(j\omega)}{1 + P(j\omega)G(j\omega)}\right| \le \delta_4(\omega), \omega \in \Omega_4$$ (2.38)

$$\text{where } |T_4(j\omega)| = \left|\frac{y(j\omega)}{d_i(j\omega)}\right| = \left|\frac{e(j\omega)}{d_i(j\omega)}\right|$$

- *Type 5: Control effort reduction specification*

$$|T_5(j\omega)| = \left|\frac{G(j\omega)}{1 + P(j\omega)G(j\omega)}\right| \le \delta_5(\omega), \omega \in \Omega_5$$ (2.39)

$$\text{where } |T_5(j\omega)| = \left|\frac{u(j\omega)}{d_e(j\omega)}\right| = \left|\frac{u(j\omega)}{d_o(j\omega)}\right| = \left|\frac{u(j\omega)}{n(j\omega)}\right| = \left|\frac{u(j\omega)}{F(j\omega)r(j\omega)}\right|$$

- *Type 6: Reference tracking specification*

$$\delta_{6\_lo}(\omega) < |T_6(j\omega)| = \left|\frac{y(j\omega)}{r(j\omega)}\right| = \left|F(j\omega)\frac{P(j\omega)G(j\omega)}{1 + P(j\omega)G(j\omega)}\right| \le \delta_{6\_up}(\omega),$$ (2.40)

$$\omega \in \Omega_6$$

or

$$\frac{|P_d(j\omega)G(j\omega)|}{|P_e(j\omega)G(j\omega)|}\frac{|1 + P_e(j\omega)G(j\omega)|}{|1 + P_d(j\omega)G(j\omega)|} \le \delta_6(\omega) = \frac{\delta_{6\_up}(\omega)}{\delta_{6\_lo}(\omega)}, \omega \in \Omega_6$$

- *Type k: General disturbance rejection specification*

$$|T_k(j\omega)| = \left|\frac{M(j\omega)}{1 + P(j\omega)G(j\omega)}\right| \le \delta_k(\omega), \omega \in \Omega_k$$ (2.41)

$$\text{where } |T_k(j\omega)| = \left|\frac{y(j\omega)}{d(j\omega)}\right| = \left|\frac{e(j\omega)}{d(j\omega)}\right| = \left|\frac{u(j\omega)}{G(j\omega)d(j\omega)}\right|$$

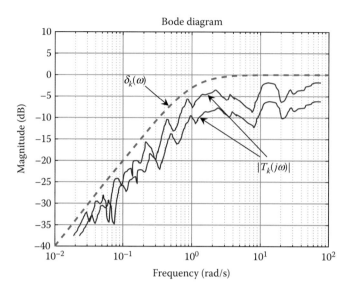

FIGURE 2.14
Specification $|T_k(j\omega)| \leq \delta_k(\omega) = |s/(s+1)|$.

Figure 2.14 shows an example of the Bode diagram for the specification $|T_k(j\omega_i)| \leq \delta_k(\omega_i)$. Following this inequality, the magnitude of the transfer function $|T_k(j\omega)|$, including all the possibilities due to the uncertainty, has to be below the limit $\delta_k(\omega)$ at every frequency of interest.

Some practical tips to define practical robust stability and performance specifications, the $\delta_k(\omega)$ functions, are included in Section 2.14.

2.7 QFT Bounds: Steps 6 through 8

See also *Bounds* window of the QFTCT and W4 in Figure 2.1.

Every plant in the ω_i template can be expressed in its polar form as $P(j\omega_i) = p\, e^{j\theta} = p\angle\theta$. Likewise, the compensator polar form is $G(j\omega_i) = g\, e^{j\varphi} = g\angle\varphi$. By substituting these polar forms in Equations 2.35 through 2.41 and rearranging the expressions, the quadratic inequalities of Equations 2.42 through 2.48 are calculated.

- *Type 1: Stability specification*

$$p^2\left(1 - \frac{1}{\delta_1^2}\right)g^2 + 2p\cos(\varphi + \theta)g + 1 \geq 0 \tag{2.42}$$

- *Type 2: Complementary sensitivity specification*

$$p^2\left(1 - \frac{1}{\delta_2^2}\right)g^2 + 2\,p\cos(\varphi + \theta)g + 1 \geq 0 \tag{2.43}$$

- *Type 3: Sensitivity or disturbances at plant output specification*

$$p^2g^2 + 2p\cos(\phi + \theta)g + \left(1 - \frac{1}{\delta_3^2}\right) \geq 0 \qquad (2.44)$$

- *Type 4: Disturbances at plant input specification*

$$p^2g^2 + 2p\cos(\phi + \theta)g + \left(1 - \frac{p^2}{\delta_4^2}\right) \geq 0 \qquad (2.45)$$

- *Type 5: Control effort reduction specification*

$$\left(p^2 - \frac{1}{\delta_5^2}\right)g^2 + 2p\cos(\phi + \theta)g + 1 \geq 0 \qquad (2.46)$$

- *Type 6: Reference tracking specification*

$$p_e^2 p_d^2\left(1 - \frac{1}{\delta_6^2}\right)g^2 + 2p_e p_d\left(p_e\cos(\phi + \theta_d) - \frac{p_d}{\delta_6^2}\cos(\phi + \theta_e)\right)g + \left(p_e^2 - \frac{p_d^2}{\delta_6^2}\right) \geq 0 \qquad (2.47)$$

- *Type k: General disturbance rejection specification*

$$p^2g^2 + 2p\cos(\phi + \theta)g + \left(1 - \frac{m^2}{\delta_k^2}\right) \geq 0 \qquad (2.48)$$

Notice that for every frequency ω_i there is a constant $\delta_k = \delta_k(j\omega_i)$, and for a fixed plant $p\angle\theta$ in the ω_i template and a fixed controller phase ϕ in $[-360°,0°]$, the unknown parameter of the inequalities is the controller magnitude g. The format of all quadratic expressions is the same:

$$I_{\omega_i}^k(p, \theta, \delta_k, \phi) = ag^2 + bg + c \geq 0 \qquad (2.49)$$

where a, b, c depend on p, θ, ϕ, and δ_k.

Now, with an appropriate algorithm—see Table 2.3 and Reference 60, the quadratic inequalities of Equations 2.42 through 2.48 are solved and translated into a set of curves on the Nichols chart for each frequency of interest and type of specification. These are the *QFT bounds*.

Notice that we have a bound $B_k(\omega_i)$ at each frequency of interest ω_i and for each specification k. The bounds can be solid or dashed lines. A solid line means that the area that meets the specification inequality is above the line, and a dashed line means that the area that meets the specification inequality is below the line.

If we have more than one specification, we need to find the intersection of all the bounds at each frequency, or worst case scenario bound. Figures 2.17 and 2.18 discuss some examples.

TABLE 2.3

Algorithm to Compute the QFT Bounds[60]

---

1. Discretize the frequency domain ω into a finite set $\Omega_k = \{\omega_i, i,...,n\}_k$.

2. Establish the uncertain plant models $\{P(j\omega)\}$ and map its boundary for each frequency $\omega_i \in \Omega_k$ on the Nichols chart: They are the n templates $\{P(j\omega_i)\}$, $i = 1,...,n$. Each one contains m points or plants: $P(j\omega_i) = \{P_r(j\omega_i) = p \angle \theta, r = 0,..., m-1\}$.

3. Select one plant as the nominal plant $P_0(j\omega_i) = p_0 \angle \theta_0$.

4. The compensator is $G(j\omega_i) = g \angle \phi$. Define a range, Φ, for the compensator's phase ϕ, and discretize it, like $\phi \in \Phi = [-360°: 5°: 0°]$.

5. First ***For-loop***: (for the frequencies $\Omega_k = \{\omega_i, i = 1,...,n\}_k$) Choose a single frequency $\omega_i \in \Omega_k$.

6.   Second ***For-loop***: (for the phases $\phi \in \Phi = [-360°: 5°: 0°]$) Choose a single controller's phase $\phi \in \Phi$.

7.   Third ***For-loop***: (for the m plants $P_r(j\omega_i)$, $r = 0,...,m-1$) Choose a single plant: $P_r(j\omega_i) = p \angle \theta$.

8.   Compute the maximum $g_{max} = g_{max}(P_r)$ and the minimum $g_{min} = g_{min}(P_r)$ of the two roots g_1 and g_2 that solve the k quadratic inequality.

9.   Choose the most restrictive of the m $g_{max}(P_r)$ and the m $g_{min}(P_r)$. Thus, $g_{max}(P)$ and $g_{min}(P)$ are obtained. They are the maximum and minimum bound points for the controller magnitude g at a phase ϕ.

10. The union of $g_{max}(P)$ and $g_{min}(P)$ along $\phi \in \Phi$ gives $g_{max} \angle \varphi$ and $g_{min} \angle \phi$, for each frequency ω, respectively.

11. The open-loop transmission is $L_0(j\omega_i) = l_0 \angle \psi_0$. Now set $l_{0\,max} \angle \psi_0 = p_0 g_{max} \angle \phi$ and $l_{0\,min} \angle \psi_0 = p_0 g_{min} \angle \phi$, being $\psi_0 = \phi + \theta_0$, $\phi = [-360°: 5°: 0]$. These are the bounds: $\{B_k(j\omega_i), \omega_i \in \Omega_k\}$.

The QFT bounds are defined from the control specifications—Equations 2.35 through 2.41, and taking into account all the model uncertainty. As a result, once the bounds are plotted in the Nichols chart, we only need to deal with the nominal plant $P_0(s)$ to find a controller $G(s)$ that meets the bounds. This is one of the great syntheses of the QFT technique. Instead of dealing with an infinite number of plants, the controller design step only needs to deal with the nominal plant $P_0(s)$, as all the model uncertainty has been already included in the QFT bounds—see also the algorithm showed in Table 2.3.

Thus, to guarantee that all the plants within the uncertainty meet a given specification at a given frequency ω_i, $L_0(\omega_i)$ must be in the correct area: above or below the bound for solid or dashed lines, respectively. As $L_0(s) = P_0(s)G(s)$, as we design an appropriate controller $G(s)$ we move (loop shape) $L_0(s)$ to the correct area at each frequency of interest—see Section 2.8. In fact, as Horowitz pointed out, an optimum controller design in the sense of no-overdesign is the one that places $L_0(\omega_i)$ on top of the bound $B_k(\omega_i)$ at each frequency of interest ω_i.

EXAMPLE 2.1 (CONT.): ARMATURE-CONTROLLED DC MOTOR. CASE A

To illustrate these concepts, let us again consider Example 2.1 (*armature-controlled DC motor*) with the uncertainty defined in *Case a*. Looking at Equations 2.35 through 2.41, we pick the following three specifications:

- *Type 1: Stability specification*

$$|T_1(j\omega)| = \left| \frac{P(j\omega)G(j\omega)}{1 + P(j\omega)G(j\omega)} \right| \le \delta_1(\omega) = W_s = 1.46$$

(2.50)

$$\omega \in [1\ 5\ 10\ 50\ 100\ 200\ 300\ 500\ 700\ 1000\ 2000\ 5000\ 10{,}000]\ \text{rad/s}$$

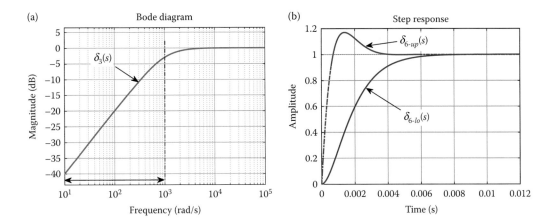

FIGURE 2.15
Control specifications for $p_{21}(s)$. (a) Sensitivity or disturbance rejection at the output of the plant: δ_3. (b) Reference tracking; δ_{6_up}, δ_{6_lo}.

which is equivalent to PM = 40.05°, GM = 4.53 dB—see Equations 2.30 and 2.31.

- *Type 3: Sensitivity or disturbances at plant output specification*—see Figure 2.15a

$$|T_3(j\omega)| = \left|\frac{1}{1+P(j\omega)G(j\omega)}\right| \leq \delta_3(\omega) = \frac{(s/a_d)}{(s/a_d)+1}; \quad a_d = 1000$$

$$\omega \in [1\ 5\ 10\ 50\ 100\ 200\ 300\ 500\ 700\ 1000]\ \text{rad/s} \tag{2.51}$$

- *Type 6: Reference tracking specification*—see Figure 2.15b

$$\delta_{6\_lo}(\omega) < |T_6(j\omega)| = \left|F(j\omega)\frac{P(j\omega)G(j\omega)}{1+P(j\omega)G(j\omega)}\right| \leq \delta_{6\_up}(\omega)$$

$$\omega \in [1\ 5\ 10\ 50\ 100\ 200\ 300\ 500\ 700\ 1000\ 2000\ 5000]\ \text{rad/s} \tag{2.52}$$

$$\delta_{6\_lo}(s) = \frac{1}{[(s/a_L)+1]^2}; \quad a_L = 1000 \tag{2.53}$$

$$\delta_{6\_up}(s) = \frac{[(s/a_U)+1]}{[(s/\omega_n)^2+(2\zeta s/\omega_n)+1]}; \quad a_U = 1000; \quad \zeta = 0.8; \quad \omega_n = \frac{1.25a_U}{\zeta} \tag{2.54}$$

Now, from these specifications—Equations 2.50 through 2.54, the plant model and the uncertainty—Equations 2.20, 2.21, and Table 2.1, and applying the algorithm introduced in Table 2.3, or the *QFTCT*, we obtain the QFT bounds for stability, sensitivity, and reference tracking—see Figure 2.16a, b, and c, respectively. Afterward, the most restrictive bound at every phase and each frequency of interest is computed to obtain the intersection of bounds—see Figures 2.16d and 2.17. In other words, for each set of bounds of the same frequency (same color), we select at each phase value (horizontal axis) the maximum vertical point of the solid-lines and the minimum vertical point of the dashed-lines.

At this point, a practical compatibility problem can arise when calculating the intersection of bounds and different-nature control specifications are simultaneously required.

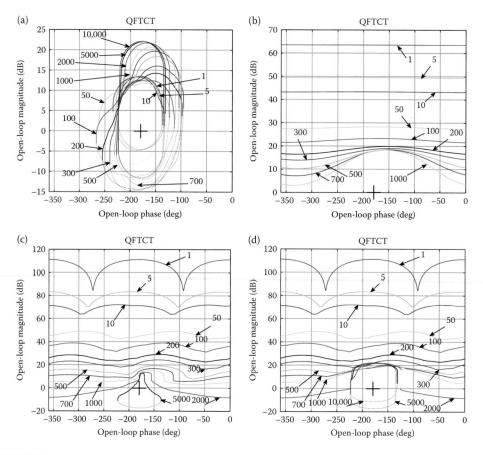

FIGURE 2.16
QFT bounds. Example 2.1, *Case a*. (a) Stability bounds, (b) sensitivity bounds, (c) reference tracking bounds, and (d) intersection of bounds.

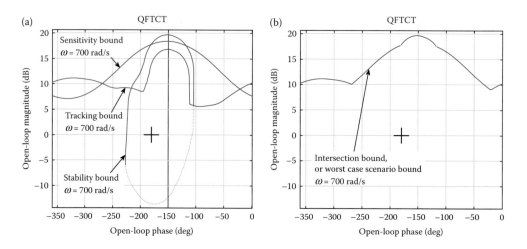

FIGURE 2.17
Example of QFT bounds for $\omega = 700$ rad/s. (a) Three bounds: stability, sensitivity, and reference tracking, (b) intersection of bounds, or most restrictive bound.

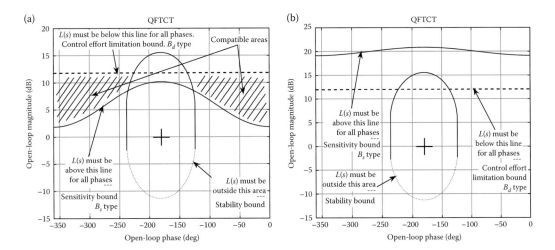

FIGURE 2.18
Compatibility of specifications (bounds) for stability, sensitivity, and control effort limitation at a given frequency. Cases: (a) compatible and (b) incompatible.

A typical case is when control effort limitations are explicitly considered along with the classical stability, disturbance rejection, or reference tracking specifications.

Graphically speaking, the control effort limitations generate a "0 to −360°" dashed-line type of bound—let's call it $B_d(j\omega_i)$, which requires $L(j\omega_i)$ to be below $B_d(j\omega_i)$ at each frequency ω_i. On the contrary, the classical stability, disturbance rejection, or reference tracking specifications typically generate a solid-line type of bound—let us call it $B_s(j\omega_i)$, which requires $L(j\omega_i)$ to be above $B_s(j\omega_i)$ at each frequency ω_i.

When $B_s(j\omega_i) < B_d(j\omega_i)$ at some phase angles, the specifications are compatible, because there are areas in the Nichols chart where we can simultaneously meet the two conditions: $L(j\omega_i) < B_d(j\omega_i)$ and $L(j\omega_i) > B_s(j\omega_i)$ at the frequency ω_i—see striped areas in Figure 2.18a. However, if $B_s(j\omega_i) > B_d(j\omega_i)$ for all phase angles, then the specifications are not compatible. In this case, it is not possible to simultaneously meet the two conditions: $L(j\omega_i) < B_d(j\omega_i)$ and $L(j\omega_i) > B_s(j\omega_i)$—see Figure 2.18b.

As we see, this graphical representation of the QFT bounds in the Nichols chart is a useful tool to analyze the compatibility of the control specifications before any controller design is attempted. In case of finding some incompatibilities, the designer can easily study each specification and the model uncertainty at every frequency, to understand the cause of such incompatibility and relax the specifications properly. For additional information about this compatibility problem, see also Reference 91.

2.8 Controller Design, $G(s)$—Loop Shaping: Step 9

See also *Controller design* window of the QFTCT and W5 in Figure 2.1.

The design of a controller for an infinite number of plants, or plants with model uncertainty, and for a set of simultaneous control specifications is a very arduous task at first. It usually requires the design and redesign of the controller for each plant and specification in a very laborious process. However, QFT gives an elegant and practical solution for this problem. It integrates the information associated to the model uncertainty and all the control specifications in a set of simple curves: the QFT bounds. This allows us to use just a single plant, the nominal plant $P_0(s)$, to design the controller $G(s)$.

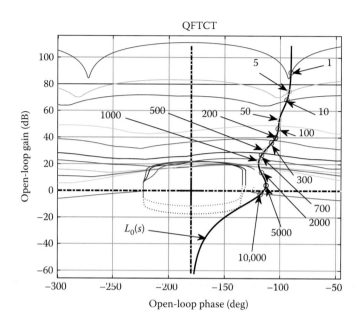

FIGURE 2.19
QFT bounds and G(s) design—loop shaping.

The design, or loop shaping, is carried out on the Nichols chart. It is done by adding poles and zeros until the nominal loop, defined as $L_0(j\omega) = P_0(j\omega)G(j\omega)$, lies near its bounds—see Figure 2.19.

As mentioned, the bounds express the plant model with uncertainty and the control specifications at each frequency. An optimal controller in the sense of QFT is obtained if $L_0(j\omega)$ lies exactly on top of the bounds at each frequency. Practically speaking a good controller design will place $L_0(j\omega)$ above the solid-line bounds and below the dashed-line bounds. This will give us also the minimum possible controller magnitude at every frequency (minimum *cost of feedback*) to meet the required specifications with all the plants within the uncertainty.

A general formulation for the controller structure $G(s)$ is expressed by the following transfer function:

$$G(s) = k_G \frac{\prod_{i=1}^{nrz}((s/z_i)+1)\prod_{i=1}^{ncz}\left((s^2/\omega_{ni}^2)+(2\zeta_i/\omega_{ni})s+1\right)}{s^r\prod_{j=1}^{mrp}((s/p_j)+1)\prod_{j=1}^{mcp}\left((s^2/\omega_{nj}^2)+(2\zeta_j/\omega_{nj})s+1\right)} \tag{2.55}$$

where k_G is the controller gain, z_i is a real zero, ω_{ni} the natural frequency and ζ_i the damping of a complex zero, and n_{rz} and n_{cz} the number of real and complex zeros, respectively. Also, p_j is a real pole, ω_{nj} the natural frequency and ζ_j the damping of a complex pole, and m_{rp} and m_{cp} the number of real and complex poles, respectively. The controller may have also some poles at the origin (integrators), with $r = 0$, 1, or 2, etc.

Some practical tips to design the controller $G(s)$, adding poles and zeros to shape the function $L_0(s) = P_0(s)G(s)$ and meet the QFT bounds, are included in Section 2.14.

EXAMPLE 2.1 (CONT.): ARMATURE-CONTROLLED DC MOTOR. CASE A

To understand these notions, let us consider again the Example 2.1 (*armature-controlled DC motor*) with the uncertainty defined in *Case a*. Looking at the intersection of bounds presented in Figure 2.16d, we design a controller (a PID with a low-pass filter) that meets these QFT bounds, as shown in Equation 2.56 and Figure 2.19.

The small circles in the line $L_0(s)$ and the associated numbers indicate the frequencies (in rad/s) for the function $L_0(s) = P_0(s)G(s)$. The objective is to move each small circle above the corresponding solid-line bounds or below the dashed-line bounds. The proposed controller $G(s)$ meets well all the bounds. The QFTCT, *Controller design* window, can be used to easily design this controller—see also Section 2.14.

$$G(s) = \frac{1100\big((s/55)+1\big)\big((s/1700)+1\big)}{s\big((s/25{,}000)+1\big)} \tag{2.56}$$

2.9 Prefilter Design, $F(s)$: Step 10

See also *Prefilter design* window of the QFTCT and W6 in Figure 2.1.

If the feedback system includes a reference tracking problem, then the best choice is to use a prefilter $F(s)$—the second degree of freedom of the control system. While the feedback controller $G(s)$ reduces the effect of the uncertainty and improves stability, disturbance rejection, and other specifications, the prefilter $F(s)$ is designed to meet reference tracking requirements. Figure 2.20 shows a typical prefilter design in the Bode diagram.

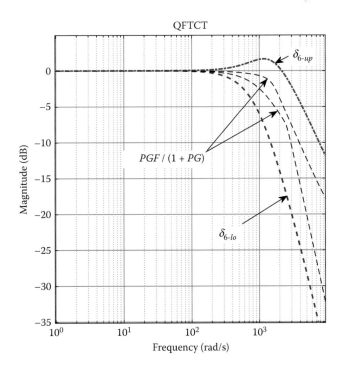

FIGURE 2.20
Prefilter $F(j\omega)$ for reference tracking specs: $\delta_{6\text{-}lo}(\omega) \leq |T_6| \leq \delta_{6\text{-}up}(\omega)$.

$\delta_{6\text{-}up}(\omega)$ and $\delta_{6\text{-}lo}(\omega)$ are the reference tracking specifications, defined as a band—see outer lines. The transfer function T_6 shows an upper and a lower edge—see the inner lines, due to the plant uncertainty. After an appropriate prefilter is designed, the band of T_6 will be between the $\delta_{6\text{-}up}(\omega)$ and $\delta_{6\text{-}lo}(\omega)$ limits: $\delta_{6\text{-}lo}(\omega) \leq |T_6| \leq \delta_{6\text{-}up}(\omega)$, with,

$$|T_6(j\omega)| = \left|\frac{y(j\omega)}{r(j\omega)}\right| = \left|\frac{P(j\omega)G(j\omega)}{1+P(j\omega)G(j\omega)}F(j\omega)\right| \qquad (2.57)$$

Some practical tips to design the prefilter $F(s)$, adding poles and zeros to modify the functions $P(s)G(s)F(s)/[1 + P(s)G(s)]$ and meet the reference tracking limits are included in Section 2.14.

NOTE: The calculation of the QFT bounds for reference tracking requires model uncertainty, as Equation 2.47 needs two plants. For this reason, in the QFTCT the reference tracking specification T_6 and the prefilter window are activated only for plants with uncertainty. Hint: If the problem deals with a fixed plant, we can always define one parameter with a very small uncertainty to activate the reference tracking specification and prefilter design.

EXAMPLE 2.1 (CONT.): ARMATURE-CONTROLLED DC MOTOR. CASE A

To understand these concepts, we consider again the Example 2.1 (*armature-controlled DC motor*) with the uncertainty defined in *Case a*.

Taking into account the controller $G(s)$—see Equation 2.56, the reference tracking specifications—see Equations 2.52 through 2.54, and the plant model—see Equations 2.20, 2.21, and Table 2.1, we design a prefilter $F(s)$ to make sure that all the input/output functions $P(s)G(s)F(s)/[1 + P(s)G(s)]$ are inside the band defined by the limits $\delta_{6\text{-}up}(\omega)$ and $\delta_{6\text{-}lo}(\omega)$. The prefilter is shown in Equation 2.58, and the input/output functions and limits in Figure 2.20.

$$F(s) = \frac{1}{(s/1100)+1} \qquad (2.58)$$

2.10 Analysis and Validation: Steps 11 through 13

See also *Analysis* window of the QFTCT and W7 in Figure 2.1.

Once the design of the controller (and prefilter if needed) is finished, it will be convenient to analyze the performance of the complete control system under different scenarios. This includes (a) frequency-domain analysis of each specification and for all the significant plants within the model uncertainty; and (b) time-domain simulations for representative cases within the uncertainty. The time-domain simulation will be performed first with the linear system, and then including all the existing nonlinear elements, if any.

EXAMPLE 2.1 (CONT.): ARMATURE-CONTROLLED DC MOTOR. CASE A

Again, to understand these notions we consider the Example 2.1 (*armature-controlled DC motor*) with the uncertainty defined in *Case a*. For the analysis and validation, we use the control block diagram shown in Figure 2.5—velocity control, with the controller $G_\omega(s) = G(s)$ in Equation 2.56, the prefilter $F_\omega(s) = F(s)$ in Equation 2.58, $H_\omega(s) = 1$, and the plant model with uncertainty—see Equations 2.20, 2.21, and Table 2.1.

The analysis of the closed-loop stability in the frequency domain is shown in Figure 2.21. The dashed line is the stability specification W_s, defined in Equation 2.50, and the solid line represents the worst case of the function $PG/(1 + PG)$ at each frequency due to the model uncertainty. The control system meets the stability specification, as the solid line is below the dashed line W_s in all cases.

The frequency-domain analysis of the sensitivity specification, or disturbance rejection at the output of the plant specification, is shown in Figure 2.22. The dashed line is the sensitivity specification $\delta_3(\omega)$, defined in Equation 2.51, and the solid line represents the worst case of the function $1/(1 + PG)$ at each frequency due to the model uncertainty. The control system meets the sensitivity specification in all cases, as the solid line is below the dashed line δ_3 from 0 to 1000 rad/s.

The time-domain analysis of the reference tracking specification is shown in Figure 2.23. The figure shows the limits $\delta_{6-up}(\omega)$ and $\delta_{6-lo}(\omega)$—see Equations 2.54 and 2.53, and the time responses $y(t)$ of 900 cases of $PGF/(1 + PG)$ to a unitary step reference $r(t)$. The control system meets the specification (is between the upper and lower limits) in all cases.

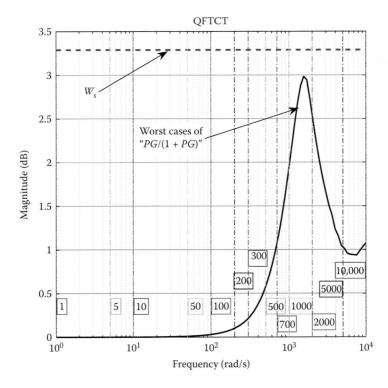

FIGURE 2.21
Stability analysis, frequency domain: W_s specification (dashed line), and worst case of $PG/(1 + PG)$ within the uncertain plants (solid line) at each frequency.

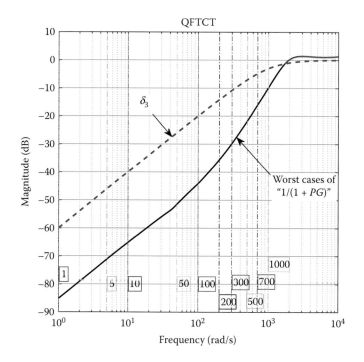

FIGURE 2.22

Sensitivity analysis, frequency domain: $\delta_3(\omega)$ specification (dashed line), and worst case of $1/(1 + PG)$ within the uncertain plants (solid line) at each frequency.

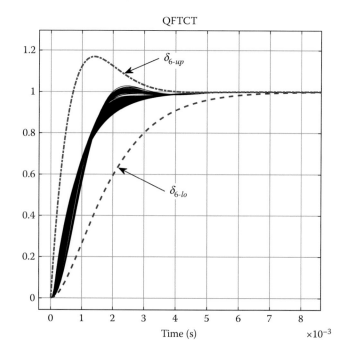

FIGURE 2.23

Reference tracking analysis, time domain: $\delta_{6\text{-}lo}$ and $\delta_{6\text{-}up}$ specs, and time responses $y(t)$ of 900 cases of $PGF/(1 + PG)$ to a unitary step reference $r(t)$.

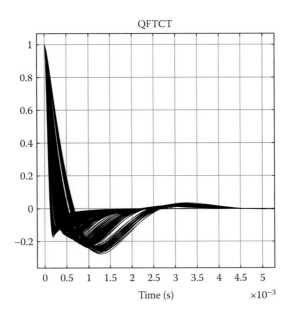

FIGURE 2.24
Sensitivity or plant output disturbance rejection: time-domain responses $y(t)$ of 900 cases of $1/(1 + PG)$ to a unitary step disturbance $d(t)$.

Finally, the time-domain analysis of the sensitivity, or disturbance rejection at the output of the plant, is shown in Figure 2.24. The figure shows the time responses $y(t)$ of 900 cases of $1/(1 + PG)$ to a unitary step disturbance $d(t)$. The control system achieves a good disturbance rejection in all cases.

EXAMPLE 2.2: AIRPLANE FLIGHT CONTROL

This section presents an additional example to illustrate the QFT methodology. The example is taken from the References 359 and 78. It consists of an automatic flight control system for an aircraft. We apply again the steps of the QFT methodology (Figure 2.1) to exemplify different aspects of the technique. The problem is a 2DOF controller as shown in Figure 2.13. The QFTCT can be used to follow the design introduced next.

Step 1. Define plant models and uncertainty: QFTCT plant definition window

The plant for the aircraft takes the form

$$P(s) = \frac{k((s/z)+1)}{s((s/p)+1)[(s/\omega_n)^2 + (2\zeta/\omega_n)s+1]} \tag{2.59}$$

where the uncertain parameters are given by

$$k \in [0.2, 2.0], \quad z \in [0.5, 0.75], \quad p \in [1.0, 10.0], \quad \omega_n \in [5.0, 6.0], \quad \zeta \in [0.8, 0.9]$$

The array of selected frequencies is:

$$\omega = [0.001\ 0.01\ 0.1\ 0.5\ 1\ 4\ 10\ 30\ 150\ 300]\ \text{rad/s}$$

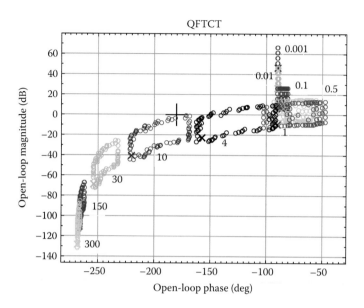

FIGURE 2.25
QFT templates for $\omega = [0.001\ 0.01\ 0.1\ 0.5\ 1\ 4\ 10\ 30\ 150\ 300]$ rad/s. Nominal plant marked with "x."

Step 2. Select nominal plant $P_0(s)$: QFTCT Plant definition window

The selected nominal plant is

$$P_0(s) = \frac{k_0((s/z_0) + 1)}{s((s/p_0) + 1)[(s/\omega_{n0})^2 + (2\zeta_0/\omega_{n0})s + 1]};$$

with $k_0 = 0.2,\quad z_0 = 0.5,\quad p_0 = 1,\quad \omega_{n0} = 5,\quad \zeta_0 = 0.8$ \hfill (2.60)

Step 3. Calculate QFT templates at given ω_i: QFTCT Templates window

The QFT templates for the plant defined in Equation 2.59 and the frequencies of interest, $\omega = [0.001\ 0.01\ 0.1\ 0.5\ 1\ 4\ 10\ 30\ 150\ 300]$ rad/s, are shown in Figure 2.25.

Step 4. Define stability specifications: QFTCT Specifications window

The desired robust stability specification is—see also Figure 2.26,

$$\left| \frac{P(j\omega)G(j\omega)}{1 + P(j\omega)G(j\omega)} \right| \leq W_s = 1.2(GM = 5.26\,\text{dB},\ PM = 49.25°)$$

$$\text{for } \omega = \begin{bmatrix} 0.001\ 0.01\ 0.1\ 0.5\ 1\ 4\ 10\ 30\ 150\ 300 \end{bmatrix}\text{rad/s} \hspace{1cm} (2.61)$$

Step 5. Define performance specifications: QFTCT Specifications window

The desired robust performance specifications are
- Sensitivity or output disturbance rejection specification—see Figure 2.27a:

$$\left| \frac{1}{1 + P(j\omega)G(j\omega)} \right| \leq \delta_{op}(\omega) = \left| \frac{0.5(j\omega)^3 + 2(j\omega)^2 + 2.8(j\omega)}{0.5(j\omega)^3 + 2(j\omega)^2 + 2.8(j\omega) + 1} \right| \hspace{0.5cm} (2.62)$$

$$\text{for } \omega = \begin{bmatrix} 0.001\ 0.01\ 0.1\ 0.5 \end{bmatrix}\text{rad/s}$$

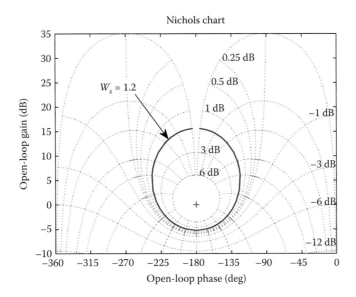

FIGURE 2.26
Stability specification in the Nichols chart: $W_s = 1.2$ (1.58 dB).

- Reference tracking specification—see Figure 2.27b:

$$\delta_{lo}(\omega) \leq \left| \frac{P(j\omega)G(j\omega)}{1+P(j\omega)G(j\omega)} F(j\omega) \right| \leq \delta_{up}(\omega) \tag{2.63}$$

where

$$\delta_{lo}(\omega) = \left| \frac{1}{0.5(j\omega)^3 + 2(j\omega)^2 + 2.8(j\omega)+1} \right| \text{ is the lower bound}$$

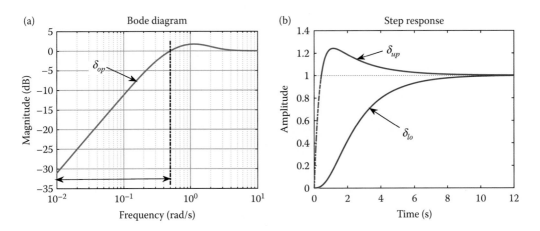

FIGURE 2.27
Control specifications for $P(s)$. (a) Sensitivity or disturbance rejection at the output of the plant: δ_{op}. (b) Reference tracking: δ_{up}, δ_{lo}.

and

$$\delta_{up}(\omega) = \left| \frac{2.8571(j\omega)+1}{0.6667\,(j\omega)^2 + 2.3333(j\omega)+1} \right| \text{ the upper bound}$$

for $\omega = \begin{bmatrix} 0.001 & 0.01 & 0.1 & 0.5 & 1 & 4 & 10 \end{bmatrix}$ rad/s

Step 6. Calculate stability QFT Bounds: QFTCT Bounds window

Using the plant model of Equation 2.59, the nominal plant of Equation 2.60 and the stability specification in Equation 2.61, we calculate the stability bounds—see Figure 2.28.

Step 7. Calculate performance QFT Bounds: QFTCTBounds window

Using the plant model of Equation 2.59, the nominal plant of Equation 2.60, the sensitivity specification in Equation 2.62, and the reference tracking specification in Equation 2.63, we calculate the sensitivity and reference tracking bounds—see Figures 2.29 and 2.30, respectively.

Step 8. Intersection of bounds & compatibility: QFTCT Bounds window

The intersection of bounds for the stability, sensitivity, and reference tracking is shown in Figure 2.31. As we have a bound solution for each frequency, the selected control specifications for the system are compatible.

Step 9. Synthetize feedback controller G(s): QFTCT Controller design window

Looking at the intersection of bounds presented in Figure 2.31, we design a controller that meets these QFT bounds, as shown in Equation 2.64 and Figure 2.32. As in the previous example, the small circles and numbers in the figure indicate the frequencies for

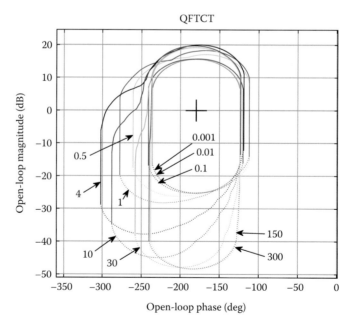

FIGURE 2.28
Stability bounds for $\omega = [0.001\ 0.01\ 0.1\ 0.5\ 1\ 4\ 10\ 30\ 150\ 300]$ rad/s.

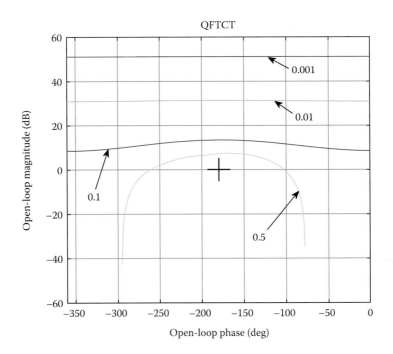

FIGURE 2.29
Sensitivity bounds for $\omega = [0.001\ 0.01\ 0.1\ 0.5]$ rad/s.

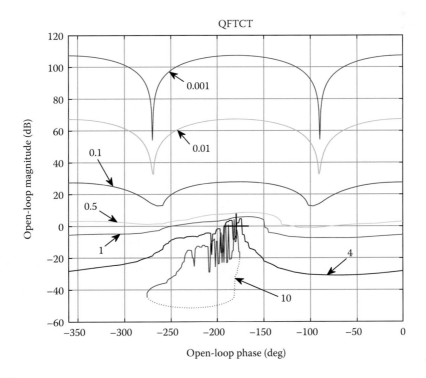

FIGURE 2.30
Reference tracking bounds for $\omega = [0.001\ 0.01\ 0.1\ 0.5\ 1\ 4\ 10]$ rad/s.

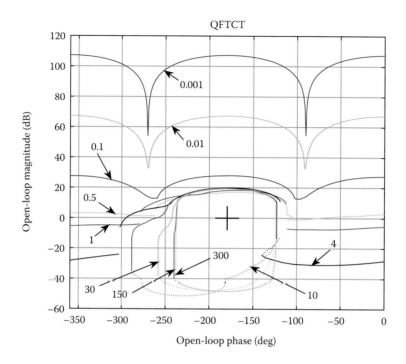

FIGURE 2.31
Intersection of bounds for $\omega = [0.001\ 0.01\ 0.1\ 0.5\ 1\ 4\ 10\ 30\ 150\ 300]$ rad/s.

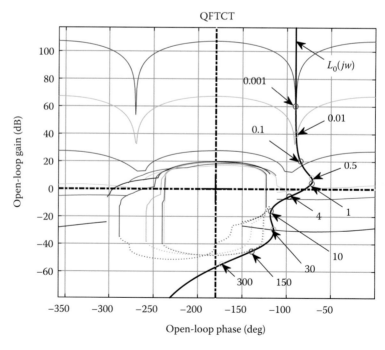

FIGURE 2.32
QFT bounds and $G(s)$ design, loop shaping.

the function $L_0(s) = P_0(s)G(s)$. The objective is to move each small circle above the solid-line bounds or below the dashed-line bounds. The proposed controller $G(s)$ meets well all the bounds. The QFTCT, *Controller design* window, can be used to easily design this controller—see also Section 2.14

$$G(s) = 5\frac{((s/4)+1)((s/13)+1)}{((s/350)+1)^2} \tag{2.64}$$

Step 10. Synthetize prefilter F(s): QFTCT Prefilter design window

Taking into account the controller $G(s)$—see Equation 2.64, the reference tracking specifications—see Equation 2.63, and the plant model—see Equation 2.59, we design a prefilter $F(s)$ to make sure that all the input/output functions $P(s)G(s)F(s)/[1 + P(s)G(s)]$ are inside the band defined by the limits $\delta_{up}(\omega)$ and $\delta_{lo}(\omega)$. The prefilter is shown in Equation 2.65, and the input/output functions and limits in Figure 2.33.

$$F(s) = \frac{1}{(s/3.5)+1} \tag{2.65}$$

Step 11. Analysis in the frequency domain: QFTCT Analysis window

For the analysis and validation, we use the 2DOF control block diagram shown in Figure 2.13, with the controller $G(s)$ in Equation 2.64, the prefilter $F(s)$ in Equation 2.65, $H(s) = 1$, and the plant model with uncertainty in Equation 2.59.

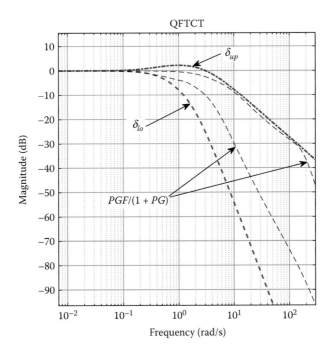

FIGURE 2.33
Prefilter $F(j\omega)$ design for $\delta_{lo}(\omega) \le |PGF/(1 + PG)| \le \delta_{up}(\omega)$.

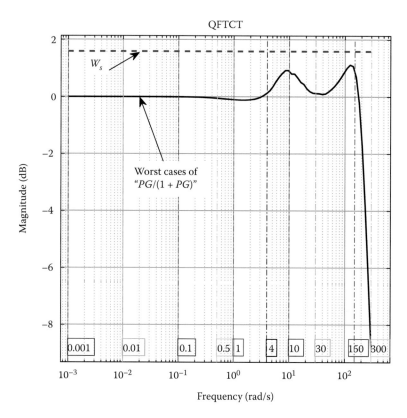

FIGURE 2.34

Stability analysis, frequency domain: W_s specification (dashed line), and worst case of $PG/(1 + PG)$ within the uncertain plants at each frequency (solid line).

The stability analysis in the frequency domain is shown in Figure 2.34, the dashed line is the stability specification W_s—Equation 2.61, and the solid line represents the worst case of the function $PG/(1 + PG)$ at each frequency and due to the model uncertainty. The control system meets the stability specification, as the solid line is below the dashed line W_s in all cases.

The analysis of the sensitivity specification in the frequency domain, or disturbance rejection at the output of the plant specification, is shown in Figure 2.35. The dashed line is the sensitivity specification $\delta_{op}(\omega)$—Equation 2.62, and the solid line represents the worst case of the function $1/(1 + PG)$ at each frequency and due to the model uncertainty. The control system meets the sensitivity specification, as the solid line is below the dashed-line δ_{op} in all cases from 0 to 0.5 rad/s.

Step 12. Analysis in the time domain (linear): QFTCT Analysis window

The analysis of the reference tracking specification in the time domain is shown in Figure 2.36. The figure shows the limits $\delta_{up}(\omega)$ and $\delta_{lo}(\omega)$—see Equation 2.63, and the time responses $y(t)$ of 500 cases of $PGF/(1 + PG)$ to a unitary step reference $r(t)$. The control system meets the specification (is between the upper and lower limits) in all cases.

Finally, the analysis in the time domain of the sensitivity, or disturbance rejection at the output of the plant specification, is shown in Figure 2.37. The figure shows the time responses $y(t)$ of 500 cases of $1/(1 + PG)$ to a unitary step disturbance $d_o(t)$. The control system achieves a good disturbance rejection in all cases.

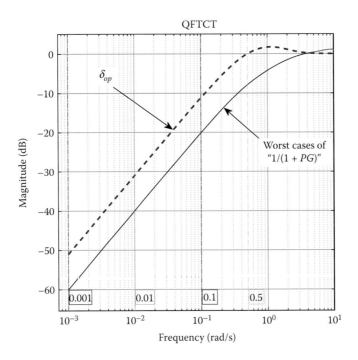

FIGURE 2.35
Sensitivity analysis, frequency domain: δ_{op} specification (dashed line), and worst case of $1/(1 + PG)$ within the uncertain plants at each frequency (solid line).

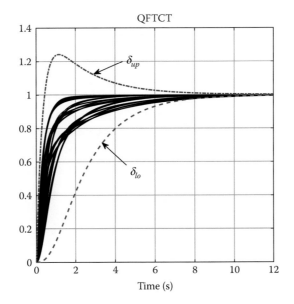

FIGURE 2.36
Reference tracking analysis, time domain: δ_{lo} and δ_{up} specifications, and time responses $y(t)$ of 500 cases of $PGF/(1 + PG)$ to a unitary step reference $r(t)$.

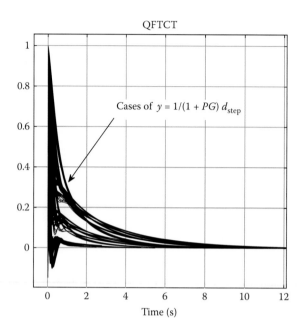

FIGURE 2.37
Sensitivity or plant output disturbance rejection analysis, time domain: time responses $y(t)$ of 500 cases of $1/(1 + PG)$ to a unitary step disturbance $d_o(t)$.

2.11 Model Matching

In some circumstances, the main objective of the control system is that the closed-loop transfer function, or complementary sensitivity function $y(s)/r(s)$, approaches a desired model $Q_0(s)$ within certain tolerances given by a function $\delta_m(\omega)$. For these cases a 3-block control structure is suggested, as shown in Figure 2.38. A complementary discussion of this structure is included in Section 9.3, Solution 9.2b.

As we see, the controller has now three blocks: the classical feedback controller $G(s)$, the prefilter, which is now the desired model $F(s) = Q_0(s)$, and a new feedforward element $G_r(s)$. Based on this diagram, the model-matching control specification is defined as

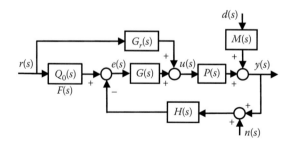

FIGURE 2.38
Model-matching structure. Controller blocks: $G(s)$, $F(s)$, and $G_r(s)$.

$$\left| Q_0(s) - \frac{y(s)}{r(s)} \right| = \left| \frac{Q_0(s) - P(s)G_r(s)}{1 + P(s)G(s)} \right| \leq \delta_m(\omega), \quad \omega \in \Omega_m \tag{2.66}$$

One of the main differences of this model-matching specification with respect to the classical reference tracking specification—see *Type 6*, Equation 2.40—is the use of phase information. The model-matching incorporates both magnitude and phase requirements $[Q_0(s) - y(s)/r(s)]$, while the classical reference tracking only uses magnitude information in the design of the prefilter $F(s)$.

The design usually starts by selecting the desired model $Q_0(s)$ for the closed-loop transfer function $y(s)/r(s)$. Then, the feedforward element $G_r(s)$ is set as,

$$G_r(s) = -P_c(s)^{-1}Q_0(s) \tag{2.67}$$

being $P_c(s)$ a central plant within the uncertainty. Subsequently, the feedback controller $G(s)$ is designed so that every possible closed-loop transfer function $y(s)/r(s)$ meets the specification given by Equation 2.66.

EXAMPLE 2.2 (CONT.): AIRPLANE FLIGHT CONTROL: FEEDBACK AND MODEL MATCHING

Consider again the flight control problem introduced in Example 2.2 to illustrate this model-matching case. The dynamics of the plant and the model uncertainty are the ones defined in Equation 2.59. The QFT templates are also the ones presented in Figure 2.25. We consider the same stability specification of the original example—Equation 2.61 and Figure 2.26, as well as the same sensitivity specification—Equation 2.62 and Figure 2.27a, and the reference tracking specification—Equation 2.63 and Figure 2.27b. In addition, we use the model-matching specification defined in Equation 2.66, with a desired model or prefilter as,

$$F(s) = Q_0(s) = \frac{1}{((s/3.5) + 1)((s/0.005) + 1)^3} \tag{2.68}$$

which is similar to the prefilter designed in Equation 2.65, with the same pole at 3.5 and three additional poles at 0.005 to have a fourth-order denominator and then a proper function $Q_0(s)/P_c(s)$. Figure 2.39 shows the Bode diagram and the step response of both prefilters—Equations 2.65 and 2.68.

Additionally, the selected model-matching specification $\delta_m(s)$ for Equation 2.66 is,

$$\delta_m(s) = \frac{(s/\gamma)}{(s/\gamma) + 1}, \text{with } \gamma = 1,$$
$$\text{and } \omega = [0.001\,0.01\,0.1\,0.5\,1\,4\,10\,30\,150\,300] \text{ rad/s} \tag{2.69}$$

Now, by selecting $P_c(s)$ as a central plant within the uncertainty

$$P_c(s) = \frac{k_c((s/z_c) + 1)}{s((s/p_c) + 1)\left[(s/\omega_{nc})^2 + (2\zeta_c/\omega_{nc})s + 1\right]},$$
$$\text{with } k_c = 1.1, z_c = 0.625, p_c = 5.5, \omega_{nc} = 5.5, \zeta_c = 0.85 \tag{2.70}$$

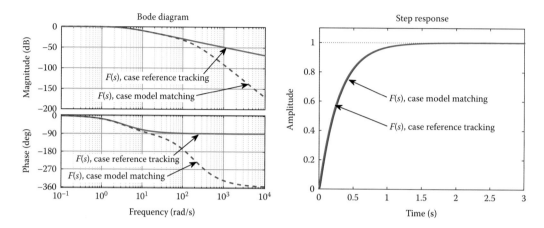

FIGURE 2.39
Prefilters $F(s)$. Solid line, as designed for the reference tracking case, Equation 2.65; and dashed line, for model matching, Equation 2.68.

the feedforward element $G_r(s)$ becomes

$$G_r(s) = \frac{Q_0(s)}{P_c(s)} = \frac{s((s/5.5)+1)[(s/5.5)^2 + (2\times0.85/5.5)s+1]}{1.1((s/0.625)+1)((s/3.5)+1)((s/0.005)+1)^3} \tag{2.71}$$

The QFT bounds for the model-matching specification Equation 2.66, with the model $Q_0(s)$, feedforward element $G_r(s)$, and model-matching objective $\delta_m(s)$ defined in Equations 2.68, 2.71 and 2.69, respectively, are shown in Figure 2.40. The intersection of

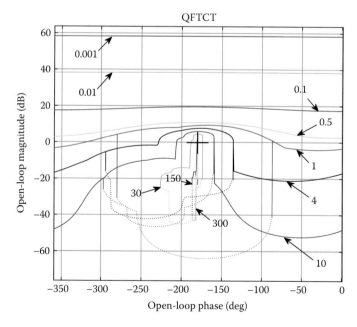

FIGURE 2.40
QFT bounds for the model-matching specification.

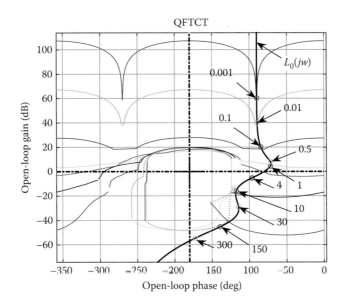

FIGURE 2.41
QFT bounds and $G(j\omega)$ design, loop shaping.

the QFT bounds of stability, sensitivity, reference tracking, and model matching, and the loop shaping of $G(s)$ are shown in Figure 2.41.

The resulting feedback controller $G(s)$ is the same as in the original flight control problem—see also Equation 2.64

$$G(s) = 5 \frac{((s/4)+1)((s/13)+1)}{((s/350)+1)^2} \tag{2.72}$$

The time-response analysis of the control system under a unitary step reference input $r(t)$ is shown in Figure 2.42. It considers the model-matching structure shown in Figure 2.38, the control functions defined in Equations 2.68, 2.71, and 2.72 for $F(s)$, $G_r(s)$ and $G(s)$, respectively, and several cases within the parametric model uncertainty defined in Equation 2.59.

For additional methodologies about this model-matching specification, see the References 138 and 121 regarding the SISO and MIMO cases, respectively.

NOTE: the QFT bounds for this model-matching case are generated with QFTCT. We create an array of functions A1 with an m.file, save it in the disk, load it from the *Plant definition* window as a transfer function array A1, and define a specification in the *Specifications* window using the case *Tk. Define by user*, with A = A1, B = 0, C = 1, D = P, and $\delta_m(s)$ as defined in Equation 2.69.

```
% m.file to create A1
% -------------------

F = tf(1,conv(conv(conv([1/3.5 1],[1/0.005 1]),[1/0.005 1]),[1/0.005 1]));
kc = 1.1; zc = 0.625; pc = 5.5; wnc = 5.5; gic = 0.85;
Pc = tf(kc*[1/zc 1],conv([1/pc 1 0],[1/wnc^2 2*gic/wnc 1]));
Gr = minreal(F/Pc);
```

```
i1m = 2; i2m = 2; i3m = 2; i4m = 2; i5m = 2;
kv = logspace(log10(0.2),log10(2),i1m);
zv = logspace(log10(0.5),log10(0.75),i2m);
pv = logspace(log10(1),log10(10),i3m);
wnv = logspace(log10(5),log10(6),i4m);
giv = logspace(log10(0.8),log10(0.9),i5m);
cc = 0;

for i1=1:i1m
  k= kv(i1);
  for i2=1:i2m
    z = zv(i2);
    for i3=1:i3m
      p = pv(i3);
      for i4=1:i4m
        wn = wnv(i4);
        for i5=1:i5m
          gi = giv(i5);
          P = tf(k*[1/z 1],conv([1/p 1 0],[1/wn^2 2*gi/wn 1]));
          cc = cc + 1;
          A1(1,1,cc) = minreal(F - P * Gr);
        end
      end
    end
  end
end

save A1 A1;
```

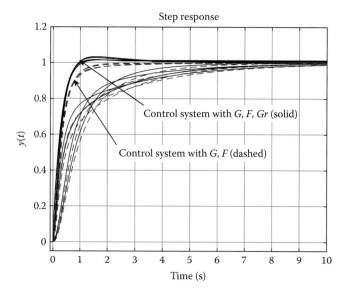

FIGURE 2.42

Comparison of model-matching control system (solid line) and the original reference tracking control system (dashed line). Time response $y(t)$ to a unitary step reference $r(t)$. Several plants within the uncertainty.

2.12 Feedforward Control

Often, the main goal of the control system is to efficiently reject the disturbances that affect the plant. For these cases, we suggest the feedback–feedforward control structure presented in Figure 2.43. A complementary discussion of this structure is included in Section 9.3, Solution 9.2a.

As we can see, the controller has now three blocks: the classical feedback controller $G(s)$, the prefilter $F(s)$ for cases with reference tracking specs, and a new feedforward element $G_f(s)$. This new block uses an additional sensor for the disturbance $d(s)$, and adds a supplementary signal to $u(s)$. Based on this diagram, the disturbance rejection specification with the feedforward element is defined as

$$\left|\frac{y(s)}{d(s)}\right| = \left|\frac{M(s)+P(s)G_f(s)}{1+P(s)G(s)}\right| \leq \delta_d(\omega), \quad \omega \in \Omega_d \tag{2.73}$$

The design usually starts by selecting a feedforward component $G_f(s)$ as

$$G_f(s) = -P_c(s)^{-1}M(s)V(s) \tag{2.74}$$

being $P_c(s)$ a central plant within the uncertainty and $V(s)$ a low-pass filter with high-frequency poles to complete a proper function for $G_f(s)$. Subsequently, the feedback controller $G(s)$ is designed so that every possible closed-loop transfer function $y(s)/d(s)$ meets the specification given by Equation 2.73.

EXAMPLE 2.2 (CONT.): AIRPLANE FLIGHT CONTROL: FEEDBACK AND FEEDFORWARD

Consider again the flight control problem introduced in Example 2.2, now adding also the feedforward component $G_f(s)$ as illustrated in Figure 2.43. The dynamics of the plant and the model uncertainty are the ones defined in Equation 2.59. The QFT templates are also the ones presented in Figure 2.25. We consider the same stability specification of the original example—Equation 2.61 and Figure 2.26, as well as the same sensitivity specification—Equation 2.62 and Figure 2.27a.

For this example, we complete the plant with a transfer function $M(s)$ from the disturbance $d(s)$ to the plant output $y(s)$ such that

$$M(s) = \frac{k_0}{s((s/p_0)+1)}; \quad \text{with} \quad k_0 = 0.2, \, p_0 = 1 \tag{2.75}$$

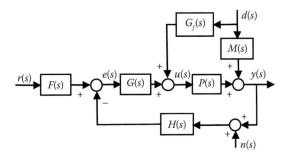

FIGURE 2.43
Feedback–feedforward structure. Controller blocks: $G(s)$, $F(s)$, and $G_f(s)$.

With this models, and selecting $P_c(s)$ as a central plant within the uncertainty as in Equation 2.70, and a low-pass filter function as $V(s) = 1/[(s/5.5) + 1]$, the feedforward component $G_f(s)$ becomes

$$G_f(s) = -P_c(s)^{-1}M(s)V(s) =$$

$$= -\frac{s((s/5.5) + 1)[(s/5.5)^2 + (2 \times 0.85/5.5)s + 1]}{1.1((s/0.625) + 1)} \frac{0.2}{s(s+1)} \frac{1}{((s/5.5) + 1)} \tag{2.76}$$

which is—see also Figure 2.44,

$$G_f(s) = -\frac{0.1818[(s/5.5)^2 + (2 \times 0.85/5.5)s + 1]}{((s/0.625) + 1)(s+1)} \tag{2.77}$$

The desired disturbance rejection specification $\delta_d(s)$ for Equation 2.73 is chosen as

$$\delta_d(s) = 0.014, \quad \text{with } \omega = [0.001\,0.01\,0.1\,0.5\,1\,4\,10]\,\text{rad/s} \tag{2.78}$$

which is more demanding than the original output disturbance rejection specification defined in Equation 2.62. The QFT bounds for the disturbance rejection specification are shown in Figure 2.45. It represents Equation 2.73, with the feedforward element $G_f(s)$ defined in Equations 2.77, the disturbance transfer function $M(s)$ of Equation 2.75, and the objective $\delta_d(s)$ defined in Equation 2.78. The intersection QFT bounds for stability, sensitivity, and disturbance rejection, and the loop shaping of $G(s)$ are shown in Figure 2.46.

After the loop shaping, the resulting feedback controller $G(s)$ is—see Section 2.14,

$$G(s) = 60\,\frac{((s/6) + 1)((s/10) + 1)((s/150) + 1)}{((s/70) + 1)((s/200) + 1)^2} \tag{2.79}$$

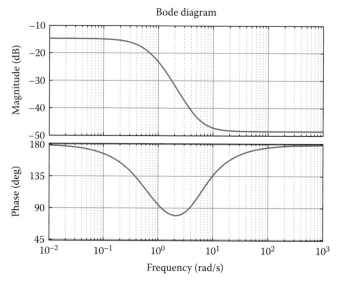

FIGURE 2.44
Bode diagram of $G_f(s)$, Equation 2.40.

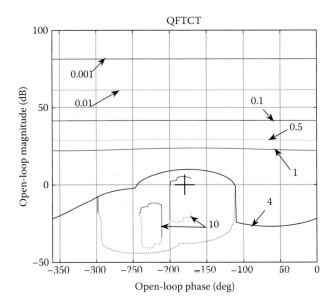

FIGURE 2.45
QFT bounds for the disturbance rejection specification including the feedforward controller—see Equations 2.73, 2.75, 2.77, and 2.78.

The time-response analysis of the control system under a unitary impulse disturbance input $d(t)$ is shown in Figure 2.47. It considers the feedback–feedforward structure shown in Figure 2.43, the control functions defined in Equations 2.79 and 2.77 for $G(s)$ and $G_f(s)$, respectively, and several cases within the parametric model uncertainty defined in Equation 2.59.

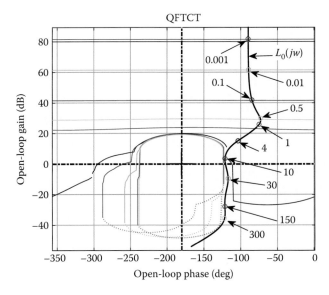

FIGURE 2.46
QFT bounds and $G(s)$ design, loop shaping.

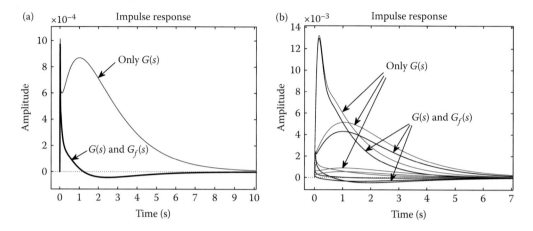

FIGURE 2.47
Time response $y(t)$ to a unitary impulse disturbance $d(t)$. (a) central plant and (b) other plants within the uncertainty. Comparison of cases: only feedback [$G(s)$] and feedback–feedforward [$G(s)$ and $G_f(s)$] structures.

For additional QFT methodologies about this feedforward specification, see the References 138 and 121 regarding the SISO and MIMO cases, respectively.

NOTE: the QFT bounds for this feedback–feedforward case are generated with QFTCT. We create an array of functions A2 with an m.file, save it in the disk, load it from the *Plant definition* window as a transfer function array A2, and define a specification in the *Specifications* window using the case *Tk. Define by user*, with A = A2, B = 0, C = 1, D = P, and $\delta_d(s)$ as defined in Equation 2.78.

```
% m.file to create A2
% -------------------

M = tf(0.2,[1 1 0]);
Gf = tf(-0.1818*[1/5.5^2 2*0.85/5.5 1],conv([1/0.625 1],[1 1]));

i1m = 2; i2m = 2; i3m = 2; i4m = 2; i5m = 2;
kv = logspace(log10(0.2),log10(2),i1m);
zv = logspace(log10(0.5),log10(0.75),i2m);
pv = logspace(log10(1),log10(10),i3m);
wnv = logspace(log10(5),log10(6),i4m);
giv = logspace(log10(0.8),log10(0.9),i5m);
cc = 0;

for i1=1:i1m
  k= kv(i1);
  for i2=1:i2m
    z = zv(i2);
    for i3=1:i3m
      p = pv(i3);
      for i4=1:i4m
        wn = wnv(i4);
        for i5=1:i5m
          gi = giv(i5);
          P = tf(k*[1/z 1],conv([1/p 1 0],[1/wn^2 2*gi/wn 1]));
          cc = cc + 1;
```

```
            A2(1,1,cc) = minreal(M + P * Gf);
        end
    end
  end
end
end

save A2 A2;
```

The feedback–feedforward solution can also be combined with the model-matching structure introduced in the previous section, as shown in Figure 2.48. In this case, the design of $F(s)$ and $G_r(s)$ for the model-matching part and of $G_f(s)$ for the feedforward part are developed independently and according to the expressions presented in Sections 2.11 and 2.12, respectively. Then, after both parts contribute to the QFT bounds—see Equations 2.66 and 2.73, the feedback controller $G(s)$ is designed (loop shaping) in the Nichols chart. A complementary discussion of this structure is included in Section 9.3, Solution 9.2c.

2.13 PID Control: Design and Tuning with QFT

Proportional integral derivative (PID) controllers are the most popular control strategy used in industry. Over 90% of control loops in real-world applications are composed of PID controllers. Nevertheless, it is quite common to find PIDs that are not the best choice for a given application, or are not tuned properly.

Frequently, control engineers decide to use a PID strategy to regulate the system without much further deliberation. However, we should ask first whether or not this is the appropriate control solution for the problem. Sometimes the plant needs a richer strategy, with some extra zeros and poles, to meet the specifications.

Yet, if the PID controller is a suitable solution, it is still important to find the appropriate methodology to tune the PID parameters.

QFT provides a powerful and easy-to-use methodology to analyze both problems: (a) to find a convenient control strategy for a given problem and (b) to tune the parameters for the controller.

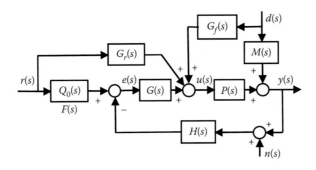

FIGURE 2.48
Feedback–feedforward model-matching structure. Controller blocks: $G(s)$, $F(s)$, $G_r(s)$, and $G_f(s)$.

The previous sections of this chapter studied how to design a QFT robust controller for a given application that presents model uncertainty and has a set of multiple control objectives or specifications. The QFT methodology is also particularly useful to design PID controllers. This section introduces a practical review of PIDs and shows how to apply the QFT methodology to design them.

A standard PID controller, with a low-pass filter for derivative part to attenuate the effect of noise, is shown in Equations 2.80 and 2.89. This standard PID form is the most common one, and is able to implement cases with either two real zeros as in Equation 2.91, or two complex zeros as in Equation 2.101. The standard PI and PD options are also shown in Equations 2.81 and 2.82, respectively

$$G_{PID}(s) = \frac{u(s)}{e(s)} = K_p \left(1 + \frac{1}{T_i s} + \frac{T_d s}{(T_d/N)s + 1} \right) \tag{2.80}$$

$$G_{PI}(s) = \frac{u(s)}{e(s)} = K_p \left(1 + \frac{1}{T_i s} \right) \tag{2.81}$$

$$G_{PD}(s) = \frac{u(s)}{e(s)} = K_p \left(1 + \frac{T_d s}{(T_d/N)s + 1} \right) \tag{2.82}$$

where K_p is the proportional gain, T_i the integral time, T_d the derivative time, N the low-pass filter constant, $e(s)$ the control error, and $u(s)$ the output of the controller.

The time-domain differential equation for a PID is composed by the three channels, proportional, integral, and derivative, as shown in Equation 2.83. When the reference $r(t)$ changes very quickly, say as a step, then the error $e(t)$ also changes very quickly, and as a consequence the derivative of the error becomes very large—see Equation 2.84.

$$u(t) = K_p \left[e(t) + \frac{1}{T_i} \int_0^t e(\tau) d\tau + T_d \frac{de(t)}{dt} \right] \tag{2.83}$$

$$K_p T_d \frac{de(t)}{dt}; \quad e(t) = r(t) - y(t) \tag{2.84}$$

The so-called *PI-D* and *I-PD Kick-off* solutions are an answer for this problem. They work well for references $r(t)$ that follow steps or are constant values. In general, we can see the PID of Equation 2.80 as an expression with three different errors—see Equations 2.85 through 2.88.

$$u(s) = K_p \left[e_p(s) + \frac{1}{T_i s} e(s) + \frac{T_d s}{1 + (T_d/N)s} e_d(s) \right] \tag{2.85}$$

with,

$$e_p(s) = b r(s) - y(s) \tag{2.86}$$

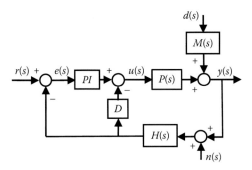

FIGURE 2.49
PI-D controller.

$$e_d(s) = c\,r(s) - y(s) \tag{2.87}$$

$$e(s) = r(s) - y(s) \tag{2.88}$$

If we select $b = 1$ and $c = 0$, we will avoid the very large initial derivative value under a step reference, while having the same response as the classical PID for a constant reference. This solution is the so-called PI-D controller, and the implementation is shown in Figure 2.49.

If we go further and select $b = 0$ and $c = 0$, we will obtain an even smoother signal under a step reference. This is the so-called I-PD controller, and the implementation is shown in Figure 2.50.

Using the 2DOF QFT methodology, we can design and implement both cases, the *PI-D* and *I-PD Kick-off* solutions. Equation 2.89 shows the feedback controller $G_{PID}(s)$ and Equation 2.90 the prefilter $F_1(s)$ as a function of b and c. By selecting $b = 1$ and $c = 0$ in Equation 2.90, we have the PI-D controller. Similarly, choosing $b = 0$ and $c = 0$ we obtain the I-PD controller—see also Figure 2.51.

$$\frac{u(s)}{e(s)} = G_{PID}(s) = \left(\frac{K_p}{T_i}\right)\frac{T_i T_d(1 + (1/N))s^2 + (T_i + (T_d/N))s + 1}{s((T_d/N)s + 1)} \tag{2.89}$$

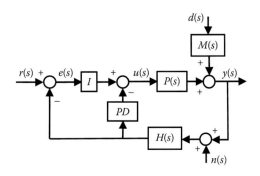

FIGURE 2.50
I-PD controller.

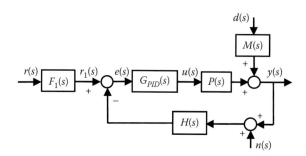

FIGURE 2.51
General 2DOF PID controller.

$$\frac{r_1(s)}{r(s)} = F_1(s) = \frac{T_iT_d(c+(b/N))s^2+(bT_i+(T_d/N))s+1}{T_iT_d(1+(1/N))s^2+(T_i+(T_d/N))s+1} \qquad (2.90)$$

These *PI-D* and *I-PD Kick-off* solutions are popular options in industrial PID controllers to weight the reference signal. However, quite often it is much better to design just a simpler prefilter *F(s)* according to the QFT-prefilter design methodology. Example 2.1 is an illustrative case of this fact, with the simple and effective prefilter of Equation 2.58.

In the presence of saturation in the actuator of the plant, the integral part of the controller can produce a wind-up problem, with an unwanted overshoot. A practical solution of this problem is presented later, in Section 7.4, with an anti-wind-up strategy.

EXAMPLE 2.1 (CONT.): ARMATURE-CONTROLLED DC MOTOR—CASE A

Consider again the Example 2.1 (*armature-controlled DC motor*) with the uncertainty defined in *Case a*. In Section 2.8, we designed a controller *G(s)*—see Equation 2.56—that met the proposed QFT bounds, as shown in Figure 2.19. This controller is in fact a PID controller with a low-pass filter for the derivative part. For convenience, we repeat the controller here again, first as a gain-zero-pole structure—see Equation 2.91, and then as the equivalent standard PID structure—see Equation 2.92.

$$G_{PID}(s) = \frac{k((s/z_1)+1)((s/z_2)+1)}{s((s/p)+1)} \qquad (2.91)$$

with $k=1100, z=55, z_2=1700, p=25{,}000$

$$G_{PID}(s) = \left(\frac{K_p}{T_i}\right)\frac{T_iT_d(1+(1/N))s^2+(T_i+(T_d/N))s+1}{s((T_d/N)s+1)} \qquad (2.92)$$

with $K_p=20.60, T_i=0.0187, T_d=5.31\times10^{-4}, N=13.28$

Figure 2.52 shows the magnitude of the controller in a Bode diagram. It starts with a slope of –20db/decade at low frequency due to the integrator (1/s), continues with two real zeros (z_1 and z_2), and finishes with a high-frequency pole (*p*).

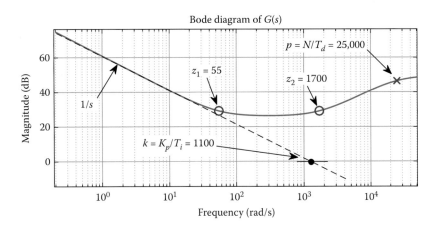

FIGURE 2.52
PID controller. Equations 2.91 or 2.92. $G_{PID}(s)$.

Format Conversion

Often, it is more convenient to design the PID controller (loop shaping) by selecting the appropriate gain, two zeros, and pole (in addition to the integrator), instead of working directly with the coefficients K_p, T_i, T_d, and N. This allows the designer to find better solutions in many cases. For these situations a conversion between the gain-zero-pole expression and the standard PID expression is needed. The conversion from the gain-zero-pole form in Equation 2.91 to the standard PID form in Equation 2.92 is easily done with the expressions in Equations 2.93 through 2.96. In this case the gain-zero-pole structure contains two real zeros, z_1, z_2.

$$T_i = \frac{1}{z_1} + \frac{1}{z_2} - \frac{1}{p} \tag{2.93}$$

$$K_p = k T_i = k \left(\frac{1}{z_1} + \frac{1}{z_2} - \frac{1}{p} \right) \tag{2.94}$$

$$T_d = \frac{1}{T_i z_1 z_2} - \frac{1}{p} = \frac{1}{((1/z_1) + (1/z_2) - (1/p)) z_1 z_2} - \frac{1}{p} \tag{2.95}$$

$$N = T_d p = \frac{p}{((1/z_1) + (1/z_2) - (1/p)) z_1 z_2} - 1 \tag{2.96}$$

If we are interested in the conversion in the opposite direction, from the standard PID structure in Equation 2.92 to the gain-zero-pole structure with two real zeros in Equation 2.91, we can use the expressions in Equations 2.97 through 2.100.

$$k = \frac{K_p}{T_i} \tag{2.97}$$

$$p = \frac{N}{T_d} \tag{2.98}$$

$$z_1 = \frac{2}{\left[(T_i + (T_d/N)) + \sqrt[2]{(T_i + (T_d/N))^2 - 4T_i T_d (1 + (1/N))}\right]} \tag{2.99}$$

$$z_2 = \frac{1}{2T_i T_d (1 + (1/N))}\left[(T_i + (T_d/N)) + \sqrt[2]{(T_i + (T_d/N))^2 - 4T_i T_d (1 + (1/N))}\right] \tag{2.100}$$

Sometimes, the numerator of the standard PID structure has two complex zeros instead of two real zeros. In this situation the equivalent gain-zero-pole structure has the form,

$$G_{PID}(s) = \frac{k\left[(s/\omega_n)^2 + (2\zeta/\omega_n)s + 1\right]}{s((s/p) + 1)} \tag{2.101}$$

Now, the conversion from the gain-zero-pole structure with two complex zeros in Equation 2.101 to the standard PID structure in Equation 2.92 is done with the expressions in Equations 2.102 through 2.105.

$$T_i = \frac{2\zeta}{\omega_n} - \frac{1}{p} \tag{2.102}$$

$$K_p = kT_i = k\left(\frac{2\zeta}{\omega_n} - \frac{1}{p}\right) \tag{2.103}$$

$$T_d = \frac{1}{T_i \omega_n^2} - \frac{1}{p} = \frac{p}{2\zeta\omega_n p - \omega_n^2} - \frac{1}{p} \tag{2.104}$$

$$N = T_d p = \frac{p^2}{2\zeta\omega_n p - \omega_n^2} - 1 \tag{2.105}$$

Finally, to go from the standard PID structure in Equation 2.92 to the gain-zero-pole structure with two complex zeros in Equation 2.101, we can use the expressions in Equations 2.106 through 2.109.

$$k = \frac{K_p}{T_i} \tag{2.106}$$

$$p = \frac{N}{T_d} \tag{2.107}$$

$$\omega_n = \frac{1}{\sqrt[2]{T_i T_d (1 + (1/N))}} \tag{2.108}$$

$$\zeta = \frac{T_i + (T_d/N)}{2\sqrt[2]{T_i T_d (1 + (1/N))}} \tag{2.109}$$

Also, as mentioned for this Example 2.1, in Section 2.9 we designed a simple prefilter $F(s)$—see Equation 2.58—that met the reference tracking specifications of Equations 2.52 through 2.54, as we saw in Figures 2.20 and 2.23.

2.14 Practical Tips

Some common questions that arise while designing a control system are how to actually choose the stability and performance specifications effectively, and how to design the feedback controller $G(s)$ and the prefilter $F(s)$. This section proposes some tips to help the designer in these important areas.

2.14.1 Selection of Specifications

2.14.1.1 Stability W_s

The stability specification has to be met at each and every frequency of the array of frequencies of interest. Then, unlike with other performance specifications, select the complete array of frequencies of interest for the stability specification.

On the other hand, there is a practical relationship between the phase margin and the overshoot of the closed-loop system time response. The larger the phase margin, the smaller the overshoot.

Some practical and frequently selected stability specifications are:

$W_s = 1.66$ (in magnitude) which is related to $PM = 35°$ and $GM = 4.09$ dB

$W_s = 1.46$ (in magnitude) which is related to $PM = 40°$ and $GM = 4.53$ dB

$W_s = 1.305$ (in magnitude) which is related to $PM = 45°$ and $GM = 4.94$ dB

2.14.1.2 Sensitivity $\delta_3(\omega)$

The rejection of disturbances that enter directly at the output of the plant is often a required objective—see d_o in Figure 2.13. The plant output disturbance rejection specification is defined as specification *Type 3*, or sensitivity, and described in Equation 2.37. For convenience, we repeat this expression below, as Equation 2.110.

As we see, the output of the plant $y(j\omega)$ is affected by the disturbance $d_o(j\omega)$ through the transfer function $T_3(j\omega)$, which is limited by the specification $\delta_3(\omega)$. Then, in the worst case scenario the change of the output of the plant due to the disturbance is: $y(j\omega) = \delta_3(\omega) \, d_o(j\omega)$.

$$|T_3(j\omega)| = \left| \frac{1}{1 + P(j\omega)G(j\omega)} \right| \leq \delta_3(\omega), \, \omega \in \Omega_3$$

(2.110)

$$\text{where } |T_3(j\omega)| = \left| \frac{y(j\omega)}{d_o(j\omega)} \right| = \left| \frac{e(j\omega)}{d_o(j\omega)} \right| = \left| \frac{e(j\omega)}{F(j\omega)r(j\omega)} \right| = \left| \frac{e(j\omega)}{n(j\omega)} \right|$$

Unlike the stability, the sensitivity specification $\delta_3(\omega)$ is usually defined for only some low to middle frequencies. Normally, we do not ask the system to reject high-frequency disturbances. In this way, we reduce high-frequency activity of the actuators and then avoid potential mechanical fatigue problems.

A practical selection of $\delta_3(\omega)$ is shown in Equation 2.111. As we can see in Figure 2.53, this expression gives a good disturbance rejection at low frequencies. The specification

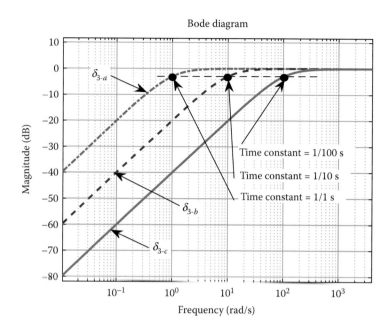

FIGURE 2.53
Sensitivity or plant output disturbance rejection specification. Examples of Equation 2.111, where $\delta_{3a,b,c}(s)$ is with $a_d = 1, 10, 100$, respectively.

has also a slope of +20 dB/decade that reaches the –3 dB level at the frequency $\omega = a_d$ (the pole), and has a 0 dB value for high frequency.

$$\delta_3(s) = \frac{(s/a_d)}{(s/a_d)+1} \tag{2.111}$$

By selecting just one parameter, the pole a_d in Equation 2.111, we can achieve different levels of disturbance rejection. The higher the parameter a_d, the more significant the attenuation of the effect of the disturbance. As an example, we present $\delta_3(s)$ with $a_d = 1$, 10, and 100. Figure 2.53 shows the Bode diagram of $\delta_3(s)$ in these three cases, and Figure 2.54 the time-domain response when a disturbance $d_o(s) = 1/s$ is introduced in the system, $y(s) = [1/(1 + P(s)G(s))] d_o(s) = \delta_3(s) d_o(s)$.

Table 2.4 shows different choices, from $a_d = 0.001$ to 10,000, and the corresponding frequency for "–20dB" attenuation and bandwidth—see Figure 2.53, and the rise time and settling time in Figure 2.54.

Additionally, Figure 2.55 shows the $\omega = [0.001, 0.05, 1]$ QFT bounds with $a_d = 1, 10$, and 100 (cases a, b, and c, respectively), and using a first-order plant $P(s) = k/[(s/p) + 1]$ with parametric uncertainty $k \in [1, 2]$, $p \in [0.1, 0.2]$.

Disturbances can enter into the system not only at the plant output, but also at many other points. Figure 2.13 showed some cases of disturbances entering at the error signal d_e, plant input d_i, plant output d_o, and at any generic point d. Section 2.6 discussed a variety of disturbance rejection specifications for all these cases—see *Types 2, 3, 4*, and *k*. The function of Equation 2.111 can also be used to define them.

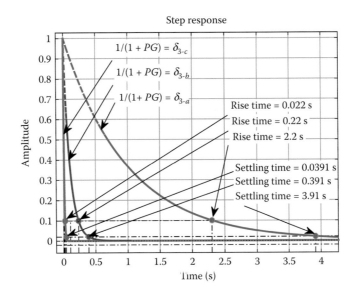

FIGURE 2.54
Step response. Examples of Equation 2.111, where $\delta_{3a,b,c}(s)$ is with $a_d = 1, 10, 100$, respectively. $y(s) = [1/(1 + P(s)G(s))]$ $d_o(s) = \delta_{3a,b,c}(s)\,d_o(s)$, with $d_o(s) = 1/s$.

TABLE 2.4

Sensitivity Specifications $\delta_3(s)$. Equation 2.111. Figures 2.53 and 2.54

a_d	Frequency (rad/s) at −20dB	Bandwidth (rad/s)	Time Constant (s)	Rise Time (s) from $y(t)$ at 10% to $y(t)$ at 90% of Steady State	Settling Time (s) from $t = 0$ to $y(t)$ at $\pm2\%$ of Steady State
0.001	0.0001	0.001	1000	2200	3910
0.01	0.001	0.01	100	220	391
0.1	0.01	0.1	10	22	39.1
1	0.1	1	1	2.2	3.91
10	1	10	0.1	0.22	0.391
100	10	100	0.01	0.022	0.0391
1000	100	1000	0.001	0.0022	0.00391
10,000	1000	10,000	0.0001	0.00022	0.000391

2.14.1.3 Reference Tracking $\delta_{6\text{-}lo}(\omega)$, $\delta_{6\text{-}up}(\omega)$

Like the sensitivity, the reference tracking specifications are usually defined for only some low-to-middle frequencies. Normally, we do not ask the system to follow a high-frequency reference. In this way, we reduce high-frequency activity of the actuators and then avoid potential mechanical fatigue problems.

A practical selection of $\delta_{6\text{-}lo}(\omega)$ and $\delta_{6\text{-}up}(\omega)$ is shown in Equations 2.112 through 2.114. As with the sensitivity, we have to select just two parameters, a_L and a_U, in these expressions. Tables 2.5 and 2.6 show different choices for a_L and a_U, from 0.001 to 10,000, and the corresponding rise time and settling time of the lower and upper specifications, $\delta_{6\text{-}lo}(\omega)$ and $\delta_{6\text{-}up}(\omega)$, respectively. Figure 2.56 shows the time response of $y(t)$ for both specifications,

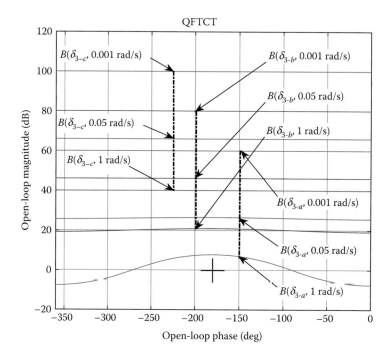

FIGURE 2.55
QFT bounds. Examples of Equation (2.111), where $\delta_{3a,b,c}(s)$ is with $a_d = 1, 10, 100$, respectively. $P(s) = k/[(s/p) + 1]$, $k \in [1, 2]$, $p \in [0.1, 0.2]$, $\omega = [0.001, 0.05, 1]$.

Equations 2.113 and 2.114, case $a_L = a_U = 1$, $\varepsilon_{lo} = 0.01$, $\varepsilon_{up} = 0.01$, $y(s) = [P(s)G(s)/(1+P(s)G(s))]$ $r(s) = \delta_6(s)\, r(s)$, and $r(s) = 1/s$. Figure 2.57 shows the Bode diagram of $\delta_{6-lo}(\omega)$ and $\delta_{6-up}(\omega)$ in this case.

$$\delta_{6\text{-}lo}(\omega) < |T_6(j\omega)| = \left|\frac{y(j\omega)}{r(j\omega)}\right| = \left|F(j\omega)\,\frac{P(j\omega)G(j\omega)}{1+P(j\omega)G(j\omega)}\right| \leq \delta_{6\text{-}up}(\omega),$$

$$\omega \in \Omega_6$$

(2.112)

TABLE 2.5

Tracking Specifications $\delta_{6-lo}(s)$. Equation 2.113. Figure 2.56. $\varepsilon_{lo} = 0.01$

a_L	Rise Time (s) from $y(t)$ at 10% to $y(t)$ at 90% of Steady State	Settling Time (s) from $t = 0$ to $y(t)$ at $\pm 2\%$ of Steady State
0.001	3360	5830
0.01	336	583
0.1	33.6	58.3
1	3.36	5.83
10	0.336	0.583
100	0.0336	0.0583
1000	0.00336	0.00583
10,000	0.000336	0.000583

TABLE 2.6

Tracking Specifications $\delta_{6\text{-}up}(s)$. Equation 2.114. Figure 2.56. $\varepsilon_{up} = 0.01$

a_U	Rise Time (s) from $y(t)$ at 10% to $y(t)$ at 90% of Steady State	Settling Time (s) from $t = 0$ to $y(t)$ at $\pm 2\%$ of Steady State
0.001	533	3230
0.01	53.3	323
0.1	5.33	32.3
1	0.533	3.23
10	0.0533	0.323
100	0.00533	0.0323
1000	0.000533	0.00323
10,000	0.0000533	0.000323

$$\delta_{6\text{-}lo}(s) = \frac{(1-\varepsilon_{lo})}{[(s/a_L)+1]^2}, \quad 0 \le \varepsilon_{lo} \tag{2.113}$$

$$\delta_{6\text{-}up}(s) = \frac{[(s/a_U)+1](1+\varepsilon_{up})}{[(s/\omega_n)^2 + (2\zeta s/\omega_n)+1]}; \quad 0 \le \varepsilon_{up}; \quad \zeta = 0.8; \quad \omega_n = \frac{1.25 a_U}{\zeta} \tag{2.114}$$

Although we are free to select the models for the reference tracking specifications $\delta_{6\text{-}lo}(\omega)$ and $\delta_{6\text{-}up}(\omega)$, we recommend to choose them so that their difference in dBs be always positive: $20 \times \log_{10}(\delta_{6\text{-}up}) - 20 \times \log_{10}(\delta_{6\text{-}lo}) \ge 0$. By doing so, the tracking bounds will be plotted in order in the Nichols chart, starting at the top of the diagram at low frequencies, and

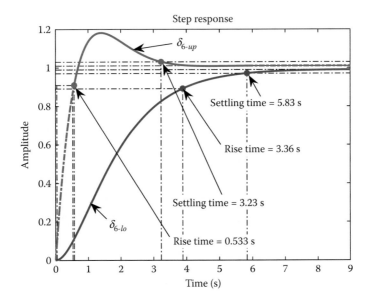

FIGURE 2.56

Step responses for Equations 2.113 and 2.114, with $a_L = a_U = 1$, $\varepsilon_{lo} = 0.01$, $\varepsilon_{up} = 0.01$. $y(s) = [P(s)G(s)/(1 + P(s)G(s))]\, r(s) = \delta_6(s)\, r(s)$, with $r(s) = 1/s$.

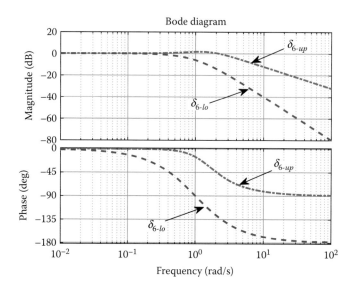

FIGURE 2.57
Bode diagram. Equations (2.113), (2.114), $a_L = a_U = 1$, $\varepsilon_{lo} = 0.01$, $\varepsilon_{up} = 0.01$.

continuing with lower and lower locations as the frequency increases. This property can be easily achieved using the proposed Equations 2.113 and 2.114.

Finally, we compare next the QFT bounds generated by the sensitivity and the reference tracking specifications. We use the expressions defined in Equation 2.111 and Equations 2.112 through 2.114, respectively, and pick $a_d = a_L = a_U = 1$, and $\varepsilon_{lo} = \varepsilon_{up} = 0.01$. We use again the simple first-order plant $P(s) = k/[(s/p) + 1]$, with the uncertainty $k \in [1, 2]$ and $p \in [0.1, 0.2]$. The QFT bounds for the frequencies $\omega = 0.2$ and 1 rad/s are shown in Figure 2.58. All of them are solid-line bounds, meaning $L_0(s)$ must be above them to meet the specs.

As we can see, at $\omega = 1$ rad/s the sensitivity bounds are dominant over the tracking bounds. However, at $\omega = 0.2$ rad/s the tracking and sensitivity bounds intersect each other at several phases (–286°, –208°, –125°, and –40°). This means that the most dominant bound can be either the tracking or the sensitivity specification, just depending on the phase.

2.14.2 Loop Shaping—Designing G(s)

Some tips to design an effective feedback controller $G(s)$ are proposed next. The tips are organized in a chronological list that the designer can follow, to find a feedback controller $G(s)$ that meets all the stability and performance specifications represented in the QFT bounds.

1. *Objective.* In the Nichols chart and from low frequency to high frequency, design the controller $G(s)$ so that at each frequency ω_i the nominal $L_0(j\omega_i) = P_0(j\omega_i)G(j\omega_i)$ is above the corresponding ω_i bound if this bound is a solid line, or below if this ω_i bound is a dashed line.

2. *Sign.* For the gain of $G(s)$, select a positive sign if the plant $P_0(s)$ has a positive sign. Select a negative sign if the plant has a negative sign. In unstable cases, this could be different—see examples L11, L12, and L13 in Section 3.6.3.

3. *Integrators.* If we need a zero steady-state error for step reference inputs, or if we look for high disturbance rejection or demanding reference tracking specifications, add one (most typical) or two integrators—see Figure 2.59, from Step 0 to Step 1.

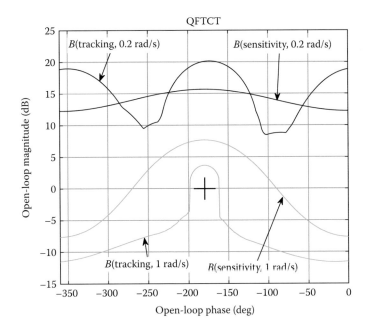

FIGURE 2.58
QFT bounds. Sensitivity: Equation 2.111 with $a_d = 1$. Tracking: Equations 2.113 and 2.114 with $a_L = a_U = 1 = 1$, $\varepsilon_{lo} = \varepsilon_{up} = 0.01$. $P(s) = k/[(s/p) + 1]$, $k \in [1, 2]$, $p \in [0.1, 0.2]$, $\omega = [0.2, 1]$ rad/s.

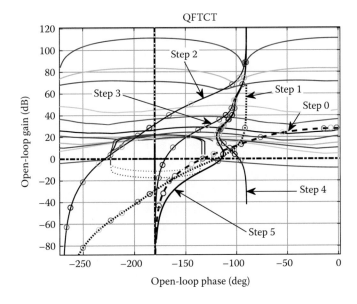

FIGURE 2.59
Loop shaping. Designing $G(s)$ of Equation 2.56, Example 2.1, in five steps: [0] original plant, [1] add an integrator, [2] change the gain to 1100, [3] add a zero at 55, [4] add a zero at 1700, [5] add a high-frequency pole at 25,000.

4. *Gain.* Tune the gain k of $G(s)$ so that $L_0(s)$ meets the lower frequency bound (or in some cases a very demanding bound)—see Figure 2.59, from Step 1 through Step 2.

$$gain = k \qquad (2.115)$$

5. *Start* at $\omega = 0$, and add zeros and poles to $G(s)$, from low frequency to high frequency.

6. *Zeros.* Add a real zero z if you need to move $L_0(s)$ to the right in the Nichols chart—see Figure 2.59, from Step 2 through Step 3, and from Step 3 through Step 4.

$$real\ zero = \left(\frac{s}{z} + 1\right) \qquad (2.116)$$

7. *Poles.* Add a real pole p if you need to move $L_0(s)$ to the left in the Nichols chart—see Figure 2.59, from Step 4 through Step 5.

$$real\ pole = \frac{1}{((s/p) + 1)} \qquad (2.117)$$

8. *Notch filters.* Add a notch filter if you need to filter out a specific frequency ω_n. The depth in magnitude of the filter at ω_n is "$20 \times \log_{10}(\zeta_1/\zeta_2)$" in dB.

$$notch\ filter = \frac{s^2 + 2\zeta_1\omega_n s + \omega_n^2}{s^2 + 2\zeta_2\omega_n s + \omega_n^2} \qquad (2.118)$$

9. *Low-pass filter* characteristic. If necessary, add high-frequency poles, as Equation 2.117 with large p, to make sure that the order of the denominator is higher than the order of the numerator. In this way, (1) we attenuate the always present high-frequency noise, and (2) we make sure that the controller is proper.

10. *Cancelations.* Sometimes it is useful to cancel some dynamics of the plant with the controller. In these cases the cancelation must be in the left-half plane (LHP).

11. *Never do.* Do not cancel any right-half plane (RHP) plant pole with a controller zero, or any RHP plant zero with a controller pole. These cancellations will always make the system unstable.

12. *Slope.* The higher the slope of the magnitude of $L_0(s) = P_0(s)G(s)$ in the Bode diagram, the larger the associated phase. According to the magnitude-phase Bode theorem,[1] a slope of $L_0(s) = -20n$ dB/decade (magnitude plot of the Bode diagram) denotes a phase of $L_0(s) = -90n$ degrees (phase plot of the Bode diagram), with $n = \pm 0,1,2....$ In this way, the steeper the slope of $L_0(s)$ (larger n), the bigger movement of $L_0(s)$ to the right in the Nichols chart, which can eventually compromise the stability (the phase margin is measured at the $-180°$ phase level). Then, a good analogy to remember is that *the amount of phase available is like the money you have in your pocket*. The possibilities of designing a good controller are related to the amount of phase of $L_0(s)$ available. Make sure you use the phase wisely, selecting the magnitude slopes intelligently.

13. *For unstable systems*, read Chapter 3 and follow guidelines of Table 3.2.
14. *For time-delay systems*, read Chapter 4.
15. *For distributed parameter systems*, read Chapter 5.
16. *For very high-performance specifications*, read Chapters 6 and 7.
17. *For multi-input multi-output systems*, read Chapter 8.
18. *For asymmetric systems*, read Chapter 9.
19. *For controller implementation*, read Chapter 10.

2.14.3 Prefilter—Designing *F*(*s*)

Finally, some tips to design a prefilter *F*(*s*) are given below. The tips are organized in a chronological list that the designer can follow, to find a prefilter that meets the reference tracking specifications.

1. *Objective*. In the Bode diagram, and from low frequency to high frequency, design the prefilter *F*(*s*) so that at each frequency ω_i the closed-loop functions $|T(j\omega)F(j\omega)|$ of all the plants within the uncertainty are between the upper $\delta_{up}(\omega)$ and lower $\delta_{lo}(\omega)$ reference tracking specs. Note: $T = PG/(1+PG)$.

2. *Gain*. The gain of the prefilter *F*(*s*) equals one.

3. *Start* at $\omega = 0$, and add zeros and poles from low frequency to high frequency.

4. *Poles*. Add a real pole *p*—see Equation 2.117, if you need to move *T*(*s*)*F*(*s*) down in the Bode diagram, from $\omega = p$ on—see Figure 2.60, from (a) to (b).

5. *Zeros*. Add a real zero *z*—see Equation 2.116, if you need to move *T*(*s*)*F*(*s*) up in the Bode diagram, from $\omega = z$ on.

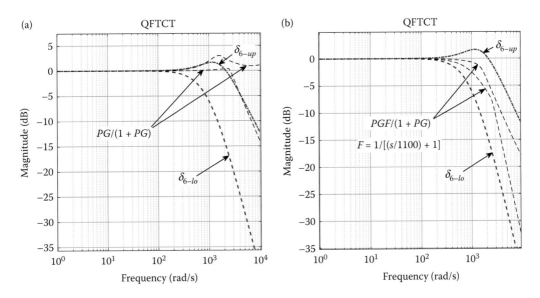

FIGURE 2.60
Prefilter design *F*(*s*), Equation 2.58, Example 2.1, in one step: (a) without prefilter and (b) adding a low-frequency pole at 1100.

2.15 Summary

This chapter presents the fundamentals of QFT. The basic steps of the QFT methodology are introduced in Sections 2.2 through 2.10 along with an illustrative example of a DC motor (Example 2.1). Afterward, an additional case of an airplane flight control system (Example 2.2) shows more details of the methodology. Finally, Sections 2.11 through 2.14 complete the chapter with model-matching control systems, feedforward control strategies, PID design and tuning, and practical control design tips.

2.16 Practice

The list shown below summarizes the collection of problems and cases included in this book that apply the control methodologies introduced in this chapter.

- *Example 2.1.* Sections 2.3 through 2.10. Armature-controlled DC motor.
- *Example 2.1.* Section 2.13. PID control. Armature-controlled DC motor.
- *Example 2.2.* Feedback solution. Airplane flight control.
- *Example 2.2.* Section 2.11. Model matching. Airplane flight control.
- *Example 2.2.* Section 2.12. Feedforward/feedback. Airplane flight control.
- *Appendix 2.* QFT Control Toolbox (QFTCT) User's guide.
- *Case study CS1.* Satellite with flexible appendages. Attitude control.
- *Case study CS2.* Wind turbine pitch control.
- *Project P1.* Appendix 1. Vehicle with active suspension system.
- *Project P2.* Appendix 1. DVD-disk head control system.
- *Project P8.* Appendix 1. Satellite with tanks partially filled.
- *Problem Q2.* Appendix 1. Control of first-order system with uncertainty.
- *Problem Q3.* Appendix 1. Control of third-order state space system.
- *Problem Q4.* Appendix 1. Field-controlled DC motor.
- *Problem Q5.* Appendix 1. Formation flying spacecraft control. Deep space.
- *Problem Q6.* Appendix 1. Helicopter control.
- *Problem Q7.* Appendix 1. Two cart problem.

3

Unstable Systems and Control Solutions

3.1 Introduction

The understanding of feedback control principles and the ability to implement them in real-world applications are among the main achievements of engineers in the last century. Moreover, the success of engineers controlling unstable plants that can hardly be operated manually opened the door to new opportunities. This was particularly important in the development of aviation. In fact, the world's first successful heavier-than-air powered airplane was intrinsically unstable. It was invented, built, tested, and controlled by the Wright brothers—see Figure 3.1.

The Wright brothers, Wilbur (1867–1912) and Orville (1871–1948), developed extraordinary mechanical skills while working in their bicycle shop in Dayton, Ohio. Among others projects, they designed new bicycles following H.J Lawson's original ideas for the so-called *"safety bicycle."*[357] Their work with these new bicycles influenced their conviction that an unstable flying machine, like an unstable bicycle, could be successfully controlled in practice.

The first flying machines designed by the brothers were simple gliders, without an engine, to understand what they defined as *"the problem of equilibrium."* They also devoted months to studying the art of flight from birds. The brothers carefully watched the movements of giant gannets, large vultures, buzzards, eagles, and small hawks in the wind. They were fascinated of how these birds smoothly modify the shape of the wings to use the wind to sail aloft, turn, rise, fall, and *control the equilibrium* under strong and light winds.

Based on these studies and experiments, the Wrights designed all kind of curved wings and tested them in a small-scale wind tunnel they built at home. They pursued a double objective: a flying machine able to react quickly enough to wind disturbances (high maneuverability), and a mechanism with sufficient control authority to regulate the machine under different wind conditions (controllability).

The brothers designed their first successful aircraft in 1903 (Figure 3.1).[357] The flying machine was open-loop unstable to increase the maneuverability. It included a small biplane wing system forward the main wings, locating its center of pressure (c.p. or lift force point of application) ahead of its center of gravity (c.g. or gravity force point of application). Nowadays, we called this aerodynamic condition static instability. Canard airplanes and the X-29 forward swept wing airplane are other examples that follow this principle to increase their maneuverability and reduce the weight (tail not needed). Using this design, the Wright brothers made the world's first heavier-than-air, powered, and controlled human flight ever, on December 17, 1903 at Kitty Hawk, North Carolina (Figure 3.1).

FIGURE 3.1
Wright brothers' unstable airplane. December 17, 1903: the World's first heavier-than-air, powered, and controlled human flight (U.S. Library of Congress).

It is important to mention that the Wright brothers' fundamental breakthrough, also described in their first U.S. patent (US-821,393), was not the invention of a flying machine, but rather, the invention of a control system able to manipulate the flying machine's surfaces and stabilize the unstable aircraft. The control system included three independent mechanisms: (1) an original *wing-warping system* for roll or lateral motion control, composed of pulleys and cables that twist the trailing edges of the wings in opposite directions; (2) a *forward elevator* for pitch or up-and-down control; and (3) a *rear rudder* for yaw or side-to-side control. These great achievements opened the door to human flight.

Other well-known examples of unstable systems include space rockets, inverted pendulums, motorcycles and bicycles, fluid instabilities, chemical and thermal instabilities, and many more. In all these cases, a careful analysis is required. This fact was very-well understood by Gunter Stein, the first recipient of the Hendrik W. Bode Lecture Award of the IEEE Control System Society (CDC, Tampa, Florida, December 1989). In his lecture, Stein highlighted Bode's critical observation that there are fundamental limitations on the performance of control systems, and that these limitations can be described by some simple integral expressions (see Bode's original book, 1945,[1] and Horowitz's book, 1963,[2]). In 2003, the *IEEE Control System Magazine* published a paper remembering the main discussions of the Stein's original Bode award lecture, including also the unstable characteristics of the Chernobyl nuclear power plant accident (1986) and of the SAAB Gripen JAS-39 airplane accident (1989).[355] This paper emphasizes three basic aspects of unstable systems:

- Unstable systems are fundamentally more difficult to control than stable ones
- Controllers for unstable systems are operationally critical
- Closed-loop systems with unstable components are only locally stable

The expressions introduced by Bode[1] state that the natural logarithm of the magnitude of the sensitivity function $S(j\omega)$ of a SISO feedback system, integrated over frequency, is constant. This constant is zero for stable systems, and positive for unstable and/or non-minimum phase (*nmp*) systems.

In practical applications, the infinite upper limit of the integrals defined by Bode is reduced to the available bandwidth Ω_a, as shown in Equations 3.1 and 3.2,[355]

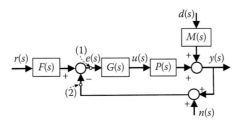

FIGURE 3.2
Closed-loop control system.

$$\int_0^{\Omega_a} \ln|S(j\omega)| \, d\omega = 0 \quad \text{for stable systems} \tag{3.1}$$

$$\int_0^{\Omega_a} \ln|S(j\omega)| \, d\omega = \pi \sum_{i=1}^{Np} \mathrm{Re}(p_i) + \pi \sum_{j=1}^{Nz} \mathrm{Re}(z_j) \quad \text{for unstable } nmp \text{ systems} \tag{3.2}$$

where $S(j\omega) = 1/[1 + P(j\omega)G(j\omega)]$ is the sensitivity function and p_i and z_j are the location of the N_p poles and N_z zeros of $P(j\omega)G(j\omega)$ in the right half plane (*RHP*).

The available bandwidth Ω_a depends on the control system technology, that is, the microprocessor, sampling rate, sensor and actuator bandwidths, hardware, etc.

The interpretation of these integrals is quite straightforward. If the controller $G(j\omega)$ reduces the sensitivity function (ln $|S(j\omega)|$ negative) at low frequencies for a much better disturbance rejection—according to $y(s) = S(s)M(s)d(s)$ and Figure 3.2, then the sensitivity must increase at high frequencies (ln $|S(j\omega)|$ positive) to meet the Bode integral's requirements: the so-called *waterbed effect*. This fact deteriorates the disturbance rejection of the control system at high frequencies. Figure 3.3a shows a typical sensitivity function $S(j\omega)$,

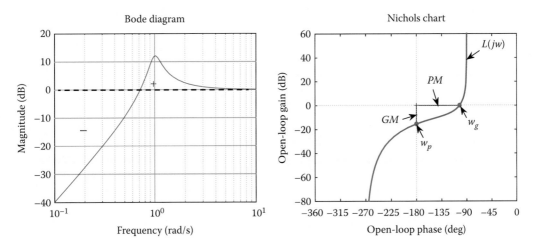

FIGURE 3.3
(a) Bode diagram of a sensitivity function: $S(j\omega) = 1/[1 + P(j\omega)G(j\omega)]$. (b) Nichols chart of $L(j\omega)$, GM and ω_p, PM and ω_g.

with values under 0 dB (magnitude <1) at low frequency, and above 0 dB (magnitude >1) at high frequency.

These intrinsic performance limitations are much more restrictive for unstable and/or non-minimum phase (*nmp*) systems than for stable systems, as the constant of the right term of the Bode's integral expressions is positive for the unstable or *nmp* systems and zero for the stable ones.

Intuitively, to stabilize a system, we need approximately $w_g > 2p_i$ and $w_g < z_j/2$, p_i and z_j being the location of the poles and zeros of $L(j\omega) = P(j\omega)G(j\omega)$ in the *RHP* respectively, and w_g the gain crossover frequency, which is the frequency where the magnitude of $L(j\omega)$ is 1 (0 dB), and being $w_g < \Omega_a$.[260,354]

3.2 Understanding Gain and Phase Margins, and W_s Circles

The gain margin (*GM*), phase margin (*PM*), and the related W_s circle are classical frequency-domain tools used to describe and quantify the stability of a closed-loop system—see Figure 3.3b. They are typically defined as control specifications or objectives to design controllers to regulate dynamic systems. Their mathematical description and practical application are part of the classical books on fundamentals of automatic control: see for instance References 259, 262, and 263.

In this section, we first discuss an intuitive understanding of these frequency-domain stability margins. Consider the closed-loop control system presented in Figure 3.2 and assume without loss of generality that the input signals are zero: $r = d = n = 0$. Now, to understand the concept of stability of a feedback system, let us introduce a periodic signal $x(t) = A \sin(w_p t)$ at point (1), with a special frequency w_p, which is the one at which all the blocks from point (1) to point (2) combined add up a phase shift of $-180°$ to the signal $x(t)$.

In this situation, as this periodic signal $x(t)$ with frequency w_p travels the closed loop, it changes its phase: first adding $-180°$ from point (1) to point (2), and then another $-180°$ from point (2) to point (1) due to the negative sign of the summer, adding up a total of $-360° = 0°$. Also, the blocks from point (1) to point (2) combined multiply the original magnitude A by a factor K, being $B = K A$, with $K = K(w_p)$. This means that after closing the loop, that is, from point (1) to point (1) back, the original signal $x(t)$ is modified to $x_{afterLoop}(t) = B \sin(w_p t)$, which has the same original phase and an amplitude B.

In this way, as $x_{afterLoop}(t)$ has the same phase as the original signal $x(t)$, the condition for stability is to have an amplitude $B < A$, which means a gain $K(w_p) < 1$, or < 0 dB. In this case, the signal $x(t)$ reduces the magnitude at every loop trip, approaching zero as it travels infinite times through the loop. On the contrary, if $K(w_p) > 1$, or > 0 dB, the signal $x(t)$ will grow unbounded, which means instability.

This is the physical interpretation for the *GM* condition for stability. It requires for the magnitude K given by all the blocks from point (1) to (1) at the frequency w_p to be $K(w_p) < 0$ dB. Thus, the *GM* is defined as $1/K(w_p)$, which is the maximum factor given by all the blocks from point (1) to (2) combined while still keeping the system stable, and being w_p the phase crossover frequency, for whom the blocks from point (1) to (2) add up a phase of $-180°$, see Figure 3.3b.

A similar interpretation for the *PM* is possible. In this case, the frequency of study w_g is the one where the blocks from point (1) to point (2) combined add up a gain $K = 1$. At this frequency w_g, also called the gain crossover frequency, we study the phase ϕ added by the loop, point (1) to point (1) back, which has to be $-180° > \phi_{11} = (\phi_{12} - 180°) \geq -360°$, or $0° > \phi_{12} \geq -180°$ for stability. This means that after traveling the loop once, the signal $x(t) = A \sin(\omega_g t)$ introduced at point (1) will be $x_{\text{afterLoop}}(t) = A \sin(\omega_g t + \phi_{11})$, with $-\pi$ rad $> \phi_{11} \geq -2\pi$ rad. In other words, after traveling the loop once, the system is stable if the signal $x(t)$ has the same original phase ($\phi_{11} = -2\pi$ rad $= 0$) or lags behind by an angle less than 2π rad or 360° (or less than π rad or 180° for ϕ_{12}, from point 1 to point 2). In this way, the *PM* is defined as the maximum phase lag given by all the blocks from point (1) to (2) combined while still keeping the system stable, see Figure 3.3b.

Putting together the *GM* (dB) and *PM* (degrees) in the Nichols Chart (NC), we can define a W_s circle around the origin (0 dB, −180°) that (1) takes into account the most demanding value of the given *GM* and *PM*, and (2) defines a constant magnitude locus for the closed-loop transfer function—see details in Chapter 2, Section 2.5.

Based on these concepts of *PMs* and *GMs*, Chapter 2 developed the U-contour or QFT stability bound in the NC, to describe the stability of a closed-loop system for plants with model uncertainty. As discussed, the condition for a stable closed-loop system is to have an open-loop transfer function $L(s) = P(s)G(s)$ that goes outside and on the right of the QFT stability bound in the NC, that is, *GM* below the 0 dB axis and *PM* right to the −180° axis, as in Figure 3.3b.

We can find, however, some special cases where this condition is not enough to assure the stability. Let us consider for instance a simple, unstable, and first-order plant $P(s) = 8/(-0.333 s + 1)$. Figure 3.4a shows the NC of the open-loop transfer function $L(s) = P(s)G_{01}(s)$ without any controller (or $G_{01}(s) = 1$). The QFT stability bound for $W_s = 1.45$ (or $GM = 4.56$ dB, $PM = 40.34°$) for all the frequencies is also shown in the figure. This case is unstable, as expected according to the stability margins ($GM = \infty$, $PM = -97.18°$) and QFT stability bound.

In order to stabilize the closed-loop system, we move the open-loop transfer function $L(s)$ to the right and outside of the QFT stability bound by adding the controller $G_{02}(s) = 1/[s (0.25 s + 1) (0.1 s + 1)]$. Figure 3.4b shows the new $L(s) = P(s)G_{02}(s)$ and the original QFT stability bound. Contrary to the first interpretation with the stability margins ($GM = \infty$ and $PM = 78.45°$ means stable system) and the QFT stability bound (L at the right and outside the stability bound means stable system), the closed-loop system with $L(s)$ represented at Figure 3.4b is still unstable. To prove it, we plot the root locus and the position

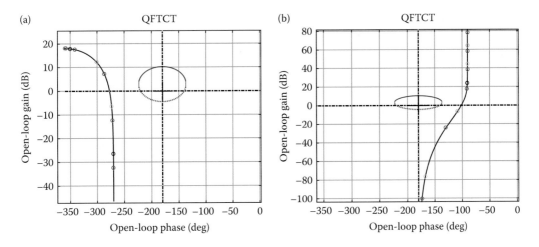

FIGURE 3.4

NC: (a) $L(s) = P(s)G_{01}(s)$, (b) $L(s) = P(s)G_{02}(s)$. In both cases, the QFT stability bound is at $W_s = 1.45$.

(a)

(b)

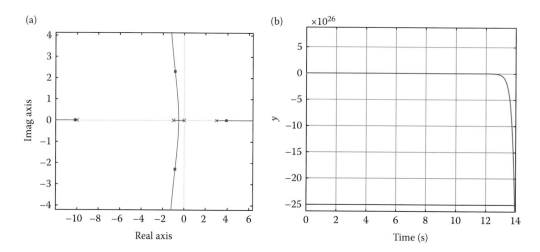

FIGURE 3.5
$L(s) = P(s)G_{02}(s)$. (a) Root locus and (b) output $y(t)$ of the close-loop system under a unitary step input $r(t)$—Figure 3.2.

of the roots of the closed-loop system (rectangle marks) in Figure 3.5a. As we can see, there is a root at $s = +4$ (at the *RHP*), which means that the closed-loop system is unstable. Also, the output signal $y(t)$ of the close-loop system under a unitary step input $r(s) = 1/s$, $y(s) = [L(s)/(1+ L(s))]\, r(s)$, is totally unbounded, as shown in Figure 3.5b.

In summary, this simple example shows a quite counterintuitive conclusion: the classical *GMs* and *PMs* and the subsequent stability analysis based on the W_s circles do not work properly in some special cases.

As a result, we need additional frequency-domain tools for a more reliable analysis of the stability of closed-loop systems. One unfailing frequency-domain tool for determining the stability of a dynamical system with feedback is the well-known Nyquist stability criterion (NSC).

Nevertheless, the NSC performs the stability analysis in the complex s-plane, and not in the NC. Also, sometimes it needs to calculate points at the infinite, which can be very complicated. To overcome these difficulties and allow the designer to use the NSC along with the QFT technique, Section 3.4 of this chapter presents a practical and comprehensive method to compute the NSC directly in the NC, and Section 3.5 shows a collection of examples.

In addition, and based on this proposed method, Section 3.6 gives guidelines to design controllers to stabilize unstable plants when dealing with frequency-domain techniques and QFT. Finally, Section 3.7 expands the example presented in this introduction, including both the stability analysis with the technique presented in Section 3.4 and the design of appropriate controllers in the NC.

3.3 The NSC[125]

The Nyquist stability criterion was introduced by the Swedish-American electrical engineer Harry Nyquist at Bell Telephone Laboratories in 1932. It is a graphical technique for determining absolute stability of a dynamical system with feedback. It is based on the Cauchy's argument principle and is calculated in the complex *s*-plane. The Nyquist criterion appears in most modern textbooks on control theory (see for instance Reference 259). It is especially useful for determining the stability of linear time invariant (*LTI*) closed-loop systems when the open-loop *L(s)* is given. It works well in any *LTI* case, even in the so-called conditionally stable systems, where the classical *GM* and *PM* analysis does not work properly. Consider the closed-loop feedback system shown in Figure 3.2 and its closed-loop transfer functions (complementary sensitivity, sensitivity, and noise transfer functions) given by

$$y(s) = \frac{P(s)G(s)F(s)}{1+P(s)G(s)}r(s) + \frac{M(s)}{1+P(s)G(s)}d(s) - \frac{P(s)G(s)}{1+P(s)G(s)}n(s) \tag{3.3}$$

with the characteristic equation

$$1 + P(s)G(s) = 1 + L(s) = 0 \tag{3.4}$$

The NSC defines a path Γ_c that clockwise encircles the complete area of the *RHP*, surrounding but not passing through any zero or pole of *L(s)*. The stability of the closed-loop system is determined by investigating how the path Γ_c is mapped by the open-loop function *L(s)* in the complex plane (see Figure 3.6). The principle of the argument states that the number of encirclements *N* in the positive direction (clockwise) around the complex point (−1,0) by the map of Γ_c, and for the frequency range $(-\infty \leq w \leq +\infty)$, equals $N = z_c - p$, where *p* is the number of poles of *L(s)* in the *RHP* and z_c the number of zeros (roots) of the characteristic equation—Equation 3.4—in the *RHP*.

The closed-loop system is stable if and only if there are not zero-pole cancelations in the *RHP* and there are no roots of the characteristic equation—Equation 3.4—in the *RHP*, that is to say: $z_c = N + p = 0$. This general condition can be reformulated with the following two rules:

- A feedback control system is stable if and only if the Nyquist contour Γ_c, when mapped in the *L(s)*-plane, does not encircle the (−1,0) point (*N* = 0) and the number of poles of *L(s)* in the *RHP* is zero (*p* = 0).

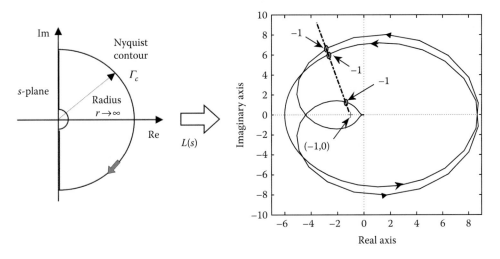

FIGURE 3.6
NSC in the complex *s*-plane. Frequency range: $-\infty \leq w \leq +\infty$. Example $N = -3$. (Adapted from Garcia-Sanz, M. 2016. *Int. J. Robust Nonlinear Control*, 26(12), 2643–2651.)

- A feedback control system is stable if and only if, for the Nyquist contour Γ_c when mapped in the $L(s)$-plane, the number of anticlockwise encirclements of the $(-1,0)$ critical point (N) equals the number of poles of $L(s)$ with positive real parts (p).

In both cases, the system is stable if $z_c = N + p = 0$. An easy way to determine the number of encirclements N is to draw a straight line out from the point $-1 + j0$ in any direction and count the net number of crossings of the map of Γ_c ($-\infty \leq \omega \leq +\infty$), with clockwise (decreasing phase) crossings being positive and counterclockwise crossings (increasing phase) negative, see Figure 3.6b.

Due to the convenience of utilizing Bode diagrams and NCs in the design of control systems in the frequency domain, several authors introduced some approximations to apply the NSC in nonpolar diagrams.[164,339,348,356] However, as Belanger states in his book,[356] *"it is sometimes tricky to count encirclements, especially if only the positive-frequency half of the Nyquist plot is given, as is the case with most software packages."* He proposes a procedure with multiple rules to count the encirclements. The method is given more ample justification in a previous paper by Vidyasagar et al.,[348] who also presents a technique to count the encirclements. However, both studies are just a short note to illustrate some common cases, but are not a comprehensive technique or a straightforward method to apply. They leave many open questions to the designer, especially in cases with poles with higher multiplicity, or poles or zeros at the origin, or multiple non-minimum phase zeros.

3.4 Nyquist Stability Criterion in the Nichols Chart[125]

This section presents the practical method introduced by the author in 2016 to study the stability of feedback systems with the Nyquist stability criterion in the Nichols chart.[125] The method can be applied to *LTI* closed-loop systems with any multiplicity of minimum and non-minimum phase zeros, stable and unstable poles, and poles at the origin, and to systems defined by nonrational functions, such as plants with time delays. The algorithm is included in the QFTCT, (click *Controller design* window, *File* menu, *Check stability* option). The MATLAB code for the algorithm can be also found in Appendix 3 and free-downloaded from the website:

http://cesc.case.edu/Stability_Nyquist_GarciaSanz.htm

The original rules of the NSC can be reformulated for the Nichols diagram as follows.[125] A feedback control system is stable if and only if:

- *Rule 1*: There are not zero-pole cancelations in the *RHP*.
- *Rule 2*: The number of poles p of the open-loop transfer function $L(s)$ in the *RHP* plus the number of encirclements N equals zero. That is to say, the number of zeros z_c of the characteristic equation is zero: $z_c = N + p = 0$.

To count the number of encirclements N directly in the NC the following technique is proposed. It divides the problem in four cases (N_a through N_d) according to Equation 3.5 and Figure 3.7.

$$N = N_a + N_b + N_c + N_d \tag{3.5}$$

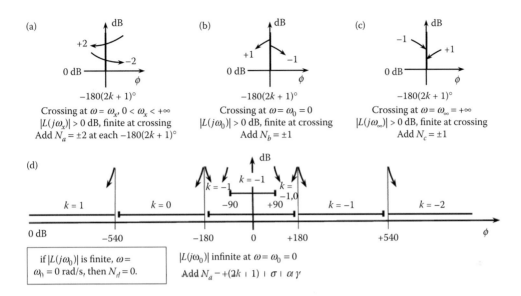

FIGURE 3.7
Rules to count the number of encirclements N in the NC: $N = N_a + N_b + N_c + N_d$.

To compute each case in the Nichols diagram, count the net number of crossings of the ray $R_0 = \{(\phi,r): \phi = -180(2k + 1)°, r > 0 \text{ dB}, k = 0, \pm 1, \pm 2,...\}$, with the left direction (decreasing phase) being positive, and right direction (increasing phase) being negative. That is,

1. *Compute N_a.* Figure 3.7a. If $L(j\omega)$ crosses the ray $R_0 = \{(\phi,r): \phi = -180°, r > 0 \text{ dB}\}$ at $\omega = \omega_x, 0 < \omega_x < +\infty$, being $|L(j\omega_x)|$ finite, then $N_a = \pm 2$ (sign according to paragraph above). For the multiple-sheeted Nichols plot, every crossing of the rays $R_0 = \{(\phi,r): \phi = -180(2k + 1)°, r > 0 \text{ dB}, k = 0, \pm 1, \pm 2,...\}$ with $|L(j\omega)|$ finite adds ± 2 to N_a (sign according to paragraph above). If there are no crossings, $N_a = 0$.

2. *Compute N_b.* Figure 3.7b. If $|L(j\omega_0)|$ is finite, $\omega = \omega_0 = 0$ rad/s, and lies on $R_0 = \{(\phi,r): \phi = -180(2k + 1)°, r > 0 \text{ dB}, k = 0, \pm 1, \pm 2,...\}$, then $N_b = \pm 1$ (sign according to paragraph above). Otherwise $N_b = 0$.

3. *Compute N_c.* Figure 3.7c. If $|L(j\omega_\infty)|$ is finite, $\omega = \omega_\infty = +\infty$ rad/s, and lies on $R_0 = \{(\phi,r): \phi = -180(2k + 1)°, r > 0 \text{ dB}, k = 0, \pm 1, \pm 2,...\}$, then $N_c = \pm 1$ (sign according to paragraph above). Otherwise $N_c = 0$.

4. *Compute N_d.* Figure 3.7d. If $|L(j\omega_0)|$ is infinite, $\omega = \omega_0 = 0$ rad/s, then

$$N_d = +2(k+1)+\sigma+\alpha\gamma \tag{3.6}$$

On the contrary, if $|L(j\omega_0)|$ is finite, $\omega = \omega_0 = 0$ rad/s, then $N_d = 0$. The parameters k, σ, α and γ are calculated as follows:

- *Parameter k.* It depends on the phase $\phi(L(\omega_1))$, being ω_1 the first frequency where $\phi(L(\omega_1)) \neq \phi(L(\omega_0))$, and the variable δ, which is

$$\delta = \text{sign}[\phi(L(\omega_0)) - \phi(L(\omega_1))] \tag{3.7}$$

and which takes the value $\delta = +1$ if the sign is positive, that is if $L(j\omega)$ goes initially to the left in the NC, or $\delta = -1$ if the sign is negative, that is if $L(j\omega)$ goes initially to the right.

The parameter k can take the following values:

1. $if -180° < \phi(L(\omega_1)) < +90° \rightarrow k = -1$ (3.8)

2. $if +90° < \phi(L(\omega_1)) < +180°$ and $\delta = -1 \rightarrow k = -1$ (3.9)

3. $if +90° < \phi(L(\omega_1)) < +180°$ and $\delta = +1 \rightarrow k = 0$ (3.10)

4. $if \phi(L(\omega_1)) \leq -180° \rightarrow k = -\text{ceil}\left[\dfrac{(\phi(L(\omega_1)) + 180)}{360}\right]$ (3.11)

5. $if \phi(L(\omega_1)) \geq +180° \rightarrow k = -\text{ceil}\left[\dfrac{(\phi(L(\omega_1)) - 180)}{360}\right]$ (3.12)

ceil(x) being a function that rounds the elements of x to the nearest integers toward infinity. k can take the values $[\ldots -2, -1, 0, +1, +2, \ldots]$ depending on the phase at ω_0 and ω_1.

- *Parameter σ. It describes the sign of the gain of the open-loop transfer function.* There are two cases:

  1. $if \text{ dcgain}(L_N) \geq 0 \text{ (positive)} \rightarrow \sigma = 0$ (3.13)

  2. $if \text{ dcgain}(L_N) < 0 \text{ (negative)} \rightarrow \sigma = 1$ (3.14)

     $L_N(j\omega)$ being the open-loop $L(j\omega)$ without poles at the origin. $\text{dcgain}(L_N) = L_N(\omega_0)$, $\omega_0 = 0$ rad/s.

- *Parameter α*. It depends on the phase $\phi(L(\omega_0))$, the variable δ and the number of poles $nP0$ at the origin. There are five cases:

  1. $if \phi(L(\omega_0)) \leq 0°$ and $\delta = +1$
     $$\rightarrow \alpha = -\text{ceil}\left(\frac{(\phi(L(\omega_0)) + 90 \times nP0)}{360}\right)$$ (3.15)

  2. $if \phi(L(\omega_0)) \leq 0°$ and $\delta = -1$
     $$\rightarrow \alpha = -\text{ceil}\left(\frac{(\phi(L(\omega_0)) + 90 \times (nP0 - 2))}{360}\right)$$ (3.16)

  3. $if \phi(L(\omega_0)) > 0°$ and $\delta = +1$
     $$\rightarrow \alpha = \text{floor}\left(\frac{(\phi(L(\omega_0)) + 90 \times (nP0 - 2))}{360}\right)$$ (3.17)

  4. $if \phi(L(\omega_0)) > 0°$ and $\delta = -1$
     $$\rightarrow \alpha = -\text{floor}\left(\frac{(\phi(L(\omega_0)) - 90 \times nP0)}{360}\right)$$ (3.18)

  5. $if \text{ order}(num.L) = \text{order}(den.L)$ and $\phi(L(\omega_\infty)) = 180°$ or $0°$
     $$\rightarrow \alpha = 0$$ (3.19)

floor(x) being a function that rounds the elements of x to the nearest integers toward minus infinity. α can take the values $[0, +1, +2, \ldots]$, depending on the phase at ω_0, ω_1, and ω_∞ and the number of poles at the origin $nP0$.

- *Parameter γ.* It is a function of the number of *LHP* and *RHP* zeros and poles of $L(s)$.

$$\gamma = 2\frac{(nZR + nPL - nZL - nPR)}{\max(|\,nZR + nPL - nZL - nPR\,|, 1)} \qquad (3.20)$$

where nZR is the number of zeros of $L(s)$ in the *RHP*, nZL is the number of zeros of $L(s)$ in the *LHP*, nPR is the number of poles of $L(s)$ in the *RHP*, and nPL is the number of poles of $L(s)$ in the *LHP*. The poles or zeros at the origin do not count in this parameter. γ can take the values $[-2, 0, +2]$, depending on the type and number of poles and zeros.

The above rules can be easily written in a simple algorithm to compute the number of encirclements N for the NSC. The method has been included in the *QFTCT for* MATLAB—see Appendix 2 and Reference 175, and the MATLAB code of the algorithm in Appendix 3. Also, the method can be applied to the Bode diagram rather than the NC. The details are left to the reader.

3.5 Examples[125]

The following examples illustrate how to apply the NSC in the NC using the technique proposed in Section 3.4. The examples include a variety of transfer functions that deal with positive and negative gains, a combination of minimum and non-minimum phase zeros, stable and unstable poles and integrators with diverse multiplicity and time-delay. As usual, the open-loop transfer function $L_j(s)$ expresses the multiplication of the plant $P_j(s)$ and the controller $G_j(s)$, so that $L_j(s) = P_j(s)G_j(s)$, $j = 1, 2, \ldots, 15$.

$$P_1(s) = \frac{140(-0.5s+1)(-0.5714s+1)(-0.6s+1)(-3.5s+1)}{s^5(-5s+1)}; \quad G_1(s) = 1 \qquad (3.21)$$

$$P_2(s) = \frac{1000(-0.5s+1)(-0.5714s+1)}{s(5s+1)(0.5s+1)(0.33s+1)(0.25s+1)}; \quad G_2(s) = 1 \qquad (3.22)$$

$$P_3(s) = \frac{-2(-0.5s+1)(0.33s+1)}{(s+1)(-5s+1)}; \quad G_3(s) = 1 \qquad (3.23)$$

$$P_4(s) = \frac{-0.25(-s+1)(10s+1)}{(-0.5s+1)(0.5s+1)}; \quad G_4(s) = 1 \qquad (3.24)$$

$$P_5(s) = \frac{-1}{s^4(-5s+1)}; \quad G_5(s) = 1 \qquad (3.25)$$

$$P_6(s) = \frac{12(-0.5s+1)(0.33s+1)}{s^2(5s+1)}; \quad G_6(s) = 1 \qquad (3.26)$$

$$P_7(s) = \frac{70(-0.5s+1)(-0.5714s+1)(-0.6s+1)}{s^3(-5s+1)}; \quad G_7(s) = 1 \tag{3.27}$$

$$P_8(s) = \frac{0.0071(-5s+1)}{s^3(-0.5s+1)(-0.5714s+1)(-0.6s+1)(-3.5s+1)}; \quad G_8(s) = 1 \tag{3.28}$$

$$P_9(s) = \frac{3 \times 10^3(-0.5s+1)(-0.5714s+1)(-0.6s+1)(-3.5s+1)}{s^6(-5s+1)} \exp(-0.8s); \quad G_9(s) = 1 \tag{3.29}$$

The first two examples (*L1* and *L2*) are associated with case N_a (Figure 3.7a). The third example (*L3*) is related to case N_b (Figure 3.7b) and the fourth example (*L4*) to case N_c (Figure 3.7c). Examples fifth to eight (*L5–L8*) deal with different options of case N_d (Figure 3.7d) and the parameters k, σ, α, and γ. The ninth example (*L9*) is a non-minimum phase unstable system with time-delay, which involves N_a and N_d. Table 3.1 and Figure 3.8 show the results of the nine examples.

TABLE 3.1

Results: $L_j(s) = P_j(s)G_j(s)$, $j = 1, 2,\dots9$. $G_j(s) = 1$

Case	N_a	N_b	N_c	N_d	k	σ	α	γ	N	P	z_c	Stab
L1	+2	0	0	+2	−1	0	+1	+2	+4	+1	+5	U
L2	+4	0	0	0	−1	0	0	+2	+4	0	+4	U
L3	0	−1	0	0	−	−	−	−	−1	+1	0	S
L4	0	0	−1	0	−	−	−	−	−1	+1	0	S
L5	0	0	0	+1	0	+1	+1	−2	+1	+1	+2	U
L6	0	0	0	+2	0	0	0	+2	+2	0	+2	U
L7	0	0	0	+2	−1	0	+1	+2	+2	+1	+3	U
L8	0	0	0	+2	+1	0	+1	−2	+2	+4	+6	U
L9	+2	0	0	+4	0	0	+1	+2	+6	+1	+7	U

Source: Garcia-Sanz, M. 2016. *Int. J. Robust Nonlinear Control*, 26(12), 2643–2651.

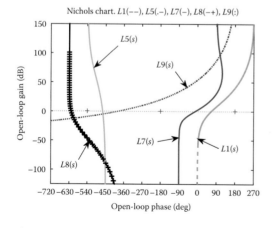
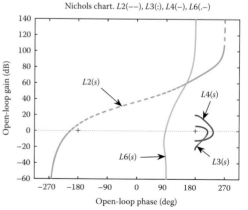

FIGURE 3.8

NCs: $L_j(s) = P_j(s)G_j(s)$, $j = 1, 2,\dots9$. (Adapted from Garcia-Sanz, M. 2016. *Int. J. Robust Nonlinear Control*, 26(12), 2643–2651.)

3.6 Guidelines to Design Controllers[125]

Based on the method proposed in Section 3.4, this new section introduces a collection of guidelines to design controllers to stabilize unstable plants. The guidelines are shown in Table 3.2. They are particularly useful for frequency-domain control design techniques, and specifically for the QFT robust control methodology.

As presented in Section 3.4, we need a number of counterclockwise encirclements, that is a negative N, to stabilize a system with a number of unstable poles in $L(s)$, that is $p > 0$. In other words, we need $N = -p$.

If p is an even number, a simple way to stabilize the plant is to design a controller according to the first case: N_a (Figure 3.7a). A negative and even N_a number can be produced by making $L(j\omega)$ to cross the rays $R_0 = \{(\phi, r): \phi = -180(2k + 1)^\circ, r > 0 \text{ dB}, k = 0, \pm 1, \pm 2, ...\}$ at a finite magnitude, at a frequency ω_x, $0 < \omega_x < +\infty$, and in the right direction. See case a_1 in Table 3.2.

If p is an odd number, two simple ways to stabilize the plant are to design controllers according to the second or third cases: N_b (Figure 3.7b) or N_c (Figure 3.7c). A negative and odd N_b number can be produced by making $L(j\omega)$ at the frequency $\omega = \omega_0 = 0$ rad/s to start at the ray $R_0 = \{(\phi, r): \phi = -180(2k + 1)^\circ, r > 0 \text{ dB}, k = 0, \pm 1, \pm 2, ...\}$ in the right direction. See case b_1 in Table 3.2. Also, a negative and odd N_c number can be produced by making $L(j\omega)$ at the frequency $\omega = \omega_\infty = +\infty$ rad/s to end at the ray $R_0 = \{(\phi, r): \phi = -180(2k + 1)^\circ, r > 0 \text{ dB}, k = 0, \pm 1, \pm 2, ...\}$ in the right direction. See case b_2 in Table 3.2. Other possibilities for p odd are also explained in Table 3.2 (see cases b_3-b_6) and illustrated in the following examples.

In addition, two useful properties to design controllers to stabilize unstable plants are presented next: the *parity interlacing property* (p.i.p.) and the *fundamental theorem of feedback control*.

3.6.1 Parity Interlacing Property

A controller $G(s)$ that stabilizes an unstable plant $P(s)$ can be itself stable (with no *RHP* poles) or unstable (with *RHP* poles).[254] This characteristic is defined by the *p.i.p.*, which says that *"a necessary and sufficient condition for the existence of a stable stabilizing compensator $G(s)$*

TABLE 3.2

Guidelines to Design a Controller $G(s)$ to Stabilize an Unstable Plant

| | | | $N = N_a + N_b + N_c + N_d = -p$ | | $\Rightarrow$ | $z_c = N + p = 0$ |
			N_a	N_b	N_c	N_d
Case (a): p even	+2,+4,+6...	a_1	−2,−4,−6...	−	−	−
Case (b): p odd	even part (0,+2,+4...)	b_0	for the even part use N_a as in case a_1			
	odd part (+1)	b_1	−	−1	−	−
		b_2	−	−	−1	−
		b_3	−2	−	−	+1 ($\sigma = 1$)
		b_4	−2	+1	−	−
		b_5	−2	−	+1	−
		b_6	add an unstable pole $\Rightarrow p$ even $\Rightarrow$ case $a1$			

Source: Garcia-Sanz, M. 2016. *Int. J. Robust Nonlinear Control*, 26(12), 2643–2651.

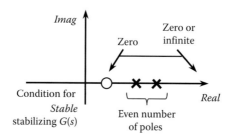

FIGURE 3.9
Parity interlacing property.

is that between each zero of the plant P(s) on the non-negative real axis (including infinity as a zero) of the s-domain, there be an even number of poles," see Figure 3.9.

The *p.i.p.* condition is a necessary and sufficient condition for the existence of a stable stabilizing compensator G(s). That is a useful condition to find controllers to stabilize unstable plants P(s) with non-minimum phase zeros. Of course, if the plant is stable, a stable compensator G(s) can always be found.

3.6.2 Fundamental Theorem of Feedback Control

It is proven that *"every plant P(s) of order n (order of the denominator) can be stabilized with a feedback compensator G(s) of order no greater than n − 1."* This is true if we consider the possibility of using not only stable compensators, but also unstable compensators if that is required.

3.6.3 Examples

The following examples illustrate how to apply the technique introduced in Section 3.6 to design controllers to stabilize unstable systems. The examples include minimum and non-minimum phase zeros and poles at the origin. Cases *L10a–L15a* are unstable plants $P_j(s)$ with $G_{ja}(s) = 1, j = 10, 11...15$. Cases *L10b–L15b* and *L15c* are the compensated stable systems with the appropriate controller $G_{jb}(s)$ or $G_{jc}(s)$. Table 3.3 and Figure 3.10 show the details.

The example $P_{10}(s)$ is an unstable plant with two *RHP* poles (see *L10a* in Figure 3.10a, and row *L10a* in Table 3.3). As $p = 2$, the number of counterclockwise encirclements needed is $N = -2$. We use case a_1 of Table 3.2, with $N_a = -2$, to design a stable controller $G_{10b}(s)$ that stabilizes the closed-loop system (see *L10b* in Figure 3.10b, and row *L10b* in Table 3.3)—Equations 3.30 and 3.31.

$$P_{10}(s) = \frac{5}{(-0.5s+1)(-0.25s+1)} \tag{3.30}$$

$$G_{10a}(s) = 1; \quad G_{10b}(s) = \frac{2 \times 10^{-5} \left(10^6 s + 1\right)(0.1s+1)}{s(0.0036s+1)} \tag{3.31}$$

The example $P_{11}(s)$ is an unstable plant with one *RHP* pole (see *L11a* in Figure 3.10a, and row *L11a* in Table 3.3). As $p = 1$, the number of counterclockwise encirclements needed is $N = -1$. We use case b_1 of Table 3.2, with $N_b = -1$, to design the controller $G_{11b}(s)$ that stabilizes the closed-loop system (see *L11b* in Figure 3.10b, and row *L11b* in Table 3.3) − Equations 3.32 and 3.33

TABLE 3.3

Results: $L_j(s) = P_j(s)G_j(s)$, $j = 10,...15$. (a) without Controller; (b) or (c) with Controller

Case	N_a	N_b	N_c	N_d	k	σ	α	γ	N	p	z_c	Stab
L10a	0	0	0	0	–	–	–	–	0	+2	+2	U
L10b	–2	0	0	0	0	0	+1	–2	–2	+2	0	S
L11a	0	0	0	0	–	–	–	–	0	+1	+1	U
L11b	0	–1	0	0	–	–	–	–	–1	+1	0	S
L12a	0	0	0	0	–	–	–	–	0	+1	+1	U
L12b	0	0	–1	0	–	–	–	–	–1	+1	0	S
L13a	0	0	0	0	–	–	–	–	0	+2	+2	U
L13b	–4	0	0	+1	0	+1	+1	–2	–3	+3	0	S
L14a	0	0	0	0	–	–	–	–	0	+1	+1	U
L14b	–2	+1	0	0	–	–	–	–	–1	+1	0	S
L15a	0	0	0	0	–	–	–	–	0	+3	+3	U
L15b	–2	–1	0	0	–	–	–	–	–3	+3	0	S
L15c	–4	0	0	0	+1	0	+2	–2	–4	+4	0	S

Source: Garcia-Sanz, M. 2016. *Int. J. Robust Nonlinear Control*, 26(12), 2643–2651.

$$P_{11}(s) = \frac{0.25}{(-0.5\ s+1)(s+1)} \tag{3.32}$$

$$G_{11a}(s) = 1; \quad G_{11b}(s) = \frac{-30(s+1)}{(0.1667\ s+1)} \tag{3.33}$$

The example $P_{12}(s)$ is an unstable non-minimum phase plant with one *RHP* pole and one *RHP* zero (see *L12a* in Figure 3.10a, and row *L12a* in Table 3.3). In this case, we have to check the *p.i.p.* condition as well. As $P_{12}(s)$ satisfies the *p.i.p.* condition and $p = 1$, the

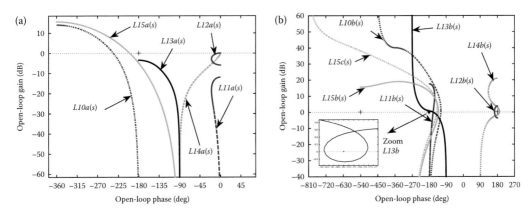

FIGURE 3.10

NCs: $L_j(s) = P_j(s)G_j(s)$, $j = 10,...15$. (a) Without controller, that is, $G_{ja}(s) = 1$—unstable systems and (b) with controller, that is, $G_{jb}(s)$ or $G_{jc}(s)$—stable systems. (Adapted from Garcia-Sanz, M. 2016. *Int. J. Robust Nonlinear Control*, 26(12), 2643–2651.)

number of counterclockwise encirclements needed is $N = -1$. We use case b_2 of Table 3.2, with $N_c = -1$, to design a stable controller $G_{12b}(s)$ that stabilizes the closed-loop system (see *L12b* in Figure 3.10b, and row *L12b* in Table 3.3)—Equations 3.34 and 3.35

$$P_{12}(s) = \frac{0.5(-s+1)}{(-0.5s+1)} \tag{3.34}$$

$$G_{12a}(s) = 1; \quad G_{12b}(s) = \frac{-1.3(0.8333s+1)}{(0.7143s+1)} \tag{3.35}$$

The example $P_{13}(s)$ is taken from Reference 254, Example 3.2. It is an unstable non-minimum phase plant with two *RHP* poles, one *RHP* zero and negative gain (see *L13a* in Figure 3.10a, and row *L13a* in Table 3.3). As $p = 2$, the number of counterclockwise encirclements needed is originally $N = -2$. However, $P_{13}(s)$ does not satisfy the *p.i.p.* condition. The plant has only one pole ($s = +3$) between the zero ($s = +2$) and the infinity, so the controller must be unstable. For this reason, we add a new unstable pole at $s = +130$, between $s = +2$ and infinity in the controller. Then, $p = 3$, and the number of counterclockwise encirclements needed is $N = -3$. We use cases a_1 and b_3 of Table 3.2, with $N_a = -4$ and $N_d = +1$, to design the controller $G_{13b}(s)$ that stabilizes the closed-loop system (see *L13b* in Figure 3.10b, and row *L13b* in Table 3.3)—Equations 3.36 and 3.37

$$P_{13}(s) = \frac{-0.6667(-0.5s+1)}{(-s+1)(-0.33s+1)} \tag{3.36}$$

$$G_{13a}(s) = 1; \quad G_{13b}(s) = \frac{0.0615\left(-17.75s^2 + 26.8750s+1\right)}{s(-0.0077s+1)} \tag{3.37}$$

Note that although this last case is a very illustrative example to understand the guidelines proposed in Table 3.2, the solution is quite fragile (see Chapter 10) and not useful from the practical point of view. In other words, small changes in the coefficients of the controller, maybe just a consequence of the digital or analog implementation, can destabilize the closed-loop system (which is the definition of fragility). For a deeper analysis of the fragility of control systems, see Section 10.4.

The example $P_{14}(s)$ is an unstable non-minimum phase plant with one *RHP* pole and one *RHP* zero (see *L14a* in Figure 3.10a, and row *L14a* in Table 3.3). As it satisfies the *p.i.p.* condition and $p = 1$, the number of counterclockwise encirclements needed is $N = -1$. We use case b_4 of Table 3.2, with $N_a = -2$ and $N_b = +1$ to design the stable controller $G_{14b}(s)$ that stabilizes the closed-loop system (see *L14b* in Figure 3.10b, and row *L14b* in Table 3.3)—Equations 3.38 and 3.39

$$P_{14}(s) = \frac{(-0.5s+1)}{(-s+1)(0.5s+1)} \tag{3.38}$$

$$G_{14a}(s) = 1; \quad G_{14b}(s) = \frac{-11.02(0.5s+1)(5.13s+1)}{\left(1.25s^2 + 37s+1\right)} \tag{3.39}$$

The last example $P_{15}(s)$ is an unstable plant with three *RHP* poles (see *L15a* in Figure 3.10a, and row *L15a* in Table 3.3). We propose two different controllers. The first controller $G_{15b}(s)$ is a stable compensator that generates $N = -3$ counterclockwise encirclements for $p = 3$. We use cases a_1 and b_1 of Table 3.2, with $N_a = -2$ and $N_b = -1$ to design this controller that stabilizes the closed-loop system (see *L15b* in Figure 3.10b, and row *L15b* in Table 3.3). The second controller $G_{15c}(s)$ is an unstable compensator that adds an additional unstable pole and generates $N = -4$ counterclockwise encirclements for $p = 4$. We use case b_6 of Table 3.2, with $N_a = -4$ (crossing of -540 and -180) to design this controller to stabilize the closed-loop system (see *L15c* in Figure 3.10b, and row *L15c* in Table 3.3) – Equations 3.40 and 3.41

$$P_{15}(s) = \frac{6}{(-0.5\,s+1)(-0.25\,s+1)(-0.1429\,s+1)} \tag{3.40}$$

$$G_{15a}(s) = 1; \quad G_{15b}(s) = \frac{-1(s+1)(0.1818\,s+1)}{(0.0033\,s+1)^2}$$

$$\tag{3.41}$$

$$G_{15c}(s) = \frac{25(s+1)(0.1429\,s+1)^3}{s(-0.1667\,s+1)(0.002\,s+1)^2}$$

3.7 Analysis of the First Case

This section develops the example presented in Section 3.2, including both the stability analysis with the technique introduced in Section 3.4 and the design of appropriate controllers in the NC introduced in Section 3.6. The plant presented in Section 3.2 has an unstable pole at the *RHP*, $s = +3$, so that

$$P_0(s) = \frac{8}{((s/-3)+1)} \tag{3.42}$$

Five controllers are propposed for this plant. The first controller $G_{01}(s)$ is presented in Equation 3.43. This is the trivial controller. The stability analysis and time response of the output $y(t)$ of the closed-loop system under a unitary step input $r(t)$ are shown in Figure 3.11. As $z_c = 1$, the system is unstable.

$$G_{01}(s) = 1 \tag{3.43}$$

The second controller $G_{02}(s)$ is the one presented in Section 3.2 and repeated in Equation 3.44. The stability analysis and time response of the output $y(t)$ of the closed-loop system under a unitary step input $r(t)$ are shown in Figure 3.12. The system is also unstable: $z_c = 1$.

$$G_{02}(s) = \frac{1}{s((s/4)+1)((s/10)+1)} \tag{3.44}$$

The third, fourth, and fifth controllers, $G_{03}(s)$, $G_{04}(s)$, and $G_{05}(s)$ are shown in Equations 3.45 through 3.47, respectively. The three controllers are able to stabilize the plant $P_0(s)$.

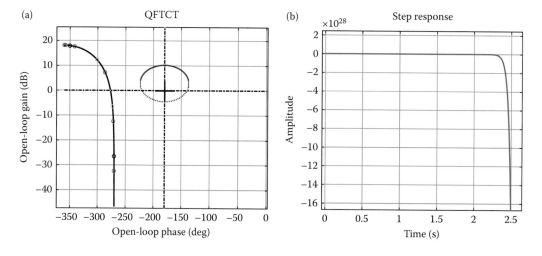

FIGURE 3.11

(a) NC: $L_{01}(s) = P_0(s)G_{01}(s)$. (b) Closed-loop time response. *Stability criterion*: **Unstable**. Rule 1: no z/p cancelation. Rule 2: $z_c = 1$, with $N = 0$, $p = 1$, and $N_a = 0$, $N_b = 0$, $N_c = 0$, $N_d = 0$.

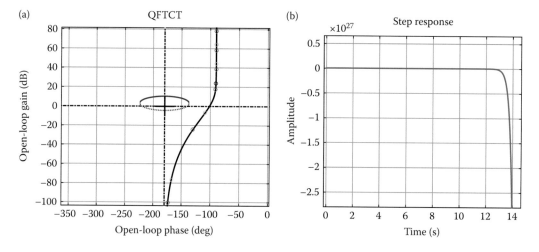

FIGURE 3.12

(a) NC: $L_{02}(s) = P_0(s)G_{02}(s)$. (b) Closed-loop time response. *Stability criterion*: **Unstable**. Rule 1: no z/p cancelation. Rule 2: $z_c = 1$, with $N = 0$, $p = 1$, and $N_a = 0$, $N_b = 0$, $N_c = 0$, $N_d = 0$, with $k = -1$, $\sigma = 0$, $\alpha = 0$, $\gamma = 2$.

Figures 3.13 through 3.15 show the stability analysis and time response of the output $y(t)$ of the closed-loop system under a unitary step input $r(t)$. In the three cases, the closed-loop system is stable: $z_c = 0$. The last case $G_{05}(s)$ is also able to achieve a zero steady-state error following a step input reference.

$$G_{03}(s) = \frac{-1}{((s/30)+1)} \tag{3.45}$$

$$G_{04}(s) = \frac{(s+1)}{((s/-1)+1)((s/100)+1)} \tag{3.46}$$

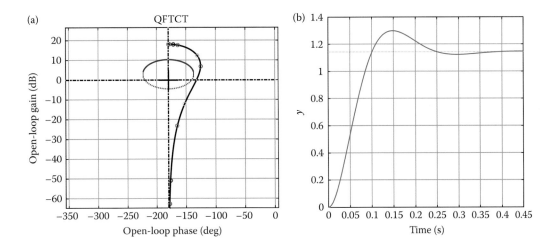

FIGURE 3.13
(a) NC: $L_{03}(s) = P_0(s)G_{03}(s)$. (b) Closed-loop time response. *Stability criterion*: **Stable**. Rule 1: no z/p cancelation. Rule 2: $z_c = 0$, with $N = -1$, $p = 1$, and $N_a = 0$, $N_b = -1$, $N_c = 0$, $N_d = 0$.

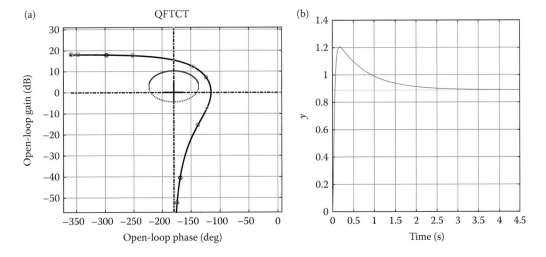

FIGURE 3.14
(a) NC: $L_{04}(s) = P_0(s)G_{04}(s)$. (b) Closed-loop time response. *Stability criterion*: **Stable**. Rule 1: no z/p cancelation. Rule 2: $z_c = 0$, with $N = -2$, $p = 2$, and $N_a = -2$, $N_b = 0$, $N_c = 0$, $N_d = 0$.

$$G_{05}(s) = \frac{-1(s+1)}{s((s/100)+1)} \tag{3.47}$$

3.8 Summary

This chapter introduced the practical and comprehensive method proposed by the author to compute the Nyquist stability criterion directly in the Nichols (magnitude/phase) chart.

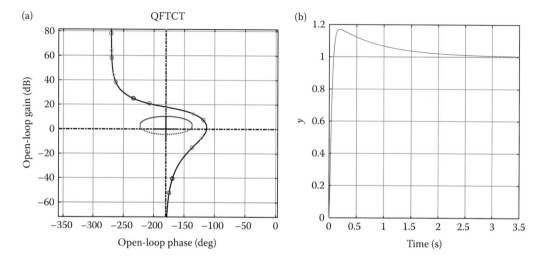

FIGURE 3.15

(a) NC: $L_{05}(s) = P_0(s)G_{05}(s)$. (b) Closed-loop time response. *Stability criterion*: **Stable**. Rule 1: no z/p cancelation. Rule 2: $z_c = 0$, with $N = -1$, $p = 1$, and $N_a = -2$, $N_b = 0$, $N_c = 0$, $N_d = 1$, with $k = 0$, $\sigma = 1$, $\alpha = 1$, $\gamma = -2$.

The method can be applied to *LTI* closed-loop systems with minimum and non-minimum phase zeros, stable and unstable poles, poles at the origin with diverse multiplicity and systems defined by nonrational functions, such as plants with time delay.

The chapter showed how to apply the method with some illustrative examples. Also, it gave guidelines to design controllers to stabilize unstable plants when dealing with frequency-domain control techniques, and in particular with the QFT robust control design methodology.

The algorithm for this method is included in the QFTCT: check *Controller design* window, *File* menu, *Check stability* option. The MATLAB code for the algorithm can be found in Appendix 3 and also free-downloaded from the website:

http://cesc.case.edu/Stability_Nyquist_GarciaSanz.htm

3.9 Practice

The list shown below summarizes the collection of problems and cases included in this book that apply the control methodologies introduced in this chapter.

- *Examples 3.1–3.9*. Section 3.5. Analysis of stability.
- *Examples 3.10–3.15*. Section 3.6. Design of control solutions to stabilize plants.
- *Example P_0*. Section 3.7. Design of multiple control solutions.
- *Appendix 3*. MATLAB code for Nyquist stability criterion in the Nichols chart.
- *Project P3*. Appendix 1. Inverted pendulum.

4

Time-Delay and Non-Minimum Phase Systems

4.1 Time-Delay Systems

Chemical and biological plants, thermal, hydraulic and pneumatic systems, and spacecraft involve transport and communication time delays.[122,6] Given a plant $P(s)$, the time-delay or dead-time t_d is defined as the time that elapses from the instant when the plant input $u(s)$ changes to the instant when the first effect of this input is seen in the plant output $y(s)$. In Laplace terms, the time-delay is represented by the exponential function e^{-st_d}. When the time-delay of a plant is comparatively larger than its time constants, an inherent bandwidth limitation makes it more difficult to achieve a satisfactory control performance.

A simple index to quantify how challenging the design of the control system is going to be is the so-called *normalized time delay*, or τ_N, which is the ratio between the *time delay* t_d and the *average residence time* T_{ar},

$$\tau_N = \frac{t_d}{T_{ar}} \tag{4.1}$$

The *average residence time* T_{ar} can be easily calculated by adding up all the time constants of the poles and the time-delay and subtracting the time constants of the zeros. Consider for instance a transfer function of a given plant like,

$$\frac{y(s)}{u(s)} = P(s) = \frac{k_p(1+T_4 s)(1+T_5 s)}{s(1+T_1 s)(1+T_2 s)(1+T_3 s)} e^{-st_d} \tag{4.2}$$

For this plant, the *average residence time* is

$$T_{ar} = T_1 + T_2 + T_3 + t_d - T_4 - T_5 \tag{4.3}$$

According to these expressions, the *normalized time delay* τ_N is always a number between zero and one ($0 \leq \tau_N \leq 1$). The larger τ_N, the more challenging the control is going to be. In large τ_N cases, a special controller design methodology, often involving predictive strategies, will be necessary to improve the system's performance.

In some situations, however, the physical limitations imposed by the time-delay can make it impossible to achieve the desired performance specifications, even with the best predictive strategies. In these circumstances, the best solution is to redesign the plant itself, reducing the time delays as much as possible.

The intrinsic limitations imposed by the time delay are explained by Bode's integral expressions presented in Chapter 3, Equations 3.1 and 3.2. Using a *Padé n*-order rational approximation for time delay,

$$e^{-st_d} \cong \frac{1 - k_1 s + k_2 s^2 - k_3 s^3 + \cdots \pm k_n s^n}{1 + k_1 s + k_2 s^2 + k_3 s^3 + \cdots + k_n s^n} \tag{4.4}$$

the Bode integral expression presented in Equation 3.2 becomes

$$\int_0^{\Omega_a} \ln |S(j\omega)| d\omega = \pi \sum_{j=1}^{n} \mathrm{Re}(z_j) \quad \text{for time delay systems} \tag{4.5}$$

where $S(j\omega) = 1/[1 + P(j\omega)G(j\omega)]$ is the sensitivity function, Ω_a is the available bandwidth, and z_j is the location of the n zeros of the *Padé n*-order rational approximation of the time-delay in the *right half plane* (RHP). As an example, for the first- and second-order *Padé* approximations, we have:

- For $n = 1$, $k_1 = t_d/2$, other $k_i = 0$, and $\mathrm{Re}(z_1) = 2/t_d$
- For $n = 2$, $k_1 = t_d/2$, $k_2 = t_d^2/12$, other $k_i = 0$, and $\mathrm{Re}(z_1) = 3/t_d$, $\mathrm{Re}(z_2) = 3/t_d$

Intuitively, to make a system with time-delay stable we need approximately $\omega_g < 1/t_d$, which is consistent with the case $\omega_g < z_j/2$ presented in Section 3.1 for *nmp* systems, and ω_g being the gain crossover frequency, which is the frequency where $|L(j\omega)| = |P(j\omega)G(j\omega)| = 1$ (0 dB).[260,354]

For open-loop stable systems, the reference tracking problem of a time-delay system can be improved substantially by introducing time-delay compensation. One of the most effective time-delay control strategies in use is the *Smith predictor* (SP), introduced by O.J.M. Smith in 1957.[338] The *SP* employs, in an inner loop, a model of the plant $\hat{P}(s)$ characterized by an estimated rational linear transfer function $\hat{P}_r(s)$ and an estimated time delay $\hat{t}_d$ to attenuate or cancel the effect of the time-delay in the real plant output $y(s)$. Then, it utilizes the un-delayed estimated output "$-y^*(s)$" as the feedback signal for control calculation. Figure 4.1 shows the control scheme of the *SP*. The real plant,

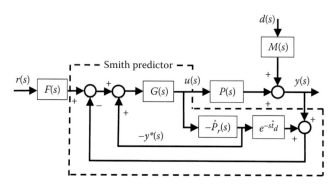

FIGURE 4.1
SP control structure.

estimated plant model, closed-loop transfer function $T(s)$, and sensitivity transfer function $S(s)$ are, respectively,

$$\text{Real plant}: \quad P(s) = P_r(s)e^{-st_d} = \frac{n_r(s)}{d_r(s)}e^{-st_d} \tag{4.6}$$

$$\text{Plant model}: \quad \hat{P}(s) = \hat{P}_r(s)e^{-s\hat{t}_d} = \frac{\hat{n}_r(s)}{\hat{d}_r(s)}e^{-s\hat{t}_d} \tag{4.7}$$

$$\frac{y(s)}{r(s)} = \frac{P(s)G(s)}{1 + \left[\hat{P}_r(s) + \left(P_r(s)e^{-st_d} - \hat{P}_r(s)e^{-s\hat{t}_d}\right)\right]G(s)}F(s) = T(s)F(s) \tag{4.8}$$

$$\frac{y(s)}{d(s)} = \frac{1 + \left(1 - e^{-s\hat{t}_d}\right)\hat{P}_r(s)G(s)}{1 + \left[\hat{P}_r(s) + \left(P_r(s)e^{-st_d} - \hat{P}_r(s)e^{-s\hat{t}_d}\right)\right]G(s)}M(s) = S(s)M(s) \tag{1.9}$$

where $P(s) = P_r(s)e^{-st_d}$ is the plant to be controlled, $\hat{P}(s) = \hat{P}_r(s)e^{-s\hat{t}_d}$ is the estimated plant model used by the SP, and $G(s)$ is the controller. Using Equations 4.8 and 4.9, the plant output is $y(s) = T(s)F(s)r(s) + S(s)M(s)d(s)$—see Figure 4.1.

If the model matches perfectly the real plant, that is, the rational linear transfer function model is $\hat{P}_r(s) = P_r(s)$ and the time-delay model is $\hat{t}_d = t_d$, then Equations 4.8 and 4.9 are simplified to

$$\frac{y(s)}{r(s)} = \frac{P_r(s)G(s)}{1 + P_r(s)G(s)}e^{-st_d}F(s) = T_r(s)e^{-st_d}F(s) \tag{4.10}$$

$$\frac{y(s)}{d(s)} = \frac{1 + (1 - e^{-s\hat{t}_d})P_r(s)G(s)}{1 + P_r(s)G(s)}M(s) = (1 - T_r(s)e^{-s\hat{t}_d})M(s) \tag{4.11}$$

being

$$T(s) = \frac{P_r(s)G(s)}{1 + P_r(s)G(s)}e^{-st_d} = T_r(s)e^{-st_d} \tag{4.12}$$

$$S(s) = 1 - T_r(s)e^{-st_d} \tag{4.13}$$

These simplifications remove the time-delay from the denominator of the closed-loop transfer function and convert the corresponding reference tracking control problem into a delay free problem: $T_r(s) = P_r(s)G(s)/[1 + P_r(s)G(s)]$.

The simplicity of this method allows the SP strategy to be implemented in low-cost digital microcontrollers, making it one of the most popular methods for compensating systems with time-delay. Indeed, the SP is available as a standard algorithm in most commercial PLCs and control devices.

However, we have to mention that although the *SP* method has the capability of transforming a time-delay problem into a delay-free problem for the reference tracking case, it still suffers significant difficulties in the disturbance rejection case, even if the plant model matches the real plant perfectly. Equations 4.11 and 4.13 and Figure 4.5b show the transfer functions for the disturbance rejection problem. The following example illustrates the advantages and difficulties of the *SP* strategy.

EXAMPLE 4.1

To illustrate the characteristics of the *SP*, consider a classical first-order plant with time-delay $P(s)$, so that,

$$P(s) = P_r(s)e^{-st_d} = \frac{9}{(2.5s+1)}e^{-0.7s}; \quad M(s) = 1 \qquad (4.14)$$

Due to the time-delay, the plant $P(s)$ is closed-loop unstable (see Chapter 3). Figure 4.2 shows the NC for $P(s)$, which does not have RHP poles and has negative stability margins. As the frequency grows, the time delay decreases the phase at a "$-(180/\pi)\,\omega\,t_d$" degrees rate, circling the point $(0\text{ dB},-180°)$ right-to-left through the positive dBs. As seen in Chapter 3, a simple method to stabilize the closed-loop system is to decrease the gain of the system with an appropriate controller $G(s)$, to circle the point $(0\text{ dB},-180°)$ right-to-left through the negative dBs. As a consequence, the closed-loop system will be stable, but slow and with a low performance due to the limitations imposed by the time-delay.

In many cases, a higher performance is required. For them, the *SP* strategy can reduce the effect of the time-delay, and allow a much more aggressive and fast controller $G(s)$. Following these ideas, we design a proportional–integral (P.I.) controller $G(s)$ with $K_p = 0.66$ and $T_i = 2$ and a unitary prefilter $F(s)$ (see the design in Example 4.1 (cont.)) such that

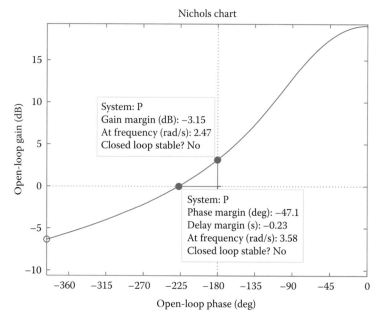

FIGURE 4.2
NC for plant $P(s)$, Equation 4.14. It is closed-loop unstable.

$$G(s) = K_p \left(1 + \frac{1}{T_i s} \right) = 0.33 \frac{2s+1}{s}; \quad F(s) = 1 \qquad (4.15)$$

The design of this controller $G(s)$ is made supposing a perfect cancellation of the effect of the time-delay, that is, the controller is designed for just the rational part of the plant, $P_r(s) = n_r(s)/d_r(s)$. If the plant model implemented in the inner loop matches the real plant, the *SP* cancels the effect of the time-delay. In this case, the stability (Figure 4.3b) and the reference tracking problem (Figure 4.4b) are excellent. On the contrary, if we apply only this controller $G(s)$ without the *SP* inner loop the system is unstable (Figure 4.3a), and the reference tracking problem does not work at all (Figure 4.4a).

Figure 4.5a shows the Bode diagram of the closed-loop transfer function $T(s)$ for the plant $P(s)$—Equation 4.14, the controller $G(s)$—Equation 4.15, and the *SP* inner loop with a perfect model matching—Equation 4.12. The curve shows an excellent reference tracking performance.

At the same time, Figure 4.5b shows the Bode diagram of the sensitivity transfer function $S(s)$ for the plant $P(s)$—Equation 4.14, the controller $G(s)$—Equation 4.15, and the *SP* inner loop with a perfect model matching—Equation 4.13. The curve shows a poor disturbance rejection performance, with some frequencies where the sensitivity amplifies the disturbances, that is, the curve is above 0 dB: $20 \log_{10}(\mathrm{abs}(S(j\omega))) > 0$ dB at some frequencies.

To overcome these disturbance rejection difficulties, several modifications of the *SP* have been proposed. See for instance: Astrom et al. (1994), Matausek and Micic (1996), Normey-Rico and Camacho (1999), etc.[351,352,353]

In addition, it is also well known that the *SP* technique is very sensitive to model-plant mismatch (plant uncertainty), either in the time delay or in the rational part of the model. Under those poor matching circumstances, stability margins may not guarantee stability even with small modeling errors. Figure 4.6 shows, for the example presented in this section, the effect of a 10% error in the gain (estimated gain = 9.9 instead of 9), of a 10% error in the time constant (estimated time constant = 2.25 instead of 2.5), and of a 10% error in the time delay (estimated time delay = 0.77 instead of 0.7). The figure shows the closed-loop system output $y(t)$ response to a unitary step input $r(t)$ at $t = 0$ s—reference tracking problem. Complementary information about this sensitivity analysis can be found in Ioannides et al. (1979), Palmor (1980), Horowitz (1983), and Yamanaka and Shimemura (1987).[342,343,345,347]

With these difficulties in mind, the next sections present the method introduced by Garcia-Sanz and Guillen[122,6] to tune the *SP* when the system presents model uncertainty in both, the rational part and the time delay. The method is composed of two algorithms. The first one is based on bandwidth frequency considerations and the second one introduces a technique to improve the design by using the QFT technique.

4.2 Robust Design of the *Smith Predictor*

This section presents the methodology introduced by the author to select the plant model for the inner loop of the *SP* when there is a model-plant mismatch in both, time delay and rational part, so that,[122,6]

$$\hat{t}_d \neq t_d \text{ and } \hat{P}_r(s) \neq P_r(s) \qquad (4.16)$$

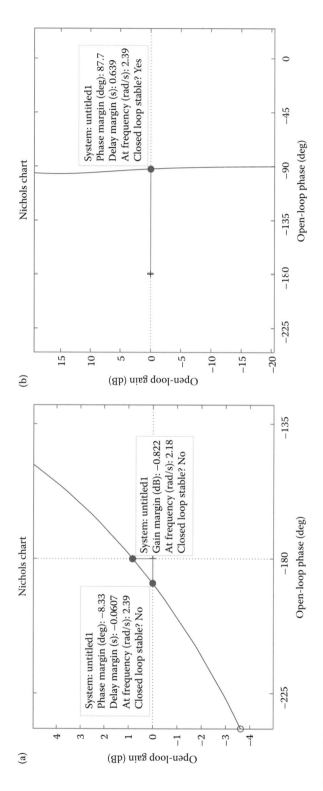

FIGURE 4.3
NC. (a) $L(s) = P_r(s)e^{-s t_d}G(s)$. (b) $L_r(s) = P_r(s)G(s)$.

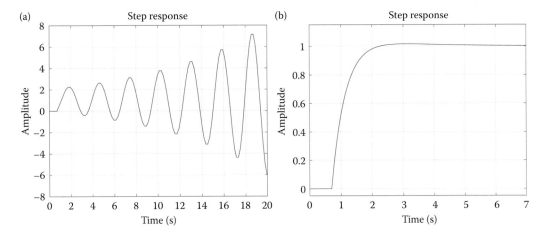

FIGURE 4.4
Closed-loop system output $y(t)$ response to a unitary step input $r(t)$ at $t = 0$ s. (a) $G(s)$ without *SP*. (b) $G(s)$ with *SP* and perfect model matching.

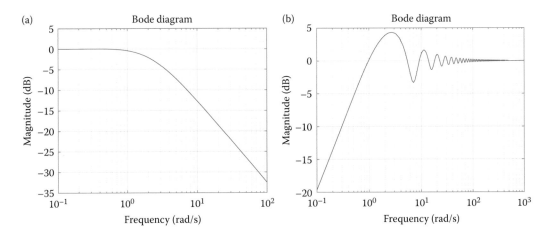

FIGURE 4.5
Magnitude plot Bode diagram. *SP* with $\hat{P}_r(s) = P_r(s)$ and $\hat{t}_d = t_d$. (a) Closed-loop transfer function $T(s) = T_r(s)e^{-st_d}$ —see Equation 4.12. (b) Sensitivity transfer function $S(s) = 1 - T_r(s)e^{-st_d}$—see Equation 4.13.

The *SP* diagram shown in Figure 4.1 is rearranged to an equivalent structure, as shown in Figure 4.7, where the expressions of the blocks $P_{eq}(s)$ and $Q(s)$ are presented in Equations 4.17 through 4.19, and where $P_r(s)$ and t_d might have uncertainty. The reference tracking problem described in Equation 4.8 is now rewritten as Equation 4.20.

$$H(s) = (1 - e^{-s\hat{t}_d}) \frac{\hat{P}_r(s)}{P_r(s)} + e^{-st_d} \tag{4.17}$$

$$Q(s) = \frac{e^{-st_d}}{H(s)} \tag{4.18}$$

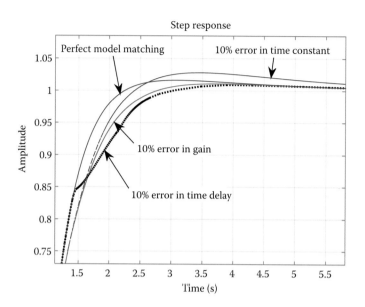

FIGURE 4.6
Closed-loop system output $y(t)$ response to a unitary step input $r(t)$ at $t = 0$ s. *SP* and $G(s)$ with: perfect model matching, 10% error in gain, 10% error in time constant, and 10% error in time delay.

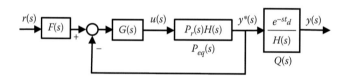

FIGURE 4.7
Equivalent diagram of *SP* for reference tracking.

$$P_{eq}(s) = P_r(s)H(s) = (1 - e^{-s\hat{t}_d})\hat{P}_r(s) + P_r(s)e^{-st_d} \tag{4.19}$$

$$\frac{y(s)}{r(s)} = \frac{P_r(s)H(s)G(s)}{1 + P_r(s)H(s)G(s)} \frac{e^{-st_d}}{H(s)} \quad F(s) = \frac{P_{eq}(s)G(s)}{1 + P_{eq}(s)G(s)} Q(s)F(s) \tag{4.20}$$

An analysis of the equivalent *SP* structure shows that, if there is no uncertainty in the model, that is to say a perfect matching or $H(s) = 1$, the time delay is eliminated from the un-delayed estimated output $y^*(s)$.

However, if there is some amount of model-plant mismatch, then the expression $H(s)$ is different from one. As a consequence, the control system is affected by the uncertainty through $H(s)$, which is in the blocks $P_{eq}(s)$ and $Q(s)$. As a result, the selection of the plant model $\hat{P}_r(s)e^{-s\hat{t}_d}$ has a significant impact on the set of values that $H(s)$ adopts over the space of uncertainty.

To address this problem, this section introduces a two-algorithm technique to tune the *SP* when the plant presents model uncertainty. The methodology gives guidelines to select the plant model for the inner loop of the *SP*. The *first algorithm* of the methodology works with the block $Q(s)$, and the *second algorithm* uses the block $H(s)$ and the QFT technique.[122,6]

4.2.1 First Algorithm

The $H(s)$ term, which appears in the $Q(s)$ block as a post-filter of the control loop (Equation 4.18 and Figure 4.7), might be responsible for some deterioration of the system performance if the magnitude of the $|Q(s)|$ block differs from "1" (0 dB) at frequencies within the system bandwidth ω_{BW} (or frequencies of interest). In this case, $y^*(s)$ would be distorted when passing through the block $Q(s)$.

Considering the complete space of parameter uncertainty and for a specified bandwidth specification, we propose the following criterion:

Criterion 1. An *SP* plant model has to be selected such that the resulting $Q(j\omega)$ does not distort $y^*(s)$ for frequencies up to ω_{BW} and for every possible plant $P_i \in P$, where $i = 1, 2, ..., n$. In other words, considering m_b the magnitude distortion limit (typically $m_b = 3$ dB) the *SP* model must satisfy

$$\left|20\log_{10}\left|Q(j\omega)\right|\right| \leq m_d, \quad 0 \leq \omega \leq \omega_{BW}, \quad \forall\left[P_r(j\omega)e^{-j\omega t_d}\right] \in P \tag{4.21}$$

The steps for the *first algorithm* (Criterion 1) are outlined in Table 4.1. It is a general procedure that can be used for any plant model with a rational part and time-delay, and for any kind of model-plant mismatch. The procedure finds the subset of plant models that could be used by the *SP* without the $Q(s)$ block causing much distortion of the output.

4.2.2 Second Algorithm

When there is a collection of possible *SP* models satisfying the *first algorithm*, then an additional degree of freedom is still available for the selection of the plant model for the *SP*. This additional degree of freedom utilizes the amount of change suffered by the equivalent-plant $P_{eq}(s)$ QFT-templates as a second criterion to select the final *SP* plant model.

The ω_i-frequency QFT-template of $P_{eq}(s)$ for two different *SP* models $\hat{P}_r(s)e^{-s\hat{t}_d}$ could significantly differ in shape and area, producing different QFT bounds and hence requiring different controllers $G(s)$. This fact introduces the question about which is the best plant model that has to be chosen from among those satisfying the *first algorithm*, to obtain the least demanding templates and ease the $G(s)$ loop-shaping. According to QFT, as a general rule, the smaller the area of the templates, the easier the compensator design becomes, and the better performance can be achieved.

TABLE 4.1

Outline for the *First Algorithm*—Criterion 1

Step	
1	Define a grid over the uncertainty of $P_{ri}(s)e^{-st_{di}} \in P$. This allows the algorithm to work with a finite set of possible plants.
2	Choose the desired closed-loop bandwidth specification ω_{BW}.
3	Select a plant $\hat{P}_{ri}(s)e^{-s\hat{t}_{di}} \in P$ as the *SP* model.
4	For this *SP* model, compute the magnitude of $Q(j\omega)$ for each plant $\in P$.
5	If the magnitude of $Q(j\omega)$ exhibits an amplification greater than m_b at frequencies lower than the desired bandwidth ω_{BW}, then that *SP* model can deteriorate the system performance. Hence it is rejected. Otherwise, the selected *SP* model passes this test.
6	Repeat from Step 3 to Step 5 for every possible *SP* model in P.

Let $\Im T(j\omega)$ represent the real-plant templates and $\Im T_{eq}(j\omega)$ represent the equivalent-plant templates when the plant $\hat{P}_r(s)e^{-s\hat{t}_d}$ is selected as the *SP* model. Also, let $A(\cdot)$ represent the area of a template on the NC; and let Ω represent the discrete set of frequencies of interest, with n_ω frequencies. The cost function I_{cost}, presented in Equation 4.22, is proposed as a measure of suitability of a specific *SP* model

$$I_{cost} = \frac{1}{n_\omega} \sum_{\omega \in \Omega} W(\omega) \frac{A(T_{eq}(j\omega))}{A(T(j\omega))} \tag{4.22}$$

This cost function is a weighted sum of normalized areas, where $W(\omega)$ represents weights that can be used to emphasize some critical frequencies, like the gain or phase crossover frequencies, resonant modes, etc. Typically, $0 \leq W(\omega) \leq 1$. An *SP* plant model that leaves every real-plant template area invariant is assigned a cost $I_{cost} = 1$ if unity weights are used. With these definitions, we propose the following criterion:

Criterion 2. From all the *SP* plant models that passed Criterion 1, select the *SP* plant model that presents the minimum cost function I_{cost}—Equation 4.22.

The steps that implement the *second algorithm* (Criterion 2) are outlined in Table 4.2. Again, this is a general procedure that can be used for any plant model with a rational part and a time-delay, and for any kind of model-plant mismatch.

The model of the plant selected for the *SP* structure with the proposed methodology avoids distortion within the operating bandwidth (*first algorithm*) and presents the least restrictive templates to the compensator design stage (*second algorithm*). Finally, though the development is made for continuous systems, the results obtained remain valid for digital control systems incorporating digital *SP*s (see Chapter 10).

EXAMPLE 4.1 (CONT.)

To illustrate and clarify the above ideas, consider again Example 4.1

$$P(s) = P_r(s)e^{-st_d} = \frac{K}{\tau s + 1} e^{-st_d} \tag{4.23}$$

which captures the essential dynamics of numerous chemical, biological, and industrial processes. In this case, the parameters for this real plant adopt values that belong to the space of parametric uncertainty given by, $K = 9.0 \pm 10\%$, $\tau = 2.5 \pm 10\%$, $t_d = 0.7 \pm 10\%$, or,

$$K \in [8.1, 9.9], \tau \in [2.25, 2.75], \text{ and } t_d \in [0.63, 0.77] \tag{4.24}$$

TABLE 4.2

Outline for the *Second Algorithm*. Criterion 2

Step	
1	Select a plant $\hat{P}_r(s)e^{-s\hat{t}_d}$ as the *SP* model from those that have successfully passed the *first algorithm*.
2	Compute the equivalent-plant templates of $P_{eq}(s)$ [see Equation 4.19] over the frequency range of interest.
3	Calculate the area of the templates, and then the cost function I_{cost} [see Equation 4.22].
4	Repeat from Steps 1 to 3 for every *SP* model that successfully passed the *first algorithm*.
5	Select the model that results in the minimum cost function I_{cost}.

Clearly, the mathematical *SP* model that better describes the plant is the first-order dynamic system with time-delay so that,

$$\hat{P}(s) = \hat{P}_r(s)e^{-s\hat{t}_d} = \frac{\hat{K}}{\hat{\tau}s+1}e^{-s\hat{t}_d} \tag{4.25}$$

Applying the First Algorithm

As is mentioned above, the *first algorithm* of the methodology is based on the analysis of the magnitude of the frequency response of $Q(j\omega)$ over the frequency range of interest and for the complete set of parameter uncertainty. Appendix 4 includes the complete MATLAB code for this example.

By fixing the desired closed-loop bandwidth to ω_{BW} (Step 2 of Table 4.1), the set of admissible *SP* models which fulfill the bandwidth specification are easily found by following Steps 3 through 6 of the *first algorithm*. The results of this example are plotted as a 3D object, where each of the three axes represent the uncertainty of every parameter τ, t_d, K (see Figures 4.8 and 4.9). The models the *SP* could adopt are those located inside the 3D figure. To obtain Figure 4.8, a grid of $5 \times 5 \times 5$ values (=125 plants) is used. Figure 4.8a shows the 3D object obtained for the bandwidth specification $\omega_{BW} = 1.0$ rad/s, Figure 4.8b for $\omega_{BW} = 2.0$ rad/s, and Figure 4.9 for $\omega_{BW} = 2.25$ rad/s. As might have been expected, the larger the bandwidth specification the smaller the set of potential candidates. Figure 4.9 also illustrates that the usual selection of a "mean" *SP* model (with the mean parameter values, shown in the figure) is not the most appropriate choice according to Criterion 1. In the limit, if the bandwidth specification is larger enough, it is presumed that there is only one possible model.

The *SP* plant models that passed the *first algorithm* for $\omega_{BW} = 2.25$ rad/s are shown in Figure 4.9 and Table 4.3. We will use these candidates in the *second algorithm* to find and select the final *SP* plant model.

Applying the Second Algorithm

Given the bandwidth specification $\omega_{BW} = 2.25$ rad/s, the *first algorithm* found the set of *SP* plant model candidates shown in Figure 4.9 and Table 4.3. Using these *SP* plant models, the *second algorithm* calculates the QFT-templates for $P_{eq}(s) = P_r(s)H(s)$ using the complete parametric uncertainty for the real plant $P(s)$ defined by Equation 4.24. Figure 4.10

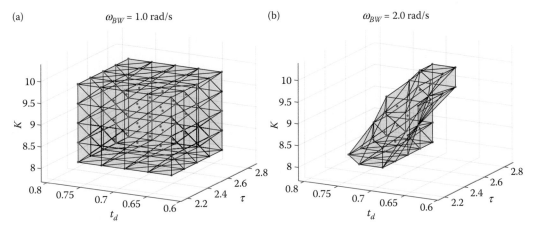

FIGURE 4.8

Plants (set of parameters) that meet *first algorithm*. (a) with $\omega_{BW} = 1.0$ rad/s and (b) with $\omega_{BW} = 2.0$ rad/s.

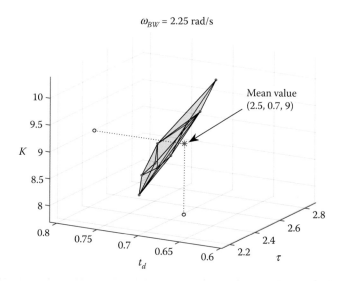

FIGURE 4.9
Plants (set of parameters) that passed *first algorithm* with $\omega_{BW} = 2.25$ rad/s. Table 4.3. See that mean value ($\tau = 2.50$, $t_d = 0.7$, $K = 9$) is outside.

TABLE 4.3

Plants (Set of Parameters) That Passed the *First Algorithm*
with $\omega_{BW} = 2.25$ rad/s (Figure 4.9)

	τ	t_d	K
1	2.375	0.735	8.10
2	2.625	0.770	8.10
3	2.750	0.770	8.10
4	2.500	0.735	8.55
5	2.625	0.735	8.55
6	2.750	0.770	8.55
7	2.750	0.735	9.00
8	2.625	0.700	9.45
9	2.750	0.700	9.90

compares these templates for $P_{eq}(s)$ with the original templates for $P(s)$. See the MATLAB code in Appendix 4.

Then, we apply Equation 4.22 and calculate the cost function I_{cost} of every *SP* plant model candidate that passed the *first algorithm* (Figure 4.9 and Table 4.3). Figure 4.11 shows the results. The candidate #6 presents the minimum value, with $I_{cost} = 0.9446$, for $W(\omega) = 1 \; \forall \omega$. Then, we select the plant model #6, with $\tau = 2.75$, $t_d = 0.77$, and $K = 8.55$ for the *SP* inner model. That is,

$$\hat{P}(s) = \hat{P}_r(s)e^{-s\hat{t}_d} = \frac{\hat{K}}{\hat{\tau}s + 1}e^{-s\hat{t}_d} = \frac{8.55}{2.75s + 1}e^{-s0.77} \qquad (4.26)$$

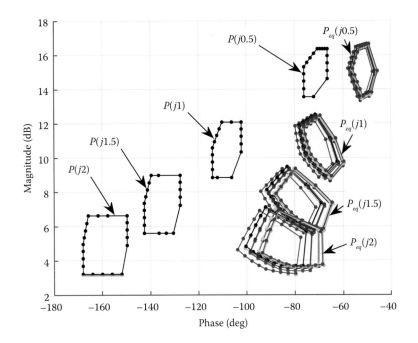

FIGURE 4.10
QFT templates for $P(s)$ and $P_{eq}(s) = P_r(s)H(s)$. Uncertainty: Equation 4.24.

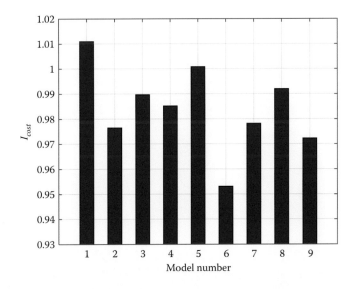

FIGURE 4.11
I_{cost} for plant models that passed *first algorithm* (plants in Table 4.3).

Finally, in order to design the controller $G(s)$ for the equivalent plant $P_{eq}(s)$ with the *SP* strategy and the selected *SP* plant model (see Figure 4.7), we propose a set of three control specifications: robust stability, robust output disturbance rejection, and robust reference tracking—see Equations 4.27 through 4.29, respectively, and Figure 4.12. Note that these specifications are proposed for $P_{eq}(s) = P_r(s)H(s)$.

- *Stability specification:*

$$\left| \frac{P_r(j\omega)H(j\omega)G(j\omega)}{1 + P_r(j\omega)H(j\omega)G(j\omega)} \right| \le W_s = 1.6(GM = 4.2\,\text{dB},\ PM = 36.4°)$$

$$\text{for } \omega = [0.001 \quad 0.005 \quad 0.01 \quad 0.05 \quad 0.1 \quad 0.5 \quad 1 \quad 5 \quad 10 \quad 50]\,\text{rad/s} \tag{4.27}$$

- *Sensitivity or output disturbance rejection specification*—see Figure 4.12a:

$$\left| \frac{1}{1 + P_r(j\omega)H(j\omega)G(j\omega)} \right| \le \delta_{op}(\omega) = \left| \frac{(j\omega)}{(j\omega) + 2} \right|$$

$$\text{for } \omega = [0.001 \quad 0.005 \quad 0.01 \quad 0.05 \quad 0.1 \quad 0.5 \quad 1]\,\text{rad/s} \tag{4.28}$$

- *Reference tracking specification*—see Figure 4.12b:

$$\delta_{lo}(\omega) < \left| \frac{P_r(j\omega)H(j\omega)G(j\omega)}{1 + P_r(j\omega)H(j\omega)G(j\omega)} F(j\omega) \right| \le \delta_{up}(\omega) \tag{4.29}$$

where,

$$\delta_{lo}(\omega) = \left| \frac{0.98}{0.1(j\omega)^2 + (j\omega) + 1} \right| \quad \text{is the lower bound}$$

and,

$$\delta_{up}(\omega) = \left| \frac{(j\omega) + 1.02}{0.2(j\omega)^2 + 0.85(j\omega) + 1} \right| \quad \text{is the upper bound}$$

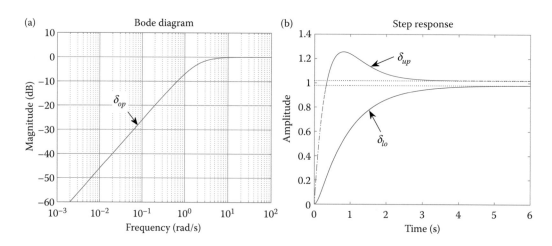

FIGURE 4.12
Control specifications for $P_{eq}(s) = P_r(s)H(s)$. (a) Disturbance rejection at the output of the plant: δ_{op}. (b) Reference tracking: δ_{up}, δ_{lo}.

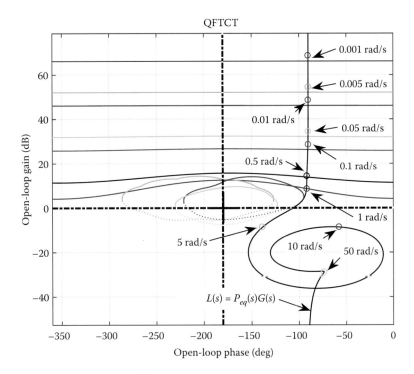

FIGURE 4.13
Loop-shaping of controller $G(s)$. $L(s) = P_{eq}(s)G(s) = P_r(s)H(s)G(s)$.

$$\text{for } \omega = [0.001 \quad 0.005 \quad 0.01 \quad 0.05 \quad 0.1 \quad 0.5 \quad 1] \text{ rad/s}$$

The QFT bounds are calculated taking into account the $P_{eq}(s)$ structure with the parameter uncertainty given by Equation 4.24, the *SP* plant model given by Equation 4.26, and the above stability, output disturbance rejection, and reference tracking specifications given by Equations 4.27 through 4.29, respectively—see Figure 4.13.

To meet the specifications and illustrate the methodology, we select a P.I. controller $G(s)$ with $K_p = 0.66$ and $T_i = 2$ (also used in Example 4.1), and a unitary prefilter $F(s)$, such that,

$$G(s) = K_p\left(1 + \frac{1}{T_i s}\right) = 0.33\frac{2s+1}{s}; F(s) = 1 \qquad (4.30)$$

The QFT-bounds and the loop shaping of $L(s) = P_{eq}(s)G(s)$ are shown in Figure 4.13. The solution includes the *SP* structure with the plant model selected with the methodology—Equation 4.26, and the QFT controller $G(s)$ presented in Equation 4.30. As Figure 4.14 shows, the control system meets the performance specifications: see Figure 4.14a for the frequency-domain analysis of the disturbance rejection specification—Equation 4.28, and Figure 4.14b for the time-domain analysis of reference tracking specification with 125 plants—Equation 4.29.

Also, Figure 4.15 shows the improvement achieved with the proposed *SP* plant model selection methodology over the classical mean value plant model selection technique with the same real plant model used in Figure 4.6.

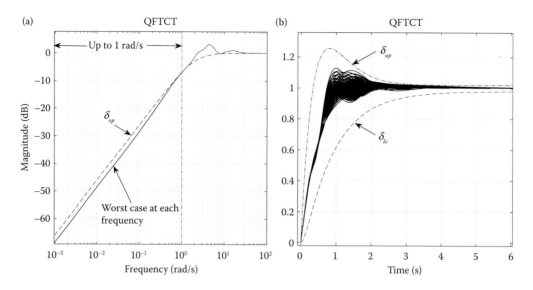

FIGURE 4.14
Analysis of controller $G(s)$ for $P_{eq}(s) = P_r(s)H(s)$. (a) Disturbance rejection at the output of the plant: δ_{op}. (b) Reference tracking: δ_{up}, δ_{lo}.

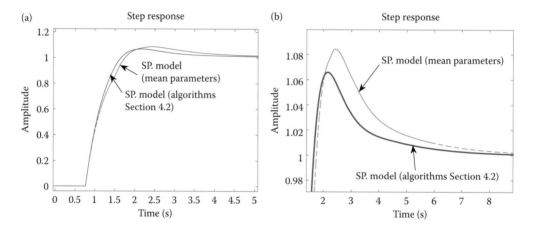

FIGURE 4.15
Plant output $y(t)$: (a) full response and (b) details. Analysis of controller $G(s)$ and *SP* with plant model $\hat{P}(s)$ according to algorithms of Section 4.2: ($\tau = 2.75$, $t_d = 0.77$, $K = 8.55$), and with plant model $\hat{P}(s)$ with mean parameters: ($\tau = 2.5$, $t_d = 0.7$, $K = 9.0$).

4.3 *Non-Minimum Phase* Systems

4.3.1 Analysis

The presence of a plant's zeros in the RHP also limits the performance achievable by closed-loop control systems. When we have RHP zeros, we say that the plant is *non-minimum phase (nmp)*. On the contrary, if there are no RHP zeros, then we say that the plant is *minimum phase (mp)*. As with the time-delay, the RHP zeros compromise the stability and reduce the control possibilities.

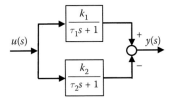

FIGURE 4.16
Example of the internal configuration of an *nmp* system.

Physically speaking, the presence of an RHP zero in the plant can be understood as the result of two parallel subsystems that interact with opposite signs within the plant. Figure 4.16 and Equations 4.31 and 4.32 show an example. The zero is in the RHP when either $[k_1 > k_2$ with $k_2\tau_1 > k_1\tau_2]$ or $[k_1 < k_2$ with $k_2\tau_1 < k_1\tau_2]$. For instance, Equation 4.33 (Example 4.2) shows an *nmp* case with one RHP zero at $z_1 = +7/11$, and Equation 4.34 (Example 4.3) shows the same case but as *mp*, with the zero $z_1 = -7/11$ at the left half plane (LHP).

$$\frac{y(s)}{u(s)} = P(s) = \frac{(k_1\tau_2 - k_2\tau_1)s + (k_1 - k_2)}{(\tau_1 s + 1)(\tau_2 s + 1)} \tag{4.31}$$

$$z_1 = \frac{-(k_1 - k_2)}{(k_1\tau_2 - k_2\tau_1)} \tag{4.32}$$

EXAMPLE 4.2: *nmp* CASE
$k_1 = 10$, $k_2 = 3$, $\tau_1 = 7$, $\tau_2 = 1$.

$$\frac{y(s)}{u(s)} = P(s) = \frac{-11s + 7}{(7s + 1)(s + 1)}, \text{ with } z_1 = \frac{7}{11} > 0 \text{ (RHP)} \tag{4.33}$$

EXAMPLE 4.3: *mp* CASE
$k_1 = 19/3$, $k_2 = -2/3$, $\tau_1 = 7$, $\tau_2 = 1$.

$$\frac{y(s)}{u(s)} = P(s) = \frac{11s + 7}{(7s + 1)(s + 1)}, \text{ with } z_1 = \frac{-7}{11} < 0 \text{ (LHP)} \tag{4.34}$$

Also, for these two cases, Figure 4.17a shows the root locus for Example 4.2—see the RHP zero, and Figure 4.17b the root locus for Example 4.3—see the zero in the LHP.

The Bode diagram for both examples is shown in Figure 4.18a. As we can see, at high frequency the *nmp* system presents a phase of $-180°$ less than the *mp* system (high frequency phase $= -270°$ instead of $-90°$). Accordingly, this is the reason for the name *nmp* or *mp* systems. The phase lag of the *nmp* system is the fact that compromises the stability and reduces the control possibilities.

Finally, the open-loop step response of both examples is shown in Figure 4.18b. The *nmp* system presents an initial inverse response, while the *mp* shows an initial direct response.

4.3.2 Control Methodology

In this section, we proposed a control methodology to regulate *nmp* systems. The technique is based on the *SP* strategy introduced in Section 4.2 for time-delay systems. In a

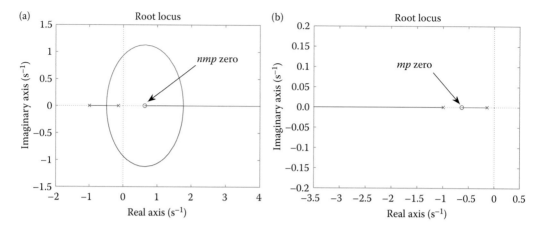

FIGURE 4.17
Root locus. (a) *nmp* case and (b) *mp* case. Examples 4.2 and 4.3.

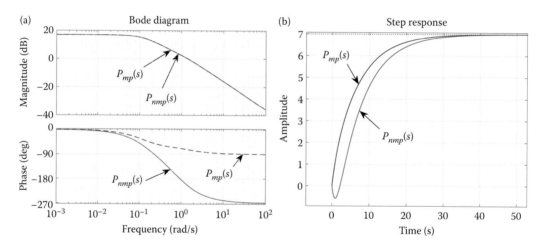

FIGURE 4.18
(a) Bode diagram and (b) step responses of *nmp* and *mp* cases. Examples 4.2 and 4.3.

similar way, we adopt a control diagram with a model of the plant in an inner loop (*internal-model based controller*) where, instead of the blocks shown in Figure 4.1 for time-delay systems, we adopt the structure presented in Figure 4.19, selecting,

$$\hat{P}_{mp}(s) \text{ instead of } P_r(s), \text{ and } APF(s) \text{ instead of } e^{-st_d} \tag{4.35}$$

As we can see in Figure 4.19, the controller employs in an inner loop an estimated *mp* plant model and an estimated *all-pass filter* (*APF*). This allows the controller to attenuate or cancel the effect of the *nmp* zero in the real plant output $y(s)$—in reference tracking specifications. Then, it utilizes the *mp* estimated output "$-y_{mp}{}^*(s)$" as the feedback signal for control calculation. The real plant and estimated plant are, respectively,

$$\text{Real plant:} \quad P(s) = \frac{(-as+1)n_r(s)}{d_r(s)} \tag{4.36}$$

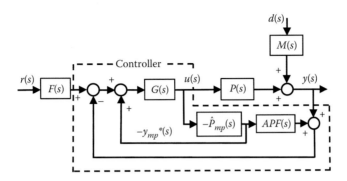

FIGURE 4.19
Extension of the *SP* concept to *nmp* systems.

$$\text{Estimated plant:} \quad \hat{P}(s) = \hat{P}_{nmp}(s) = \hat{P}_{mp}(s)APF(s) \tag{4.37}$$

with,

$$\text{Minimum phase plant model:} \quad \hat{P}_{mp}(s) = \frac{(\hat{a}s+1)\hat{n}_r(s)}{\hat{d}_r(s)} \tag{4.38}$$

$$\text{All-pass filter:} \quad APF(s) = \frac{(-\hat{a}s+1)}{(\hat{a}s+1)} \tag{4.39}$$

Now, based on these expressions, and similar to Equations 4.8 and 4.9, the closed-loop transfer function $T(s)$ and the sensitivity transfer function $S(s)$ are

$$\frac{y(s)}{r(s)} = \frac{P(s)G(s)}{1 + \left[\hat{P}_{mp}(s) + \left(P(s) - \hat{P}_{mp}(s)APF(s)\right)\right]G(s)}F(s) = T(s)F(s) \tag{4.40}$$

$$\frac{y(s)}{d(s)} = \frac{1 + (1 - APF(s))\hat{P}_{mp}(s)G(s)}{1 + \left[\hat{P}_{mp}(s) + \left(P(s) - \hat{P}_{mp}(s)APF(s)\right)\right]G(s)}M(s) = S(s)M(s) \tag{4.41}$$

For open-loop stable systems, the tracking problem of these *nmp* systems can be improved substantially by introducing this RHP zero compensation—see Equation 4.40.

Notice also that the *APF* defined in Equation 4.39 coincides with the time-delay *Padé* approximation of first order—see Equation 4.4. For this reason, for cases where there is a model-plant mismatch in both, the *nmp* zero and the rational part,

$$\hat{a} \neq a \text{ and } \hat{P}_{mp}(s) \neq P_{mp}(s) \tag{4.42}$$

the selection of the plant model for the inner loop of the controller can be done according to the same methodology introduced in Section 4.2, *first* and *second* algorithms. The details are left to the reader.

4.4 Summary

The presence of time-delay limits the performance achievable by closed-loop control systems. For open-loop stable systems, the tracking problem of a time-delay system can be improved substantially by introducing time-delay compensation. One of the most popular and effective time-delay control strategies in use is the *SP*.

It is well known, however, that the *SP* may be very sensitive to model-plant mismatch, resulting in a poor performance when model uncertainty is present. This chapter introduced a method to design *SPs* when the plant is not precisely known. The method includes two criteria. The first one, presented as *first algorithm*, is based on bandwidth frequency considerations. It finds the set of model candidates for the plant so that, if the *SP* adopts one of them, the desired bandwidth is not reduced by the effect of the parameter uncertainty. The second criterion, presented as *second algorithm*, introduces some guidelines to improve the design of the *SP* by using the QFT technique.

An illustrative example shows in detail how to effectively apply the methodology step by step to a plant with time-delay and model uncertainty. The MATLAB code for the algorithms for this example is presented in Appendix 4.

Finally, the presence of plant's zeros in the RHP (*nmp* systems) also limits the performance achievable by closed-loop control systems. This chapter also analyzed the main properties of these systems. Also, it introduced a methodology similar to the one proposed for the *SP*, to design an *internal-model based controller* when the plant is not precisely known. For open-loop stable systems, the tracking problem of these *nmp* systems can be improved substantially by introducing this RHP zero compensation.

4.5 Practice

The list shown below summarizes the collection of problems and cases included in this book that apply the control methodologies introduced in this chapter.

- Example 4.1. Sections 4.1 and 4.2. First-order plant with delay and *SP*.
- Examples 4.2 and 4.3. Section 4.3. *nmp* and *mp* systems.
- Appendix 4. MATLAB code for Example 4.1.
- *Project P6*. Appendix 1. Central heating system with long-time delays.

5

Distributed Parameter Systems

5.1 Introduction

The physical systems we have studied in the previous chapters are lumped systems. A lumped system is one in which its elements are considered as concentrated at singular points in space, and all the dependent variables of interest are a function of time alone, that is, $f(t)$. A typical example is an electrical circuit with resistors R, capacitors C, and inductances L. The dependent variables are the current $i(t)$ and the voltage $v(t)$, which are a function of time alone. Generally, lumped systems are described with a set of ordinary differential equations (ODEs). In this electrical circuit example, we can describe the system with the well-known ODEs

$$i(t) = C\frac{dv(t)}{dt} \ ; \quad i(t) = R\,v(t); \quad v(t) = L\frac{di(t)}{dt} \tag{5.1}$$

On the contrary, a distributed parameter system (DPS) is one in which the dependent variables are functions of both time and spatial variables, that is, $g(t, x, y, z, \ldots)$. In this case, the elements are distributed in space (x, y, z). A classic DPS example is an electrical line, where the resistance, capacity, and inductance are not constant but a function of the line length x, that is, $R(x)$, $C(x)$, and $L(x)$. Then, the dependent variables, current and voltage, are a function of time and distance: $i(t,x)$ and $v(t,x)$. In these DPS cases, we need a set of partial differential equations (PDEs) to describe the system.

Examples for DPSs include flexible robots, heat transfer problems, large mechanical structures, spacecraft with flexible appendages, electrical transmission lines, etc. Due to the high complexity and computational cost of solving PDEs, DPSs are often approximated by much simpler lumped systems and the corresponding ODEs. When this lumped approximation is not enough to catch the DPS characteristics, methods based on PDEs are required. In the last two decades, a significant number of papers about control of DPSs have been proposed. They include a variety of topics like stability, controllability, observability, sensor location, estimation, and optimal control of DPSs. However, at the same time, there are still very few practical approaches for DPS control design, and even fewer for DPS robust control.

This chapter introduces a practical quantitative robust control technique to design one-point feedback controllers for DPSs with uncertainty. The method considers the spatial distribution of the relevant points where the inputs and outputs of the control system are applied (actuators, sensors, disturbances, and control objectives), and a new set of transfer functions (TFs) that describe the relationships between those distributed inputs and outputs.

Based on the definition of such distributed TFs, the classical robust stability and robust performance specifications are extended to the DPS case, and a new set of quadratic inequalities are defined for the DPS QFT bounds. As a result, we propose an advanced control method able to deal with uncertainty in both the model and the spatial distribution of the inputs and the outputs. The chapter also includes a well-known DPS heat transfer example to illustrate the use and simplicity of the proposed methodology.

5.2 Modeling Approaches for PDE[7]

DPSs can be described with PDEs. According to the classical book written by Farlow,[358] a general second-order linear PDE in two variables is of the form

$$AU_{xx} + BU_{xy} + CU_{yy} + DU_x + EU_y + FU = G \tag{5.2}$$

where $U_{xx} = \partial^2 U/\partial x^2$, $U_{xy} = \partial^2 U / \partial x \partial y$, $U_x = \partial U/\partial x$, $U_y = \partial U/\partial y$, etc.

This equation is written here with the two independent variables x and y. Note, however, that in many problems one of the two variables stands for time, and hence the case is written in terms of x and t. Following Equation 5.2, there is a basic classification to linear PDE involving three types of equations. They are:

1. *Parabolic equations,* that satisfy the property $B^2 - 4AC = 0$ (as for heat flow and diffusion processes)
2. *Hyperbolic equations,* that satisfy the property $B^2 - 4AC > 0$ (as for wave equations)
3. *Elliptic equations,* that satisfy the property $B^2 - 4AC < 0$ (as for some mechanical systems)

Some very well-known examples of these three types are: the heat exchange equation for parabolic type $\{\partial T/\partial t = K\, \partial^2 T/\partial x^2\}$; the one-dimensional (1D) wave equation for hyperbolic type $\{\partial^2 T/\partial t^2 = K\, \partial^2 T/\partial x^2\}$; and the Euler–Lagrange equation for the elliptic type $\{\partial^2 T/\partial t^2 + \partial^2 T/\partial x^2 = 0\}$.

Frequently, for parabolic systems an easy model approximation is to use equivalent electrical circuits (see Figures 5.1 and 5.5 and Example 5.1). Generally speaking, the set of differential equations describing the dynamic performance of these kinds of DPSs can be formulated in terms of *through variables* (I) and *across variables* (V)—see References 230 and

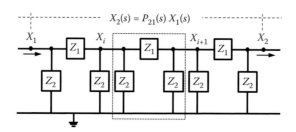

FIGURE 5.1
Π-equivalent electrical model.

236. The through variable (I) can represent an electrical current, a mechanical force or torque, a heat flow rate, etc. The across variable (V) can represent an electrical voltage difference, a mechanical linear or angular velocity difference, or a temperature difference, etc.

The equivalent electrical model P_{x2x1} (Figure 5.1) is defined with Π (Pi) elements in series between the points of interest X_1 and X_2. The more Π elements introduced, the larger the bandwidth where the model is accurate enough to be used. The minimum amount of elements to be considered can be studied with the Anderson–Parks criteria.[346]

For the three types of PDEs, another general approach of P_{x2x1} can be obtained by getting the solution of the PDE. This is not a trivial task, and often it leads to either an irrational solution or a very complex TF. In these cases, it is convenient to transform the irrational solutions into a series of sums or products by using some mathematical approaches such as the Taylor series, Fourier series, or Weirstrass factorization (see, e.g., Equation 5.37). As before, the number of components of the sums or products can be studied with the Anderson–Parks criteria.[346]

5.3 Generalized DPS Control System Structure[7,137]

As is seen in the previous chapters, the classical 2DOF general feedback structure includes a loop compensator $G(s)$ and a prefilter $F(s)$ in series with the feedback loop (Figure 5.2). Based on this structure, the following well-known equations for the plant output $y(s)$, control signal $u(s)$, and signal error $e(s)$ are obtained—see also Equations 2.32 through 2.34.

$$y = \frac{PG}{1+PGH}Fr + \frac{1}{1+PGH}(PG\,d_e + P\,d_i + d_o + M\,d) - \frac{PGH}{1+PGH}n \qquad (5.3)$$

$$u = \frac{G}{1+PGH}Fr + \frac{G}{1+PGH}(d_e - HP\,d_i - H\,d_o - HM\,d) - \frac{GH}{1+PGH}n \qquad (5.4)$$

$$e = \frac{1}{1+PGH}Fr - \frac{H}{1+PGH}(PG\,d_e - P\,d_i - d_o - M\,d) - \frac{H}{1+PGH}n \qquad (5.5)$$

where $P(s)$ denotes the plant with uncertainty, $H(s)$ the dynamics for the sensor and filters, $M(s)$ the dynamics of the effect of the generic disturbances $d(s)$ over the plant output, and $r(s)$, $n(s)$, $d_e(s)$, $d_i(s)$, $d_o(s)$ represent the reference signal for tracking, sensor noise, and

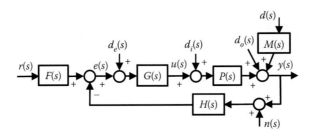

FIGURE 5.2
Classical 2DOF feedback structure.

external disturbances at the error signal, plant input and plant output, respectively. The Laplace variable s is omitted in the equations for simplicity. To deal with DPSs, this chapter proposes a generalization of this classical 2DOF feedback structure. In particular, we propose to include the information of the spatial location (x_a, x_d, x_s, and x_o) of the points where the actuator, disturbances, sensor, and control objectives are applied, respectively, as shown in Figure 5.3.

Based on these points, a PDE TF can be defined between every two points of interest. These PDE TFs vary with the relative position, topology, and spatial distribution of the points of interest. Now, in a similar way to Equations 5.3 through 5.5, the dynamics of the DPS system is explained by

$$y_{xs} = P_{xs\,xd}\,u_{xd} + P_{xs\,xa}\,u_{xa} \tag{5.6}$$

$$y_{xo} = P_{xo\,xd}\,u_{xd} + P_{xo\,xa}\,u_{xa} \tag{5.7}$$

$$u_{xa} = G[F\,r - H(y_{xs} + n) + d_e] + d_i \tag{5.8}$$

$$u_{xd} = M\,d + d_o \tag{5.9}$$

$$e = F\,r - H(y_{xs} + n) \tag{5.10}$$

where P_{x2x1} denote the Laplace TF between the input x_1 ($x_1 = x_a$ or x_d) and the output x_2 ($x_2 = x_s$ or x_o). The signals u_{xd}, u_{xa}, e, y_{xo}, and y_{xs} represent the external disturbances, the actuator, the error, the control objective, and the sensor output, respectively.

Using Equations 5.6 through 5.10, and after some straightforward algebraic manipulation, the expressions for the DPS 2DOF feedback structure of Figure 5.3 are

$$y_{xs} = \frac{P_{xs\,xa}}{1 + P_{xs\,xa}GH}[G(F\,r + d_e - H\,n) + d_i] + \frac{P_{xs\,xd}}{1 + P_{xs\,xa}GH}u_{xd} \tag{5.11}$$

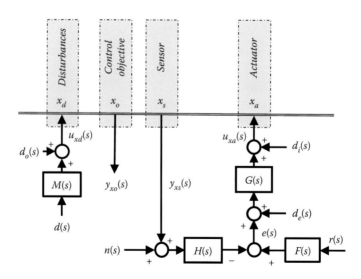

FIGURE 5.3
Generalized 2DOF DPS control system structure.[137]

$$y_{xo} = \frac{P_{xo\,xa}}{1+P_{xs\,xa}GH}[G(Fr+d_e-Hn)+d_i] + \left[P_{xo\,xd} - \frac{P_{xo\,xa}P_{xs\,xd}GH}{1+P_{xs\,xa}GH}\right]u_{xd} \quad (5.12)$$

$$u_{xa} = \frac{1}{1+P_{xs\,xa}GH}[G(Fr+d_e-Hn)+d_i] - \frac{GHP_{xs\,xd}}{1+P_{xs\,xa}GH}u_{xd} \quad (5.13)$$

$$e = \frac{1}{1+P_{xs\,xa}GH}(Fr-Hn) - \frac{HP_{xs\,xa}}{1+P_{xs\,xa}GH}(Gd_e+d_i) - \frac{HP_{xs\,xd}}{1+P_{xs\,xa}GH}u_{xd} \quad (5.14)$$

Note that these TFs depend on the compensator G, prefilter F, sensor dynamics H, and the spatial distribution and topology through the four TFs $P_{xs\,xa}$, $P_{xo\,xa}$, $P_{xs\,xd}$, and $P_{xo\,xd}$, and where $u_{xd} = Md + d_o$.

The new four TFs $\{P_{xs\,xa}, P_{xo\,xa}, P_{xs\,xd}, P_{xo\,xd}\}$ describe the relationship between inputs and outputs at the points of interest (x_a, x_d, x_s, x_o). These TFs are a representation of the PDEs that preserve the distributed spatial configuration of the problem. Furthermore, the implementation of the problem with these rational expressions allows the designer to use classical control theory tools related to Laplace TFs.

Remark 5.1

A useful characteristic of the proposed methodology is the fact that the points of interest $(x_a, x_d, x_s,$ and $x_o)$ may not be fixed at a particular location. They can slowly move along in space or else the current location be uncertain. In other words, the application of the actuator, disturbance, sensor, and control objective may not be defined at a single point but at a distribution of points.

Remark 5.2

A potential problem to be addressed when applying this control design methodology is the definition of the TFs $\{P_{xs\,xa}, P_{xo\,xa}, P_{xs\,xd}, \text{ and } P_{xo\,xd}\}$, which being rational must also comply with the distributed nature of the system. See Section 5.2 for details.

Remark 5.3

Note that the 2DOF DPS control structure proposed here (Figure 5.3) is a generalization of the classical lumped control diagram (Figure 5.2). Consider for instance the case where the sensor, the control objective, and the application of the disturbance d are at the same point, that is, $x_s = x_o = x_d$ in Figure 5.3. Then, the TFs turn out to be $P = P_{xs\,xa} = P_{xo\,xa}$ and $P_{xo\,xd} = P_{xs\,xd} = 1$. Hence, substituting these equivalences in Equation 5.12, the control objective expression becomes

$$y_{xo} = y_{xs} = \frac{P_{xs\,xa}}{1+P_{xs\,xa}GH}[G(Fr+d_e-Hn)+d_i] + \frac{1}{1+P_{xs\,xa}GH}(Md+d_o) \quad (5.15)$$

which matches the classical lumped system shown in Figure 5.2 and Equation 5.3, being the lumped TF for the system plant $P = P_{xs\,xa}$.

5.4 Extension of QFT to DPS[7,137]

In the early 1980s, Professor Horowitz proposed to Kannai an interesting mathematical task: the analysis of the causality and stability of linear systems described by partial differential operators.[137,344] Some years later, and based on these preliminary results, Horowitz and Azor[130,131] developed the very first QFT approach to DPSs. The method was a two-variable generalization (complex variables s_1 and s_2) of the QFT design technique developed for lumped uncertain plants. It required a three-dimensional (3D) Bode diagram and a double Laplace transform for the 1D or x-space problem; a four-dimensional (4D) Bode diagram and a triple Laplace transform for the two-dimensional (2D) or xy-space problem; and a five-dimensional (5D) Bode diagram and a quadruple Laplace transform for the 3D or xyz-space problem. The method also required the implementation of a nonrealistic continuous feedback loop at every single point of the spatial distribution.[133]

Some years later, and based on new results achieved by Horowitz, Kannai, and Kelemen,[134,135] a one-point feedback approach to DPSs was presented. The methodology was more realistic, dealing with only a single output point at a time. However, no guidelines to solve the practical system difficulties were included.

Meanwhile, Chait *et al.*[132] presented a new Nyquist stability criterion for DPS, as well as studied the "spillover" effects of the truncation of DPS models; and Hedge and Nataraj[136] developed a two-point feedback approach.

In October 2000, at the Public University of Navarre (Spain), and during one of his last public presentations, Professor Horowitz challenged Professor García-Sanz and his PhD students to explore new ideas related to DPS and QFT.[137] This chapter is a consequence of these original conversations with Professor Horowitz, and explains the resulting methodology to design a one-point feedback QFT controller for DPSs with uncertainty. The previous section just described the proposed 2DOF control structure for DPS; this section introduces the extension of QFT to DPS; and next section includes an illustrative example.

As for the lumped approach – see Section 2.7 and Equations 2.35 through 2.41 from Equations 2.32 to 2.34, the DPS-generalized TFs $T_k(j\omega_i)$ extracted from Equations 5.11 to 5.14 and Figure 5.3 are shown in Table 5.1 as inequality performance specifications—see Equations 5.16 through 5.25. Subsequently, as we did with lumped systems—see Equations 2.42 through 2.48, they are translated into quadratic inequalities in Table 5.2—see Equations 5.26 through 5.35.

When extending the lumped QFT methodology to the DPS case, one of the main differences encountered is the fact that the set of plants $P_{x2\,x1}\,(j\omega_i) = \{P_{x2\,x1}\,(j\omega_i)\ \omega_i \in \Omega_k\}$ not only includes parametric and nonparametric uncertainty related to physical characteristics of the model, but also uncertainty related to the location of the inputs and outputs x_a, x_d, x_s, and x_o.

Now, the QFT nominal plant corresponds to $P_{xs\,xa0} \in \mathscr{P}_{xs\,xa}(j\omega_i)$, and the templates and QFT bounds consider the uncertainty related to the complete DPS model, with the TFs: $\{\mathscr{P}_{xs\,xa}$, $\mathscr{P}_{xo\,xa}$, $\mathscr{P}_{xo\,xd}$, and $\mathscr{P}_{xs\,xd}\}$.

Remark 5.4

Based on Remark 5.1, as this proposed methodology can deal with uncertainty at the location of the points x_a, x_d, x_s, and x_o, the QFT templates of $P_{x1\,x2}$ are expanded to include this spatial uncertainty.

TABLE 5.1

Feedback specifications for DPS when $H = 1$.

k	Inequalities	Equations
1	$\displaystyle \left| T_1(j\omega) \right| = \left| \frac{y_{xo}(j\omega)}{F(j\omega)r(j\omega)} \right| = \left| \frac{y_{xo}(j\omega)}{n(j\omega)} \right| = \left| \frac{y_{xo}(j\omega)}{d_e(j\omega)} \right|$ $\displaystyle = \left| \frac{P_{xoxa}(j\omega)G(j\omega)}{1 + P_{xsxa}(j\omega)G(j\omega)} \right| \leq \delta_1(\omega),\ \omega \in \Omega_1$	(5.16)
2	$\displaystyle \left| T_2(j\omega) \right| = \left| \frac{y_{xo}(j\omega)}{u_{xd}(j\omega)} \right| = \left| P_{xoxd}(j\omega) - \frac{P_{xoxa}(j\omega)P_{xsxd}(j\omega)G(j\omega)}{1 + P_{xsxa}(j\omega)G(j\omega)} \right| \leq \delta_2(\omega),\ \omega \in \Omega_2$	(5.17)
3	$\displaystyle \left| T_3(j\omega) \right| = \left| \frac{u_{xa}(j\omega)}{F(j\omega)r(j\omega)} \right| = \left| \frac{u_{xa}(j\omega)}{d_e(j\omega)} \right| = \left| \frac{u_{xa}(j\omega)}{n(j\omega)} \right|$ $\displaystyle = \left| \frac{G(j\omega)}{1 + P_{xsxa}(j\omega)G(j\omega)} \right| \leq \delta_3(\omega),\ \omega \in \Omega_3$	(5.18)
4	$\displaystyle \left| T_4(j\omega) \right| = \left| \frac{u_{xa}(j\omega)}{u_{xd}(j\omega)} \right| = \left| \frac{P_{xsxd}(j\omega)G(j\omega)}{1 + P_{xsxa}(j\omega)G(j\omega)} \right| \leq \delta_4(\omega),\ \omega \in \Omega_4$	(5.19)
5	$\displaystyle \delta_{5lo}(\omega) \leq \left| T_5(j\omega) \right| = \left| \frac{y_{xo}(j\omega)}{r(j\omega)} \right| = \left| F(j\omega)\frac{P_{xoxa}(j\omega)G(j\omega)}{1 + P_{xsxa}(j\omega)G(j\omega)} \right| \leq \delta_{5up}(\omega),\ \omega \in \Omega_5$ $\displaystyle \frac{\left| P_{xoxad}(j\omega)G(j\omega) \right|}{\left| P_{xoxae}(j\omega)G(j\omega) \right|} \frac{\left| 1 + P_{xsxae}(j\omega)G(j\omega) \right|}{\left| 1 + P_{xsxad}(j\omega)G(j\omega) \right|} \leq \delta_5(\omega) = \frac{\delta_{5up}(\omega)}{\delta_{5lo}(\omega)},\ \omega \in \Omega_5$	(5.20)
6	$\displaystyle \left| T_6(j\omega) \right| = \left| \frac{e(j\omega)}{F(j\omega)r(j\omega)} \right| = \left| \frac{e(j\omega)}{n(j\omega)} \right| = \left| \frac{u_{xa}(j\omega)}{d_i(j\omega)} \right|$ $\displaystyle = \left| \frac{1}{1 + P_{xsxa}(j\omega)G(j\omega)} \right| \leq \delta_6(\omega),\ \omega \in \Omega_6$	(5.21)
7	$\displaystyle \left| T_7(j\omega) \right| = \left| \frac{e(j\omega)}{u_{xd}(j\omega)} \right| = \left| \frac{y_{xs}(j\omega)}{u_{xd}(j\omega)} \right| = \left| \frac{P_{xsxd}(j\omega)}{1 + P_{xsxa}(j\omega)G(j\omega)} \right| \leq \delta_7(\omega),\ \omega \in \Omega_7$	(5.22)
8	$\displaystyle \left| T_8(j\omega) \right| = \left| \frac{e(j\omega)}{d_e(j\omega)} \right| = \left| \frac{y_{xs}(j\omega)}{F(j\omega)r(j\omega)} \right| = \left| \frac{y_{xs}(j\omega)}{n(j\omega)} \right| = \left| \frac{y_{xs}(j\omega)}{d_e(j\omega)} \right|$ $\displaystyle = \left| \frac{P_{xsxa}(j\omega)G(j\omega)}{1 + P_{xsxa}(j\omega)G(j\omega)} \right| \leq \delta_8(\omega),\ \omega \in \Omega_8$	(5.23)
9	$\displaystyle \left| T_9(j\omega) \right| = \left| \frac{e(j\omega)}{d_i(j\omega)} \right| = \left| \frac{y_{xs}(j\omega)}{d_i(j\omega)} \right| = \left| \frac{P_{xsxa}(j\omega)}{1 + P_{xsxa}(j\omega)G(j\omega)} \right| \leq \delta_9(\omega),\ \omega \in \Omega_9$	(5.24)
10	$\displaystyle \left| T_{10}(j\omega) \right| = \left| \frac{y_{xo}(j\omega)}{d_i(j\omega)} \right| = \left| \frac{P_{xoxa}(j\omega)}{1 + P_{xsxa}(j\omega)G(j\omega)} \right| \leq \delta_{10}(\omega),\ \omega \in \Omega_{10}$	(5.25)

TABLE 5.2

Quadratic inequalities for DPS with $H = 1$.

k	Inequalities	Equations
1	$\left(p_{xsxa}^2 - \dfrac{p_{xoxa}^2}{\delta_1^{\,2}}\right)g^2 + 2p_{xsxa}\cos(\phi + \theta_{xsxa})g + 1 \geq 0$	(5.26)
2	$\left[p_{xsxa}^2 - \dfrac{\left(p_{xsxa}\,p_{xoxd}\right)^2 + \left(p_{xoxa}\,p_{xsxd}\right)^2 - \dfrac{2p_{xsxa}\,p_{xoxd}\,p_{xoxa}\,p_{xsxd}\cos\left(-\theta_{xsxa}-\theta_{xoxd}+\theta_{xoxa}+\theta_{xsxd}\right)}{\delta_2^{\,2}}}{\delta_2^{\,2}}\right]g^2$ $+ \left(2p_{xsxa}\cos(\phi+\theta_{xsxa}) - 2p_{xoxd}\left[\begin{array}{l}p_{xsxa}\,p_{xoxd}\cos\left(\phi+\theta_{xsxa}\right) - \\ p_{xoxa}\,p_{xsxd}\cos\left(-\phi-\theta_{xsxd}+\theta_{xoxd}-\theta_{xoxa}\right)\end{array}\right]\Big/\delta_2^{\,2}\right)g$ $+ \left(1 - \dfrac{p_{xoxd}^{\,2}}{\delta_2^{\,2}}\right) \geq 0$	(5.27)
3	$\left(p_{xsxa}^2 - \dfrac{1}{\delta_3^{\,2}}\right)g^2 + 2p_{xsxa}\cos(\phi + \theta_{xsxa})g + 1 \geq 0$	(5.28)
4	$\left(p_{xsxa}^2 - \dfrac{p_{xsxd}^2}{\delta_4^{\,2}}\right)g^2 + 2p_{xsxa}\cos(\phi + \theta_{xsxa})g + 1 \geq 0$	(5.29)
5	$\left(p_{xoxa_e}^2 p_{xsxa_d}^2 - \dfrac{p_{xoxa_d}^2 p_{xsxa_e}^2}{\delta_5^{\,2}}\right)g^2$ $+ 2\left(p_{xoxa_e}^2 p_{xsxa_d}\cos\left(\phi+\theta_{xsxa_d}\right) - \dfrac{p_{xoxa_d}^2 p_{xsxa_e}\cos\left(\phi+\theta_{xsxa_e}\right)}{\delta_5^{\,2}}\right)g$ $+ \left(p_{xoxa_e}^2 - \dfrac{p_{xoxa_d}^2}{\delta_5^{\,2}}\right) \geq 0$	(5.30)
6	$p_{xsxa}^2 g^2 + 2p_{xsxa}\cos(\phi + \theta_{xsxa})g + \left(1 - \dfrac{1}{\delta_6^{\,2}}\right) \geq 0$	(5.31)
7	$p_{xsxa}^2 g^2 + 2p_{xsxa}\cos\left(\phi + \theta_{xsxa}\right)g + \left(1 - \dfrac{p_{xsxd}^2}{\delta_7^{\,2}}\right) \geq 0$	(5.32)
8	$p_{xsxa}^2\left(1 - \dfrac{1}{\delta_8^{\,2}}\right)g^2 + 2p_{xsxa}\cos\left(\phi + \theta_{xsxa}\right)g + 1 \geq 0$	(5.33)
9	$p_{xsxa}^2 g^2 + 2p_{xsxa}\cos\left(\phi + \theta_{xsxa}\right)g + \left(1 - \dfrac{p_{xsxa}^2}{\delta_9^{\,2}}\right) \geq 0$	(5.34)
10	$p_{xsxa}^2 g^2 + 2p_{xsxa}\cos\left(\phi + \theta_{xsxa}\right)g + \left(1 - \dfrac{p_{xoxa}^2}{\delta_{10}^{\,2}}\right) \geq 0$	(5.35)

Remark 5.5

When controlling any DPS by a one-point feedback structure, some limitations due to the spatial distribution arise. In particular, and because of the position of $P_{xo\ xd}$ in Equation 5.12, the compensator G cannot control the effect of the disturbance u_{xd} at the output y_{xo}. Also, for the same reason, there is no solution to the DPS problem if $P_{xo\ xd}$ has poles in the RHP.

EXAMPLE 5.1: HEAT CONDUCTION, DISTRIBUTED TEMPERATURE

This section applies the QFT DPS control design methodology introduced in this chapter to a well-known heat transmission problem. Additional information can be found in Reference 134.

Description

Consider the heat transmission problem with a temperature-distributed prismatic plant shown in Figure 5.4, whose PDE model is

$$
\begin{aligned}
&\frac{\partial T}{\partial t} - \frac{\partial^2 T}{\partial x^2} = v(x,t), \quad x \in (0,\pi), \quad t \geq 0 \\
&T(x,0) = 0, \quad x \in [0,\pi] \\
&T(0,t) = T(\pi,t) = 0; \quad v(x,t) = (\delta_{t=0})(\delta_{x=xi})
\end{aligned}
\tag{5.36}
$$

where $y_{xs}(x,t) = T(x,t)$ and $u_{xa}(x,t) = v(x,t)$ are the temperature and heat input, respectively. It is assumed that the sensor and the actuator points match up at $x_s = x_a$, and disturbances are applied at the control objective point x_o and at the sensor point x_s. In this example, we design a one-feedback loop to achieve robust stability and robust disturbance rejection and reference tracking specifications at x_o and x_s.

Modeling

This heat transmission problem, described by the PDE in Equation 5.36 and Figure 5.4, can also be modeled as the irrational TF shown in Equation 5.37, which represents at

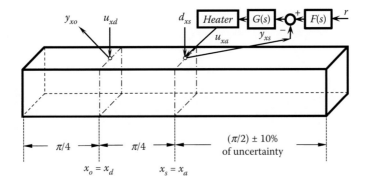

FIGURE 5.4
Heat transmission problem. Temperature-distributed prismatic plant.

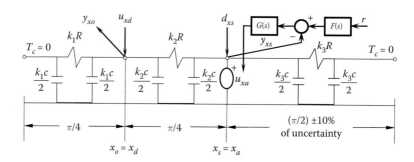

FIGURE 5.5
Π-equivalent electrical model.

time t the behavior of the temperature $T(x,t)$ at the point x of the distributed prismatic plant under a heat input $v(x_i,t)$ at the point x_i—see also Kelemen et al.[134]

$$P_{xxi}(s) = \frac{\left(e^{\sqrt{s}(\pi-x)} - e^{-\sqrt{s}(\pi-x)}\right)\left(e^{\sqrt{s}\,x_i} - e^{-\sqrt{s}\,x_i}\right)}{2\sqrt{s}\left(e^{\sqrt{s}\,\pi} - e^{-\sqrt{s}\,\pi}\right)} \tag{5.37}$$

As detailed in Section 5.2, the irrational solution calculated for the problem can be either transformed into a sum of terms or into a product of terms. In this case, since Equation 5.36 is a parabolic equation, an equivalent electrical model of the plant is chosen (see Figure 5.5). Solving the resultant system for a number of *Π*-element cases, the respective groups of rational TFs P_{xoxd}, P_{xoxa}, P_{xsxd}, P_{xsxa} can be obtained. The parameters used in this example are the distances k_1, k_2, and k_3 [m], the resistance per meter $R = 1/kA$, and the capacitance per meter $c = \rho\,C_p\,A$, where ρ is the density (kg/m³), k is the thermal conductivity (W/(m °C)), C_p is the specific heat capacity (J/(kg °C)), and A is the section area (m²). This electric circuit can be easily solved applying classical circuit techniques, resulting in the following expression $[Y] = [Q]^{-1}\,[U]$:

$$\begin{bmatrix} y_{xo} \\ y_{xs} \end{bmatrix} = \begin{bmatrix} \dfrac{1}{k_1R} + \dfrac{1}{\dfrac{2}{k_1cs}} + \dfrac{1}{k_2R} + \dfrac{1}{\dfrac{2}{k_2cs}} & -\dfrac{1}{k_2R} \\[2em] -\dfrac{1}{k_2R} & \dfrac{1}{k_2R} + \dfrac{1}{\dfrac{2}{k_2cs}} + \dfrac{1}{k_3R} + \dfrac{1}{\dfrac{2}{k_3cs}} \end{bmatrix}^{-1} \begin{bmatrix} u_{xd} + \dfrac{T_c}{k_1R} \\[1.5em] u_{xa} + \dfrac{T_c}{k_3R} + d_{xs} \end{bmatrix} \tag{5.38}$$

Calculating the inverse matrix, $[P] = [Q]^{-1}$, we have $[Y] = [P]\,[U]$, or

$$\begin{bmatrix} y_{xo} \\ y_{xs} \end{bmatrix} = \begin{bmatrix} P_{xo\,xd} & P_{xo\,xa} \\ P_{xs\,xd} & P_{xs\,xa} \end{bmatrix} \begin{bmatrix} u_{xd} + \dfrac{T_c}{k_1R} \\[1.5em] u_{xa} + \dfrac{T_c}{k_3R} + d_{xs} \end{bmatrix} \tag{5.39}$$

where the transfer matrix $[P]$ contains the elements

$$P_{xo\,xd} = 2Rk_1\sigma[Rck_2(k_2k_3 + k_3^2)s + 2(k_2 + k_3)] \tag{5.40}$$

$$P_{xo\,xa} = 4Rk_1k_3\sigma \tag{5.41}$$

$$P_{xs\,xd} = 4Rk_1k_3\sigma \tag{5.42}$$

$$P_{xs\,xa} = 2Rk_3\sigma[Rck_1(k_1k_2 + k_2^2)s + 2(k_1 + k_2)] \tag{5.43}$$

with

$$\sigma = \frac{1}{a_1s^2 + a_2s + a_3} \tag{5.44}$$

$$a_1 = R^2c^2k_1k_2k_3(k_1k_2 + k_1k_3 + k_2k_3 + k_2^2) \tag{5.45}$$

$$a_2 = 2Rc(k_1^2k_2 + k_1^2k_3 + k_1k_2^2 + 2k_1k_2k_3 + k_1k_3^2 + k_2^2k_3 + k_2k_3^2) \tag{5.46}$$

$$a_3 = 4(k_1 + k_2 + k_3) \tag{5.47}$$

Now, introducing the fixed parameters $k_1 = \pi/4$, $k_2 = \pi/4$, and $T_c = 0$, and the uncertain parameters $k_3 \in [0.45\pi, 5.5\pi]$, $R \in [0.9, 1.1]$, and $c \in [0.9, 1.1]$, we calculate the uncertain plants and generate the QFT templates.

The MATLAB code (*mfile*) that calculates the plants, first in symbolic math (Q and P), and then as TFs with parametric uncertainty (Pxoxd, Pxsxd, Pxoxa, and Pxsxa), is presented in Appendix 5. After saving them in the disk, these TFs are introduced in the QFTCT, plant definition window, *load transfer function array*, as shown in Figure 5.6a.

In addition, in order to introduce the specification y_{xo}/u_{xd}, the TF B1=PxsxaPxoxd-PxoxaPxsxd is also included. This specification is introduced later by using the "*define by user*" option in the specifications window of the QFTCT, with A = Pxoxd, B = B1, C = 1, D = P.

The system plant in the plant definition window is $P_{xsxa}(s)$. The QFT templates for the system plant and frequencies of interest are shown in Figure 5.6b.

Control Specifications

The system performance specifications are selected according to the following requirements:

1. Stability at x_s and x_0
   It is defined by the most restrictive condition of the two expressions

$$\left|\frac{P_{xs\,xa}G}{1 + P_{xs\,xa}G}\right| \le 1.6, 0.0005 \le \omega \le 1000 \text{ rad/s} \tag{5.48}$$

$$\left|\frac{P_{xo\,xa}G}{1 + P_{xs\,xa}G}\right| \le 1.6, \ 0.0005 \le \omega \le 1000 \text{ rad/s} \tag{5.49}$$

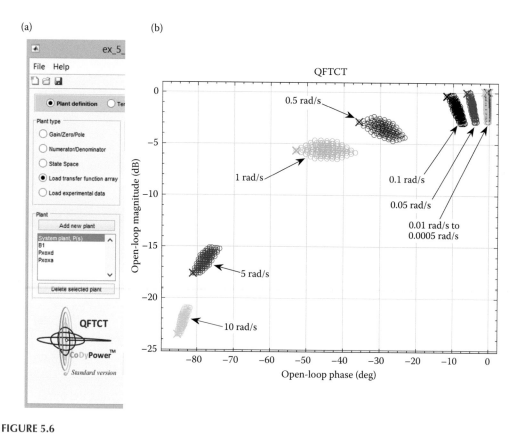

FIGURE 5.6
QFTCT. (a) Plant definition window. TFs imported from disk. (b) Templates window: $P_{xs\,xa}(s)$ system plant templates.

2. Disturbance rejection at x_0

$$\left|\frac{y_{xo}}{u_{xd}}\right| = \left|P_{xo\,xd} - \frac{P_{xo\,xa}\,P_{xs\,xd}\,G}{1 + P_{xs\,xa}G}\right| \leq 0.9\left|P_{xo\,xd}\right|,\ 0.005 \leq \omega \leq 0.1\,\text{rad/s} \tag{5.50}$$

3. Disturbance rejection at x_s

$$\left|\frac{y_{xs}}{d_{xs}}\right| = \left|\frac{1}{1 + P_{xs\,xa}G}\right| \leq \delta_o(\omega) = \left|\frac{700s}{700s + 2}\right|,\ 0.0005 \leq \omega \leq 10\,\text{rad/s} \tag{5.51}$$

4. Reference tracking specification at x_s.

$$\delta_{lo}(\omega) \leq \left|\frac{P_{xs\,xa}G}{1 + P_{xs\,xa}G}F\right| \leq \delta_{up}(\omega)\ 0.0005 \leq \omega \leq 1\,\text{rad/s} \tag{5.52}$$

where

$$\delta_{lo}(\omega) = \left|\frac{0.98}{20(j\omega)^2 + 250(j\omega) + 1}\right|\ \text{is the lower bound}$$

and

$$\delta_{up}(\omega) = \left| \frac{20(j\omega)+1.02}{20(j\omega)^2+100(j\omega)+1} \right| \text{ is the upper bound}$$

Controller Design

The robust stability bounds for Equation 5.48 are obtained from the quadratic inequalities given by Equation 5.33. The robust stability bounds for Equation 5.49 are calculated from the quadratic inequalities given by Equation 5.26. The robust disturbance rejection bounds for Equations 5.50 and 5.51 are found from the quadratic inequalities given by Equations 5.27 and 5.31, respectively. The robust reference tracking bounds for Equation 5.52 are obtained using the quadratic inequalities given by Equation 2.47, with $P(s) = P_{xsxa}(s)$ in Equation 2.40. The nominal open-loop is $L_1 = G_1 P_{xsxa0}$.

The $G_1(s)$ controller is designed using a standard loop shaping QFT technique—see Equation 5.53. Figure 5.7 shows the NC for the loop TF L_1 with the proposed compensator $G_1(s)$.

$$G_1(s) = \frac{0.0043((s/0.004)+1)}{s((s/100)+1)} \tag{5.53}$$

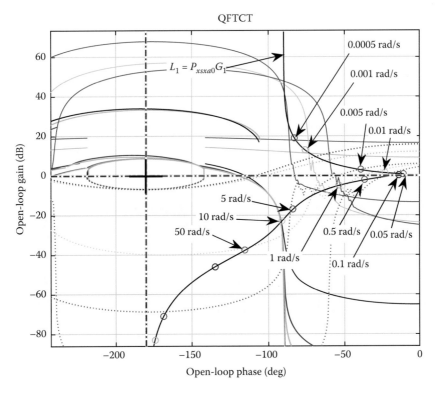

FIGURE 5.7
Loop shaping for the controller G_1, with the open-loop TF: $L_1 = P_{xsxa0} \, G_1$.

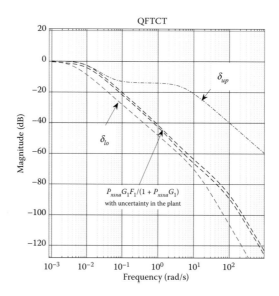

FIGURE 5.8
Prefilter design for the plant P_{xsxa} with uncertainty, the compensator G_1 and the prefilter F_1. Reference tracking specs according to Equation 5.52.

Finally, a prefilter $F_1(s)$ is designed to meet the reference tracking specifications. The expression is shown in Equation 5.54 and the Bode diagram in Figure 5.8.

$$F_1(s) = \frac{((s/0.0021)+1)((s/2.9)+1)}{((s/0.0045)+1)((s/0.009)+1)} \tag{5.54}$$

Simulations

Figures 5.9 through 5.11 show the results at the points x_s and x_o. They are obtained applying the controller $G_1(s)$ and the prefilter $F_1(s)$ to 125 plants within the space of

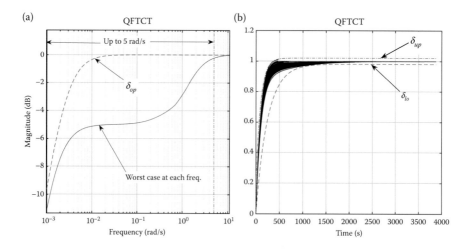

FIGURE 5.9
Results at point x_s with G_1 and F_1. (a) Disturbance rejection y_{xs}/d_{xs}, Equation 5.51. (b) Reference tracking y_{xs}/r, Equation 5.52, with unitary step input r.

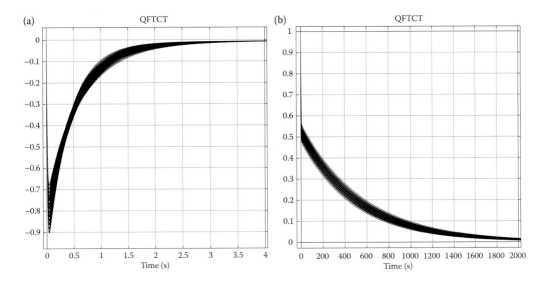

FIGURE 5.10

Results at point x_s with G_1 and F_1. Disturbance rejection y_{xs}/d_{xs}, Equation 5.51. Output y_{xs} after (a) unitary impulse input d_{xs} and (b) unitary step input d_{xs}.

parametric uncertainty (see Appendix 5). The system is closed-loop stable and meets well the specifications for the disturbance rejection and reference tracking problems in all cases.

Note

Each rational TF used in this problem (P_{xoxd}, P_{xoxa}, P_{xsxd}, and P_{xsxa}) is composed of a single Π-element (see Figure 5.5). In other problems, a more complex modeling, involving multiple Π-elements for each TF, can be necessary. The same controller design technique

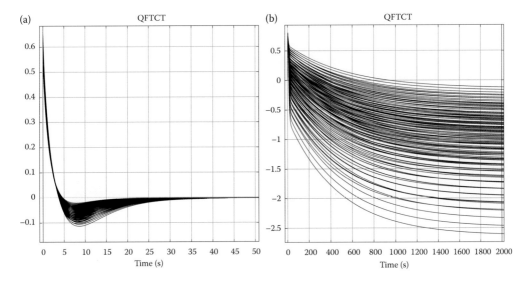

FIGURE 5.11

Results at point x_o with G_1 and F_1. Disturbance rejection y_{xo}/u_{xd}, Equation 5.50. Output y_{xo} after (a) unitary impulse input u_{xd} and (b) unitary step input u_{xd}.

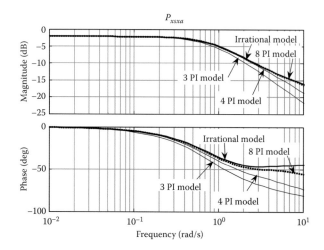

FIGURE 5.12

Bode diagram for $P_{xs\mu0}$ according to the irrational model Equation 5.37, and the rational electric circuit approach given by 3, 4, and 8 Π-elements. (Adapted from Garcia-Sanz, M. and Houpis, C.H. 2012. *Wind Energy Systems: Control Engineering Design (Part I: QFT Control, Part II: Wind Turbine Design and Control).* A CRC Press book, Taylor & Francis Group, Boca Raton, FL; Garcia-Sanz, M., Huarte, A., and Asenjo, A. 2007. *Int. J. Robust Nonlinear Control,* 17(2–3), 135–153.)

can be applied in these cases with TFs with more Π-elements. Figure 5.12 shows how the equivalent electrical models for 3, 4, and 8 Π-elements approximate the irrational expression of Equation 5.37. According to Anderson and Parks,[346] the number of Π-elements is valid for the frequencies $\omega < 2/(Rc)$, where R and c are, respectively, the resistance and capacitance per length in each Π-element.

5.5 Summary

There is no need for very complex methodologies to design controllers for DPS. On the contrary, using the practical approach presented in this chapter, classical QFT methods based on new bound definitions can be applied to solve DPS control problems. A spatial distribution of the location of the relevant points was considered, allowing the designer to place the actuator, disturbances, sensor, and control objectives in any spatial location of interest, including location uncertainty. From this topology, new stability and performance specifications, TFs, and quadratic inequalities for DPS were introduced. The simplicity and excellent practical results of the methodology are illustrated with a well-known PDE example.

5.6 Practice

The list shown below summarizes the collection of problems and cases included in this book that apply the control methodologies introduced in this chapter.

- Example 5.1. Heat transmission system with distributed temperature.
- *Appendix 5.* MATLAB code for Example 5.1.

6

Gain Scheduling/Switching Control Solutions

6.1 Introduction

This chapter introduces a hybrid methodology to design nonlinear robust control systems able to go beyond the classical linear limitations. Combining robust controller designs and stable switching, the new system optimizes the time response of the plant by fast adaptation of the controller parameters during the transient response according to certain rules based on the amplitude of the error. The methodology is based on both a graphical frequency-domain stability criterion for switching linear systems and the use of the robust QFT control system design technique.

Switching control has demonstrated to be an efficient tool in achieving tight performance specifications in control systems.[329,334] The way to reach this enhancement is by designing various parallel controllers with different characteristics, and continuously selecting among them the optimum one that controls the system best (Figure 6.1). Performance specifications that are not achievable by a simple linear controller, as the limitation theory predicts,[330] can be attained through suitable switching rules.

One of the main issues in switching control techniques is that the system stability is not assured *a priori*, even if the switching is made between stable controllers. This is the reason why most of the current literature about switching systems is still devoted to stability issues. See References 326–336 for general results about stability criteria and applications to some practical cases.

This chapter introduces a nonlinear switching robust controller design methodology, including a graphical frequency-domain criterion to ensure the system stability and several illustrative examples.

6.2 System Stability Under Switching[7,163]

Figure 6.1 shows the general scheme to be used with the switching control system. A set of controllers is designed and a supervisor selects the most suitable one, depending on the state of the system and/or some external parameters.

One of the main difficulties found when switching techniques are applied is that, in general, the system stability is not assured, even if switching is made between stable

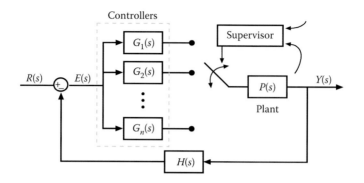

FIGURE 6.1

Switching control scheme. (Adapted from Garcia-Sanz, M. and Houpis, C.H. 2012. *Wind Energy Systems: Control Engineering Design (Part I: QFT Control, Part II: Wind Turbine Design and Control).* A CRC Press book, Taylor & Francis Group, Boca Raton, FL.)

controllers. Some extra conditions must be met to assure stability. In particular, it is proven that a system represented by

$$\dot{x}(t) = A(t)x(t), \quad A(t) \in A = \{A_1, ..., A_m\}, \quad A_i \text{ Hurwitz} \tag{6.1}$$

with arbitrary switching within the set of matrices A is exponentially stable if and only if there exists a common Lyapunov function (*CLF*) for all A_i in the set A.[328]

It also has been proven that the existence of a common quadratic Lyapunov function (*CQLF*) is a sufficient condition for exponential stability.[331] In this context, the main issue in linear switching systems is finding conditions under which the existence of a *CQLF* is assured. In particular, the circle criterion (see Section 7.2) provides the necessary and sufficient conditions for the existence of a *CQLF* for two systems in companion form,[326,327,335] that is, the systems

$$\dot{x}(t) = A x(t) \tag{6.2}$$

$$\dot{x}(t) = (A - g \Delta^T) x(t) \tag{6.3}$$

with

$$
A = \begin{bmatrix}
0 & 1 & 0 & \cdots & 0 & 0 \\
0 & 0 & 1 & \cdots & 0 & 0 \\
0 & 0 & 0 & \cdots & 0 & 0 \\
\vdots & \vdots & \vdots & \ddots & \vdots & \vdots \\
0 & 0 & 0 & \cdots & 0 & 1 \\
-e_0 & -e_1 & -e_2 & \cdots & -e_{n-2} & -e_{n-1}
\end{bmatrix}, \quad
g = \begin{bmatrix} 0 \\ 0 \\ 0 \\ \vdots \\ 0 \\ 1 \end{bmatrix}, \quad
\Delta = \begin{bmatrix} \Delta e_0 \\ \Delta e_1 \\ \Delta e_2 \\ \vdots \\ \Delta e_{n-2} \\ \Delta e_{n-1} \end{bmatrix} \tag{6.4}
$$

have a *CQLF* if and only if

$$1 + \mathrm{Re}\{\Delta^T (sI - A)^{-1} g\} > 0, \quad s = j\omega, \quad \text{for all frequency } \omega \tag{6.5}$$

This chapter considers stability for arbitrary switching between two closed-loop systems with transfer functions $T_1(s) = L_1(s)/[1 + L_1(s)]$ and $T_2(s) = L_2(s)/[1 + L_2(s)]$, both stable. For these two transfer functions, $L_1(s) = P(s)G_1(s)$ and $L_2(s) = P(s)G_2(s)$ are proper. Additionally, $L_1(s)$ and $L_2(s)$ have the same number of poles, and the same number of zeros. The switching takes place by changing the gain and the position of the poles and zeros. The open-loop transfer functions for both systems are:

$$L_1(s) = \frac{b_{n-1}s^{n-1} + \cdots + b_0}{s^n + a_{n-1}s^{n-1} + \cdots + a_0} = \frac{N(s)}{D(s)} \tag{6.6}$$

and

$$L_2(s) = \frac{(b_{n-1} + \Delta b_{n-1})s^{n-1} + \cdots + (b_0 + \Delta b_0)}{s^n + (a_{n-1} + \Delta a_{n-1})s^{n-1} + \cdots + (a_0 + \Delta a_0)} = \frac{N(s) + \Delta N(s)}{D(s) + \Delta D(s)} \tag{6.7}$$

For the sake of clarity, and without losing generality, in this analysis a general expression where the order of the numerator is one less than that of the denominator is used. If it is not the case, then some b_i coefficients can be zero. The closed-loop transfer functions are

$$T_1(s) = \frac{L_1(s)}{1 + L_1(s)} = \frac{N(s)}{D(s) + N(s)} \tag{6.8}$$

and

$$T_2(s) = \frac{L_2(s)}{1 + L_2(s)} = \frac{N(s) + \Delta N(s)}{D(s) + \Delta D(s) + N(s) + \Delta N(s)} \tag{6.9}$$

where the characteristic equations are

$$D(s) + N(s) = s^n + e_{n-1}s^{n-1} + \cdots + e_1 s + e_0 \tag{6.10}$$

and

$$D(s) + \Delta D(s) + N(s) + \Delta N(s) = s^n + (e_{n-1} + \Delta e_{n-1})s^{n-1} + \cdots + (e_0 + \Delta e_0) \tag{6.11}$$

with $e_i = a_i + b_i$ and $\Delta e_i = \Delta a_i + \Delta b_i$.

Using these expressions for the coefficients e_i and Δe_i, the matrices A, g, and Δ are defined. Now the circle criterion is applied to guarantee stability under arbitrary switching as

$$(s\mathbf{I} - A)^{-1} = s^n + e_{n-1}s^{n-1} + \cdots + e_1 s + e_0$$

and

$$\Delta^T g = \Delta e_{n-1}s^{n-1} + \cdots + \Delta e_1 s + \Delta e_0$$

Equation 6.5 becomes

$$1 + \text{Re}\left\{\frac{\Delta e_{n-1}s^{n-1} + \cdots + \Delta e_1 s + \Delta e_0}{s^n + e_{n-1}s^{n-1} + \cdots + e_1 s + e_0}\right\} > 0, \quad s = j\omega \tag{6.12}$$

for all frequency ω. After some simple manipulation,

$$\text{Re}\left\{1 + \frac{\Delta e_{n-1}s^{n-1} + \cdots + \Delta e_1 s + \Delta e_0}{s^n + e_{n-1}s^{n-1} + \cdots + e_1 s + e_0}\right\} = \text{Re}\left\{\frac{N(s) + \Delta N(s) + D(s) + \Delta D(s)}{N(s) + D(s)}\right\}$$

$$= \text{Re}\left\{\frac{(N(s) + \Delta N(s) + D(s) + \Delta D(s))\,/\,(N(s) + \Delta N(s))}{(N(s) + D(s))\,/\,N(s)}\left(\frac{N(s) + \Delta N(s)}{N(s)}\right)\right\}$$

$$= \text{Re}\left\{\frac{1 + L_2(s)}{1 + L_1(s)}\left(\frac{D(s) + \Delta D(s)}{D(s)}\right)\right\} \tag{6.13}$$

the condition can be expressed in the following form:

$$\text{Re}\left\{\frac{1 + L_2(s)}{1 + L_1(s)}\left(\frac{D(s) + \Delta D(s)}{D(s)}\right)\right\} > 0, \quad s = j\omega \tag{6.14}$$

for all frequency ω.

As this formulation of the circle criterion is applied to open-loop transfer functions, due to symmetry, it is sufficient to check it at only positive frequencies. The Equation 6.14 condition is then equivalent to

$$\left|\arg\{1 + L_2(j\omega)\} - \arg\{1 + L_1(j\omega)\} + \arg\left\{\frac{D(j\omega) + \Delta D(j\omega)}{D(j\omega)}\right\}\right| < \frac{\pi}{2} \tag{6.15}$$

for all $\omega \geq 0$.

By denoting the following expressions:

$$\varphi_{12}(\omega)^\circ = \left|\arg\{1 + L_2(j\omega)\} - \arg\{1 + L_1(j\omega)\}\right| \tag{6.16}$$

$$\alpha(\omega)^\circ = \left|\arg\left\{\frac{D(j\omega) + \Delta D(j\omega)}{D(j\omega)}\right\}\right| \tag{6.17}$$

and using the triangle inequality, a sufficient condition for Equation 6.15 is:

$$\varphi_{12}(\omega)^\circ < 90^\circ - \alpha(\omega)^\circ \tag{6.18}$$

degrees for all $\omega \geq 0$.

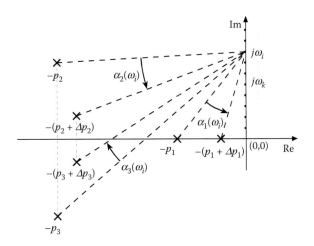

FIGURE 6.2
Complex plane $\alpha(s)$ for a system with three switching poles. $\alpha(\omega_i)^n - \alpha_1(\omega_i)^n + \alpha_2(\omega_i)^n + \alpha_3(\omega_i)^n$. (Adapted from Garcia-Sanz, M. and Houpis, C.H. 2012. *Wind Energy Systems: Control Engineering Design (Part I: QFT Control, Part II: Wind Turbine Design and Control)*. A CRC Press book, Taylor & Francis Group, Boca Raton, FL.)

With this result, the criterion can be applied graphically to both the Complex plane and the Nichols diagram. In the first case, the criterion is expressed by requiring $L_1(j\omega)$ and $L_2(j\omega)$ to be inside of an arc of $[90° - \alpha(\omega)°]$ degrees around the critical point $(-1,0)$ at each frequency.

Similarly, in the Nichols diagram the condition based upon angles is easily checked by plotting $[1 + L_1(j\omega)]$ and $[1 + L_2(j\omega)]$, and the distance of $\varphi_{12}(\omega)$ on the horizontal axis at each frequency. To assure stability, this distance should be less than $[90° - \alpha(\omega)°]$ degrees.

It is noted that the function $\alpha(\omega)$ can also be expressed according to Equation 6.19, so that it may be considered as a measurement of the change of the controller poles, as shown in Figure 6.2. Consequently, the larger the movement made by the poles, the larger the conservativeness introduced by the triangle inequality of Equation 6.18. Note that for each frequency ω_i there is a different angle $\alpha(\omega_i)°$,

$$\alpha(\omega)° = \left| \arg\left\{ \frac{D(j\omega) + \Delta D(j\omega)}{D(j\omega)} \right\} \right| =$$

$$\left| \arg\left\{ \frac{\prod_{j=1}^{n}(j\omega + p_j + \Delta p_j)}{\prod_{j=1}^{n}(j\omega + p_j)} \right\} \right| = \left| \sum_{j=1}^{n} \arg\{j\omega + p_j + \Delta p_j\} - \arg\{j\omega + p_j\} \right| \quad (6.19)$$

At this point two questions arise. Firstly, the criterion presented above is applied to switching between two isolated controllers with the same structure and the same plant model. However, it is possible that the designer may desire to do switching among more than two systems, or even among an infinite number of systems, which can also be considered as a *linear parameter varying* (LPV) system, where the controller varies continuously. Secondly, real systems present uncertainty, so the criterion must be modified in some manner to take the uncertainty into account. The next discussion undertakes both issues.

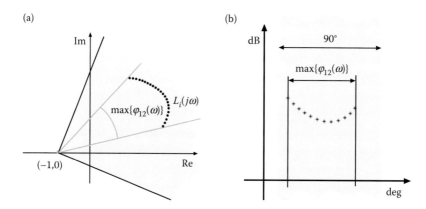

FIGURE 6.3
Criterion for continuous switching: (a) Complex plane, plotting $L_i(j\omega)$; (b) Nichols plot, plotting $[1 + L_i(j\omega)]$. (Adapted from Garcia-Sanz, M. and Houpis, C.H. 2012. *Wind Energy Systems: Control Engineering Design (Part I: QFT Control, Part II: Wind Turbine Design and Control).* A CRC Press book, Taylor & Francis Group, Boca Raton, FL.)

If the switching is made among a set of several controllers, the criterion must be satisfied for each pair of controllers. Checking this condition may be an impossible task if there is more than one pole moving, because the angle α is different for each pair of controllers. For this reason, consider the case of a controller whose parameters change continuously with the error, and allow the gain and zeros to be the variables. Then, the angle $\alpha(\omega)°$ is null for every frequency, and the only condition to be satisfied is that the angle between any two possible systems $L_i(j\omega)$ and the critical point $(-1,0)$ is less than 90°. Moreover, under this premise the conservativeness introduced in Equation 6.18 vanishes. The condition can be checked graphically with a grid of the possible open-loop systems that the controller variation can generate, as shown in Figure 6.3a. The maximum angle $\varphi_{12}(\omega)°$ must be contained in a 90° arc from $(-1,0)$. In the Nichols plot, the way to apply the criterion is to draw the grid of possible $1 + L_i(j\omega)$ systems, and check that the maximum horizontal distance is less than 90°. Figure 6.3b illustrates the criterion on the Nichols plot.

Using similar arguments, it is also easy to deal with uncertainty. For an uncertain system, the *template* $\Im P(j\omega)$ is the area in the Nichols chart of the possible plants within the uncertainty at the frequency ω_i. If the system is governed by a switching controller, each point of the template can change its position due to the switching. From this point of view, switching can be considered as a mechanism that modifies the position and the shape of the templates of $[1 + L_i(j\omega)]$. To be sure that the switching is stable, the above criterion must be applied to the whole template.

It has been traditionally considered in robust control theory that uncertainty changes the plant slowly in comparison with the system dynamics. If the switching laws depend on the state of the system, then the switching is much faster than changes due to uncertainty. Consequently, for each point of the departure template there is only one corresponding point in the arrival template. Furthermore, it is assumed that uncertainty does not affect the angle α.

Then, the Nichols chart is a very clear diagram to test the stability of the uncertain switching system. Figure 6.4 shows the templates of $[1 + L_i(j\omega)]$ and the application of the method. If, during the displacement of each point of the first template to its corresponding point of the second one, the maximum horizontal distance between any two points of this path is less than 90°, the stability condition is satisfied at that particular frequency. Although it is a laborious task, usually it is not necessary to check each point of the template at each frequency, because the whole set of templates are much closer together than the critical distance.[7,163]

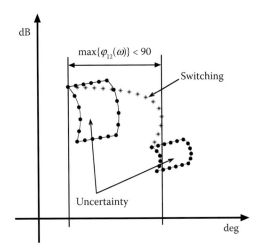

FIGURE 6.4
Templates of $|1 + L_i(j\omega)|$ and stability criterion on the Nichols chart. (Adapted from Garcia-Sanz, M. and Houpis, C.H. 2012. *Wind Energy Systems: Control Engineering Design (Part I: QFT Control, Part II: Wind Turbine Design and Control)*. A CRC Press book, Taylor & Francis Group, Boca Raton, FL.)

6.3 Methodology

Quantitative feedback theory has demonstrated to be an excellent tool to deal with several, often conflicting, control specifications. As it is shown in the previous chapters, it is a transparent design technique that allows the designer to consider all the specifications simultaneously and in the same plot. The QFT philosophy permits the design of a controller that satisfies all the performance specifications for every plant within the uncertainty, and using (quantifying) the minimum amount of feedback and the minimum controller gain at every frequency of interest.

Based on the analysis presented in the previous section, a switching control methodology is proposed here to go beyond the classical performance limitations imposed by linear systems. The methodology combines the QFT robust control design method and the stable switching technique presented in the preceding section.

Specifically, we propose to use the error amplitude as the switching signal, so that, when the output is far from the reference (large error), the system needs to be more stable, and also faster, but precision is not so necessary. Conversely, when there is a small error, some amount of stability margin can be sacrificed in order to increase the low-frequency gain, and therefore to also increase the precision and the disturbance rejection. These ideas, combined with the QFT method, lead to the design procedure (Method 1), as presented by the following four steps:

Method 1
$L_i(s)$, $i = 1,2,\ldots,n$, *with gain and zeros variation, constant poles*
For switching among many controllers (two or more) and for systems with uncertainty.

- *Step 1. Preliminary controller*: Obtain a preliminary linear controller for the system by applying the QFT robust control methodology, dealing with parametric and/or nonparametric uncertainty (templates), time-domain and frequency-domain specifications (QFT-bounds), and applying loop-shaping techniques, as shown in the previous chapters.

- *Step 2. Two controllers: One for small errors, one for large errors:* Use the preliminary QFT controller as the starting point to design two complementary controllers with the same structure, where the gain and the zeros can vary freely, but the poles remain unchanged. The characteristics of these two complementary controllers can be related with the error amplitude. As Boris Lurie[256] explains, when the error is large, the bandwidth must increase to obtain a fast response, but the loop gain does not need to be high. Additionally, when the error is small, the bandwidth is reduced to avoid the effects of noise, while the low frequencies gain is increased to minimize the jitter and the reference tracking error. In terms of loop-shaping, these rules are (1) for small errors, increase the low-frequency gain and bring nearer the zeros and (2) for large errors, decrease the low-frequency gain and move the zeros further away. Apart from this, reasonable stability margins must be maintained, although they can be considerably reduced for the small error situation if necessary.

- *Step 3. Stability analysis:* The robustness of the extreme designs guarantees that both linear systems are stable for every plant within the uncertainty. However, it is necessary to apply the criterion presented in the previous section (see Equation 6.18 with $\alpha = 0$) to assure that the switching between both controllers is also stable. One advantage of this graphical criterion is that it gives information about the frequencies where conditions are not satisfied, so that the designer can go back to Step 2 and change the extreme controllers in this region if necessary.

- *Step 4. Switching function:* The switching function relates the error amplitude with the position of the controller parameters. Select the most appropriate switching function from the time-domain simulation with each controller (see Example 6.3 in the next section for more information).

Method 2
$L_i(s)$, $i = 1,2$, *with variation in gain, poles, and zeros*
For switching among two controllers and for systems without uncertainty.

The stability is checked graphically, Equations 6.18 and 6.19, in both, the Complex plane or the Nichols diagram. In the first case, the criterion is expressed by requiring $L_1(j\omega)$ and $L_2(j\omega)$ to be inside of an arc of $[90° - \alpha(\omega)°]$ degrees around the point $(-1,0)$ at each frequency (Figure 6.3a); and in the second case, plotting $[1 + L_1(j\omega)]$ and $[1 + L_2(j\omega)]$, the distance $\varphi_{12}(\omega)$ on the horizontal axis at each frequency should be less than $[90° - \alpha(\omega)°]$ degrees (Figure 6.3b).[7,163]

6.4 Examples

To validate the theory introduced in the previous sections, four illustrative switching examples are presented here: three based on Shorten et al.[335] and one based on a benchmark problem presented at the American Control Conference in 1992 (the classical two-mass-spring system).[349,350]

EXAMPLE 6.1: THERE DOES NOT EXIST A *CQLF*

Example 6.1, Case 1.

$L_i(s)$ with gain and poles variation, constant zeros. Method 2.

Consider two systems given by the matrices,

$$A_1 = \begin{bmatrix} 0 & 1 & 0 \\ 0 & 0 & 1 \\ -1 & -2 & -3 \end{bmatrix}, \quad A_2 = \begin{bmatrix} 0 & 1 & 0 \\ 0 & 0 & 1 \\ -2 & -3 & -1 \end{bmatrix} \tag{6.20}$$

According to the methodology presented by Shorten et al.[335], the stability under switching between these two systems cannot be assured because A_1A_2 has two real negative eigenvalues: $eig(A_1A_2) = [1.0, -1.0, -2.0]$.

The same result can be found applying the methodology introduced in Sections 6.2 and 6.3. Let us assume that A_1 and A_2 represent two closed-loop systems, being the other state-space matrices: $B_1 = B_2 = [-1\ 0\ 0]^T$, $C_1 = C_2 = [0\ 1\ 1]$, and $D_1 = D_2 = [0]$. The open-loop transfer functions $L_1(s)$ and $L_2(s)$, which are defined by the closed-loop systems A_1 and A_2, have different gain and poles but the same zero, such that,

$$L_1(s) = \frac{(s+1)}{s(s+2.618)(s+0.382)} \tag{6.21}$$

$$L_2(s) = \frac{2(s+1)}{s(s+0.5 \pm 0.8666j)} \tag{6.22}$$

Figure 6.5 shows Equation 6.18 results at every frequency. The switching produces a gain increase and a change of two real poles by a pair of complex conjugate ones. Then, the stability is analyzed with Method 2, Section 6.3. The condition described in Equation 6.18 fails in the range of frequencies $1\ \text{rad/s} \le w \le 1.67\ \text{rad/s}$. The MATLAB code used in this case is in Appendix 6.

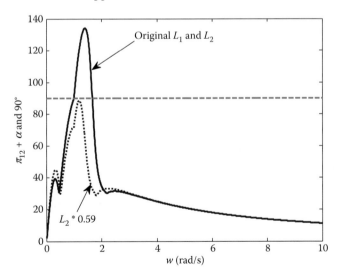

FIGURE 6.5

Example 6.1, Case 1. Equation 6.18, original L_1 & L_2, and with 0.59 L_2. (Adapted from Garcia-Sanz, M. and Houpis, C.H. 2012. *Wind Energy Systems: Control Engineering Design (Part I: QFT Control, Part II: Wind Turbine Design and Control)*. A CRC Press book, Taylor & Francis Group, Boca Raton, FL.)

A simple solution to make the system stable under switching is to reduce the gain of the second controller by multiplying the second transfer function $L_2(s)$ by 0.59—see also Figure 6.5.

Example 6.1, Case 2
$L_i(s)$ with gain and zeros variation, constant poles. Method 1
 As a second case, consider another two systems given by the matrices,

$$A_1 = \begin{bmatrix} -3 & -2 & -1 \\ 1 & 0 & 0 \\ 0 & 1 & 0 \end{bmatrix}, \quad A_2 = \begin{bmatrix} -1 & -3 & -2 \\ 1 & 0 & 0 \\ 0 & 1 & 0 \end{bmatrix} \tag{6.23}$$

According to the methodology presented by Shorten et al.[335], the stability under switching between these two systems cannot be assured. Again, the matrices $A_1 A_2$ has the same two real negative eigenvalues: eig($A_1 A_2$) = [1.0, –1.0, –2.0].
 Once more, we can find the same result applying the methodology introduced in Sections 6.2 and 6.3. Let us assume again that A_1 and A_2 represent two closed-loop systems, being the other state-space matrices: $B_1 = B_2 = [1\ 0\ 0]^T$, $C_1 = [3\ 2\ 0]$, $C_2 = [1\ 3\ 1]$, and $D_1 = D_2 = [0]$. Now, the two open-loop transfer functions for the switching $L_i(j\omega)$, $i = 1,2$, have different gain and different zeros but the same poles, such that,

$$L_1(s) = \frac{3s(s+0.6667)}{(s+1)(s-0.5\pm0.8660j)} \tag{6.24}$$

$$L_2(s) = \frac{(s+2.618)(s+0.382)}{(s+1)(s-0.5\pm0.8660j)} \tag{6.25}$$

As the changes are only in the zeros and the gain, $\alpha(\omega)$ is null for all frequencies (Method 1, Section 6.3). In this case, as is shown in Figure 6.6, the condition of

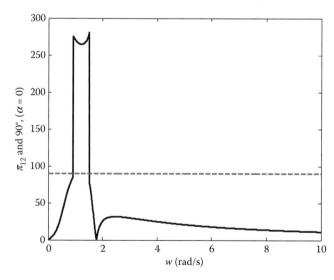

FIGURE 6.6
Example 6.1, Case 2. Equation 6.18. (Adapted from Garcia-Sanz, M. and Houpis, C.H. 2012. *Wind Energy Systems: Control Engineering Design (Part I: QFT Control, Part II: Wind Turbine Design and Control)*. A CRC Press book, Taylor & Francis Group, Boca Raton, FL.)

Equation 6.18 fails in the range of frequencies 0.9 rad/s $\leq \omega \leq$ 1.52 rad/s. The MATLAB code used in this case is also in Appendix 6.

EXAMPLE 6.2: THERE EXIST A *CQLF*

Again, according to the methodology presented by Shorten et al.[335], the stability under switching between the two systems given by the matrices

$$A_1 = \begin{bmatrix} -3 & -2 & -1 \\ 1 & 0 & 0 \\ 0 & 1 & 0 \end{bmatrix}, \quad A_2 = \begin{bmatrix} -3 & -1 & -1 \\ 1 & 0 & 0 \\ 0 & 1 & 0 \end{bmatrix} \tag{6.26}$$

is assured because $A_1 A_2$ has no real negative eigenvalues.

This result, based on the $A_1 A_2$ eigenvalues[335], can also be found by applying the methodology introduced in Sections 6.2 and 6.3. Again it is assumed that these matrices represent two closed-loop systems, being the other state-space matrices: $B_1 = B_2 = [1 \ 0 \ 0]^T$, $C_1 = C_2 = [0 \ 0 \ 1]$, and $D_1 = D_2 = [0]$. Now, the two open-loop transfer functions for the switching are

$$L_1(s) = \frac{1}{s(s+2)(s+1)} \tag{6.27}$$

$$L_2(s) = \frac{1}{s(s+2.618)(s+0.382)} \tag{6.28}$$

As it is shown in Figure 6.7, the condition of Equation 6.18 is fulfilled for every frequency of interest, so the system under switching is stable. The MATLAB code used in this case is also in Appendix 6.

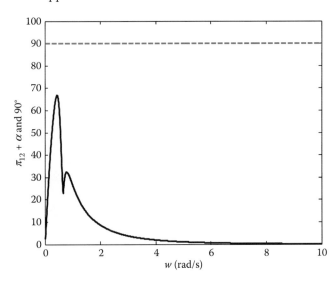

FIGURE 6.7
Example 6.2, Equation 6.18. (Adapted from Garcia-Sanz, M. and Houpis, C.H. 2012. *Wind Energy Systems: Control Engineering Design (Part I: QFT Control, Part II: Wind Turbine Design and Control)*. A CRC Press book, Taylor & Francis Group, Boca Raton, FL.)

EXAMPLE 6.3: A NEW SOLUTION FOR THE ACC'92 BENCHMARK PROBLEM

This example presents and improves one of the numerous solutions published for the classical two-mass-spring system (Figure 6.8), also known as the ACC'92 benchmark problem.[349,350]

The benchmark consists of two carts with masses m_1 and m_2 connected by a spring with stiffness k. The control force $u(t)$ acts on the first cart. The position of the second cart $x_2(t)$ is the variable to be controlled, which results in a non-collocated control problem. The system can be represented in state-space as,

$$
\begin{bmatrix} \dot{x}_1 \\ \dot{x}_2 \\ \dot{x}_3 \\ \dot{x}_4 \end{bmatrix} = \begin{bmatrix} 0 & 0 & 1 & 0 \\ 0 & 0 & 0 & 1 \\ -k/m_1 & k/m_1 & 0 & 0 \\ k/m_2 & -k/m_2 & 0 & 0 \end{bmatrix} \begin{bmatrix} x_1 \\ x_2 \\ x_3 \\ x_4 \end{bmatrix} + \begin{bmatrix} 0 \\ 0 \\ 1/m_1 \\ 0 \end{bmatrix} (u+w_1) + \begin{bmatrix} 0 \\ 0 \\ 0 \\ 1/m_2 \end{bmatrix} w_2
$$

$$
y = x_2 + v
$$

$$(6.29)$$

being $x_1(t)$ and $x_2(t)$ the positions of the carts 1 and 2, respectively, $x_3(t)$ and $x_4(t)$ the velocities of the carts 1 and 2, respectively, $u(t)$ the control signal or force that acts on the first cart, $y(t)$ the sensor output, $w_1(t)$ and $w_2(t)$ the disturbances acting on the carts 1 and 2, respectively, $v(t)$ the sensor noise, and $x_2(t)$ is the variable to be controlled.

Assuming a nominal plant $m_1 = m_2 = 1$ and $k = 1$, the transfer function between the plant input $u(s)$ and the plant output $x_2(s)$ is,

$$
\frac{x_2(s)}{u(s)} = P(s) = \frac{1}{s^2(s^2+2)}
$$

$$(6.30)$$

The original linear controller proposed in Reference 349 is

$$
G_{fix}(s) = \frac{2246.3(s+0.237)(s^2-0.681s+1.132)}{(s+33.19)(s+11.79)(s^2+4.95s+7.563)}
$$

$$(6.31)$$

Applying now the switching strategy introduced in this chapter (Section 6.3, Method 1), the original results presented that Reference 349 for the benchmark are improved. The switching solution proposed here is composed of two channels. The first one, $G_{sw1}(s)$, is for the small error regime, with one pole placed closer to the origin. The second one, $G_{sw2}(s)$, is for the large error regime, with one zero placed closer to the origin. The threshold value in the error is selected based on simulations with the original

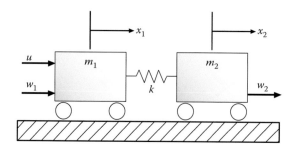

FIGURE 6.8
Two-mass–spring system ($m_1 = m_2 = 1, k = 1$) (Adapted from Garcia-Sanz, M. and Houpis, C.H. 2012. *Wind Energy Systems: Control Engineering Design (Part I: QFT Control, Part II: Wind Turbine Design and Control)*. A CRC Press book, Taylor & Francis Group, Boca Raton, FL.).

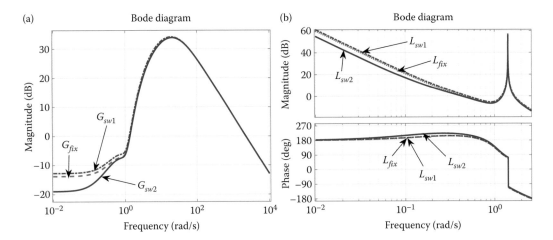

FIGURE 6.9
Bode diagrams. (a) G_{fix}, G_{sw1}, and G_{sw2}. (b) L_{fix}, L_{sw1}, and L_{sw2}.

controller. Figure 6.9 shows the Bode diagrams of the controllers G_{fix}, G_{sw1}, and G_{sw2}, and the open-loop transfer functions $L_{fix} = P\,G_{fix}$, $L_{sw1} = P\,G_{sw1}$, and $L_{sw2} = P\,G_{sw2}$. Figure 6.10a shows the Nichols chart of L_{fix}, L_{sw1}, and L_{sw2}. Figure 6.10b shows the Nichols chart of $(1 + L_{sw1})$ and $(1 + L_{sw2})$ to check the stability according to Section 6.2, which is clearly stable under switching. The switching controller is,

$$G_{sw}(s) = \begin{cases} G_{sw1}(s) = \dfrac{2246.3\,(s+0.237)(s^2 - 0.681s + 1.132)}{(s+33.19)(s+10.5)(s^2 + 4.95s + 7.563)} & \text{for} \quad |e(t)| \le 0.5 \\[3mm] G_{sw2}(s) = \dfrac{2246.3\,(s+0.13)(s^2 - 0.681s + 1.132)}{(s+33.19)(s+11.79)(s^2 + 4.95s + 7.563)} & \text{for} \quad |e(t)| > 0.5 \end{cases}$$

(6.32)

Finally, Figures 6.11 through 6.13 show the improvement with respect to the design introduced in Reference 349. The response of the system to a change in the reference signal [$r(t)$ = unitary step input at $t = 5$ s] is shown in Figure 6.11. The response to a 0.5-magnitude, 1-s-period pulse disturbances, w_2 on mass m_2 at $t = 50$ s and w_1 on mass m_1 at $t = 100$ s, are shown in Figures 6.12 and 6.13, respectively.

6.5 Summary

A practical methodology to design robust controllers that works under a switching mechanism is presented in this chapter. The method is capable of optimizing performance and stability simultaneously, going beyond the classical linear limitations and giving a solution for the well-known robustness-performance trade-off. Based on the frequency-domain approach, the method combines a new graphical stability criterion for switching linear systems and the robust QFT technique. The new formulation is applied here to four illustrative examples.

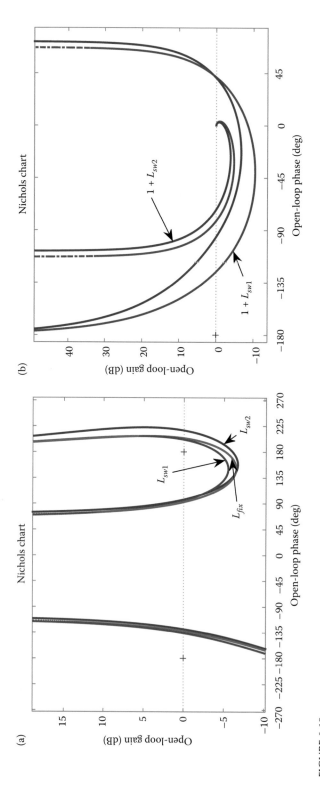

FIGURE 6.10
Nichols chart. (a) L_{fix}, L_{sw1}, and L_{sw2}; (b) $1 + L_{sw1}$ and $1 + L_{sw2}$.

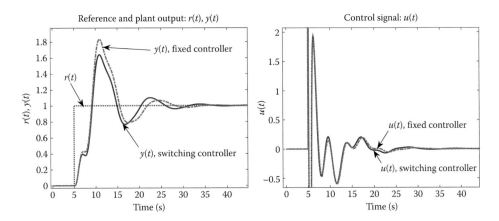

FIGURE 6.11
Response to a unit step reference tracking at $t = 5$ s.

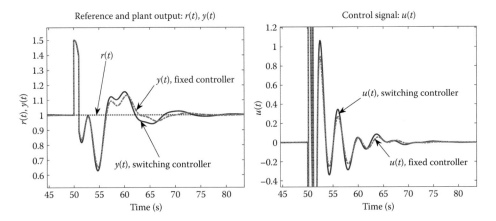

FIGURE 6.12
Response to 0.5 pulse disturbance w_2 on m_2 at $t = 50$ s.

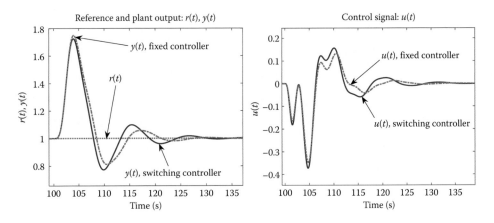

FIGURE 6.13
Response to 0.5 pulse disturbance w_1 on m_1 at $t = 100$ s.

6.6 Practice

The list shown below summarizes the collection of problems and cases included in this book that apply the control methodologies introduced in this chapter.

- *Examples 6.1 and 6.2.* Section 6.4. Systems with gain, zeros, and poles variations.
- *Example 6.3.* Section 6.4. Two-mass-spring benchmark system.
- *Appendix 6.* MATLAB code for *Examples 6.1 through 6.3.*

7

Nonlinear Dynamic Control

7.1 Introduction

Chapter 6 developed a practical methodology to design control systems composed of a set of linear transfer functions or controllers that are switched in real-time following any given switching function. The method gives sufficient conditions for stability under switching and is capable of optimizing performance and stability simultaneously. It goes beyond the classical linear limitations while giving a solution for the well-known robustness/performance trade-off. Based on the frequency domain, the method combines a graphical stability criterion for linear systems that work under switching and the robust QFT.

Extending these results to more complex systems, this new chapter introduces an advanced control design methodology to deal with systems with one or even several nonlinearities, either in the plant, or in the controller, or in both.

Reviewing the fundamentals, Section 7.2 presents the circle stability criterion to study systems with one bounded and static or even time-variant nonlinearity.

Based on this stability criterion, Section 7.3 introduces a practical engineering methodology to design controllers for systems with one nonlinearity in the plant (e.g., saturation or dead-zone in the actuator), and the same nonlinearity deliberately introduced in the controller. The method, known here as *nonlinear dynamic control—one nonlinearity*, is applied to the design of an anti-windup strategy for a classical PID controller in Section 7.4.

Section 7.5, *Nonlinear dynamic control—several nonlinearities*, expands the methodology to systems with more than one nonlinearity. In this case, the nonlinear blocks can be either in the plant (such as saturation, dead-zone, relay, hysteresis, or friction), and/or in the controller (to go beyond the classical linear limitations). Finally, Example 7.2 applies this new methodology to a field-controlled DC motor case.

7.2 The Circle Stability Criterion

The circle stability criterion is a generalization of the Nyquist criterion for linear systems with one bounded, static or even time-varying nonlinearity, $N_0(u,t)$. It gives *sufficient conditions* for absolute stability. It assumes that the nonlinearity is bounded between an upper (k_2) and a lower (k_1) limit, as Figure 7.1 and Equation 7.0.[230,354]

$$
\begin{aligned}
&\text{for } u = 0 \Rightarrow N_0(0,t) = 0 \\
&\text{for } u \neq 0 \Rightarrow k_1 \leq N_0(u,t) \leq k_2
\end{aligned}
\tag{7.0}
$$

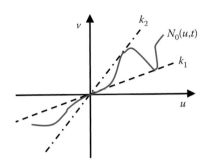

FIGURE 7.1
The nonlinearity, $v = N_0(u,t)\, u$, and its bounds k_2 and k_1, with $k_2 \geq k_1 \geq 0$.

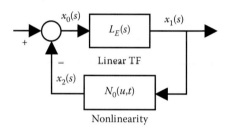

FIGURE 7.2
Representation as a feedback system with one linear function and one nonlinearity.

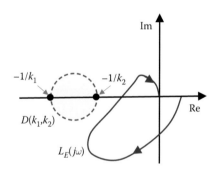

FIGURE 7.3
Conditions for stability according to circle criterion.

Considering a control system represented as a feedback connection of one equivalent linear transfer function $L_E(s)$ and one nonlinear element $N_0(u,t)$, as shown in Figure 7.2, and assuming that the nonlinearity N_0 satisfies the conditions described in Equation 7.0, the system is absolutely stable (*sufficient condition*) if one of the following conditions is satisfied—see also Figure 7.3:

Case 1. If $0 \leq k_1 \leq k_2$, the Nyquist plot of $L_E(j\omega)$, $-\infty \leq \omega \leq \infty$, does not enter the disk $D(k_1,k_2)$ and encircles it n times in the counterclockwise direction, n being the number of poles of $L_E(j\omega)$ with positive real part.

Case 2. If $0 = k_1 \leq k_2$, $L_E(j\omega)$ has no poles in the *RHP* and the Nyquist plot of $L_E(j\omega)$, $-\infty \leq \omega \leq \infty$, lies at the right of the vertical line defined by Re $= -1/k_2$.

Case 3. If $k_1 \leq 0 \leq k_2$, $L_E(j\omega)$ has no poles in the *RHP* and the Nyquist plot of $L_E(j\omega)$, $-\infty \leq \omega \leq \infty$, lies in the interior of the disk $D(k_1, k_2)$.

If, $k_2 = k_1 = 1$, the nonlinearity converts into a linear element (see Figure 7.1), the circle becomes the point "−1" (see Figure 7.3), and the circle criterion turns into the Nyquist stability criterion (see Chapter 3 for more details).

7.3 Nonlinear Dynamic Control: One Nonlinearity

Real systems contain nonlinearities of diverse characteristics. It is common to have position and velocity saturation limitations in the plant actuators, or sometimes dead zones, backlash, friction, etc. In this section, we study a methodology to design control systems with one nonlinearity.

Consider the system shown in Figure 7.4, where there is a nonlinearity N_0 at the input of the plant. To deal with this problem, we could measure the output of the nonlinearity and use this information in an inner-loop of the control system, as shown in Figure 7.4. Instead, we could also introduce the model of this nonlinearity $\hat{N}_0$ in the inner-loop of the control system, as shown in Figure 7.5. This solution, known as *Nonlinear Dynamic Controller (NDC)*, is equivalent to Figure 7.4.[256]

$\hat{N}_0$ being the model of the nonlinearity N_0, $P(s)$ the plant, $G_1(s)$ the feedback controller, and $A(s)$ the inner-loop controller, from Figure 7.5 we have

$$\frac{u(s)}{e(s)} = \frac{G_1(s)}{1 + A(s)(1 - \hat{N}_0)} \tag{7.1}$$

$$y(s) = \frac{P(s)N_0 G_1(s)}{1 + A(s)(1 - \hat{N}_0) + P(s)N_0 G_1(s)}[F(s)r(s) - n(s)] +$$
$$+ \frac{[1 + A(s)(1 - \hat{N}_0)]M(s)}{1 + A(s)(1 - \hat{N}_0) + P(s)N_0 G_1(s)}d(s) \tag{7.2}$$

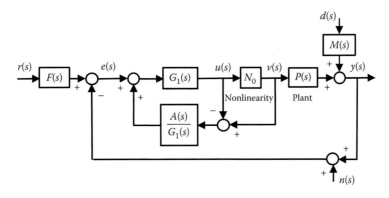

FIGURE 7.4
Control system with an *NDC*.

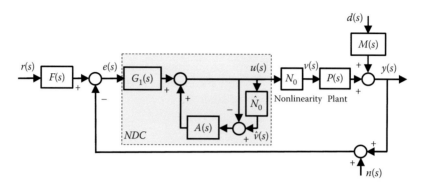

FIGURE 7.5
Equivalent control system with the NDC.

Obviously, in the trivial case, if the model of the nonlinearity matches the real nonlinearity and is in the linear zone $\hat{N}_0 = N_0 = 1$, then Equation 7.2 becomes the well-known expression and diagram given by Equation 7.3 and Figure 7.6

$$y(s) = \frac{P(s)G_1(s)}{1+P(s)G_1(s)}[F(s)r(s)-n(s)] + \frac{M(s)}{1+P(s)G_1(s)}d(s) \tag{7.3}$$

In general, the NDC methodology can deal with any bounded nonlinearity N_0, like a dead zone, backlash, saturation, etc. The only restriction, imposed by the circle stability criteria, is that if we have more than one nonlinearity, all of them should be combined in one block N_0.

In order to apply the circle stability criterion, we consider a perfect nonlinear model $\hat{N}_0 = N_0$, and rearrange Figure 7.5 as Figure 7.7.

Now, being $L_1(s) = P(s)G_1(s)$, the equivalent linear transfer function $L_E(s)$ in Figure 7.7, calculated as it is in Figure 7.2, results as

$$\frac{x_1(s)}{-x_2(s)} = \frac{x_1(s)}{x_0(s)} = L_E(s) = \frac{L_1(s)-A(s)}{1+A(s)} \tag{7.4}$$

Note that in Figure 7.7, the transfer function from the input $x_2(s)$ to the output $x_1(s)$, which is the one within the dashed envelope, is "$-L_E(s)$." This is consistent with the $L_E(s)$, N_0 and negative feedback sign $[x_0(s) = -x_2(s)]$ shown in Figure 7.2.

Design Methodology (NDC: One Nonlinearity)
The design of an NDC—one nonlinearity case, Figure 7.5—is composed of the following sequential steps:

- **Step A.** *Control element $G_1(s)$ and prefilter $F(s)$.* Use the standard QFT methodology to design a feedback controller $G_1(s)$ and a prefilter $F(s)$ to regulate the plant $P(s)$ with model uncertainty and to achieve robust stability and robust performance specifications. At this step, we consider the plant without nonlinearities (or working with small signals), as the control system in Figure 7.6. The design must be Nyquist stable—see Chapter 3.

- **Step B.** *Inner-loop and control element $A(s)$.* Include the inner-loop for the NDC as in Figure 7.5. Find the model for the nonlinearity, $\hat{N}_0 = N_0$, and $L_E(s)$ as Equation 7.4

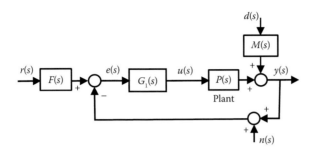

FIGURE 7.6
Equivalent control system with $\hat{N}_0 = N_0 = 1$.

for Figure 7.7. Design the transfer function $A(s)$ with the same number of integrators than the controller $G_1(s)$, and with the appropriate zeros and poles so that $L_1(s)$ dominates at low frequency $[|L_1(j\omega)| \gg |L_E(j\omega)|$ at $\omega_{\text{low}}]$ and $L_E(s)$ dominates at high frequency $[|L_E(j\omega)| \gg |L_1(j\omega)|$ at $\omega_{\text{high}}]$—see Figure 7.15.

- **Step C.** *System stability. Circle criterion.* Rearrange the control system as it is in Figure 7.7 or Figure 7.2, combining all the nonlinearities (real and model) in one single block. Calculate the equivalent transfer function $L_E(s)$, as Equation 7.4 for Figure 7.7. The system has to be stable according to the circle criterion, as presented in Section 7.2. The method gives *sufficient conditions* for stability.

- **Step D.** *Simulation and discussion.* Simulate the complete *NDC* system, with all the nonlinearities, inner-loop and model uncertainties. Check all the stability and performance specifications.

Additional Notes

Remark 7.1

The *NDC* methodology requires $L_1(s)$ to be stable according to the Nyquist criterion and $L_E(s)$ to be stable according to the circle criterion.

Remark 7.2

As a result, the *NDC* improves classical linear performance. It is less restrictive than other classical methods, because it allows $L_1(s)$ to be just Nyquist stable, instead of circle stable

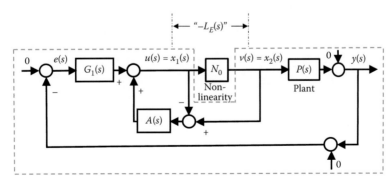

FIGURE 7.7
Equivalent control system to calculate $L_E(s)$.

(which is the condition for L_E). This fact permits $L_1(s)$ to be type 2 or 3 (2 or 3 integrators) if necessary. The circle criterion is more restrictive because typically it only accepts the system to be type 0 or 1.

Remark 7.3

If the system (plant or controller) contains more than one nonlinearity, we must lump all the nonlinearities together by block diagram manipulation and obtain the overall nonlinearity $N_0(u,t)$.

Remark 7.4

The methodology can deal with time-variant nonlinearities $N_0(u,t)$ as long as they are bounded within the sector: $k_1 \leq N_0(u,t) \leq k_2$.

7.4 Anti-Windup Solution for PID Controllers

The *NDC* shown in Figures 7.4 and 7.5 can also be used to provide an anti-windup solution to the classical PID controllers when the nonlinearity N_0 is the actuator saturation. A classical anti-windup solution with an inner-loop around the integral part is shown in Figure 7.8.[249]

This is equivalent to the *NDC* presented in Figure 7.5, which is also Figure 7.9 with the following expressions for $G_1(s)$ and $A(s)$:

$$G_1(s) = G_{PID}(s) = K_p \left(1 + \frac{1}{T_i s} + \frac{T_d s}{(T_d/N)s + 1} \right) \tag{7.5}$$

$$A(s) = \frac{k_a}{s} \tag{7.6}$$

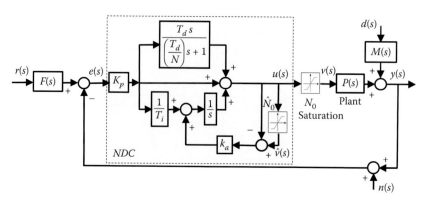

FIGURE 7.8
PID controller with an anti-windup solution.

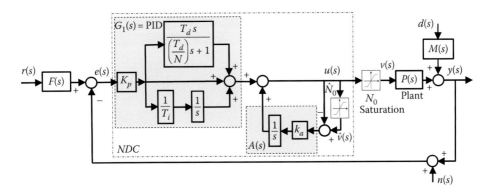

FIGURE 7.9
PID controller with anti-windup solution as *NDC*.

EXAMPLE 7.1: PID CONTROLLER WITH AN ANTI-WINDUP SOLUTION

To illustrate the discussion, consider a third-order plant with parametric uncertainty. The following subsections describe the steps to apply the design methodology. The control system is based on Figure 7.9. The plant to be controlled is described by the following third-order expression:

$$P(s) = \frac{k_1}{((1/\mu_1)s + 1)^3} \tag{7.7}$$

with the parametric uncertainty, $k_1 \in [0.5, 1.5]$, $\mu_1 \in [0.83, 1.25]$. We use the first values of the parameter intervals ($k_1 = 0.5$, $\mu_1 = 0.83$) for the nominal plant $P_0(s)$, and the central values ($k_1 = 1$, $\mu_1 = 1$) for the central plant $P_c(s)$. The disturbance transfer function is $M(s) = 1$. The actuator saturation limits are: $u_{max} = +1.10$, $u_{min} = -1.10$, and the slope of the central part is one—see Figure 7.10.

Step A: Control Element $G_1(s)$ and Prefilter $F(s)$

The control specifications required for this example include the robust stability, disturbance rejection at the output of the plant, and reference tracking objectives presented in Equations 7.8 through 7.10, respectively.

- *Stability specification:*

$$\left| \frac{P(j\omega)G_1(j\omega)}{1 + P(j\omega)G_1(j\omega)} \right| \le W_s = 1.93 \ (GM = 3.62 \ \text{dB}, PM = 30.02°) \tag{7.8}$$

for $\omega = [0.001 \quad 0.005 \quad 0.01 \quad 0.05 \quad 0.1 \quad 0.5 \quad 1 \quad 2 \quad 3 \quad 5 \quad 10 \quad 50 \quad 100]$ rad/s

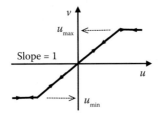

FIGURE 7.10
Actuator saturation. Slope = 1. $u_{max} = +1.10$, $u_{min} = -1.10$.

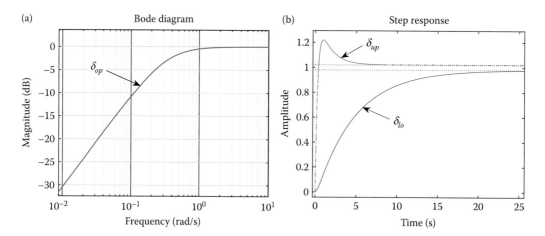

FIGURE 7.11
Control specifications for $P(s)$. (a) Disturbance rejection at the output of the plant: δ_{op}. (b) Reference tracking: δ_{up}, δ_{lo}.

- *Sensitivity or output disturbance rejection specification—see Figure 7.11a:*

$$\left| \frac{1}{1 + P(j\omega)G_1(j\omega)} \right| \leq \delta_{op}(\omega) = \left| \frac{3(j\omega)}{3(j\omega) + 1} \right|$$

$$\text{for } \omega = [0.001 \quad 0.005 \quad 0.01 \quad 0.05 \quad 0.1 \quad 0.5] \text{ rad/s}$$

(7.9)

- *Reference tracking specification—see Figure 7.11b:*

$$\delta_{lo}(\omega) \leq \left| \frac{P(j\omega)G_1(j\omega)}{1 + P(j\omega)G_1(j\omega)} F(j\omega) \right| \leq \delta_{up}(\omega)$$

(7.10)

where

$$\delta_{lo}(\omega) = \left| \frac{0.98}{1.4(j\omega)^2 + 5(j\omega) + 1} \right| \text{ is the lower bound}$$

and

$$\delta_{up}(\omega) = \left| \frac{2(j\omega) + 1.02}{0.4(j\omega)^2 + 1.7(j\omega) + 1} \right| \text{ is the upper bound}$$

$$\text{for } \omega = [0.001 \quad 0.005 \quad 0.01 \quad 0.05 \quad 0.1 \quad 0.5] \text{ rad/s}$$

The QFT bounds are calculated taking into account the $P(s)$ transfer function with the parameter uncertainty given by Equation 7.7, and the stability, output disturbance rejection and reference tracking specifications given by Equations 7.8 through 7.10, respectively—see Figure 7.11.

To meet the specifications, we design a PID controller $G_1(s)$ with a low-pass filter for the derivative part, as shown in Equation 7.11. The parameters of the PID (standard

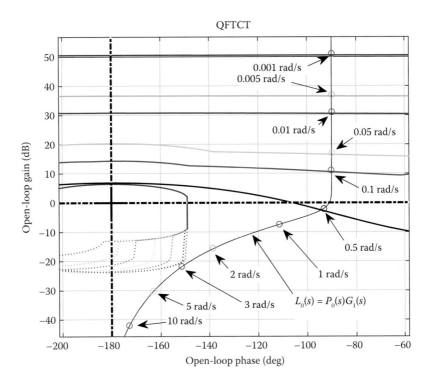

FIGURE 7.12
Loop-shaping of controller $G_1(s)$. $L_0(s) = P_0(s)G_1(s)$.

form) are: $K_p = 2.5177$, $T_i = 3.5967$, $T_d = 1.1088$, and $N = 336$. The QFT bounds and the loop shaping of $L_0(s) = P_0(s)G_1(s)$ meeting all the specifications are shown in Figure 7.12.

$$G_1(s) = G_{PID}(s) = K_p\left(1 + \frac{1}{T_i s} + \frac{T_d s}{(T_d/N)s + 1}\right)$$

$$= 2.5177\left(1 + \frac{1}{3.5967s} + \frac{1.1088s}{0.0033s + 1}\right) = 0.7\frac{(s/0.5)^2 + (2 \times 0.9/0.5)s + 1}{s((1/300)s + 1)} \quad (7.11)$$

The prefilter element $F(s)$ is designed with the QFTCT as a unitary gain, for the plant $P(s)$ and the controller $G_1(s)$. Figure 7.13 shows the upper and lower spec limits and the closed-loop transfer function with this design. The prefilter is

$$F(s) = 1 \quad (7.12)$$

Figure 7.14 shows how the control system meets all the performance specifications. Figure 7.14a is the frequency domain analysis of the disturbance rejection specification—Equation 7.9, and Figure 7.14b the time domain analysis of the reference tracking specification with 125 plants—Equation 7.10. The design is Nyquist stable.

Step B: Inner-Loop and Control Element A(s)

The function $A(s)$ of the *NDC* is designed so that: (1) $A(s)$ has the same number of integrators as $G_1(s)$, and (2) $L_1(s) = P(s)G_1(s)$ dominates at low frequencies and $L_E(s) = [L_1(s) - A(s)]/[1 + A(s)]$ dominates at high frequencies. Equation 7.13 shows the expression for $A(s)$.

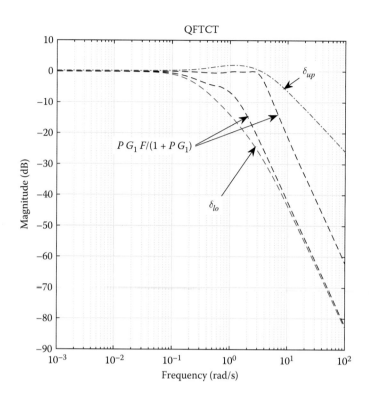

FIGURE 7.13
Design of prefilter $F(s)$ for $P(s)$ and $G_1(s)$.

Figure 7.15a shows $G_1(s)$ and $A(s)$, and Figure 7.15b compares the magnitudes of $L_1(s)$ and $L_E(s)$ using the nominal plant $P_0(s)$.

$$A(s) = \frac{k_a}{s} = \frac{0.9}{s} \tag{7.13}$$

Step C: System Stability. Circle Criterion

In order to analyze the absolute stability of the system according to the circle criterion (Section 7.2), we have first to arrange the control system of Figure 7.9 as a feedback connection of one linear transfer function $L_E(s)$ and one nonlinear element $N_0(u,t)$, as it was shown in Figure 7.2. As discussed in Equation 7.4, considering a perfect nonlinear model $\hat{N}_0 = N_0$ the equivalent linear transfer function $L_E(s)$ is

$$L_E(s) = \frac{L_1(s) - A(s)}{1 + A(s)}, \text{ being } L_1(s) = P(s)G_1(s) \tag{7.14}$$

As we see, the function $L_E(s)$ depends on the plant $P(s)$, which varies according to the parametric uncertainty—see Equation 7.7. After a careful inspection of the function $L_E(s)$ at all the cases defined by the parametric uncertainty, we appreciate that it does not have poles in the *RHP*.

On the other hand, the nonlinearity N_0 of the system is the actuator saturation that satisfies the expressions:

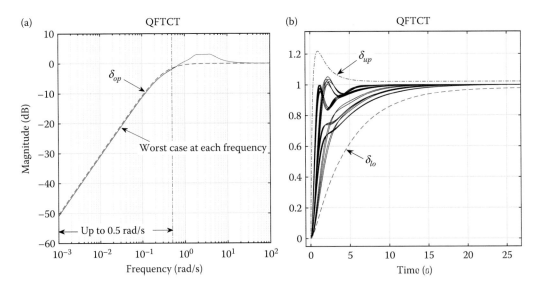

FIGURE 7.14
Analysis of controller $G_1(s)$ and prefilter $F(s)$ for $P(s)$. (a) Disturbance rejection at plant output: δ_{op}. (b) Reference tracking: δ_{up}, δ_{lo}.

$$\text{for } u = 0 \Rightarrow N_0(0) = 0$$
$$\text{for } u \neq 0 \Rightarrow k_1 \leq N_0(u) \leq k_2, k_1 = 0, k_2 = 1 \tag{7.15}$$

Now, according to *Case 2* (Section 7.2), the *sufficient condition* for stability of the closed-loop system is that the Nyquist plot of $L_E(j\omega)$ does not enter or enclose the circle D defined by $(-1/k_1, -1/k_2)$, which in this problem is a vertical line at Re $= -1$ (circle of infinite radius, from $-\infty$ to -1). Figure 7.16 shows $L_E(j\omega)$, $0 \leq \omega \leq \infty$, for 100 plants within the parametric uncertainty, and the circle (vertical line). As all the functions $L_E(j\omega)$ lie at the right of the vertical line Re $= -1$ (outside the circle), the closed-loop system is absolutely stable for all the cases within the parametric uncertainty.

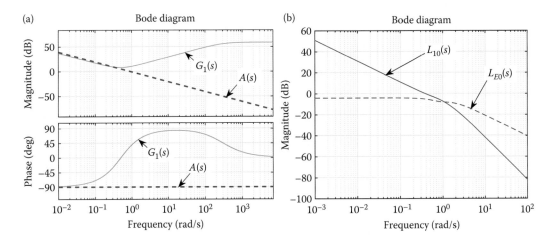

FIGURE 7.15
Bode diagrams. (a) $G_1(s)$ and $A(s)$. (b) $L_{10}(s)$ and $L_{E0}(s)$.

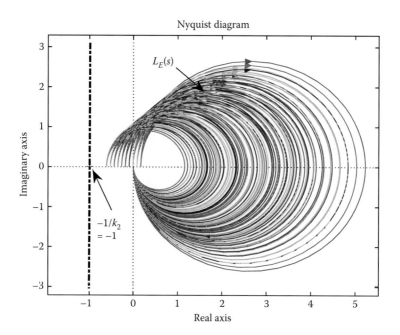

FIGURE 7.16
Circle criterion for 100 plants within the parametric uncertainty in $L_E(s)$, N_0 = saturation, and circle with $k_1 = 0$, $k_2 = 1$.

Step D: Simulation and Discussion

Finally, Figures 7.17 and 7.18 show the results of the simulation of the closed-loop system presented in Figure 7.9, with the central plant $P(s) = P_c(s)$—Equation 7.7, the QFT/PID controller $G_1(s)$—Equation 7.11, and the prefilter $F(s)$—Equation 7.12. The reference input starts at $r(t) = 0$ and at $t = 1$ s jumps to $r(t) = 1$ (as a unitary step). The disturbance

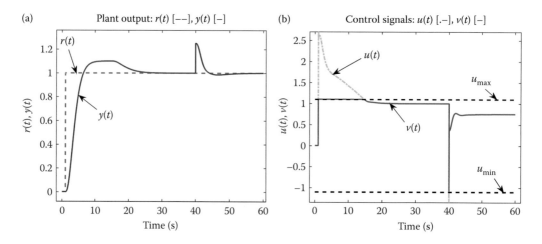

FIGURE 7.17
Simulation of closed-loop system with central plant $P_c(s)$, PID controller $G_1(s)$, $F(s)$, and $A(s) = 0$. (a) Output $y(t)$ response to a unitary step input $r(t)$ at $t = 1$ s and a 0.25 step disturbance $d(t)$ at $t = 40$ s. (b) Controller output $u(t)$ and signal $v(t)$ after the actuator saturation.

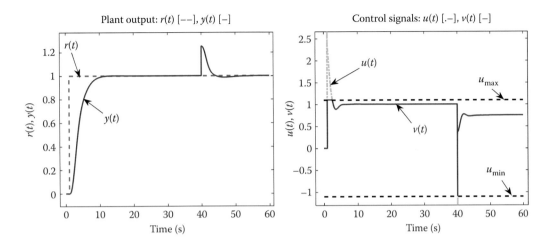

FIGURE 7.18

Simulation of closed-loop system with central plant $F_c(s)$, PID controller $G_1(s)$, $F(s)$ and $A(s) = 0.9/s$. (a) Output $y(t)$ response to a unitary step input $r(t)$ at $t = 1$ s and a 0.25 step disturbance $d(t)$ at $t = 40$ s. (b) Controller output $u(t)$ and signal $v(t)$ after the actuator saturation.

starts at $d(t) = 0$ and at $t = 40$ s jumps to $d(t) = 0.25$ (as a step). In both figures, Figures 7.17 and 7.18, the first part shows the reference input $r(t)$ and the plant output $y(t)$, and the second part the controller output $u(t)$ and the signal $v(t)$ after the actuator saturation. The first figure, Figure 7.17, shows the results without the *NDC*, that is to say, $A(s) = 0$. The second figure, Figure 7.18, shows the results with the *NDC*, that is to say, with $A(s)$ as defined in Equation 7.13. As we see, the QFT/PID works well in both cases, and for the reference tracking and disturbance rejection specifications. Moreover, the controller with the additional *NDC* inner-loop improves the performance, eliminating the windup problem and achieving a much better reference tracking without any overshoot.

7.5 Nonlinear Dynamic Control: Several Nonlinearities

This section extends the methodology introduced in Sections 7.3 and 7.4 to systems with more than one nonlinearity. The nonlinearities can be either a part of the controller, or a part of the plant, or both. In the first case, the introduction of nonlinearities in the controller is a way to achieve high performance, allowing the system to go beyond the classical linear limitations. In the second case, the nonlinearities are a part of the plant and can typically include saturation, dead zone, relay, hysteresis, friction, backlash, etc.

As discussed in Sections 7.2 and 7.3, when we can lump the nonlinearities of the system into just one single block, we can then apply the circle criterion to analyze the stability. In this case, the methodology presented in Sections 7.3 and 7.4 is completely valid. However, if the system presents more than one nonlinearity, and these nonlinearities cannot be combined in a single block, then the circle criterion and the corresponding method presented in Sections 7.3 and 7.4 cannot be applied. This section presents a solution for these more general cases. It is based on the describing function (*DF*) technique. The method can deal with any kind of nonlinearity, like single valued, double valued, with memory, etc.

7.5.1 Describing Functions

It is well known that the model or transfer function of a linear system does not depend on the amplitude of the input signal applied to the system.[230,256] On the contrary, when we deal with nonlinear systems, the corresponding model is typically a function of the amplitude of the input signal. For these cases, in this section, we calculate the model of such nonlinearities with the *Describing Function (DF) technique.*

By definition, the *DF* of a nonlinearity N_0 is an LTI transfer function approximation that depends on the amplitude of the applied input. In other words, given a sine wave input $u(t)$ of amplitude E and frequency ω, $u(t) = E \sin(\omega t)$, the output $v(t)$ of N_0 will be also a sine wave of the same frequency ω but with a scaled amplitude and shifted phase, $v(t) = N_0 u(t) \approx DF\, u(t) = Q \sin(\omega t + \phi)$, being

$$N_0 \approx DF(E, j\omega) = \frac{B_1 + jA_1}{E} \tag{7.16}$$

where B_1 and A_1 depend on the type of nonlinearity, as we will see in the following subsections, and where E is the amplitude of the input to the nonlinearity.

The *DF* method needs to satisfy the *low-pass filter conjecture*, which states that the system has a low-pass filter characteristic that attenuates the harmonics (frequency components higher than the fundamental frequency) of the input. This condition can be easily satisfied in real control systems. Errors due to the violation of this condition can be fixed by adding some extra phase in the controller.

In the following paragraphs, we present the *DF* models of some practical nonlinearities that will be either introduced in the controller to achieve high-performance or found in real-world plants. The expressions B_1 and A_1 are the real and imaginary parts of Equation 7.16, respectively. The input of the nonlinearity is $u(t) = E \sin(\omega t)$, E being the amplitude.

- *Nonlinear element N_{01}*: This is a useful nonlinearity with hysteresis for switching between two controllers. It can be implemented as channel 1 of the controller, and combined with the nonlinear element N_{02} for channel 2. The graphical description is presented in Figure 7.19, and the *DF* model in Equations 7.17 and 7.18.

$$\text{If } E < \delta_l, \text{ then } DF(E) = 1 \tag{7.17}$$

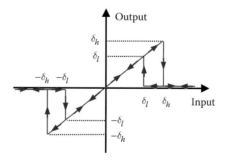

FIGURE 7.19
Nonlinear element N_{01}.

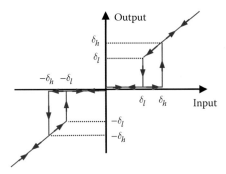

FIGURE 7.20
Nonlinear element N_{02}.

$$\text{If } E > \delta_h, \text{ then } DF(E) = \frac{B_1 + jA_1}{E}, \text{ with} \qquad (7.18)$$

$$B_1 = 0$$

$$A_1 = \frac{1}{\pi E}\left[\delta_h^2 - \delta_l^2\right]$$

- *Nonlinear element N_{02}*: This is the complement of N_{01}. It can be implemented as channel 2 of the controller, and combined with the nonlinear element N_{01} for channel 1. The graphical description is presented in Figure 7.20, and the *DF* model in Equations 7.19 and 7.20. Graphically, $N_{01} + N_{02} = 1$.

$$\text{If } E < \delta_l, \text{ then } DF(E) = 0 \qquad (7.19)$$

$$\text{If } E > \delta_h, \text{ then } DF(E) = \frac{B_1 + jA_1}{E}, \text{ with} \qquad (7.20)$$

$$B_1 = E$$

$$A_1 = -\frac{1}{\pi E}\left[\delta_h^2 - \delta_l^2\right]$$

- *Nonlinear element N_{03}*: This is a useful nonlinearity without hysteresis for switching between two controllers. It can be implemented as channel 1 of the controller, and combined with the nonlinear element N_{04} for channel 2. The graphical description is presented in Figure 7.21, and the *DF* model in Equations 7.21 and 7.22:

$$\text{If } E < \delta, \text{ then } DF(E) = 1 \qquad (7.21)$$

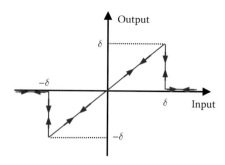

FIGURE 7.21
Nonlinear element N_{03}.

$$\text{If } E > \delta, \text{ then } DF(E) = \frac{B_1 + jA_1}{E}, \text{ with} \tag{7.22}$$

$$B_1 = E - \left(\frac{\delta}{\pi}\right)\sqrt{1 - \left(\frac{\delta}{E}\right)^2}$$

$$A_1 = 0$$

- *Nonlinear element N_{04}*: This is the complement of N_{03}. It can be implemented as channel 2 of the controller, and combined with the nonlinear element N_{03} for channel 1. The graphical description is presented in Figure 7.22 and the *DF* model in Equations 7.23 and 7.24. Graphically, $N_{03} + N_{04} = 1$.

$$\text{If } E < \delta, \text{ then } DF(E) = 0 \tag{7.23}$$

$$\text{If } E > \delta, \text{ then } DF(E) = \frac{B_1 + jA_1}{E}, \text{ with} \tag{7.24}$$

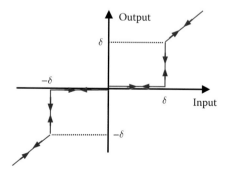

FIGURE 7.22
Nonlinear element N_{04}.

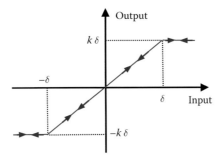

FIGURE 7.23
Nonlinear element N_{05}: saturation.

$$B_1 = \left(\frac{\delta}{\pi}\right)\sqrt{1 - \left(\frac{\delta}{E}\right)^2}$$

$$A_1 = 0$$

- *Nonlinear element N_{05}:* This is the saturation nonlinearity. The graphical description is presented in Figure 7.23, and the *DF* model in Equations 7.25 and 7.26.

$$\text{If } E < \delta, \text{ then } DF(E) = k \qquad (7.25)$$

$$\text{If } E > \delta, \text{ then } DF(E) = \frac{B_1 + jA_1}{E}, \text{ with} \qquad (7.26)$$

$$B_1 = \frac{2kE}{\pi}\left[\sin^{-1}\left(\frac{\delta}{E}\right) + \left(\frac{\delta}{E}\right)\sqrt{1 - \left(\frac{\delta}{E}\right)^2}\right]$$

$$A_1 = 0$$

- *Nonlinear element N_{06}:* This is the dead-zone nonlinearity. The graphical description is in Figure 7.24, and the *DF* model in Equations 7.27 and 7.28. Note that $N_{05} = k - N_{06}$.

$$\text{If } E < \delta, \text{ then } DF(E) = 0 \qquad (7.27)$$

$$\text{If } E > \delta, \text{ then } DF(E) = \frac{B_1 + jA_1}{E}, \text{ with} \qquad (7.28)$$

$$B_1 = kE - \frac{2kE}{\pi}\left[\sin^{-1}\left(\frac{\delta}{E}\right) + \left(\frac{\delta}{E}\right)\sqrt{1 - \left(\frac{\delta}{E}\right)^2}\right]$$

$$A_1 = 0$$

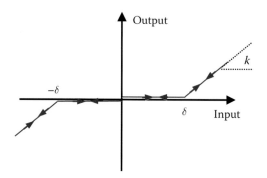

FIGURE 7.24
Nonlinear element N_{06}: dead zone.

- *Nonlinear element N_{07}:* This is the relay nonlinearity. The graphical description is presented in Figure 7.25, and the *DF* model in Equations 7.29 and 7.30.

$$\text{For all } E, DF(E) = \frac{B_1 + jA_1}{E}, \text{with}$$

$$B_1 = \frac{4\alpha}{\pi} E \tag{7.29}$$

$$A_1 = 0 \tag{7.30}$$

- *Nonlinear element N_{08}:* This is the on–off nonlinearity with hysteresis. The graphical description is presented in Figure 7.26, and the *DF* model in Equations 7.31 and 7.32.

$$\text{If } E < \delta, \text{then } DF(E) = 0 \tag{7.31}$$

$$\text{If } E > \delta, \text{then } DF(E) = \frac{B_1 + jA_1}{E}, \text{with} \tag{7.32}$$

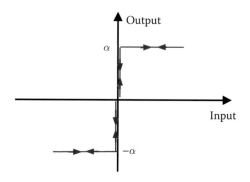

FIGURE 7.25
Nonlinear element N_{07}: relay.

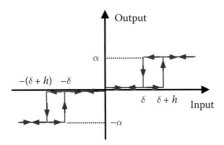

FIGURE 7.26
Nonlinear element N_{08}: on–off with hysteresis.

$$B_1 = \frac{2\alpha}{\pi}\left[\sqrt{1-\left(\frac{\delta+h}{E}\right)^2} + \sqrt{1-\left(\frac{\delta}{E}\right)^2}\right]$$

$$A_1 = -\frac{2\alpha}{\pi}\left(\frac{h}{E}\right)$$

7.5.2 Isolines

As mentioned before, the described function DF is a function of the amplitude of the input E and the frequency ω.[256] As in linear systems, we can plot the open-loop transfer function L of the system in the NC, but now including also the DFs. These lines L are the so-called isolines. They are calculated cascading from right to left (or output to input) the linear and nonlinear DF elements, like for instance: $L = P_1\,DF_{N1}\,P_2\,[C_1\,DF_{N3} + (C_2 + C_3\,DF_{N4})]$. Figure 7.27 shows an example.

The system is considered stable if all the isolines satisfy the Nyquist stability criterion, as introduced in Chapter 3. However, note that as the DF technique is an approximation these results can be still confusing in some cases, mainly due to the phase uncertainty caused by the potential harmonics. Nevertheless, providing a safety margin of about 20°, the NDC should be adequate for many practical applications.

Design Methodology (NDC: Several Nonlinearities)
The design of an NDC—several nonlinearities case is composed of the following sequential steps:

- **Step A.** *Controller structure.* Select the controller topology, including k parallel channels to switch ($k = 1$ to number of channels n), inner-loops, linear transfer functions, preferred nonlinearities for the controller, and switching functions. We typically use the *error* as the signal to switch between the channels, being active/ aggressive controllers for small errors and moderate controllers for large errors.
- **Step B.k.** *Control element for kth channel, $G_k(s)$ and prefilter $F_k(s)$.* Use the standard QFT methodology to design feedback controllers $G_k(s)$ and prefilters $F_k(s)$ to regulate the plant $P(s)$ with model uncertainty, and to achieve robust stability and performance specifications at each selected channel or switched zone "k." At this step, we consider the plant and the controllers without nonlinearities (or working

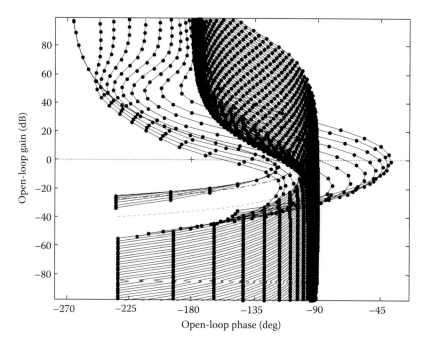

FIGURE 7.27

Isolines L in the NC, for input error E from 10^{-7} to 10^{2}.

with small signals), as the control system in Figure 7.6. The design must be Nyquist stable for each case—see Chapter 3.

- **Step C.** *Inner-loops and control elements A(s).*

  C.1. Select the inner-loop structure, either (a) with one inner-loop after the addition of all the channels $k = 1$ to n, or (b) with one inner-loop in each channel. Use the first option if all the controllers $G_k(s)$ have the same number of integrators, and the second one if the controllers $G_k(s)$ have a different number of integrators.

  C.2. Include the inner-loop for the *NDC* as in Figure 7.5. Find the model for the nonlinearity, $\hat{N}_0 = N_0$. Design the function $A(s)$ with the same number of integrators than the controller $G_k(s)$. Also, design $A(s)$ so that $L_k(s)$ dominates at low frequency $[|L_k(j\omega)| \gg |L_{Ek}(j\omega)|$ at $\omega_{\text{low}}]$ and $L_{Ek}(s)$ dominates at high frequency $[|L_{Ek}(j\omega)| \gg |L_k(j\omega)|$ at $\omega_{\text{high}}]$—see example in Figure 7.15.

  C.3. Rearrange the control system as it is in Figure 7.7 or 7.2, combining all the nonlinearities (real and model) in one single block. Calculate the equivalent transfer function $L_{Ek}(s)$, as Equation 7.4 for Figure 7.7. The system has to be stable according to the circle criterion, as presented in Section 7.2.

- **Step D.** *Isolines. Stability analysis for the complete system.* Calculate the *DF* of every nonlinearity, either in the controller or in the plant. Write an algorithm to calculate the isolines cascading the linear and nonlinear elements. Include the *DF* of each nonlinearity at their specific position in the loop, from the input (right of the expression) to the output (left of the expression); for instance: $L = P_1 \, DF_{N1} \, P_2 \, [G_1 \, DF_{N2} + (G_2 + G_3 \, DF_{N3})]$. See Example 7.2 and Appendix 7. Then, for the nominal plant, and for every amplitude of the error E (input) and every frequency of interest ω, plot the isoline in the NC. Include also the QFT stability bounds for the nominal plant. Check the stability with the isolines and according to the Nyquist criterion for NCs introduced in Chapter 3 and the QFT stability bounds. The method gives *sufficient conditions* for stability.

- **Step E.** *Simulation and discussion.* Simulate the complete NDC system, with all the nonlinearities, parallel channels, inner-loops, switching cases, and model uncertainties. Check all the robust stability and performance specifications.

Additional Notes

Remark 7.5

If the nonlinearity N_0 is independent of time, $N_0 = N_0(u)$, and $k_1 \leq N_0(u) \leq k_2$, then the *DF* also satisfies $k_1 \leq DF(E) \leq k_2$ and hence the graph of $-1/DF(E)$ lies entirely within the circle $D(-1/k_1, -1/k_2)$ of the circle criterion. In other words, the *DF* method and the circle criterion are consistent.

EXAMPLE 7.2: NDC WITH SEVERAL NONLINEARITIES

To understand the *NDC* methodology for several nonlinearities, consider an illustrative example composed of a conventional field-controlled DC motor system and its third-order plant model with parametric uncertainty given by

$$\frac{y(s)}{u(s)} = P(s) = \frac{K_m}{s(Js+D)(L_a s + R_a)} = \frac{(k/(ab))}{s((1/a)s+1)((1/b)s+1)} \tag{7.33}$$

where $y(s)$ is the angle of the shaft of the DC motor to be controlled, $u(s)$ is the voltage applied to the input, J is the inertia of the rotating elements, D is the viscous friction, K_m is the motor-torque constant, R_a is the resistance, and L_a is the inductance of the motor armature. As a first approximation, it is possible to combine these parameters into $k = K_m/(L_a J)$ as the gain, $1/a = J/D$ as the mechanical time constant, and $1/b = L_a/R_a$ as the electrical time constant of the system.

For a small motor, the parameters and the associate uncertainty for the system are: $k \in [610, 1050]$, $a \in [1, 15]$, $b \in [150, 170]$. We select the nominal plant $P_0(s)$ as the one with $k = 610$, $a = 1$, $b = 150$. In addition, we consider a saturation constraint in the actuator, being the upper and lower limits, respectively, $u_{max} = +1.10$, $u_{min} = -1.10$, and the slope of the central part unitary—see also Figure 7.10.

Step A: Controller Structure

The structure proposed for the NDC is shown in Figure 7.28. It consists of two parallel channels, $G_1(s) N_1$ and $G_2(s) N_2$, and an anti-windup nonlinear inner-loop with a linear transfer function $A(s)$ and a saturation model $\hat{N}_0$. $G_1(s) N_1$ is the aggressive channel and $G_2(s) N_2$ the moderate channel.

N_1 and N_2 are the nonlinearities defined, respectively, in Figures 7.19 and 7.20, that is, $N_1 = N_{01}$ with Equations 7.17 and 7.18, and $N_2 = N_{02}$ with Equations 7.19 and 7.20—see also Figure 7.29. The parameters selected for these two nonlinearities are: $\delta_h = 0.15$ and $\delta_l = 0.075$. Note that $N_{01} + N_{02} = 1$.

As a summary, the system has four nonlinearities: the actuator saturation N_0, the actuator saturation model $\hat{N}_0$ and the two control channels N_1 and N_2.

Step B.1: Control Elements for First Channel, $G_1(s)$ and Prefilter $F_1(s)$: Aggressive

The control specifications selected for the first channel include robust stability, aggressive disturbance rejection at the output of the plant and reference tracking objectives, as presented in Equations 7.34 through 7.36, respectively.

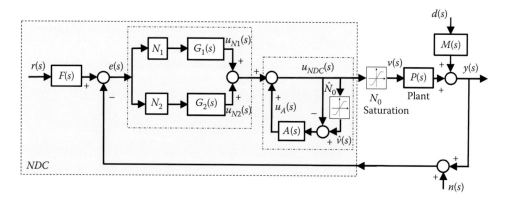

FIGURE 7.28
NDC block diagram: two channels and one common inner-loop.

- Stability specification:

$$\left|\frac{P(j\omega)G_1(j\omega)}{1+P(j\omega)G_1(j\omega)}\right| \leq W_{s1} = 1.46\,(GM = 4.53\,\text{dB}, PM = 40.05°) \tag{7.34}$$

$$\text{for } \omega = \begin{bmatrix} 0.001 & 0.005 & 0.01 & 0.05 & 0.1 & 0.2 & 0.5 & 0.7 \\ 1 & 2 & 3 & 5 & 10 & 50 & 100 & 500 & 1000 \end{bmatrix} \text{rad/s}$$

- Sensitivity or output disturbance rejection specification—see Figure 7.30a:

$$\left|\frac{1}{1+P(j\omega)G_1(j\omega)}\right| \leq \delta_{op1}(\omega) = \left|\frac{0.07(j\omega)}{0.07(j\omega)+1}\right|$$

$$\text{for } \omega = \begin{bmatrix} 0.001 & 0.005 & 0.01 & 0.05 & 0.1 & 0.2 & 0.5 & 0.7 \\ 1 & 2 & 3 & 5 & 10 & 50 \end{bmatrix} \text{rad/s} \tag{7.35}$$

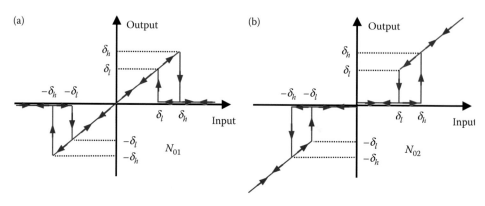

FIGURE 7.29
Nonlinear elements for the *NDC* of Figure 7.28: (a) $N_1 = N_{01}$. (b) $N_2 = N_{02}$.

- Reference tracking specification—see Figure 7.30b:

$$\delta_{lo1}(\omega) \leq \left| \frac{P(j\omega)G_1(j\omega)}{1+P(j\omega)G_1(j\omega)} F_1(j\omega) \right| \leq \delta_{up1}(\omega) \tag{7.36}$$

where

$$\delta_{lo1}(\omega) = \left| \frac{0.98}{0.0625(j\omega)^2 + 1.05(j\omega) + 1} \right| \text{ is the lower bound}$$

and

$$\delta_{up1}(\omega) = \left| \frac{2(j\omega) + 1.02}{0.4(j\omega)^2 + 1.7(j\omega) + 1} \right| \text{ is the upper bound}$$

for $\omega = [0.001 \quad 0.005 \quad 0.01 \quad 0.05 \quad 0.1 \quad 0.2 \quad 0.5 \quad 0.7 \quad 1]$ rad/s

The QFT bounds are calculated taking into account the $P(s)$ transfer function with the parameter uncertainty given by Equation 7.33, and the robust stability, output disturbance rejection and reference tracking specifications given by Equations 7.34 through 7.36, respectively—see Figure 7.30.

To meet the specifications, we design a PID with a low-pass filter for the $G_1(s)$ controller, as shown in Equation 7.37. The QFT bounds and the loop shaping of $L_{10}(s) = P_0(s)G_1(s)$ meeting all the specifications are shown in Figure 7.31.

$$G_1(s) = 89 \frac{((1/1.2)s + 1)((1/5)s + 1)}{s((1/500)s + 1)} \tag{7.37}$$

The prefilter element $F_1(s)$ is also designed with the QFTCT, for the plant $P(s)$, and for the controller $G_1(s)$. Figure 7.32 shows the design. The expression for the prefilter is

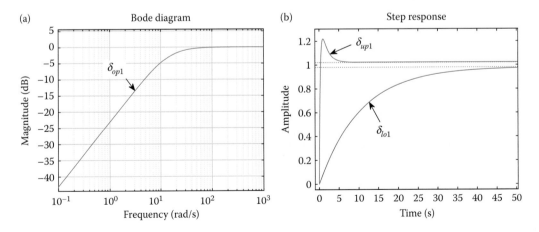

FIGURE 7.30

Control specifications for $P(s)$. (a) Disturbance rejection at the output of the plant: δ_{op1}. (b) Reference tracking: δ_{up1}, δ_{lo1}. Aggressive channel.

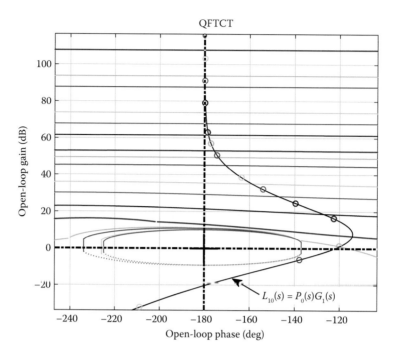

FIGURE 7.31
Loop-shaping of controller $G_1(s)$. $L_{10}(s) = P_0(s)G_1(s)$.

$$F(s) = F_1(s) = \frac{((1/21)s + 1)((1/48)s + 1)}{((1/4)s + 1)((1/8)s + 1)} \tag{7.38}$$

Figure 7.33 shows how the control system (G_1, F_1) meets the performance specifications. Figure 7.33a is the frequency domain analysis of the disturbance rejection specification—Equation 7.35, and Figure 7.33b the time domain analysis of the reference tracking specification with 125 plants—Equation 7.36. The design is Nyquist stable.

Step B.2: Control Elements for Second Channel, $G_2(s)$ and Prefilter $F_2(s)$: Moderate

The control specifications selected for the second channel include robust stability, moderate disturbance rejection at the output of the plant and reference tracking objectives, as presented in Equations 7.39 through 7.41, respectively.

- Stability specification:

$$\left| \frac{P(j\omega)G_2(j\omega)}{1 + P(j\omega)G_2(j\omega)} \right| \leq W_{s2} = 1.46 \ (GM = 4.53 \text{ dB}, PM = 40.05°)$$

$$\text{for } \omega = \begin{bmatrix} 0.001 & 0.005 & 0.01 & 0.05 & 0.1 & 0.2 & 0.5 & 0.7 \\ 1 & 2 & 3 & 5 & 10 & 50 & 100 & 500 & 1000 \end{bmatrix} \text{rad/s} \tag{7.39}$$

- Sensitivity or output disturbance rejection specification—see Figure 7.34a:

$$\left| \frac{1}{1 + P(j\omega)G_2(j\omega)} \right| \leq \delta_{op2}(\omega) = \left| \frac{10(j\omega)}{10(j\omega) + 1} \right|$$

$$\text{for } \omega = [0.001 \ \ 0.005 \ \ 0.01 \ \ 0.05 \ \ 0.1 \ \ 0.2 \ \ 0.5 \ \ 0.7 \ \ 1] \text{ rad/s} \tag{7.40}$$

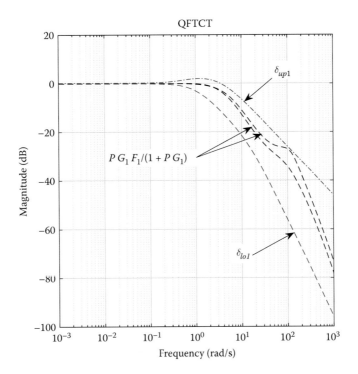

FIGURE 7.32
Design of prefilter $F_1(s)$ for $P(s)$ and $G_1(s)$.

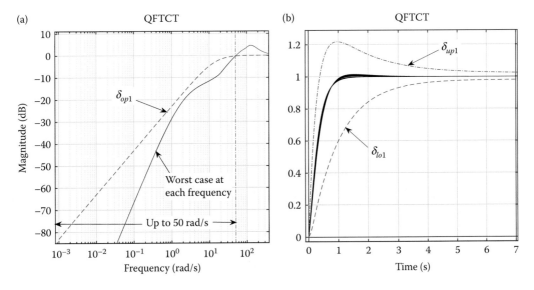

FIGURE 7.33
Analysis of controller $G_1(s)$ and prefilter $F_1(s)$ for $P(s)$. (a) Disturbance rejection at plant output: δ_{op1}. (b) Reference tracking: δ_{up1}, δ_{lo1}.

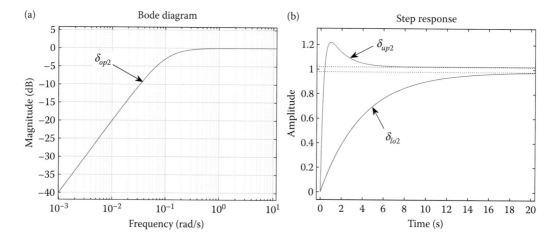

FIGURE 7.34
Control specifications for $P(s)$. (a) Disturbance rejection at the output of the plant: δ_{op2}. (b) Reference tracking: δ_{up2}, δ_{lo2}. Moderate channel.

- Reference tracking specification—see Figure 7.34b:

$$\delta_{lo2}(\omega) \leq \left| \frac{P(j\omega)G_2(j\omega)}{1 + P(j\omega)G_2(j\omega)} F_2(j\omega) \right| \leq \delta_{up2}(\omega) \tag{7.41}$$

where

$$\delta_{lo2}(\omega) = \left| \frac{0.98}{0.001(j\omega)^2 + 4(j\omega) + 1} \right| \text{ is the lower bound}$$

and

$$\delta_{up2}(\omega) = \left| \frac{2(j\omega) + 1.02}{0.4(j\omega)^2 + 1.7(j\omega) + 1} \right| \text{ is the upper bound}$$

for $\omega = [0.001\quad 0.005\quad 0.01\quad 0.05\quad 0.1\quad 0.2\quad 0.5\quad 0.7\quad 1]$ rad/s

The QFT bounds are calculated taking into account the $P(s)$ transfer function with the parameter uncertainty given by Equation 7.33, and the robust stability, output disturbance rejection and reference tracking specifications given by Equations 7.39 through 7.41, respectively—see Figure 7.34.

To meet the specifications, we also design PID with a low-pass filter for the $G_2(s)$ controller, as shown in Equation 7.42. The QFT bounds and the loop shaping of $L_{20}(s) = P_0(s)G_2(s)$ meeting all the specifications are shown in Figure 7.35.

$$G_2(s) = 0.0035 \frac{((1/0.002)s + 1)((1/3)s + 1)}{s((1/10)s + 1)} \tag{7.42}$$

The prefilter element $F_2(s)$ is also designed with the QFTCT, for the plant $P(s)$, and for the controller $G_2(s)$. Figure 7.36 shows the design. The expression for the prefilter is

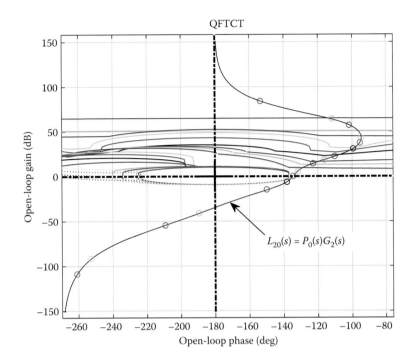

FIGURE 7.35
Loop-shaping of controller $G_2(s)$. $L_{20}(s) = P_0(s)G_2(s)$.

$$F(s) = F_2(s) = F_1(s) \tag{7.43}$$

Figure 7.37 shows how the control system (G_2, F_2) meets the performance specifications. Figure 7.37a is the frequency domain analysis of the disturbance rejection specification—Equation 7.40, and Figure 7.37b the time domain analysis of the reference tracking specification with 125 plants—Equation 7.41. The design is Nyquist stable.

The purpose of implementing two channels, $G_1(s)$ and $G_2(s)$, in the controller is to improve the system performance beyond the linear limitations. The element $G_1(s)$ is designed as an active/aggressive controller to work with small errors (see $N_1 = N_{01}$) and the element $G_2(s)$ as a moderate controller to work with large errors (see $N_2 = N_{02}$). Figure 7.38 compares both dynamics over the frequencies of interest.

Step C: Inner-Loop and Control Element A(s)

$A(s)$ is designed so that: (1) There is a common inner-loop for all the channels— see Figure 7.28; (2) $A(s)$ has the same number of integrators than $G_1(s)$ and $G_2(s)$; (3) $L_{10}(s) = P_0(s)G_1(s)$ dominates at low frequencies and $L_{E10}(s) = [L_{10}(s) - A(s)]/[1 + A(s)]$ dominates at high frequencies; and (4) $L_{20}(s) = P_0(s)G_2(s)$ dominates at low frequencies and $L_{E20}(s) = [L_{20}(s) - A(s)]/[1 + A(s)]$ dominates at high frequencies. $A(s)$ is shown in Equation 7.44. Figure 7.39a shows $G_1(s)$, $G_2(s)$, and $A(s)$. Figure 7.39b compares $L_{10}(s)$, $L_{E10}(s)$, $L_{20}(s)$, and $L_{E20}(s)$ using the nominal plant $P_0(s)$.

$$A(s) = \frac{k_a}{s} = \frac{4.5}{s} \tag{7.44}$$

Note that we have rearranged the control system of Figure 7.28 for each channel as a feedback connection of one linear transfer function $L_E(s)$ and one nonlinear

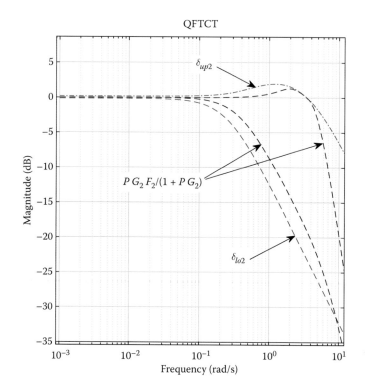

FIGURE 7.36
Design of prefilter $F_2(s)$ for $P(s)$ and $G_2(s)$.

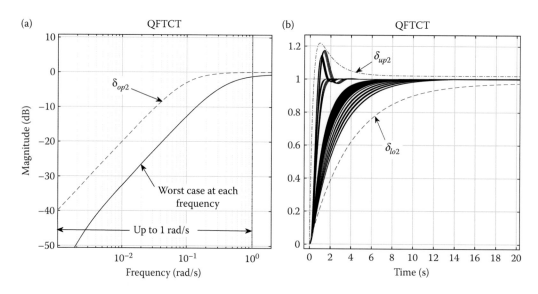

FIGURE 7.37
Analysis of controller $G_2(s)$ and prefilter $F_2(s)$ for $P(s)$. (a) Disturbance rejection at plant output: δ_{op2}. (b) Reference tracking: δ_{up2}, δ_{lo2}.

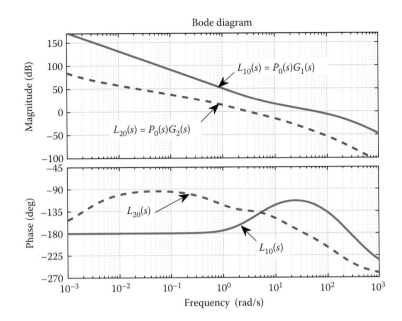

FIGURE 7.38
Channel comparison: $L_{10}(s) = P_0(s)G_1(s)$, $L_{20}(s) = P_0(s)G_2(s)$.

element $N_0(u,t)$, as it was shown in Figure 7.2. The equivalent linear transfer functions $L_{E1}(s)$ and $L_{E2}(s)$ follow Equation 7.4, considering a perfect nonlinear model $\hat{N}_0 = N_0$: $L_{E1}(s) = [L_1(s) - A(s)]/[1 + A(s)]$, $L_{E2}(s) = [L_2(s) - A(s)]/[1 + A(s)]$, being $L_1(s) = P(s)G_1(s)$ and $L_2(s) = P(s)G_2(s)$.

Now, according to the circle criterion *Case 2* (Section 7.2), the *sufficient condition* for stability of the closed-loop system in each channel, calculated independently, is that $L_E(j\omega)$ has no RHP poles and the Nyquist plot of $L_E(j\omega)$ does not enter or enclose the circle D defined by $(-1/k_1, -1/k_2)$, which in this problem is a vertical line at Re $= -1$ (circle of infinite radius, from $-\infty$ to -1). Figure 7.40a shows $L_{E1}(j\omega)$, $0 \le \omega \le \infty$, for 64 plants within the parametric uncertainty and the circle (vertical line), and Figure 7.40b shows $L_{E2}(j\omega)$, $0 \le \omega \le \infty$, also for 64 plants within the parametric uncertainty and the circle (vertical line).

As all the functions $L_{E2}(j\omega)$ lie at the right of the vertical line Re $= -1$ (outside the circle), the closed-loop system for the second channel is absolutely stable for all the cases within the parametric uncertainty.

In the first channel, however, some cases of $L_{E1}(s)$ cross the vertical line at Re $= -1$ (circle), and then violate the circle stability criterion—see Figure 7.40a. These cases correspond to the plants with $1 \le a < 4.5$. Note that the circle criterion gives only sufficient conditions. This means that if it is satisfied, the system is stable. On the contrary, if it is not satisfied (as in L_{E1} for cases with $1 \le a < 4.5$), we cannot conclude whether the system is stable or not.

The complete analysis of the stability, with all the nonlinearities involved, will be performed with the isoline method in the next step.

Step D: Isolines—Stability Analysis for the Complete System

The *DFs* for every nonlinearity, both in the controller and in the plant, are given by the expressions:

- $N_0 = N_{05}$, actuator saturation—Equations 7.25 and 7.26
- $\hat{N}_0 = N_0 = N_{05}$, controller actuator saturation model—Equations 7.25 and 7.26

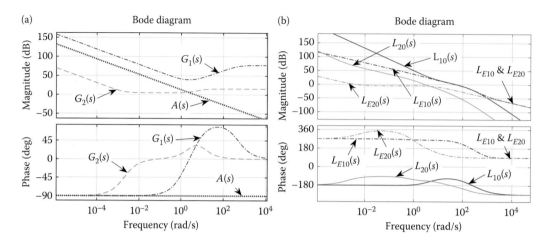

FIGURE 7.39
Bode diagrams. (a) $G_1(s)$, $G_2(s)$, $A(s)$. (b) $L_{10}(s)$, $L_{E10}(s)$, $L_{20}(s)$, $L_{E20}(s)$.

- $N_1 = N_{01}$, controller first channel nonlinearity—Equations 7.17 and 7.18
- $N_2 = N_{02}$, controller second channel nonlinearity—Equations 7.19 and 7.20

The isolines are calculated cascading the linear and nonlinear elements. They include the *DF* of each nonlinearity at their specific position in the loop, from the input (right of the expression) to the output (left of the expression). That is,

$$L = P(s)DF_{N0}\left(\frac{1}{1+A(s)DF_{\hat{N}0}}\right)[G_1(s)DF_{N1}+G_2(s)DF_{N2}] \qquad (7.45)$$

The algorithm that calculates the isolines for every amplitude E of the input and frequency ω is included in Appendix 7. Figure 7.41 represents in the NC the

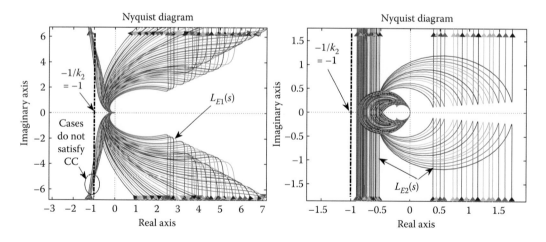

FIGURE 7.40
Circle criterion for 64 plants within the parametric uncertainty, $L_{E1}(s)$ and $L_{E2}(s)$, $N_0 =$ saturation, circle with $k_1 = 0$, $k_2 = 1$.

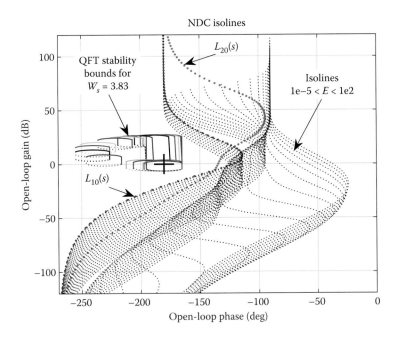

FIGURE 7.41
Isolines in the NC, for error E from 10^{-5} to 10^2, ω from 10^{-4} to 10^5 rad/s, and QFT stability bounds for $W_s = 3.83$ (or $PM = 15°$).

isolines L—Equation 7.45, for the input errors $10^{-5} \leq E \leq 10^2$ and for the frequencies $10^{-4} \leq \omega \leq 10^5$ rad/s. The figure is calculated for the nominal plant—Equation 7.33. It also shows the QFT stability bounds found for this nominal plant and $W_s = 3.83$ (or $PM = 15°$). As there is no RHP poles and the isolines do not enter into the QFT bounds, the *NDC* system is *stable* for all the cases within the model uncertainty. The method gives *sufficient conditions* for stability—see also Chapter 3.

Step E: Simulations and Discussion

Finally, Figures 7.42 and 7.43 show the results of the simulation of the closed-loop system presented in Figure 7.28, with the nominal plant $P(s) = P_0(s)$—Equation 7.33 and the actuator saturation, the QFT controllers $G_1(s)$ and $G_2(s)$—Equations 7.37 and 7.42, the prefilter $F(s)$—Equations 7.38 and 7.43, the function $A(s)$—Equation 7.44, and the three controller nonlinearities $\hat{N}_0$, N_1, and N_2.

The reference input starts at $r(t) = 0$ and at $t = 1$ s jumps to $r(t) = 1$ (as a unitary step). The disturbance starts at $d(t) = 0$ and at $t = 40$ s jumps to $d(t) = 0.25$ (as a step).

Figure 7.42 shows the reference input $r(t)$, the plant output $y_1(t)$ of only the first channel and without the nonlinearity (or $N_1 = 1$, $N_2 = 0$), the plant output $y_2(t)$ of only the second channel and without the nonlinearity (or $N_1 = 0$, $N_2 = 1$), and the plant output of the complete *NDC* $y_{NDC}(t)$. In all cases, the actuator saturation, controller anti-windup inner-loop, and prefilter are present. As we can see in the figure, the *NDC* improves significantly the performance in both, the reference tracking problem—Figure 7.42a, and the disturbance rejection problem—Figure 7.42b.

Note that the plant used in the simulation is the nominal plant $P_0(s)$, with $k = 610$, $a = 1$, $b = 150$. This case did not satisfy the circle criterion for $L_{E1}(s)$—see Figure 7.40a, but met the isoline stability criterion—see Figure 7.41.

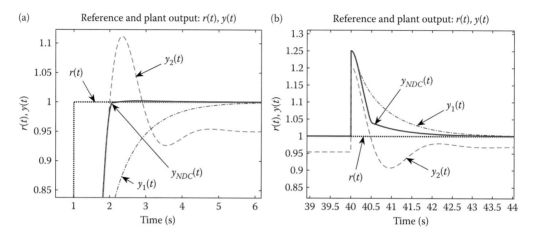

FIGURE 7.42
NDC results with nominal plant $P_0(s)$. (a) Reference tracking. (b) Disturbance rejection.

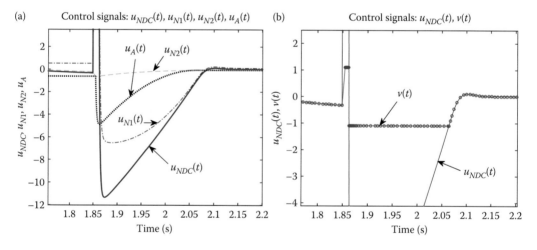

FIGURE 7.43
NDC results with nominal plant $P_0(s)$. (a) $u_{NDC}(t) = u_{N1}(t) + u_{N2}(t) + u_A(t)$. (b) $u_{NDC}(t)$ and $v(t)$.

Finally, Figure 7.43 shows the control signals for the case of the complete *NDC* presented in Figure 7.42, including the first and second channels $u_{N1}(t)$ and $u_{N2}(t)$, the inner-loop anti-windup signal $u_A(t)$, the complete *NDC* control signal $u_{NDC}(t)$, which is the addition $u_{NDC}(t) = u_{N1}(t) + u_{N2}(t) + u_A(t)$, and the signal after the saturation of the actuator $v(t)$.

7.6 Summary

This chapter has introduced the *nonlinear dynamic control* (NDC) methodology. It is a practical control design method based on QFT that can deal with one or several nonlinearities in system, either in the plant and/or deliberately introduced in the controller to go beyond the linear limitations and achieve a very high performance.

7.7 Practice

The list shown below summarizes the collection of problems and cases included in this book that apply the control methodologies introduced in this chapter.

- Example 7.1. Section 7.4. System with actuator saturation. PID with anti-windup.
- Example 7.2. Section 7.5. Field-controlled DC motor with saturation. PIDs within an NDC structure.
- Appendix 7. MATLAB code for Examples 7.1 and 7.2.
- Case study CS4. Radio telescope servo system

8

Multi-Input Multi-Output Systems: Analysis and Control

8.1 Introduction

Control of multi-input multi-output MIMO systems with model uncertainty is often a difficult problem. Contrary to the single-input single-output SISO systems we have studied so far, the structure of a MIMO plant is composed of some internal channels that connect the inputs and outputs of the plant in more than one way. This fact introduces challenging interactions among the inputs and outputs. To illustrate this problem, let us consider the conventional counter-flow heat exchanger shown in Figure 8.1—Example 8.1.

Heat exchangers are key components of many energy, chemical, and biological systems. Food pasteurization processes are a practical example of this case, where the main objective is to increase the temperature of the product (milk, juice, etc.) to a certain value to eliminate the potential bacteria, while maintaining the temperature of the secondary fluid at a given level for efficiency purposes.

With this objective, the main components of the pasteurization process shown in Figure 8.1 are a counter-flow heat exchanger, a tank with a secondary fluid, an internal electrical heater, an open hydraulic circuit that brings the product into the heat exchanger, and a close hydraulic circuit controlled by a variable peristaltic pump that provides the hot fluid to the heat exchanger. The sensors of the system are T_1 and T_2, which are the temperatures (plant outputs) to be controlled. At the same time, the two actuators of the system (plant inputs) are the heat flux Q produced by the electrical resistor inside the tank, and the volume flow rate N generated by the peristaltic pump in the hot fluid line.

To simplify the case, the product line is held at a constant volume flow rate. As we can see in Figure 8.1a, a change in the pump flow rate N is able to change both temperatures, T_1 and T_2. Specifically, a decrease of N decreases T_1 and T_2, and an increase of N increases T_1 and T_2. This is described by the transfer functions $p_{11}(s)$ and $p_{21}(s)$, respectively, shown in Figure 8.1b.

Similarly, a variation of the heat flux Q is able to change both temperatures, T_1 and T_2. Specifically, a decrease of Q decreases T_1 and T_2, and an increase of Q increases T_1 and T_2. This is also described by the transfer functions $p_{12}(s)$ and $p_{22}(s)$, respectively, also shown in Figure 8.1b. In other words, we can write

$$\begin{bmatrix} T_1(s) \\ T_2(s) \end{bmatrix} = \begin{bmatrix} p_{11}(s) & p_{12}(s) \\ p_{21}(s) & p_{22}(s) \end{bmatrix} \begin{bmatrix} N(s) \\ Q(s) \end{bmatrix} \tag{8.1}$$

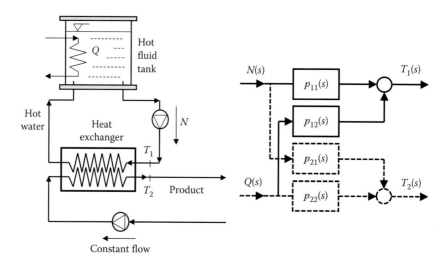

FIGURE 8.1
Counter-flow heat exchanger: a 2 × 2 MIMO system—Example 8.1.

In general, naming $y_1(s) = T_1(s)$, $y_2(s) = T_2(s)$, $u_1(s) = N(s)$, and $u_2(s) = Q(s)$, we have

$$\begin{bmatrix} y_1(s) \\ y_2(s) \end{bmatrix} = \begin{bmatrix} p_{11}(s) & p_{12}(s) \\ p_{21}(s) & p_{22}(s) \end{bmatrix} \begin{bmatrix} u_1(s) \\ u_2(s) \end{bmatrix}, \quad \text{or} \quad y(s) = P(s)u(s) \tag{8.2}$$

For the syntax, note that the order of the subscripts in a transfer function $p_{ab}(s)$ is read from right to left, "b" being the input channel and "a" the output channel. That is to say: $y_a(s) = p_{ab}(s)\, u_b(s)$.

A quick analysis of this plant model brings new important questions beyond the SISO systems. For instance, in order to control the temperatures T_1 and T_2 we have to pick the most appropriate actuator for each one, to close the two control loops. In other words, to control T_1, should we select the flow rate N or the heat flux Q? Similarly, to control T_2, should we select the flow rate N or the heat flux Q? This is the so-called *pairing problem*. A solution for this pairing problem is presented in Example 8.1 and Section 8.3.2. For this heat exchanger, we selected N to control T_1 and Q to control T_2, as shown in Figures 8.2 and 8.3.

Once this pairing problem is solved, another new important question is to quantify the coupling or interaction between the loops. In Figure 8.2 this interaction is related to the transfer function $p_{12}(s)$—from $Q(s)$ to $T_1(s)$, and to the transfer function $p_{21}(s)$—from $N(s)$ to $T_2(s)$. This new question is also addressed in Section 8.3.2.

Moving forward, once we selected the pairing and quantified the interaction among the loops, and based on this interaction, we have to decide the most appropriate controller structure. Considering a controller expression as the matrix shown in Equation 8.3,

$$\begin{bmatrix} N(s) \\ Q(s) \end{bmatrix} = \begin{bmatrix} u_1(s) \\ u_2(s) \end{bmatrix} = \begin{bmatrix} g_{11}(s) & g_{12}(s) \\ g_{21}(s) & g_{22}(s) \end{bmatrix} \begin{bmatrix} e_1(s) \\ e_2(s) \end{bmatrix}, \quad \text{or} \quad u(s) = G(s)e(s) \tag{8.3}$$

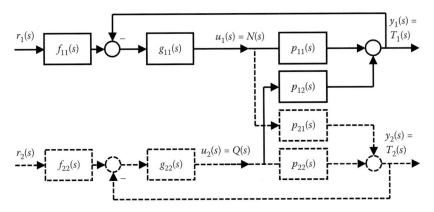

FIGURE 8.2
Diagonal controller for the 2 × 2 heat exchanger MIMO system.

we can select between a diagonal controller—see Figure 8.2, or a non-diagonal or full-matrix controller—see Figure 8.3, $e_1(s)$ and $e_2(s)$ being the errors: $e_1(s) = f_{11}(s) \, r_1(s) - T_1(s)$, and $r_2(s) = f_{22}(s) \, r_2(s) - T_2(s)$.

Sections 8.5 and 8.6 present two methodologies (called *Method 1* and *Method 2*, respectively) to design non-diagonal MIMO QFT controllers.

Other new important problems in MIMO systems are the poles and zeros of the MIMO plant, the stability of the MIMO plant, potential non-minimum phase transmission zeros, and the directionality of inputs and outputs of the MIMO plant. Section 8.3 also discusses these new problems.

EXAMPLE 8.1: A HEAT EXCHANGER

To illustrate the discussion, consider again the counter-flow heat exchanger and pasteurization process showed in Figure 8.1 and Equations 8.1 and 8.2. As presented, the plant inputs (actuators) are $u_1(s) = N(s)$ and $u_2(s) = Q(s)$, and the plant outputs (sensors) $y_1(s) = T_1(s)$ and $y_2(s) = T_2(s)$. The following subsections describe (a) the plant definition, (b) the input–output pairing, (c) the control specifications, (d) the design of the independent controller g_{11}, (e) the design of the independent controller g_{22}, and (f) the MIMO plant simulation.

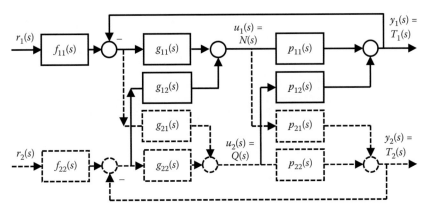

FIGURE 8.3
Non-diagonal controller for the 2 × 2 heat exchanger MIMO system.

1. *Plant Definition.* The elements of the transfer matrix for the heat exchanger—see Equations 8.1 and 8.2—are described by the following first-order models:

$$p_{11}(s) = \frac{k_{11}}{\tau_{11}s+1}; \quad p_{12}(s) = \frac{k_{12}}{\tau_{12}s+1}; \quad p_{21}(s) = \frac{k_{21}}{\tau_{21}s+1}; \quad p_{22}(s) = \frac{k_{22}}{\tau_{22}s+1} \tag{8.4}$$

with the parametric uncertainty

$$k_{11} \in [6.3,7.7], \ \tau_{11} \in [3.6,4.4], \ k_{12} \in [8.9,9.1], \ \tau_{12} \in [4.9,5.1]$$
$$k_{21} \in [2.9,3.1], \ \tau_{21} \in [7.9,8.1], \ k_{22} \in [4.5,5.5], \ \tau_{22} \in [1.8,2.2]$$

and the nominal plant: $k_{11} = 7$, $\tau_{11} = 4$, $k_{12} = 9$, $\tau_{12} = 5$, $k_{21} = 3$, $\tau_{21} = 8$, $k_{22} = 5$, $\tau_{22} = 2$.

2. *Input–Output Pairing.* The relative gain analysis (RGA) of the nominal plant presented in Equations 8.2 and 8.4—see Section 8.3.2 for details—gives the following Λ matrix:

$$\Lambda = \boldsymbol{P}_0 \otimes (\boldsymbol{P}_0^{-1})^T = \begin{bmatrix} 7 & 9 \\ 3 & 5 \end{bmatrix} \otimes \left(\begin{bmatrix} 7 & 9 \\ 3 & 5 \end{bmatrix}^{-1} \right)^T = \begin{bmatrix} 4.3750 & -3.3750 \\ -3.3750 & 4.3750 \end{bmatrix} \begin{matrix} y_1 \\ y_2 \end{matrix} \tag{8.5}$$

$$\begin{matrix} u_1 \qquad u_2 \end{matrix}$$

$\boldsymbol{P}_0$ being the nominal plant at $s = 0$. Based on these results, we pick the elements $\lambda_{11} = 4.3750$ and $\lambda_{22} = 4.3750$, which means to pair the temperature $T_1(s)$ with the pump flow rate $N(s)$, or $[y_1(s) - u_1(s)]$, and the temperature $T_2(s)$ with the heater $Q(s)$, or $[y_2(s) - u_2(s)]$. Figure 8.2 shows the loops based on this selection.

3. *Control Specifications.* The first control loop includes a compensator $g_{11}(s)$ to regulate the actuator $u_1(s) = N(s)$ and control the output $y_1(s) = T_1(s)$ of the plant $p_{11}(s)$. The second control loop includes a compensator $g_{22}(s)$ to regulate the actuator $u_2(s) = Q(s)$ and control the output $y_2(s) = T_2(s)$ of the plant $p_{22}(s)$—see Figure 8.2. The control specifications for both loops are the same, including stability, output disturbance rejection, and reference tracking objectives, as shown below.

a. Stability specification:

$$\left| \frac{p_{ii}(j\omega)g_{ii}(j\omega)}{1+p_{ii}(j\omega)g_{ii}(j\omega)} \right| \leq W_s = 1.01(GM = 5.98 \text{ dB}, PM = 59.3°), i = 1,2 \tag{8.6}$$

for $\omega = [0.001 \ 0.005 \ 0.01 \ 0.05 \ 0.1 \ 0.5 \ 1 \ 5 \ 10 \ 50 \ 100 \ 500 \ 1000] \text{ rad/s}$

b. Sensitivity or output disturbance rejection specification—See Figure 8.4a:

$$\left| \frac{1}{1+p_{ii}(j\omega)g_{ii}(j\omega)} \right| \leq \delta_{op}(\omega) = \left| \frac{(j\omega)}{(j\omega)+1} \right|, i = 1,2 \tag{8.7}$$

for $\omega = [0.001 \ 0.005 \ 0.01 \ 0.05 \ 0.1 \ 0.5 \ 1 \ 5 \ 10] \text{ rad/s}$

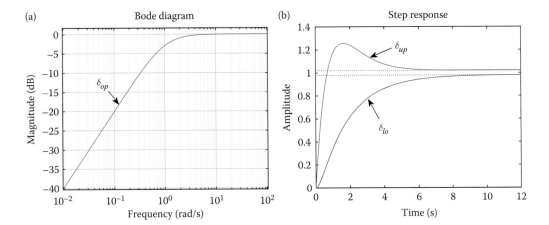

FIGURE 8.4
Control specifications for $p_{11}(s)$ and $p_{22}(s)$. (a) Disturbance rejection at the output of the plant: δ_{op}. (b) Reference tracking: δ_{up}, δ_{lo}.

    c.   Reference tracking specification—See Figure 8.4b:

$$\delta_{lo}(\omega) \leq \left| \frac{p_{ii}(j\omega)g_{ii}(j\omega)}{1 + p_{ii}(j\omega)g_{ii}(j\omega)} f_{ii}(j\omega) \right| \leq \delta_{up}(\omega), i = 1, 2 \tag{8.8}$$

where,

$$\delta_{lo}(\omega) = \left| \frac{0.98}{0.4(j\omega)^2 + 2(j\omega) + 1} \right| \text{ is the lower bound}$$

and,

$$\delta_{up}(\omega) = \left| \frac{2(j\omega) + 1.02}{0.8(j\omega)^2 + 1.7(j\omega) + 1} \right| \text{ the upper bound}$$

for $\omega = [0.001\ 0.005\ 0.01\ 0.05\ 0.1\ 0.5\ 1\ 5\ 10]$ rad/s

4. *Controller $g_{11}(s)$—Independent.* The QFT bounds are calculated taking into account the $p_{11}(s)$ transfer function with the parametric uncertainty given by Equation 8.4, and the stability, output disturbance rejection and reference tracking specifications given by Equations 8.6 through 8.8, respectively—see Figure 8.4. To meet the specifications, we select a PI compensator with a filter for the $g_{11}(s)$ controller and a unitary prefilter $f_{11}(s)$, such that,

$$g_{11}(s) = 0.18 \frac{((1/0.2)s + 1)}{s((1/390)s + 1)}; \quad f_{11}(s) = 1 \tag{8.9}$$

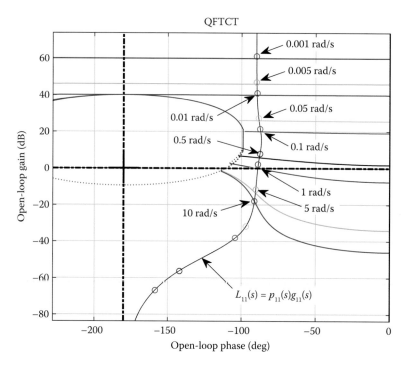

FIGURE 8.5
Loop-shaping of controller $g_{11}(s)$. $L_{11}(s) = p_{11}(s)\, g_{11}(s)$.

The QFT bounds and the loop shaping of $L_{11}(s) = p_{11}(s)g_{11}(s)$ are shown in Figure 8.5. The solution includes the QFT controller $g_{11}(s)$ presented in Equation 8.9. As Figure 8.6 shows, the control system meets all the performance specifications: see Figure 8.6a for the frequency-domain analysis of the disturbance rejection specification—Equation 8.7, and Figure 8.6b for the time-domain analysis of the reference tracking specification with 125 plants—Equation 8.8.

5. *Controller* $g_{22}(s)$—*Independent.* For the second control loop, the QFT bounds are calculated taking into account the $p_{22}(s)$ transfer function with the parametric uncertainty given by Equation 8.4, and the stability, output disturbance rejection and reference tracking specifications given by Equations 8.6 through 8.8, respectively—see Figure 8.4. To meet the specifications, we also select a PI compensator with a filter for the $g_{22}(s)$ controller and a unitary prefilter $f_{22}(s)$, such that

$$g_{22}(s) = 0.25 \frac{((1/35)s + 1)}{s((1/190)s + 1)}; \quad f_{22}(s) = 1 \tag{8.10}$$

The QFT bounds and the loop shaping of $L_{22}(s) = p_{22}(s)g_{22}(s)$ are shown in Figure 8.7. The solution includes the QFT controller $g_{22}(s)$ presented in Equation 8.10. As Figure 8.8 shows, the control system meets all the performance specifications: see Figure 8.8a for the frequency-domain analysis of the disturbance rejection specification—Equation 8.7, and Figure 8.8b for the time-domain analysis of the reference tracking specification with 125 plants—Equation 8.8.

6. *MIMO Plant Simulation and Discussion.* Although both controllers $g_{11}(s)$ and $g_{22}(s)$ achieve an excellent performance controlling each plant $p_{11}(s)$ and $p_{22}(s)$

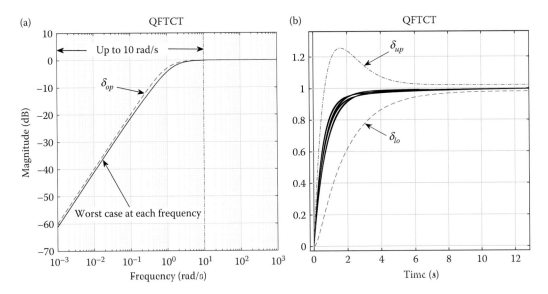

FIGURE 8.6
Analysis of controller $g_{11}(s)$ and prefilter $f_{11}(s)$ for $p_{11}(s)$. (a) Disturbance rejection at plant output: δ_{op}. (b) Reference tracking: δ_{up}, δ_{lo}.

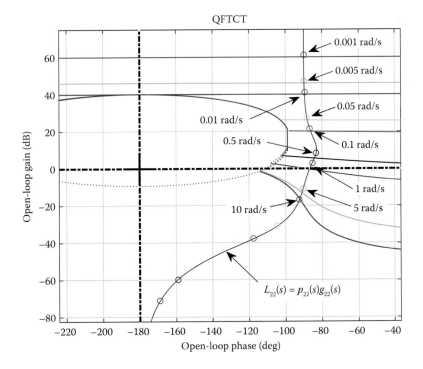

FIGURE 8.7
Loop-shaping of controller $g_{22}(s)$. $L_{22}(s) = p_{22}(s)\, g_{22}(s)$.

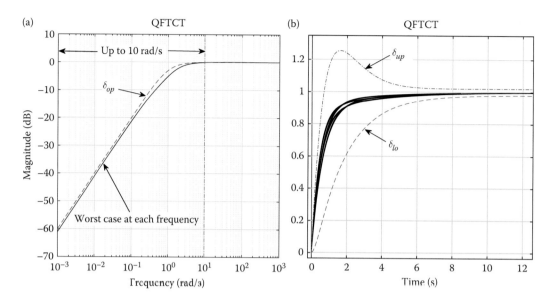

FIGURE 8.8

Analysis of controller $g_{22}(s)$ and prefilter $f_{22}(s)$ for $p_{22}(s)$. (a) Disturbance rejection at plant output: δ_{op}. (b) Reference tracking: δ_{up}, δ_{lo}.

independently—see Figures 8.6 and 8.8, when we implement them in the actual heat exchanger (the complete MIMO system) the results are quite different. Figure 8.9 shows the simulation this complete case, with the two loops of Figure 8.2, with the nominal plant defined in Equation 8.4, and with the independent controllers designed in Equations 8.9 and 8.10. The first reference starts at $r_1 = 0$ and changes to $r_1 = 1$ at $t = 20$ s. The second reference starts at $r_2 = 0$ and changes to $r_2 = 1$ at $t = 170$ s. Even though the reference tracking is still excellent, there is a disturbance at the second loop at $t = 20$ s due to the coupling with the first loop. Additionally, there is another important disturbance at the first loop at $t = 170$ s due to the coupling with the second loop. The following sections of this chapter address this problem and introduce several methodologies to design robust QFT controllers able to reduce the coupling among the loops of the MIMO system.

8.2 Formulation for $n \times n$ Systems

In general, a MIMO system is normally composed of a set of internal subsystems $p_{ij}(s)$ that connect some of the inputs (u_j, $j = 1, 2,...,n$) with more than one output (y_i, $i = 1, 2,...,n$), or some of the outputs with more than one input of the plant—see Figure 8.10.

The mathematical description of a linear-time-invariant (LTI) MIMO system may be expressed as a set of ordinary differential equations, as a state space representation, or as a transfer matrix description (P matrix).

In the first case, the plant model consists of n coupled *LTI* differential equations, with n inputs (u_j, $j = 1, 2, ...,n$) and n outputs (y_i, $i = 1, 2, ...,n$), so that

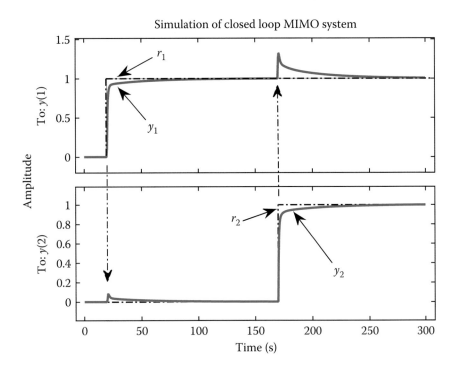

FIGURE 8.9
Closed-loop MIMO system. Figure 8.2. Inputs: $r_1(s)$, $r_2(s)$. Outputs: $y_1(s)$, $y_2(s)$. $P = [p_{11}(s)\ p_{12}(s);\ p_{21}(s)\ p_{22}(s)]$, Equation 8.4, nominal plant. $G = [g_{11}(s)\ 0;\ 0\ g_{22}(s)]$, $F = [f_{11}(s)\ 0;\ 0\ f_{22}(s)]$, Equations 8.9 and 8.10. Independent.

$$\alpha_{11}(s)y_1(s) + \cdots + \alpha_{1n}(s)y_n(s) = \beta_{11}(s)u_1(s) + \cdots + \beta_{1n}(s)u_n(s)$$
$$\alpha_{21}(s)y_1(s) + \cdots + \alpha_{2n}(s)y_n(s) = \beta_{21}(s)u_1(s) + \cdots + \beta_{2n}(s)u_n(s)$$
$$\cdots \qquad\qquad (8.11)$$
$$\alpha_{n1}(s)y_1(s) + \cdots + \alpha_{nn}(s)y_n(s) = \beta_{n1}(s)u_1(s) + \cdots + \beta_{nn}(s)u_n(s)$$

where $\alpha_{ij}(s)$ and $\beta_{ij}(s)$ are polynomials in s, $y_1(s)$ to $y_n(s)$ are the outputs, and $u_1(s)$ to $u_n(s)$ are the inputs. Grouping in two matrices, $\boldsymbol{\alpha}(s)$ and $\boldsymbol{\beta}(s)$,

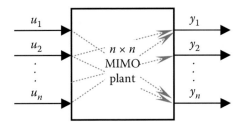

FIGURE 8.10
An $n \times n$ MIMO system.

$$
\alpha(s) = \begin{vmatrix} \alpha_{11}(s) & \alpha_{12}(s) & \cdots & \alpha_{1n}(s) \\ \alpha_{21}(s) & \alpha_{22}(s) & \cdots & \alpha_{2n}(s) \\ \vdots & \vdots & & \vdots \\ \alpha_{n1}(s) & \alpha_{n2}(s) & \cdots & \alpha_{nn}(s) \end{vmatrix} \quad \beta(s) = \begin{vmatrix} \beta_{11}(s) & \beta_{12}(s) & \cdots & \beta_{1n}(s) \\ \beta_{21}(s) & \beta_{22}(s) & \cdots & \beta_{2n}(s) \\ \vdots & \vdots & & \vdots \\ \beta_{n1}(s) & \beta_{n2}(s) & \cdots & \beta_{nn}(s) \end{vmatrix} \tag{8.12}
$$

the set of differential equations given by Equation 8.11 can be expressed as,

$$
\underset{n \times n}{\alpha(s)} \begin{bmatrix} y_1(s) \\ y_2(s) \\ \vdots \\ y_n(s) \end{bmatrix}_{n \times 1} = \underset{n \times n}{\beta(s)} \begin{bmatrix} u_1(s) \\ u_2(s) \\ \vdots \\ u_n(s) \end{bmatrix}_{n \times 1}. \tag{8.13}
$$

Now, pre-multiplying by $\alpha(s)^{-1}$, with $\alpha(s)$ nonsingular, Equation 8.13 can then be written as

$$
y(s) = P(s)u(s) \tag{8.14}
$$

$P(s)$ being the $n \times n$ plant transfer function matrix (TFM, third representation option).

$$
P(s) = \alpha(s)^{-1}\beta(s) = \begin{vmatrix} p_{11}(s) & p_{12}(s) & \cdots & p_{1n}(s) \\ p_{21}(s) & p_{22}(s) & \cdots & p_{2n}(s) \\ \vdots & \vdots & \ddots & \vdots \\ p_{n1}(s) & p_{n2}(s) & \cdots & p_{nn}(s) \end{vmatrix} \tag{8.15}
$$

In the second representation option, the state-space case, the system can be written as

$$
\begin{aligned}
\dot{x}(t) &= A\,x(t) + B\,u(t) \\
y(t) &= C\,x(t) + D\,u(t)
\end{aligned} \tag{8.16}
$$

where t denotes time, A, B, C, and D are constant matrices of $m \times m$, $m \times n$, $n \times m$, and $n \times n$ dimension, respectively, and where x is the $m \times 1$ vector of state variables, y the $n \times 1$ vector of outputs, and u the $n \times 1$ vector of inputs. Once translated into the Laplace domain, the $n \times n$ plant TFM $P(s)$ (third representation option) can be again evaluated as

$$
y(s) = P(s)u(s) \tag{8.17}
$$

with

$$
P(s) = C[sI - A]^{-1}B + D \tag{8.18}
$$

being I the $m \times m$ identity matrix.

The plant matrix $P(s) = [p_{ij}(s)]$ is a member of the set $P = \{P(s)\}$ of possible plant matrices within the model uncertainty. In addition, the controller $G(s)$ and the prefilter $F(s)$ are now matrices as well, as expressed in Equation 8.19.

$$G(s) = \begin{bmatrix} g_{11}(s) & g_{12}(s) & \cdots & g_{1n}(s) \\ g_{21}(s) & g_{22}(s) & \cdots & g_{2n}(s) \\ \vdots & \vdots & \ddots & \vdots \\ g_{n1}(s) & g_{n2}(s) & \cdots & g_{nn}(s) \end{bmatrix}; \quad F(s) = \begin{bmatrix} f_{11}(s) & f_{12}(s) & \cdots & f_{1n}(s) \\ f_{21}(s) & f_{22}(s) & \cdots & f_{2n}(s) \\ \vdots & \vdots & \ddots & \vdots \\ f_{n1}(s) & f_{n2}(s) & \cdots & f_{nn}(s) \end{bmatrix} \quad (8.19)$$

The closed-loop transfer function of the control system, from the vector of inputs $r(s)$ to the vector of outputs $y(s)$, is

$$y(s) = [I + P(s)G(s)]^{-1} P(s)G(s)F(s)r(s) \quad (8.20)$$

where the closed-loop system control ratio matrix $T(s)$ is

$$T(s) = [I + P(s)G(s)]^{-1} P(s)G(s)F(s) = \begin{bmatrix} t_{11}(s) & \cdots & t_{1n}(s) \\ \vdots & \ddots & \vdots \\ t_{n1}(s) & \cdots & t_{nn}(s) \end{bmatrix} \quad (8.21)$$

and where each element $t_{ij}(s) = y_i(s)/r_j(s)$ relates the *j*th *input* to the *i*th *output*.

To appreciate the difficulty of the design problem, Equation 8.22 shows the complex expression of $t_{11}(s)$ for just a 3×3 plant $P(s)$, a pure diagonal controller $G(s)$, and a pure diagonal prefilter $F(s)$. The Laplace variable s is omitted in the equation for simplicity.

$$\begin{aligned}
t_{11} = \{ & (p_{11}f_{11}g_{11})(1 + p_{22}g_{22})(1 + p_{33}g_{33} - p_{23}p_{32}g_{22}g_{33}) \\
& -(p_{21}f_{11}g_{11})[p_{12}g_{22}(1 + p_{33}g_{33}) - p_{32}p_{13}g_{22}g_{33}] \\
& +(p_{31}f_{11}g_{11})[p_{23}p_{12}g_{22}g_{33} - (1 + p_{22}g_{22})p_{13}g_{33}]\}/ \\
& \{(1 + p_{11}g_{11})[(1 + p_{22}g_{22})(1 + p_{33}g_{33}) - p_{23}p_{32}g_{22}g_{33}] \\
& -p_{21}g_{11}[p_{12}g_{22}(1 + p_{33}g_{33}) - p_{32}p_{13}g_{22}g_{33}] \\
& +p_{31}g_{11}[p_{12}p_{23}g_{22}g_{33} - p_{13}g_{33}(1 + p_{22}g_{22})]\}
\end{aligned} \quad (8.22)$$

For the 3×3 system, there are $n^2 = 9$ transfer functions t_{ij} like the one shown in Equation 8.22. Additionally, the expressions may have parametric uncertainty in the nine plant transfer functions $p_{ij}(s)$. The robust control design objective is a system which behaves as desired for the entire range of uncertainty. This requires finding three $f_{ij}(s)$ and three to nine $g_{ij}(s)$ such that each $t_{ij}(s)$ stays within its acceptable region defined by the specifications no matter how the $p_{ij}(s)$ may vary. Clearly, this is a difficult problem. Even the stability problem alone, ensuring that the characteristic polynomial (the denominator of Equation 8.22) has no factors in the right-half plane (RHP) for all possible $p_{ij}(s)$, is difficult. However, as we will see in the next sections, the QFT design methodology systematizes and simplifies the manner of achieving a satisfactory design.

8.3 MIMO Systems—Description and Characteristics

Two of the main characteristics that define a MIMO system are the input and output direc-tionality—different vectors to actuate u and to measure y, and the coupling among control loops—each input u_i can affect some outputs y_i, and each output can be affected by several inputs. This problem, which is known as interaction or coupling, makes the control system design less intuitive since any change in one loop interferes with the rest of the plant loops.

The systems considered from now on are supposed to be linearizable, at least within a range of operating conditions, as with most of physical real problems. As mentioned before, we model the MIMO system as an $n \times n$ matrix of transfer functions $P(s) = [p_{ij}(s)]$, also called the plant TFM, which relates the n input variables [manipulated variables, $u_j(s)$ with $j = 1,..., n$] with the n output variables [controlled variables, $y_i(s)$ with $i = 1,..., n$], so that $[y_i(s)] = P(s) [u_j(s)]$.

In general, the MIMO TFM $P(s)$ can be rectangular. However, most of the related lit-erature deals with square systems, that is, with the same number of inputs and outputs. If it is not the case for the plant under study, there exist different procedures that can be followed, such as using weighting matrices which reduce the system to a square effective plant matrix,[6] leaving some outputs (inputs) uncontrolled (not used), or looking for inde-pendent extra inputs or outputs, depending on which one is in excess.[252]

Multivariable systems have aroused great interest within the control community and many design techniques have been developed. This is not only because of their mathematical and computational challenge (derived from the matrix representation), but also due to inherent features that do not appear in SISO systems. The particular nature of MIMO systems poses additional difficulties to control design such as the above mentioned input/output directionality and coupling, the potential transmission RHP zeros, system closed-loop stability, etc.; all with the intrinsic uncertainty of real-world applications.

8.3.1 Loop Coupling and Controller Structure

One of the most distinctive aspects of MIMO plants is the existence of coupling among the control loops. Thus, one input (manipulated variable) can affect various outputs (con-trolled variables), and one output can be affected by several inputs. Consequently, when we apply a control signal to one of the plant inputs, more than one plant output is affected. This fact makes hard to predict the control action that is simultaneously needed in order to get the desired performance at all the outputs of the plant.

The first and easiest way that comes to mind when dealing with a MIMO system is to reduce it to a set of SISO problems, ignoring the system interactions. This is known as *decentralized control*.[260] In this case, each input is responsible for only one output and the resulting controller $G(s)$ is *diagonal*. Finding a suitable input–output pairing becomes there-fore essential for decentralized control. However, this approach is only valid in systems where the coupling among loops is not significant, which unfortunately is not the case for many real applications. In other approaches the goal is to remove, or at least greatly reduce, the effects of the interaction before performing a decentralized control of the somehow decoupled plant, as if they are independent input–output pairs.

In any case, it is necessary to quantify the amount of coupling present in the system. Many of the MIMO design techniques, particularly the sequential ones, strongly depend on the correct selection and pairing of inputs and outputs at the beginning of the design procedure. Determining the controller structure is also crucial. This means deciding whether the multivariable system can be divided into several SISO or smaller MIMO subsystems, and establishing the off-diagonal compensators needed if a populated matrix controller is to be designed, avoiding non-required extra controllers. This issue becomes extremely complex in the presence of large coupling and has generated great interest within the control community, as shown by the numerous related works as References 246, 294, 297, 298, 299, 305, 309, and 311. Nevertheless, too often only the extreme controller structures (the fully centralized [fully populated matrix] and the fully decentralized [set of SISO loops]) are discussed.[302]

8.3.2 Interaction Analysis

An extensive amount of work on the way of quantifying the system interaction can be found in the literature.[245,260] One of the most popular techniques is the *Relative Gain Analysis*, or *Relative Gain Array* (RGA), defined by Bristol as a matrix of relative gains Λ based on the steady-state gains of the plant:[265]

$$\Lambda = \begin{bmatrix} \lambda_{11} & \lambda_{12} & \cdots & \lambda_{1n} \\ \lambda_{21} & \lambda_{22} & \cdots & \lambda_{2n} \\ \vdots & \vdots & \cdots & \vdots \\ \lambda_{n1} & \lambda_{n2} & \cdots & \lambda_{nn} \end{bmatrix} \tag{8.23}$$

The elements λ_{ij} that constitute this matrix are dimensionless and represent the relation between two gains of the system:

$$\lambda_{ij} = \frac{K_{OFF}}{K_{ON}} \tag{8.24}$$

where K_{OFF} is the open-loop gain between the input j and the output i when the rest of loops are open, while K_{ON} is the open-loop gain between the same input j and output i when the remaining loops are working in automatic mode, that is, they are closed.

A practical way of computing the RGA is through the following matrix expression:

$$\Lambda = P_0 \otimes (P_0^{-1})^T \tag{8.25}$$

where P_0 is the $n \times n$ matrix that represents the steady-state ($\omega = 0$) model. Its elements are determined by applying the final value theorem to the transfer functions $P(s)$ under a unitary step input ($1/s$):

$$P_0 = P(s = 0) = \lim_{s \to 0} s P(s)\left(\frac{1}{s}\right) = \lim_{s \to 0} P(s) \tag{8.26}$$

The operator [$\otimes$] denotes element-by-element multiplication (Hadamard or Schur product). The RGA provides a scaling independent measure of the coupling among loops, and also useful information on how to select the pairing of variables.[239] Its elements λ_{ij} give information about the interaction among the control loops, so that:

- $\lambda_{ij} = 1 \Rightarrow$ The closure of the rest of loops does not change the influence of the input j on the output i. Hence the ij loop is decoupled from the rest of the system and can be treated as a SISO subsystem.

- $\lambda_{ij} = 0 \Rightarrow$ There is no influence of the manipulated variable j over the control variable i.

- $0 < \lambda_{ij} < 1 \Rightarrow$ When the rest of loops are closed, the gain between the input j and the output i increases, that is, $K_{ON} > K_{OFF}$.

- $\lambda_{ij} < 0 \Rightarrow$ At the closure of the remaining loops ($\neq ij$), the gain of the ij system changes its sign. This is a very sensitive situation, where the operation of the other loops changes the sign of the ij loop. In this case, if we do the pairing ij, we say that the system does not have integrity.

- $\lambda_{ij} > 1 \Rightarrow$ When the rest of loops are closed, the gain between the input j and the output i decreases, that is, $K_{OFF} > K_{ON}$.

- $\lambda_{ij} > 10 \Rightarrow$ In this case, the ij loop is very sensitive to modeling errors and to small variations in the gain. Pairings of variables with large RGA values are undesirable.

According to the meaning of the RGA elements outlined above, it is desired to pair variables (u_j with y_i) so that λ_{ij} is *positive and close to one*. This means that the gain from the input u_j to output y_i is not very much affected by the other loops.

On the other hand, a pairing corresponding to $0 < \lambda_{ij} < 1$ means that the other loops reinforce the gain of the ij loop. Also, a pairing corresponding to $\lambda_{ij} > 1$ means that the other loops reduce the gain of the ij loop. Finally, negative values of λ_{ij} are undesirable because this means that the steady-state gain in the ij loop changes the sign when the other loops are closed.

Notice also that the RGA matrix has two useful properties: (a) the λ_{ij} elements are dimensionless, which means that they are ready to compare channels of very different nature (pressure, temperature, concentrations, etc); and (b) the addition of all the elements of a column or all the elements of a row are always one: $\sum_j \lambda_{ij} = \sum_i \lambda_{ij} = 1$.

Given its importance, the RGA method has been the subject of multiple revisions and research. For instance, although originally defined for the steady-state gain, the RGA was extended to a frequency-dependent definition and used to assess the interaction at frequencies other than zero.[239,260,279,300] In most cases, it is the value of RGA at frequencies close to crossover which is the most important one, and both the gain and the phase are to be taken into account. For a detailed analysis of the plant, RGA is considered as a function of frequency:

$$\text{RGA}(j\omega) = P(j\omega) \otimes (P^{-1}(j\omega))^T \tag{8.27}$$

where $P(j\omega)$ is a frequency-dependent matrix.

Further information on how to perform the pairing is available in Reference 239, and different properties of the RGA can be consulted in References 260, 265, 294, 303, 304, 308, and 312.

Other measures of interaction that exist in the literature are: the *Block Relative Gain*;[298,301,306] the *Relative Disturbance Gain*;[295,296,303,304] or the *Generalized Relative Disturbance Gain*.[296]

EXAMPLE 8.2

Given the 3×3 MIMO plant at $s = 0$,

$$
\begin{bmatrix} y_1 \\ y_2 \\ y_3 \end{bmatrix} = \begin{bmatrix} 2.662 & 8.351 & 8.351 \\ 0.3816 & -0.5586 & -0.5586 \\ 0 & 11.896 & -0.3511 \end{bmatrix} \begin{bmatrix} u_1 \\ u_2 \\ u_3 \end{bmatrix} \tag{8.28}
$$

The RGA matrix is

$$
\Lambda = P_0 \otimes (P_0^{-1})^T = \begin{array}{ccc} u_1 & u_2 & u_3 \\ \begin{bmatrix} 0.318 & 0.0195 & \mathbf{0.663} \\ \mathbf{0.682} & 0.00913 & 0.309 \\ 0 & \mathbf{0.971} & 0.0287 \end{bmatrix} & \begin{array}{c} y_1 \\ y_2 \\ y_3 \end{array} \end{array} \tag{8.29}
$$

and then, the selected pairing is $(y_1 - u_3)$, $(y_2 - u_1)$, $(y_3 - u_2)$.

8.3.3 Multivariable Poles and Zeros

Due to the abovementioned interaction among loops, the poles and zeros of a multivariable system may differ from what could be deduced from direct observation of the elements of the plant TFM.[245] In fact, the pole positions can be inferred from the matrix elements $p_{ij}(s)$, but not their multiplicity, which is of great importance when applying Nyquist-like stability theorems in the presence of RHP poles. Regarding the multivariable zeros (also known as transmission zeros), neither the position nor the multiplicity can be derived from direct observation of $p_{ij}(s)$. These multivariable zeros present a transmission-blocking property, since they provoke the loss of rank of the plant TFM (see example below).

Thus, it is necessary to determine the effective poles and zeros of a MIMO system, for example, by using the so-called Smith-McMillan form,[264] as Rosenbrock first suggested.[229,231,270,274] Alternative definitions for transmission zeros can be found in the References 232, 272, 273, 275, and 281. Further information on this issue is available in the References 229, 237, 245, and 266. Regarding the RGA method introduced in the previous section, the following theorem can be useful to find RHP transmission zeros.

Theorem

Given TFM $P(s)$ with stable elements $p_{ij}(s)$ with no poles or zeros at the origin $s = 0$, and assuming that $\lambda_{ij}(\infty)$ is finite and not zero, if $\lambda_{ij}(0)$ and $\lambda_{ij}(\infty)$ have opposite signs, then one of the following sentences is true: (a) the element $p_{ij}(s)$ has an RHP zero, or (b) the matrix $P(s)$ has an RHP zero.

EXAMPLE 8.3

Given the MIMO plant,

$$
P(s) = \begin{bmatrix} \dfrac{4(s-1)}{(s+1)} & \dfrac{s}{(s+2)} \\ \dfrac{-6}{(s+1)} & \dfrac{(s-2)}{(s+1)} \end{bmatrix} \tag{8.30}
$$

taking the common denominator,

$$P(s) = \frac{1}{(s+1)(s+2)} \begin{bmatrix} 4(s-1)(s+2) & s(s+1) \\ -6(s+2) & (s-2)(s+2) \end{bmatrix} \tag{8.31}$$

the poles of $P(s)$ are at $s = -1$ and $s = -2$.

For the zeros, first we calculate the inverse $P(s)^{-1}$, which is

$$P(s)^{-1} = \frac{1}{(s^3 + 0.5s^2 - 2.5s + 4)}$$
$$\times \begin{bmatrix} 0.25s^3 + 0.25s^2 - s - 1 & -0.25s^3 - 0.5s^2 - 0.25s \\ 1.5s^2 + 4.5s + 3 & s^3 + 2s^2 - s - 2 \end{bmatrix} \tag{8.32}$$

Now, the poles of the inverse are the zeros of the original $P(s)$. Then, the zeros of $P(s)$ are $s = -2.32$, $s = 0.91 + 0.95j$, and $s = 0.91 - 0.95j$. Note that these zeros cannot be derived from direct observation of $p_{ij}(s)$ in Equation 8.30.

EXAMPLE 8.4

Given the MIMO plant,

$$P(s) = \frac{1}{(5s+1)} \begin{bmatrix} (s+1) & (s+4) \\ (s+1) & (2s+2) \end{bmatrix} \tag{8.33}$$

There is one pole at $s = -0.2$.

For the zeros, first we calculate the inverse $P(s)^{-1}$, which is

$$P(s)^{-1} = \frac{1}{(s-2)(s+1)}$$
$$\times \begin{bmatrix} (10s+2)(s+1) & (-5s^2 - 21s - 4) \\ (-5s-1)(s+1) & (5s+1)(s+1) \end{bmatrix} \tag{8.34}$$

Again, the poles of the inverse are the zeros of the original $P(s)$. Then, the zeros of $P(s)$ are $s = 2$, $s = -1$. Note that once again these zeros cannot be derived from direct observation of $p_{ij}(s)$ in Equation 8.33.

In this case, plant at $s = 0$ is

$$P_0 = P(s=0) = \begin{bmatrix} 1 & 4 \\ 1 & 2 \end{bmatrix} \tag{8.35}$$

and then the RGA at $s = 0$,

$$\Lambda_0 = P_0 \otimes (P_0^{-1})^T = \begin{bmatrix} -1 & 2 \\ 2 & -1 \end{bmatrix} \tag{8.36}$$

At the same time, the plant at $s = \infty$ is

$$P_\infty = P(s = \infty) = \begin{bmatrix} 0.2 & 0.2 \\ 0.2 & 0.4 \end{bmatrix} \tag{8.37}$$

and the RGA at $s = \infty$,

$$\Lambda_\infty = P_\infty \otimes \left(P_\infty^{-1}\right)^T = \begin{bmatrix} 2 & -1 \\ -1 & 2 \end{bmatrix} \tag{8.38}$$

In other words, $\lambda_{11}(0) = -1$, and $\lambda_{11}(\infty) = 2$. According to the theorem introduced in this section, as $\lambda_{11}(0)$ and $\lambda_{11}(\infty)$ have different signs, and because the elements of the diagonal of $P(s)$ do not have RHP zeros, then the matrix $P(s)$ must have an RHP zero. This is consistent with the calculations above that found an RHP zero at $s = 2$.

8.3.4 Directionality

Directionality is one of the most essential differences between MIMO and SISO plants.[243,260] A given direction is a combination of input signal values: for instance $[u_1, u_2, u_3] = [4\ 1\ 3]$ has the same direction as $[u_1, u_2, u_3] = [8\ 2\ 6]$, which is $2 \times [4\ 1\ 3]$. Inherently, MIMO systems present spatial (directional) and frequency dependency. Basically, they respond differently to input signals lying in distinct directions. As a result, the relationship between the open-loop and closed-loop properties of the feedback system is less obvious. This directionality is completely in accordance with the TFM representation for MIMO systems.

8.3.5 Gain and Phase

The concept of gain of a system is somehow easy to translate to MIMO plants through the *Singular Value Decomposition* (SVD) of the TFM,[244,260,280,282] which provides the plant gain at each particular frequency with respect to the main directions, which are determined by the corresponding singular vectors.

However, the extension of the notion of phase, as understood in scalar systems, is not so straightforward. Several attempts have been made to define a multivariable phase, such as can be seen in References 238, 243, 287, and 289. On the other hand, transmission zeros contribute with extra phase lag in some directions, but not in others.[285] Generally speaking, the change imposed by a MIMO system upon a vector signal can be observed in the magnitude, the direction, and the phase.[243]

EXAMPLE 8.1 (CONT.)

From the heat exchanger case, Example 8.1, the MIMO plant $P(s)$—see Equation 8.4—with the nominal parameters and at steady state $s = 0$ is

$$P_0 = P(s = 0) = \begin{bmatrix} 7 & 9 \\ 3 & 5 \end{bmatrix} \tag{8.39}$$

SVD for P_0 is:

$$SVD(P_0) \Rightarrow \underbrace{\begin{bmatrix} -0.891 & -0.454 \\ -0.454 & 0.891 \end{bmatrix}}_{u} \underbrace{\begin{bmatrix} 12.791 & 0 \\ 0 & 0.625 \end{bmatrix}}_{\sigma} \underbrace{\begin{bmatrix} -0.594 & -0.804 \\ -0.804 & 0.594 \end{bmatrix}^H}_{v^H}$$

(8.40)

being $P_0 = U \sigma V^H$ (the exponent H is for the conjugate transpose).

This means that $\sigma_{max} = 12.791$ is the largest gain of the MIMO system, and it is for an input in the direction: $\bar{v} = \begin{bmatrix} -0.594 \\ -0.804 \end{bmatrix}$, and $\sigma_{min} = 0.625$ is the smallest gain of the MIMO system, and it is for an input in the direction: $\underline{v} = \begin{bmatrix} -0.804 \\ 0.594 \end{bmatrix}$.

Also, the condition of the matrix is $\gamma = \sigma_{max}/\sigma_{min} = 12.791/0.625 = 20.45$.

8.3.6 Effect of Poles and Zeros

The effect of multivariable poles and zeros strongly depends on directionality as well. That is, their nature is only perceptible for particular directions. So, the TFM transmittance gets unbounded when the matrix is evaluated at a pole, but only in the directions determined by the residue matrix at the pole. Likewise, transmission zeros exert their blocking influence provided the TFM is evaluated at the zero, and the input signal lies in the corresponding null-space.[243]

8.3.7 Disturbance and Noise Signals

Because of directionality, disturbance and noise signals generally do not equally affect all the loops. In general, they have more influence on some loops than on others. Depending on the disturbance direction, that is, the direction of the system output vector resulting from a specific disturbance, some disturbances may be easily rejected, while others may not. The disturbance direction can have an influence in two ways: (1) through the magnitude of the manipulated variables needed to cancel the effect of the disturbance at steady-state, independently of the designed controllers, and (2) through its effect on closed-loop performance of the controlled outputs. To address this issue, Skogestad and Morari defined the *Disturbance Condition Number*.[303,304] It measures the magnitude of the manipulated variables needed to counteract a disturbance acting in a particular direction relative to the "best" possible direction.

8.3.8 Uncertainty

Uncertainty, present in all real-world systems, adds complexity to MIMO systems, especially in the crossover frequency region. Indeed, uncertainty is one of the reasons (together with the presence of disturbances, and the original instability of the plant if that is the case) why feedback is necessary in control systems.

There exist multiple sources of uncertainty (model/plant mismatch), for instance:

- The model is known only approximately or is inaccurately identified.
- The model varies because of a change in the operating conditions (experimental models are accurate for a limited range of operating conditions), wear of components, nonlinearities, etc.

- Measurement devices are not perfect and their resolution range may be limited.
- The structure or order of the system is unknown at high frequencies.
- The plant model is sometimes simplified to carry out the controller design, and the neglected dynamics are considered as uncertainty.
- Other events such as sensor and actuator failures, changes in the control objectives, the switch from automatic to manual (or the other way around) in any loop, inaccuracy in the implementation of the control laws, etc.

The uncertainty can be characterized as *unstructured* when the only available knowledge is the loop location, the stability, and a frequency-dependent magnitude of the uncertainty. The weights used for that magnitude (or bound) are generally stable and minimum phase to avoid additional problems, and multiplicative weights are usually preferred. This description is useful for representing unmodeled dynamics, particularly in the high-frequency range, and small nonlinearities. Different ways of expressing the unstructured uncertainty mathematically and their corresponding properties are available in the book by Skogestad and Postlethwaite.[260]

Nevertheless, unstructured uncertainty is often a poor assumption for MIMO plants. It can sometimes lead to highly conservative designs since the controller has to face events that, in fact, are not likely to exist. On the one hand, errors on particular model parameters, such as mode shapes, natural frequencies, damping values, etc., are highly structured. This is the so-called *parametric uncertainty*. Likewise, parameter errors arising in linearized models are correlated, that is, they are not independent. On the other hand, uncertainty that is unstructured at a component level becomes structured when analyzed at a system level.

Thus, in all those cases, it is more convenient to use *structured uncertainty*. Several approaches can be followed to represent this type of uncertainty. For example, a diagonal block can be utilized,[290,291] or a straightforward and accurate representation of the uncertain elements can be performed by means of the plant templates (which are particularly useful for parametric uncertainty). As we saw in Section 2.4, the templates describe the set of possible frequency responses of a plant at each frequency.[6,7] Indeed, the QFT robust control theory can quantitatively handle both types of uncertainty, structured, and unstructured.

Alternative approaches for describing uncertainty are also available, but so far its practicality is somehow limited for controller design. An example is the assumption of a probabilistic distribution (e.g., normal, uniform) for parametric uncertainty.

As for the rest of system features, uncertainty in MIMO systems also displays directionality properties. One loop may contain substantially more uncertainty due to unmodeled dynamics or parameter variations than other loops. Added to this, and again because of directionality, uncertainty at the plant input or output has a different effect. Primarily, input uncertainty is usually a diagonal perturbation, since in principle there is no reason to assume that the perturbations in the manipulated variables are correlated. This uncertainty represents errors on the relative variation rather than on the absolute value.[303,304]

8.3.9 Stability

Stability of MIMO systems is also a crucial point in the design process. In the literature, and depending on the design methodology applied, different ways of assessing the feedback system stability exist.

One of the main approaches is the *generalized Nyquist stability criterion,* in its direct and inverse version.[229,278] It places an encirclement condition on the Nyquist plot of the determinant of the return difference matrix.[190] However, it is necessary to get a diagonally dominant system for this criterion to be practical. This is achieved by means of pre-compensation. The designer is helped in this task by the Gershgorin and Ostrowski bands—see References 229, 231, and 245, or by Mees' theorem—see Reference 288. This stability criterion is mainly used in nonsequential classical methodologies, for example, the Inverse Nyquist Array[267] and Direct Nyquist Array.[229,231] By contrast, sequential classical techniques do not make a direct use of it. Proofs of the multivariable Nyquist stability criterion have been given from different viewpoints. See for instance References 235, 271, 277, and 284.

An alternative way of checking stability is by means of the *Smith-McMillan poles.*[264] This approach is applied in classical sequential methodologies through stability conditions such as those defined by De Bedout and Franchek for non-diagonal sequential techniques.[109]

A completely different strategy is adopted by synthesis techniques, which make use of stability robustness results such as the *small-gain theorem.*[233] This states that a feedback loop composed of stable operators will remain stable if the product of all the operator gains is smaller than unity. The theorem is applied to systems with unstructured uncertainty. When the phases of perturbations, rather than their gains, can be bounded, the *small-phase theorem* can be used.[289] However, the main drawback of this approach is the highly conservative results it may provide. In the presence of structured uncertainty, results based on the *structured singular value* SSV can be used instead.[291]

8.4 MIMO QFT Control—Overview

As is indicated in previous chapters, QFT is an engineering control design methodology which explicitly emphasizes the use of feedback to simultaneously reduce the effects of plant uncertainty and satisfy performance specifications. It is deeply rooted in classical frequency response analysis involving Bode diagrams, template manipulations, and Nichols charts. It relies on the observation that feedback is principally needed when the plant presents model uncertainty or when there are uncertain disturbances acting on the plant.

Model uncertainty, frequency-domain specifications, and desired time-domain responses translated into frequency-domain tolerances lead to the so-called Horowitz-Sidi bounds—see Section 2.7. These bounds serve as a guide for shaping the nominal loop transfer function $L(s) = P(s)G(s)$, which involves the selection of gain, poles, and zeros to design the appropriate controller $G(s)$—see Section 2.8. On the whole, the QFT main objective is to synthesize (loop shape) a simple, low-order controller with minimum bandwidth, which satisfies the desired performance specifications for all the possible plants due to the model uncertainty. The use of CAD tools has made the MIMO QFT controller design much simpler—see for instance the QFTCT for MATLAB developed by the author and used in this book (Appendix B).

The first proposal for MIMO QFT design was made by Horowitz in his first book in 1963,[2] where he pointed out the possibility of using diagonal controllers for quantitative design. This was divided into different frequency ranges: for the low-frequency interval the controller gain generally needs to be high and is easily determined. As for the medium and high-frequency bands, he suggested the progressive tuning loop-by-loop sorted in increasing order. A more systematic and precise approach was later introduced by Shaked et al. in 1976.[276] However, no proof of convergence to a solution was provided.

The first rigorous MIMO QFT methodology was developed by Horowitz in 1979.[92] This nonsequential technique translates the original $n \times n$ MIMO problem with uncertainty into n multi-input single-output (MISO) systems with uncertainty, disturbances, and specifications derived from the initial problem. The coupling is then treated as a disturbance at the plant input, and the individual solutions guarantee the whole multivariable solution. This is assured by the application of the Schauder's fixed point theorem. This theory maps the desired fixed point on the basis of unit impulse functions.

As before, the design applies a different approach to each major frequency range. Loops are designed as basically noninteracting (BNI) at low frequency, whereas in the middle and high-frequency range the design is based on the effect of the noise at the plant input, especially in problems with significant uncertainty.

On the whole, the first Horowitz's method is a direct technique oriented toward MIMO plants with uncertainty. It also considers a trade-off among the control loops at each range of frequency. Nevertheless, the type of plant which can be dealt with is constrained in several ways, and the method places necessary conditions depending on the system size, which hampers its application to high-order systems. In addition, it presents potential overdesign and may generate highly conservative designs. Additional references on this methodology and its applications are available in References 93, 94, 95, and 177.

An improvement of the preceding technique was also provided by Horowitz with a sequential procedure.[96] There exist some similarities between this technique and the SRD method by Mayne,[269,283] such as the fact that the resulting controller is diagonal or that they proceed as if each input–output pair was a standard SISO system with loop interaction behaving as an external disturbance. Besides, both methods incorporate the effects of each loop once it is designed into the subsequent loop designs.

Nevertheless, the main difference is that Horowitz's methodology relies on a factorization of the return difference matrix which is based on the inverse of the plant TFM. By using the inverse plant, a much simpler relationship between the closed-loop and the open-loop TFMs is obtained. One of Horowitz's major contributions with this technique is that he dealt with the problem of robust stability by considering parametric uncertainty.

The stability proof for Horowitz's improved method was provided in the work by Yaniv and Horowitz,[98] and De Bedout and Franchek.[109] By and large, the method constituted a great step forward in MIMO QFT design techniques. First, as abovementioned, parametric uncertainty was considered. Second, the Schauder's fixed point theorem was no longer needed. Third, the limitation related to the system size from the first method was avoided. Finally, it reduced the conservativeness of the former method by using the concept of equivalent plant (which takes into account the controllers that were previously designed). All in all, the second method is a much more powerful technique (although obviously more complicated than other classical approaches), and the physical sense is kept all along the procedure.

Different authors, such as Nwokah, Yaniv, and Horowitz again, made some improvements of these first two MIMO QFT design methods in subsequent works.[97–99] A detailed compilation of the above techniques is presented in the book by Garcia-Sanz and Houpis.[7]

An alternative approach to MIMO QFT methodologies was presented by Park, Chait, and Steinbuch in 1994, who developed a direct technique. In other words, the inversion of the plant matrix was not required anymore, which therefore simplified the design process to some extent.[100]

The methodologies outlined so far only deal with the problem of designing a diagonal controller. Nevertheless, there exist potential benefits in the use of full-matrix compensators. Horowitz already commented that the use of diagonal controllers was established just to simplify the theoretical development, but that in practice it could be convenient to consider the off-diagonal elements as well. These terms could then be used to reduce the level of coupling in open-loop, and therefore reduce the amount of feedback needed in the diagonal compensators to fulfill the required specifications.[92]

Furthermore, as Franchek, Herman, and Nwokah demonstrated,[313] non-diagonal compensators can be used for ensuring that no SISO loop introduces extra unstable poles into the subsequent loops in sequential procedures based on the inverse plant domain. As a result, the minimum crossover frequency needed to achieve closed-loop stability can be reduced in these succeeding loops. In other words, the actuation bandwidth requirements can be relaxed. Additionally, specific integrity objectives can be achieved, allowing the design of fault-tolerant MIMO systems. In the case of Horowitz's diagonal sequential improved method, however, it is not possible to remove the unstable poles originally present in those subsequent loops, and a more general design technique is necessary for that purpose.[109] On the other hand, diagonal compensators are limited for the correction of the plant directionality. There even exist cases where a diagonal or triangular controller cannot stabilize the system.[109]

On balance, the designer has greater flexibility to design the MIMO feedback control system when using fully populated controllers. But the introduction of such non-diagonal controllers poses two main issues: the way of determining the off-diagonal compensators and the need for suitable stability conditions. In systems controlled by a full-matrix compensator, the property of diagonal dominance is not necessary. The Gershgorin circles become too conservative in that case and the stability test gets more complicated. As a result, different stability results are needed. Sufficient stability conditions for non-diagonal sequential procedures have been defined by De Bedout and Franchek.[109]

Regarding the determination of the off-diagonal compensators, different techniques have been proposed. The first attempt in non-diagonal MIMO QFT was proposed by Horowitz and coworkers,[23,95] who suggested the pre-multiplication of the plant by a full matrix. Yaniv[102] presented a procedure where a non-diagonal de-coupler is applied as a pre-compensator and a classical diagonal controller is designed afterward. Therein, the main objective becomes the improvement of the system bandwidth.

A different approach was adopted by Boje and Nwokah.[106,191] They used the Perron–Frobenius root as a measure of interaction and of the level of triangularization of the uncertain plant. The full-matrix pre-compensator is accordingly designed to reduce the coupling before designing a diagonal QFT controller.

On the other hand, Franchek and collaborators[310,313] proposed a non-diagonal sequential procedure. They made use of the Gauss elimination technique[293] to introduce the effects of the controllers previously designed by means of a recursive expression. Integrity considerations are also included. The controller is then divided into three parts with differentiated roles in the design process. The technique achieves the reduction of the required bandwidth with respect to previous classical sequential techniques. Additionally, De Bedout and Franchek established sufficient stability conditions for non-diagonal sequential procedures.[109]

At the same time, and following Horowitz's original ideas,[96] García-Sanz and collaborators extended the sequential methodology to the design of fully populated MIMO controllers.[6,7,110,114,115,206,208,209,211] In this case, the role of the non-diagonal terms is simultaneously analyzed for the fundamental cases of reference tracking, disturbance rejection at plant input, and disturbance rejection at plant output. The compensators are aimed at the reduction of the coupling on the basis of defined coupling matrices, which are accordingly minimized. Additionally, this method has been proved to be a reliable design tool in real applications from different fields, as heat exchangers,[205] robotics,[114] vehicles,[207] industrial furnaces,[206] wastewater treatment plants,[208,212] spacecraft flying in formation (NASA-JPL),[209] spacecraft with flexible appendages (ESA-ESTEC),[211] and wind turbines.[7,202] This method is presented in Sections 8.5 and 8.8.

Some years later, Garcia-Sanz and Eguinoa introduced a reformulation of the previous full-matrix QFT robust control methodology for MIMO plants with uncertainty.[119] The new methodology includes a generalization of the previous non-diagonal MIMO QFT techniques. It avoids former hypotheses of diagonal dominance and simplifies the calculations for the off-diagonal elements and the method itself. It also reformulates the classical matrix definition of MIMO specifications by designing a new set of loop-by-loop QFT bounds on the Nichols chart, with necessary and sufficient conditions, giving explicit expressions to share the load among the loops of the MIMO system to achieve the matrix specifications. And all for stability, reference tracking, disturbance rejection at plant input and output, and noise attenuation problems. This method is presented in Sections 8.6 and 8.9.

Regarding the field of nonsequential MIMO QFT techniques, it is to be remarked the approach by Jayasuriya, Kerr and coworkers.[112,113,117,204,210] Stability conditions have also been established within this framework.[112,116]

Other approaches have also been introduced for particular types of MIMO systems. For example, there are results on non-minimum phase (nmp) MIMO plants.[143] It is noted that not all the $n \times n$ transfer functions have to suffer the limitations imposed by nmp behavior.[141] The MIMO system has the capacity to relocate the RHP zeros in those loops that are less important, while the critical loops are kept as minimum-phase. Likewise, some research has been done for unstable and strongly nmp MIMO systems, for example, the X-29 aircraft.[142,178,210] One interesting suggestion is the *singular-G method*,[142,178] which makes use of a singular compensator—that is, with a determinant equal to zero, which implies that one output is dependent from the rest of outputs. In this way, the technique allows easing the nmp problem and the instability in the MIMO system, and simultaneously achieving good performance.

8.5 Non-Diagonal MIMO QFT—Method 1[7,110,114,115,206,208,209,211,212]

A non-diagonal (fully populated) matrix compensator allows the designer more design flexibility to regulate MIMO systems than the classical diagonal controller structure. Two methodologies (*Method 1* and *Method 2*) to design non-diagonal matrix compensators (fully populated G) for MIMO systems with model uncertainty are presented in this chapter.

Method 1, introduced in the section, studies three classical cases: reference tracking, external disturbance rejection at plant input, and external disturbance rejection at plant output. The method analyses the role played by the non-diagonal compensator elements g_{ij} $(i \neq j)$ by means of three coupling matrices (C_1, C_2, and C_3) and a quality function η_{ij}. It quantifies the amount of interaction among the control loops, and proposes a sequential design methodology for the fully populated matrix compensator that yields n equivalent tracking SISO systems and n equivalent disturbance rejection SISO systems.[110,114,115,206,208,209,211,212] As a result, the off-diagonal elements g_{ij} $(i \neq j)$ of the compensator matrix reduce (or cancel if there is no uncertainty) the level of coupling between loops, and the diagonal elements g_{kk}

regulate the system with less bandwidth requirements than the diagonal elements of the diagonal G methods.

8.5.1 The Coupling Matrix

Coupling among the control loops is one of the main challenges in MIMO systems. This subsection defines an index (the coupling matrix) that allows the designer to quantify the loop interaction in the MIMO system. Following Horowitz's ideas, consider an $n \times n$ linear multivariable system—see Figure 8.11, composed of a plant P, a fully populated matrix compensator G, and a prefilter F. These matrices are defined as follows:

$$P = \begin{bmatrix} p_{11} & p_{12} & \cdots & p_{1n} \\ p_{21} & p_{22} & \cdots & p_{2n} \\ \vdots & \vdots & \ddots & \vdots \\ p_{n1} & p_{n2} & \cdots & p_{nn} \end{bmatrix}; \quad G = \begin{bmatrix} g_{11} & g_{12} & \cdots & g_{1n} \\ g_{21} & g_{22} & \cdots & g_{2n} \\ \vdots & \vdots & \ddots & \vdots \\ g_{n1} & g_{n2} & \cdots & g_{nn} \end{bmatrix};$$

$$F = \begin{bmatrix} f_{11} & f_{12} & \cdots & f_{1n} \\ f_{21} & f_{22} & \cdots & f_{2n} \\ \vdots & \vdots & \ddots & \vdots \\ f_{n1} & f_{n2} & \cdots & f_{nn} \end{bmatrix}$$

(8.41)

Notice that all the matrices, elements and signals are Laplace functions. For the sake of clarity, the dependence on (s) is omitted in this section. Figure 8.11 shows the plant input disturbance transfer function P_{di}, and the plant output disturbance transfer function P_{do}, where $\mathcal{P} \in P$, and P is the set of possible plants due to uncertainty. The reference vector r' and the external disturbance vectors at plant input d_i' and plant output d_o' are the inputs of the system. The output vector y represents the variables to be controlled. The plant inverse P^{-1}, denoted by P^* in this chapter, is presented in the following format:

$$P^{-1} = P^* = \left[p_{ij}^* \right] = \Lambda + B = \begin{bmatrix} p_{11}^* & 0 & 0 \\ 0 & \cdots & 0 \\ 0 & 0 & p_{nn}^* \end{bmatrix} + \begin{bmatrix} 0 & \cdots & p_{1n}^* \\ \cdots & 0 & \cdots \\ p_{n1}^* & \cdots & 0 \end{bmatrix}$$

(8.42)

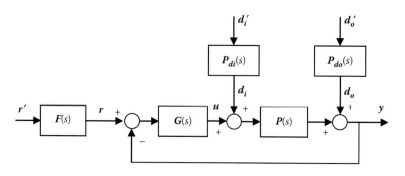

FIGURE 8.11
Structure of a 2DOF MIMO System. (Adapted from Garcia-Sanz, M. and Houpis, C.H. 2012. *Wind Energy Systems: Control Engineering Design (Part I: QFT Control, Part II: Wind Turbine Design and Control)*. A CRC Press book, Taylor & Francis Group, Boca Raton, FL.)

and the compensator matrix is broken up into two parts as follows:

$$G = G_d + G_b = \begin{bmatrix} g_{11} & 0 & 0 \\ 0 & \ldots & 0 \\ 0 & 0 & g_{nn} \end{bmatrix} + \begin{bmatrix} 0 & \ldots & g_{1n} \\ \ldots & 0 & \ldots \\ g_{n1} & \ldots & 0 \end{bmatrix} \tag{8.43}$$

Note that now we use the symbol Λ for the diagonal part, and B for the balance of P^*; and that G_d is the diagonal part and G_b is the balance of G.

The following three subsections introduce an index to quantify the loop interaction in the three classical cases: reference tracking, external disturbances at the plant input, and the external disturbances at the plant output. In this chapter, the index is called the *coupling matrix* C and, depending on the case, shows three different notations: C_1, C_2, and C_3, respectively. The use of these coupling matrices enables the achievement of essentially n equivalent tracking SISO systems and n equivalent disturbance rejection SISO systems.

8.5.2 Tracking

The transfer function matrix (TFM) of the control system for the reference tracking problem without any external disturbance ($d_i{}' = 0$, $d_o{}' = 0$ in Figure 8.11) is written as shown in Equation 8.44,

$$y = (I + PG)^{-1} PGr = T_{y/r}r = T_{y/r}Fr' \tag{8.44}$$

Using Equations 8.42 and 8.43, Equation 8.44 is rewritten as

$$T_{y/r}r = \left(I + \Lambda^{-1}G_d\right)^{-1} \Lambda^{-1}G_d r + \left(I + \Lambda^{-1}G_d\right)^{-1} \Lambda^{-1}[G_b r - (B + G_b)T_{y/r}r] \tag{8.45}$$

which is the closed-loop TFM, and represents another way to describe the same idea introduced by Horowitz as

$$y_{ij} = t_{ij}^{y/r} r_j = \left(t_{r_{ii}} + t_{c_{ij}}\right)r_j, i,j = 1,2,\ldots,n$$

where

$$t_{r_{ii}} = w_{ii}g_{ii}; \quad t_{c_{ij}} = w_{ii}c_{ij}; \quad w_{ii} = \frac{1}{p_{ii}^* + g_{ii}}; \quad c_{ij} = -\sum_{k \neq i} t_{kj}\, p_{ik}^*, k = 1,2,\ldots,n$$

An analysis of Equation 8.45 reveals that it can be broken up into two parts as follows:

1. *A diagonal term T_{y/r_d} given by,*

$$T_{y/r\_d} = \left(I + \Lambda^{-1}G_d\right)^{-1} \Lambda^{-1}G_d \tag{8.46}$$

that represents a pure diagonal structure. Note that it does not depend on the non-diagonal part of the plant inverse B nor on the non-diagonal part of the compensator G_b. It is equivalent to n reference tracking SISO systems formed by plants equal to the elements of Λ^{-1} when the n corresponding parts of a diagonal G_d control them, as shown in Figure 8.12a. In this figure $t_i(s)$ represents the closed-loop control ratio.

2. *A non-diagonal term T_{y/r_b} given by*

$$T_{y/r\_b} = \left(I + \Lambda^{-1}G_d\right)^{-1}\Lambda^{-1}\left[G_b - (B + G_b)T_{y/r}\right] = \left(I + \Lambda^{-1}G_d\right)^{-1}\Lambda^{-1}C_1 \qquad (8.47)$$

which represents a non-diagonal structure. It is equivalent to the same n previous systems with cross-coupling (internal) disturbances coming from the references of the other loops $c_{1ij}\, r_j$—see Figure 8.12b. In Equation 8.47, the matrix C_1 is the only part that depends on the non-diagonal parts of both the plant inverse B and the compensator G_b. Hence, it represents the *coupling matrix C* of the equivalent system for reference tracking problems, or

$$C_1 = G_b - (B + G_b)T_{y/r} = \begin{bmatrix} 0 & c_{1_{12}} & \cdots & c_{1_{1m}} \\ c_{1_{21}} & 0 & \cdots & c_{1_{2m}} \\ \vdots & \vdots & \ddots & \vdots \\ c_{1_{m1}} & c_{1_{m2}} & \cdots & 0 \end{bmatrix} \qquad (8.48)$$

Each element $c_{1_{ij}}$ of this matrix C_1 obeys

$$c_{1ij} = g_{ij}(1 - \delta_{ij}) - \sum_{k=1}^{n}\left(p_{ik}^* + g_{ik}\right)t_{kj}(1 - \delta_{ik}) \qquad (8.49)$$

(a)

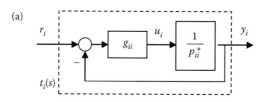

(b)

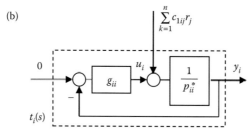

FIGURE 8.12
i-th equivalent decoupled SISO systems: (a) reference tracking, (b) disturbance at plant input. (Adapted from Garcia-Sanz, M. and Houpis, C.H. 2012. *Wind Energy Systems: Control Engineering Design (Part I: QFT Control, Part II: Wind Turbine Design and Control)*. A CRC Press book, Taylor & Francis Group, Boca Raton, FL.)

where δ_{ik} (and δ_{ij}) is the Kronecker delta that is defined as

$$\delta_{ik} = \begin{cases} \delta_{ik} = 1 \Leftrightarrow k = i \\ \delta_{ik} = 0 \Leftrightarrow k \neq i \end{cases} \qquad (8.50)$$

8.5.3 Disturbance Rejection at Plant Input

The transfer matrix from the external disturbance d_i' at the plant input to the plant output y is written as shown in the following equation—see Figure 8.11,

$$y = (I + PG)^{-1} P d_i = T_{y/di} d_i = T_{y/di} P_{di} d_i' \qquad (8.51)$$

Using Equations 8.42 and 8.43, Equation 8.51 is rewritten as

$$T_{y/di} d_i = \left(I + \Lambda^{-1} G_d\right)^{-1} \Lambda^{-1} d_i - \left(I + \Lambda^{-1} G_d\right)^{-1} \Lambda^{-1} \left[(B + G_b) T_{y/di}\right] d_i \qquad (8.52)$$

Again, from Equation 8.52 it is possible to define two different terms as follows
  1. *A diagonal term* T_{y/di_d} *given by,*

$$T_{y/di\_d} = \left(I + \Lambda^{-1} G_d\right)^{-1} \Lambda^{-1} \qquad (8.53)$$

that is equivalent to n regulator SISO systems, as shown in Figure 8.13a.
  2. *A non-diagonal term* T_{y/di_b} *given by,*

$$T_{y/di\_b} = \left(I + \Lambda^{-1} G_d\right)^{-1} \Lambda^{-1} (B + G_b) T_{y/di} = \left(I + \Lambda^{-1} G_d\right)^{-1} \Lambda^{-1} C_2 \qquad (8.54)$$

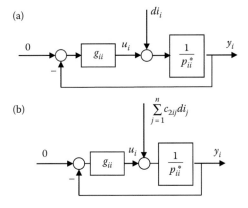

FIGURE 8.13
i-th equivalent decoupled SISO systems: (a) external disturbances at plant input, (b) internal disturbances. (Adapted from Garcia-Sanz, M. and Houpis, C.H. 2012. *Wind Energy Systems: Control Engineering Design (Part I: QFT Control, Part II: Wind Turbine Design and Control)*. A CRC Press book, Taylor & Francis Group, Boca Raton, FL.)

which represents a non-diagonal structure, equivalent to the same n previous systems with external disturbances $c_{2ij}\, di_j$ at the plant input, as shown in Figure 8.13b. In Equation 8.54, the matrix C_2 comprises the coupling. It represents the *coupling matrix* of the equivalent system for external disturbance rejection at the plant input problems, and is given by

$$C_2 = (B + G_b)T_{y/di} \tag{8.55}$$

Each element c_{2ij} of this matrix obeys,

$$c_{2ij} = \sum_{k=1}^{n} \left(p_{ik}^* + g_{ik} \right) t_{kj}(1 - \delta_{ik}) \tag{8.56}$$

where δ_{ki} is the Kronecker delta defined in Equation 8.50.

8.5.4 Disturbance Rejection at Plant Output

The transfer matrix from the external disturbance d_o' at the plant output to the output y is written as shown in Equation 8.57—see Figure 8.11,

$$y = (I + PG)^{-1} d_o = T_{y/do}\, d_o = T_{y/do}\, P_{do}\, d_o' \tag{8.57}$$

Using Equations 8.42 and 8.43, and repeating the procedure of the previous subsections, Equation 8.57 is rewritten as

$$T_{y/do} d_o = \left(I + \Lambda^{-1}G_d \right)^{-1} d_o + \left(I + \Lambda^{-1}G_d \right)^{-1} \Lambda^{-1}\left[B - (B + G_b)T_{y/do} \right] d_o \tag{8.58}$$

From Equation 8.58, it is also possible to define two terms:

1. *A diagonal term* T_{y/do_d} *given by*

$$T_{y/do\_d} = \left(I + \Lambda^{-1}G_d \right)^{-1} \tag{8.59}$$

that is equivalent to the n regulator SISO systems shown in Figure 8.14a.

2. *A non-diagonal term* T_{y/do_b} *given by*

$$T_{y/do\_b} = \left(I + \Lambda^{-1}G_d \right)^{-1} \Lambda^{-1}\left[B - (B + G_b)T_{y/do} \right] = \left(I + \Lambda^{-1}G_d \right)^{-1} \Lambda^{-1} C_3 \tag{8.60}$$

which represents a non-diagonal structure, and is equivalent to the same n previous systems with external disturbances $c_{3ij}\, do_j$ at the plant input, as shown in Figure 8.14b. In Equation 8.60, the matrix C_3 comprises the coupling. It represents the *coupling matrix* of the equivalent system for external disturbance rejection for the plant output problems and is given by,

$$C_3 = B - (B + G_b)T_{y/do} \tag{8.61}$$

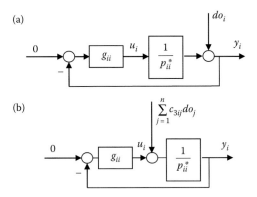

FIGURE 8.14
i-th equivalent decoupled SISO systems: (a) external disturbances at plant output, (b) internal disturbances. (Adapted from Garcia-Sanz, M. and Houpis, C.H. 2012. *Wind Energy Systems: Control Engineering Design (Part I: QFT Control, Part II: Wind Turbine Design and Control).* A CRC Press book, Taylor & Francis Group, Boca Raton, FL.)

Each element of the coupling matrix, c_{3ij} obeys,

$$c_{3ij} = p_{ij}^*(1 - \delta_{ij}) - \sum_{k=1}^{n} (p_{ik}^* + g_{ik}) t_{kj} (1 - \delta_{ik}) \tag{8.62}$$

where δ_{ik} (and δ_{ij}) is the Kronecker delta as defined in Equation 8.50.

8.5.5 The Coupling Elements

In order to design a MIMO compensator that reduces the level of coupling, we have to study the influence of every non-diagonal element g_{ij} on the coupling elements c_{1ij}, c_{2ij}, and c_{3ij}, as defined by Equations 8.49, 8.56, and 8.62, respectively. To quantify the coupling effects, these elements are simplified by applying the following hypothesis.

Hypothesis H1

$$\left|(p_{ij}^* + g_{ij})t_{jj}\right| \gg \left|(p_{ik}^* + g_{ik})t_{kj}\right|, \quad \text{for} \quad k \neq j, \quad \text{and in the bandwidth of } t_{jj} \tag{8.63}$$

Note that this is a common situation in many systems, as it is desirable for the diagonal elements t_{jj} to be much larger than the non-diagonal elements t_{kj}, once the pairing of the most convenient variables have been applied. Thus,

$$|t_{jj}| \gg |t_{kj}|, \quad \text{for} \quad k \neq j, \quad \text{and in the bandwidth of } t_{jj} \tag{8.64}$$

Now, using Hypothesis H1, Equations 8.49, 8.56, and 8.62, which describe the coupling elements in the tracking problem, the disturbance rejection at the plant input and disturbance rejection at plant output, respectively, are rewritten as

$$c_{1ij} = g_{ij} - t_{jj}(p_{ij}^* + g_{ij}); \quad i \neq j \tag{8.65}$$

$$c_{2ij} = t_{jj}\left(p_{ij}^* + g_{ij}\right); \quad i \neq j \tag{8.66}$$

$$c_{3ij} = p_{ij}^* - t_{jj}\left(p_{ij}^* + g_{ij}\right); \quad i \neq j \tag{8.67}$$

Also, taking the elements t_{jj} from the equivalent system derived from Equations 8.46, 8.53, and 8.59, we have

$$t_{jj} = \frac{g_{jj}\left(p_{ij}^*\right)^{-1}}{1 + g_{jj}\left(p_{ij}^*\right)^{-1}} \tag{8.68}$$

$$t_{jj} = \frac{\left(p_{ij}^*\right)^{-1}}{1 + g_{jj}\left(p_{ij}^*\right)^{-1}} \tag{8.69}$$

$$t_{jj} = \frac{1}{1 + g_{jj}\left(p_{ij}^*\right)^{-1}} \tag{8.70}$$

and then

$$c_{1ij} = g_{ij} - \frac{g_{jj}\left(p_{ij}^* + g_{ij}\right)}{\left(p_{jj}^* + g_{jj}\right)}; \quad i \neq j \tag{8.71}$$

$$c_{2ij} = \frac{\left(p_{ij}^* + g_{ij}\right)}{\left(p_{jj}^* + g_{jj}\right)}; \quad i \neq j \tag{8.72}$$

$$c_{3ij} = p_{ij}^* - \frac{p_{jj}^*\left(p_{ij}^* + g_{ij}\right)}{\left(p_{jj}^* + g_{jj}\right)}; \quad i \neq j \tag{8.73}$$

8.5.6 The Optimal Non-Diagonal Compensator

As stated previously, the purpose of non-diagonal compensators is to reduce the coupling effect to achieve the desired loop performance specifications. The optimum non-diagonal compensators for the three cases (tracking and disturbance rejection at the plant input and output) are obtained making the coupling elements c_{1ij}, c_{2ij}, and c_{3ij} of Equations 8.71 through 8.73 equal to zero.

Note that in these three expressions the elements p_{ij}^* and p_{jj}^* have uncertainty. In general every uncertain plant p_{ij}^* can be a plant represented by the family:

$$\left\{p_{ij}^*\right\} = p_{ij}^{*N}(1 + \Delta_{ij}), \quad 0 \leq |\Delta_{ij}| \leq \Delta p_{ij}^*, \quad \text{for} \quad i,j = 1,\ldots,n \tag{8.74}$$

where p_{ij}^{*N} is the selected nominal plant for the non-diagonal controller expression, and Δp_{ij}^* is the maximum of the non-parametric uncertainty radii $|\Delta_{ij}|$—see Figure 8.15b. Note also that Δp_{ij}^* depends on the selection of p_{ij}^{*N}.

The selected plants p_{ij}^{*N} and p_{jj}^{*N} that are chosen for the optimum non-diagonal compensator must comply with the following rules:

1. If the uncertain parameters of the plants show a uniform probability distribution—see Figure 8.15a, which is typical in robust control, then the elements p_{ij}^* and p_{jj}^* for the optimum non-diagonal compensator are the plants p_{ij}^{*N} and p_{jj}^{*N}. These plants minimize the maximum of the nonparametric uncertainty radii Δp_{ij}^* and Δp_{jj}^* that comprise the plant templates—see Figure 8.15b.

2. If the uncertain parameters of the plants show a nonuniform probability distribution—see Figure 8.15c, then the elements p_{ij}^* and p_{jj}^* for the optimum non-diagonal compensator are the plants p_{ij}^{*N} and p_{jj}^{*N}, whose set of parameters maximize the area of the probability distribution in the regions $[a_{ij} - \varepsilon, a_{ij} + \varepsilon]$ and $[a_{jj} - \varepsilon, a_{jj} + \varepsilon]$ ($\forall$ parameter $a_{ij}, b_{ij} \ldots, a_{jj}, b_{jj} \ldots$), respectively.

These rules of selection are analyzed again in Section 8.5.7, where the coupling effects with the optimum non-diagonal compensator are computed. By setting Equations 8.71, 8.72, and 8.73 equal to zero and using Equation 8.74, the optimum non-diagonal compensator for each case is obtained—see Equations 8.75, 8.76, and 8.77.

8.5.6.1 Tracking

$$g_{ij}^{opt} = F_{pd}\left(g_{jj} \frac{p_{ij}^{*N}}{p_{jj}^{*N}}\right), \quad \text{for} \quad i \neq j \tag{8.75}$$

8.5.6.2 Disturbance Rejection at Plant Input

$$g_{ij}^{opt} = F_{pd}\left(-p_{ij}^{*N}\right), \quad \text{for } i \neq j \tag{8.76}$$

8.5.6.3 Disturbance Rejection at Plant Output

$$g_{ij}^{opt} = F_{pd}\left(g_{jj} \frac{p_{ij}^{*N}}{p_{jj}^{*N}}\right), \quad \text{for } i \neq j \tag{8.77}$$

The function $F_{pd}(A)$ means in every case a proper stable and minimum phase function made from the dominant poles and zeros of the expression A.

8.5.7 The Coupling Effects

The rules of Section 8.5.6 are utilized for choosing the plants p_{ij}^{*N} and p_{jj}^{*N}. These plants are inserted into Equations 8.75 through 8.77 in order to obtain the respective g_{ij}^{opt}, which are in turn utilized for determining the minimum achievable coupling effects given by Equations 8.78, 8.80, and 8.82. In a similar manner, the maximum coupling effects for the diagonal compensator matrix case, given by Equations 8.79, 8.81, and 8.83, are computed by substituting $g_{ij} = 0$ ($i \neq j$) into the coupling expressions of Equations 8.71 through 8.73, respectively.

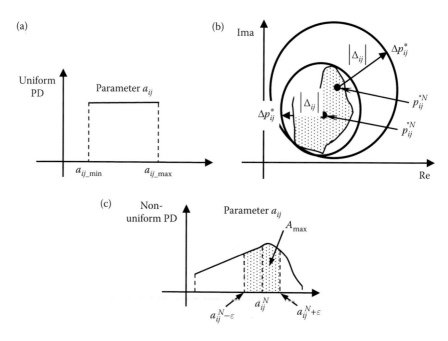

FIGURE 8.15
(a) and (c) Probability distribution of the parameter a_{ij}; (b) two possibilities of the maximum nonparametric uncertainty radii Δp_{ij}^*. (Adapted from Garcia-Sanz, M. and Houpis, C.H. 2012. *Wind Energy Systems: Control Engineering Design (Part I: QFT Control, Part II: Wind Turbine Design and Control)*. A CRC Press book, Taylor & Francis Group, Boca Raton, FL.)

8.5.7.1 Tracking

$$\left| c_{1ij} \right|_{g_{ij}=g_{ij}^{opt}} = \left| \psi_{ij}(\Delta_{jj} - \Delta_{ij})g_{jj} \right| \tag{8.78}$$

$$\left| c_{1ij} \right|_{g_{ij}=0} = \left| \psi_{ij}(1 + \Delta_{ij})g_{jj} \right| \tag{8.79}$$

8.5.7.2 Disturbance Rejection at Plant Input

$$\left| c_{2ij} \right|_{g_{ij}=g_{ij}^{opt}} = \left| \psi_{ij}\Delta_{ij} \right| \tag{8.80}$$

$$\left| c_{2ij} \right|_{g_{ij}=0} = \left| \psi_{ij}(1 + \Delta_{ij}) \right| \tag{8.81}$$

8.5.7.3 Disturbance Rejection at Plant Output

$$\left| c_{3ij} \right|_{g_{ij}=g_{ij}^{opt}} = \left| \psi_{ij}(\Delta_{ij} - \Delta_{jj})g_{jj} \right| \tag{8.82}$$

$$\left| c_{3ij} \right|_{g_{ij}=0} = \left| \psi_{ij}(1 + \Delta_{ij})g_{jj} \right| \tag{8.83}$$

where

$$\psi_{ij} = \frac{p_{ij}^{*N}}{(1+\Delta_{jj})p_{jj}^{*N} + g_{jj}}$$ (8.84)

and where the uncertainty is

$$0 \le |\Delta_{ij}| \le \Delta p_{ij}^{*}, \quad 0 \le |\Delta_{jj}| \le \Delta p_{jj}^{*}, \quad \text{for} \quad i,j = 1,\ldots,n$$

The coupling effects, calculated for the pure diagonal compensator cases, result in three expressions (8.79), (8.81), and (8.83) that still present a nonzero value, even if $(p_{ij}^{*N}, p_{jj}^{*N})$ is selected so that the actual plant mismatching disappears ($\Delta_{ij} = 0$ and $\Delta_{jj} = 0$). However, the coupling effects obtained with the optimum non-diagonal compensators (see Equations 8.78, 8.80, and 8.82) tend to zero when the mismatching disappears ($\Delta_{ij} = 0$ and $\Delta_{jj} = 0$).

8.5.8 Quality Function of the Designed Compensator

Figure 8.16 shows the coupling bands for a common system. The $|c_{ij}|_{g_{ij}=0}$ case, computed from Equations 8.79, 8.81, or 8.83, is the upper limit. The lower limit represents the minimum coupling effect $|c_{ij}|_{g_{ij}=g_{ij}^{opt}}$, obtained with the optimum element g_{ij}^{opt} from Equations 8.78, 8.80, or 8.82. This allow us to define a quality function η_{ij} for a given non-diagonal compensator g_{ij} ($i \ne j$), so that

$$\eta_{ij}(\%) = 100 \left\{ \frac{\log_{10}\left[\left[\max\left\{|c_{ij}|_{g_{ij}=0}\right\}/\max\left\{|c_{ij}|_{g_{ij}=g_{ij}}\right\}\right]\right]}{\log_{10}\left[\left[\max\left\{|c_{ij}|_{g_{ij}=0}\right\}/\max\left\{|c_{ij}|_{g_{ij}=g_{ij}^{opt}}\right\}\right]\right]} \right\}$$ (8.85)

This quality function becomes a proximity measure of the coupling effect c_{ij} to the minimum achievable coupling effect. Thus, the function is useful to quantify the amount of

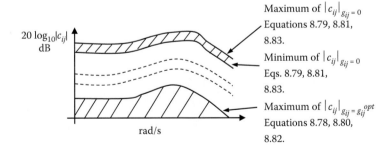

FIGURE 8.16
Coupling effect bands with different non-diagonal compensators. (Adapted from Garcia-Sanz, M. and Houpis, C.H. 2012. *Wind Energy Systems: Control Engineering Design (Part I: QFT Control, Part II: Wind Turbine Design and Control)*. A CRC Press book, Taylor & Francis Group, Boca Raton, FL.)

loop interaction and to design the non-diagonal compensators. If η_{ij} is closed to 100%, then the coupling effect is a minimum and the g_{ij} compensator tends to be the optimum one. A suitable non-diagonal compensator maximizes the quality function of Equation 8.85.

8.5.9 Design Methodology

With all the preliminary analysis, this section presents the steps we have to follow to design the MIMO QFT controller according to Method 1. The design method is a sequential procedure by closing the loops. It only needs to fulfill Hypothesis H1—see Equations 8.63 and 8.64. For more details, Section 8.8 presents an example of the application of this methodology.

Step A: Controller Structure
First, find the most appropriate pairing for the inputs and outputs of the plant according to RGA technique—see Section 8.3.2 and Equations 8.23 through 8.27.

Second, find the appropriate structure for the controller matrix $G(s)$. Typically, we need off-diagonal elements $g_{ij}(s)$ for the controller $G(s)$ in the positions ij where the loop coupling is significant. As a practical rule, the coupling between the loops i and j is considered important when the absolute value of the corresponding RGA element is greater than 1.2 (or less than 0.8), or $|\lambda_{ij}| \geq 1.2$ (or $|\lambda_{ij}| \leq 0.8$). To keep the solution as simpler as possible, start with a triangular controller matrix $G(s)$ if possible, and then add further elements if necessary.

Third, arrange the matrix P^* so that $q_{11} = (1/p_{11}^*)$ has the smallest bandwidth, $q_{22} = (1/p_{22}^*)$ the next smallest bandwidth, and so on. The sequential compensator design technique, as described in Figure 8.17, is composed of the steps: A, $B.i.1$ to $B.i.3$, $i = 1$ to n (one for each column "i"), and C—see also Figure 8.19.

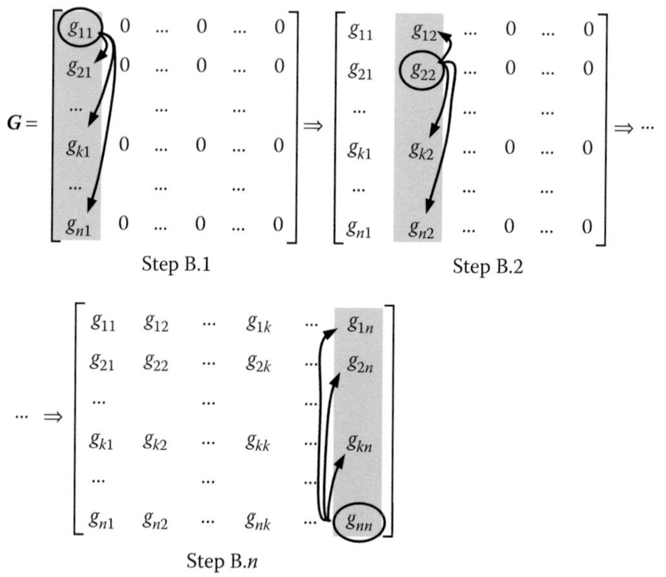

FIGURE 8.17
Steps *B.i*, $i = 1$ to n columns, of the sequential design technique. (Adapted from Garcia-Sanz, M. and Houpis, C.H. 2012. *Wind Energy Systems: Control Engineering Design (Part I: QFT Control, Part II: Wind Turbine Design and Control).* A CRC Press book, Taylor & Francis Group, Boca Raton, FL.)

Step B.i.1: Design of the Diagonal Controller Elements $g_{ii}(s)$
The design of the element g_{ii} is calculated using the standard QFT loop-shaping technique for the inverse of the equivalent plant $q_{ii} = (1/p_{ii}^{*e})$ and to achieve robust stability and robust performance specifications. The equivalent plant satisfies the recursive relationship[313] given by Equation 8.86.

$$\left[p_{ii}^{*e}\right]_k = \left[p_{ii}^*\right]_{k-1} - \frac{\left(\left[p_{i(i-1)}^*\right]_{k-1} + \left[g_{i(i-1)}\right]_{k-1}\right)\left(\left[p_{(i-1)i}^*\right]_{k-1} + \left[g_{(i-1)i}\right]_{k-1}\right)}{\left[p_{(i-1)(i-1)}^*\right]_{k-1} + \left[g_{(i-1)(i-1)}\right]_{k-1}}; \quad i \geq k; \quad \left[\boldsymbol{P}^*\right]_{k=1} = \boldsymbol{P}^* \quad (8.86)$$

This equation is an extension for the non-diagonal case of the recursive expression proposed by Horowitz[96] as the *Improved Design Technique*.

At this point, the design has also to meet two stability conditions:[109] (a) $L_{ii}(s) = [p_{ii}^{*e}(s)]^{-1}g_{ii}(s)$ has to satisfy the Nyquist encirclement condition and (b) no RHP pole-zero cancellations have to occur between $g_{ii}(s)$ and $q_{ii}(s) = [p_{ii}^{*e}(s)]^{-1}$—see Chapter 3.

If the system requires reference tracking specifications as $a_{ii}(j\omega) \leq \left|t_{ii}^{y/r}(j\omega)\right| \leq b_{ii}(j\omega)$, since $t_{ii}^{y/r} = t_{rii} + t_{cii}$, the tracking bounds b_{ii} and a_{ii} are corrected to take into account the cross-coupling specification τ_{cii}, so that (see also Equation 8.45)

$$b_{ii}^c = b_{ii} - T_{cii}, \quad a_{ii}^c = a_{ii} + T_{cii} \quad (8.87)$$

$$t_{cii} = w_{ii}c_{ii} \leq \tau_{cii} \quad (8.88)$$

$$a_{ii}^c(j\omega) \leq \left|t_{rii}(j\omega)\right| \leq b_{ii}^c(j\omega) \quad (8.89)$$

These are the same corrections proposed by Horowitz[92,96] for their original MIMO QFT methods. However, for the non-diagonal compensator these corrections are less demanding. The coupling expression $t_{cii} = w_{ii}c_{ii}$ is now minor as compared to the diagonal methods—for instance compare Equations 8.78 and 8.79. That is the off-diagonal elements g_{ij} $(i \neq j)$ of the matrix compensator attenuate or cancel the cross coupling effects. This results in the diagonal elements g_{ii} of the non-diagonal method requiring a smaller bandwidth than the diagonal elements of the diagonal compensator methods.

Step B.i.2: Design of the Diagonal Prefilter Elements $f_{ii}(s)$
The design of a prefilter $F(s)$ is necessary in case of reference tracking specifications. As one of the objectives of the full-matrix controller $G(s)$ is to reduce the loop interaction, the prefilter $F(s)$ can be matrix diagonal. The $f_{ii}(s)$ prefilter element is designed for the equivalent plant $q_{ii}(s)$, and for the diagonal controller $g_{ii}(s)$ designed in the previous *Step B.i.1*.

Step B.i.3: Design of the Non-Diagonal Controller Elements $g_{ji}(s)$
The $(n-1)$ non-diagonal elements $g_{ji}(j \neq i, j = 1,2,...,n)$ of the j-th compensator column are designed to minimize the cross-coupling terms c_{ji} given by Equations 8.71 through 8.73. The optimum compensator elements described in Equations 8.75 through 8.77 are utilized in order to achieve this goal. After we finish the Steps *B.i, i = 1* to *n* columns, we proceed with *Step C*.

Step C: Final Checks

Once the design of $G(s)$ is finished, the system has to meet two additional stability conditions:[109] (c) no Smith-McMillan pole-zero cancellations have to occur between $P(s)$ and $G(s)$ and (d) no Smith-McMillan pole-zero cancellations have to occur in $|P^*(s) + G(s)|$.

Also, although very remote, the possibility of introducing RHP transmission zeros due to the controller design theoretically exists. This undesirable situation cannot be detected until the whole multivariable system design is completed. To avoid this problem, the multivariable zeros of $P(s)G(s)$ are determined using the Smith-McMillan form over the set of possible plants P due to the uncertainty. If there are no new RHP zeros apart from those that might already be present in $P(s)$, the method concludes. Otherwise, we rectify this undesirable situation by modifying the non-diagonal elements placed in the last column of the matrix $G(s)$, according to the Smith-McMillan expressions.[115]

8.5.10 Some Practical Issues

The sequential non-diagonal MIMO QFT techniques introduced in this chapter arrive at a robust stable closed-loop system if, for each $P \in \mathcal{P}$,[109]

1. Each $L_i(s) = (p_{ii}^{*e})^{-1} g_{ii}(s)$, $i = 1, \ldots, n$, satisfies the Nyquist encirclement condition,
2. No RHP pole-zero cancellations occur between $g_{ii}(s)$ and $(p_{ii}^{*e})^{-1}$, $i = 1, \ldots, n$,
3. No Smith-McMillan pole-zero cancellations occur between $P(s)$ and $G(s)$, and
4. No Smith-McMillan pole-zero cancellations occur in $|P^*(s) + G(s)|$.

On the other hand, as mentioned, the resulting matrix PG should be checked to ensure that RHP transmission zeros or unstable modes have not been introduced by the new compensator elements g_{kk} or g_{ik}, which would obviously cause an unnecessary loss of control performance. If these *n.m.p.* (non-minimum phase) zeros appear due to the designed compensator elements, supplementary constraints in the determinant of PG should be imposed to re-calculate the compensator.

Incidentally, arbitrarily picking the wrong order of the loops to be designed can result in the nonexistence of a solution. This may occur if the solution process is based on satisfying an upper limit of the phase margin frequency ω_ϕ for each loop. To avoid that potential problem, as has been introduced in *Step A* of the methodology, loop i having the smallest phase margin frequency has to be chosen as the first loop to be designed. The loop that has the next smallest phase margin frequency is next, and so on.[6,7]

Finally, notice that the calculation of the equivalent plant $q_{ii} = (1/p_{ii}^{*e})$ needs the inverse of the plant matrix, and usually introduces some exact pole-zero cancellations. This operation can be precisely done with symbolic mathematic tools, but could be erroneous when using numerical calculations due to computer round errors. In MATLAB, the command $q_{ii} = minreal(q_{ii}, tolMR)$ helps removing the extra zeros and poles erroneously introduced—see Appendix 8.

8.6 Non-Diagonal MIMO QFT—Method 2[7,119]

Method 2 is a formal reformulation and generalization of *Method 1*. It avoids the former hypotheses of diagonal dominance and simplifies the calculations for the off-diagonal elements. The method also reformulates the classical matrix definition of MIMO specifications by designing a new set of loop-by-loop QFT bounds on the Nichols chart, with necessary and sufficient conditions. It also gives explicit expressions to share the load among the loops of the MIMO system to achieve the matrix specifications. And all for stability, reference tracking, disturbance rejection at plant input and output, and noise attenuation problems.[119]

One of the most significant differences observed between H-infinity control techniques and MIMO QFT techniques is the way of defining the performance specifications. While classical techniques deal with a matrix definition of specifications (e.g., $||T(j\omega)||_\infty < \delta_1$ or $||S(j\omega)||_\infty < \delta_2$), QFT describes the specifications in terms of a loop-by-loop definition (e.g., $|t_{ii}(j\omega)| < \delta_r(\omega)$ or $|s_{ii}(j\omega)| < \delta_k(\omega)$). This fact is analyzed in Section 8.6.5, where the matrix MIMO specifications are formally translated into a set of loop-by-loop QFT bounds on the Nichols chart, with necessary and sufficient conditions.

Sections 8.6.1 through 8.6.5 introduce the reformulation of the full-matrix MIMO QFT technique. In particular, Sections 8.6.1 through 8.6.4 present the new methodology itself, and Section 8.6.5 presents the technique to translate the classical matrix specifications into the QFT methodology. Additionally, Section 8.9 presents a detailed example of the application of the methodology.

8.6.1 Non-Diagonal MIMO QFT Reformulation

The first objective of the non-diagonal elements (g_{ij}, $i \neq j$) of the full-matrix controller $G(s)$ is to minimize the loop interactions (coupling). They act as feedforward functions, cancelling some of the dynamics of the non-diagonal elements (p_{ij}, $i \neq j$) of the plant matrix $P(s)$. As a consequence, they are sensitive to model uncertainty. But their inclusion relieves the design of the diagonal compensators by reducing the amount of feedback (bandwidth) necessary to fulfill the performance specifications. They can also be required to achieve internal stability in some systems where diagonal or triangular controllers are not enough.[92] By contrast, the diagonal elements of the controller (g_{ii}) act as feedback functions, minimizing the effect of the uncertainty (sensitivity) and fulfilling the performance and stability specifications. They also have to overcome the uncertainty of the non-diagonal plant elements (p_{ij}, $i \neq j$), minimizing the residual coupling that the non-diagonal elements of the controller (g_{ij}, $i \neq j$) have not cancelled.

Taking these considerations as a starting point, the reformulation presented in this section for the full-matrix controller design methodology adopts a four step procedure (*A* through *D*). The dominant characteristic of the system will determine the way of designing the off-diagonal compensators: reference tracking or disturbance rejection at plant output (Case 1, Section 8.6.2), or disturbance rejection at plant input (Case 2, Section 8.6.3). Of course, independently of which case is chosen, any type of specification (reference tracking, disturbance rejection at plant input and output, noise attenuation, etc.) can be introduced if required when it comes to design the diagonal controller elements, as is usual within the QFT framework. The distinction is therefore just based on the role assigned to the off-diagonal compensators.

8.6.2 Case 1: Reference Tracking and Disturbance Rejection at Plant Output

Consider an $n \times n$ linear multivariable system—see Figure 8.11 with $d_i' = 0$, composed of a plant P, a fully populated matrix controller $G = G_\alpha G_\beta$, a prefilter F, and a plant output disturbance transfer function P_{do}, where $\mathcal{P} \in \mathcal{P}$, and P is the set of possible plants due to uncertainty, and,

$$
P = \begin{bmatrix} p_{11} & p_{12} & \cdots & p_{1n} \\ p_{21} & p_{22} & \cdots & p_{2n} \\ \cdots & \cdots & \cdots & \cdots \\ p_{n1} & p_{n2} & \cdots & p_{nn} \end{bmatrix}; \quad
F = \begin{bmatrix} f_{11} & f_{12} & \cdots & f_{1n} \\ f_{21} & f_{22} & \cdots & f_{2n} \\ \cdots & \cdots & \cdots & \cdots \\ f_{n1} & f_{n2} & \cdots & f_{nn} \end{bmatrix}
\tag{8.90}
$$

$$
G = \begin{bmatrix} g_{11} & g_{12} & \cdots & g_{1n} \\ g_{21} & g_{22} & \cdots & g_{2n} \\ \cdots & \cdots & \cdots & \cdots \\ g_{n1} & g_{n2} & \cdots & g_{nn} \end{bmatrix} = G_\alpha \, G_\beta;
$$

$$
G_\alpha = \begin{bmatrix} g_{11}^\alpha & g_{12}^\alpha & \cdots & g_{1n}^\alpha \\ g_{21}^\alpha & g_{22}^\alpha & \cdots & g_{2n}^\alpha \\ \cdots & \cdots & \cdots & \cdots \\ g_{n1}^\alpha & g_{n2}^\alpha & \cdots & g_{nn}^\alpha \end{bmatrix}; \quad
G_\beta = \begin{bmatrix} g_{11}^\beta & 0 & \cdots & 0 \\ 0 & g_{22}^\beta & \cdots & 0 \\ \cdots & \cdots & \cdots & \cdots \\ 0 & 0 & \cdots & g_{nn}^\beta \end{bmatrix}
\tag{8.91}
$$

Note that for the sake of clarity, the dependence on (s) is omitted in this section. The reference vector r' and the external disturbance vector at plant output d_o' are the inputs of the system. The output vector y is the variable to be controlled. The plant inverse P^* is

$$
P^{-1} = P^* = \left[p_{ij}^* \right] = \begin{bmatrix} p_{11}^* & p_{12}^* & \cdots & p_{1n}^* \\ p_{21}^* & p_{22}^* & \cdots & p_{2n}^* \\ \vdots & \vdots & \ddots & \vdots \\ p_{n1}^* & p_{n2}^* & \cdots & p_{nn}^* \end{bmatrix}
\tag{8.92}
$$

8.6.2.1 Methodology

Step A: Controller Structure
In the same way as in *Method 1* (Section 8.5.9, *Step A*), by using RGA[265] over the frequencies of interest—see Equation 8.93,[260] and taking into account the requirement of minimum complexity for the controller, the method identifies the input–output pairing and the most appropriate structure for the matrix compensator, that is, the required off-diagonal and diagonal elements, which correspond with the significant values of the RGA matrix.[211]

$$
RGA(j\omega) = P(j\omega) \otimes (P^{-1}(j\omega))^T
\tag{8.93}
$$

where $\otimes$ denotes element-by-element multiplication (see Section 8.3.2).

Step B: Design of G_α
The fully populated matrix controller G is composed of two matrices: $G = G_\alpha G_\beta$—see Equation 8.91. The main objective of the pre-compensator G_α is to diagonalize the plant P as much as possible. The initial expression used to determine G_α is

$$G_\alpha = \left[g_{ij}^\alpha\right] = \hat{P}^{-1}\hat{P}_{diag} = \left[\frac{\hat{p}_{jj}\,\hat{\Delta}_{ji}}{\hat{\Delta}}\right]_{ij} \tag{8.94}$$

where $\hat{P}$ is a plant matrix selected within the uncertainty, $\hat{P}_{diag}$ is its diagonal part, $\hat{\Delta}$ is the determinant of the $\hat{P}$ matrix, and $\hat{\Delta}_{ji}$ is the *ji*th cofactor of the $\hat{P}$ matrix. The plant matrix $\hat{P}$ is selected so that the expression of the extended matrix in Equation 8.95 presents the closest form to a diagonal matrix, nulling the off-diagonal terms as much as possible.

$$P^x = PG_\alpha = \left[p_{ij}^x\right] \tag{8.95}$$

The starting point for the design of G_α is the well-known ideal decoupling method.[268] Further discussion on this subject is given in Remark 8.1. However, instead of performing it directly, and to avoid some of its drawbacks, the interest is focused on approximating the frequency response of the matrix in Equation 8.94 over the frequencies of interest and the uncertainty. Thus, the $[g_{ij}^\alpha]$ compensators are shaped following the mean value in magnitude and phase at every frequency of the region plotted by Equation 8.94 within the uncertainty. Due to this uncertainty, no exact cancellation is achieved. The residual coupling is managed through the design of the diagonal feedback controller G_β. Note that since the feedback compensators in G_β are designed by robust MIMO QFT, and not as a decentralized control system, the role of the controller G_α is not to achieve an exact decoupling, but to ease the design of G_β. That is, to reduce the amount of feedback needed to achieve the robust performance specifications.

Besides, this approach allows modifying, when necessary, the final form of the controller G_α in Equation 8.94 so that:

- No RHP or imaginary axis pole-zero cancellation occurs between P and G_α or its elements,
- No RHP transmission elements (Smith-McMillan) are introduced by the controller G_α,
- The relative difference of the number of poles and zeros in each element of the G_α matrix is the same as in Equation 8.94 in order to ease the design of the G_β controller,
- The RGA of the system is improved, looking for positive and close-to-one diagonal elements λ_{ii} of the RGA matrix. That is, the pre-compensator G_α decouples the system to some extent, which is its main goal.

Step C.i.1: Design of the Diagonal Controller Elements g_{ii}^β
After determining G_α, the method proceeds with the design of a diagonal matrix G_β that fulfills the desired robust stability and robust performance specifications for the extended plant $P^x = PG_\alpha$. Its inverse, the P^{x*} matrix in Equation 8.96, is reorganized so that $(p_{11}^{x*})^{-1}$

has the smallest bandwidth, $(p_{22}^{x*})^{-1}$ the next smallest bandwidth, and so on. This eases the existence of a solution and will avoid unnecessary overdesign related to the order in which loops are designed.[6,7]

$$\boldsymbol{P}^{x*} = (\boldsymbol{P}^x)^{-1} = \left[p_{ij}^{x*}\right] = \begin{bmatrix} p_{11}^{x*} & p_{12}^{x*} & \cdots & p_{1n}^{x*} \\ p_{21}^{x*} & p_{22}^{x*} & \cdots & p_{2n}^{x*} \\ \vdots & \vdots & \ddots & \vdots \\ p_{n1}^{x*} & p_{n2}^{x*} & \cdots & p_{nn}^{x*} \end{bmatrix} \qquad (8.96)$$

The compensators g_{ii}^{β} (from $i = 1$ to n) are calculated by using the sequential (loop-by-loop) standard QFT loop-shaping methodology for the inverse of the equivalent extended plant: $q_{xii} = [p_{ii}^{x*e}]_i^{-1}$. This satisfies the recursive relationship of Equation 8.97—see also Reference 313 with $g_{ij} = 0$, $i \neq j$, and takes into account the compensator elements previously designed—this is the so-called *sequential method*.

$$\left[p_{ij}^{x*e}\right]_k = \left[p_{ij}^{x*e}\right]_{k-1} - \frac{\left[p_{i(k-1)}^{x*e}\right]_{k-1} \left[p_{(k-1)j}^{x*e}\right]_{k-1}}{\left[p_{(k-1)(k-1)}^{x*e}\right]_{k-1} + g_{(k-1)(k-1)}^{\beta}} ; \quad i,j \geq k; \quad \left[\boldsymbol{P}^{x*e}\right]_{k=1} = \boldsymbol{P}^{x*} \qquad (8.97)$$

The presence of model uncertainty reduces the diagonalization effect of the precompensator $\boldsymbol{G}_{\alpha}$ over $\boldsymbol{P}$. This diagonalization is a real cancellation only when the uncertainty of the plant $\boldsymbol{P}$ is exactly at the working point $\hat{\boldsymbol{P}}$, and $\boldsymbol{P}\boldsymbol{G}_{\alpha} = \hat{\boldsymbol{P}}\hat{\boldsymbol{P}}^{-1}\hat{\boldsymbol{P}}_{diag} = \hat{\boldsymbol{P}}_{diag}$. If the plant $\boldsymbol{P}$ is working on a different point, the extended plant that the compensator $\boldsymbol{G}_{\beta}$ sees includes off-diagonal elements as well. Consequently, the performance specifications (disturbance rejection) for the elements g_{ii}^{β} have to be more demanding in order to avoid the residual coupling. This is an easy solution when using this MIMO QFT methodology, which takes into account the coupling effect loop-by-loop.[6,7]

Step C.i.2: Design of the Diagonal Prefilter Elements f_{ii}
The design of a prefilter $\boldsymbol{F}$ is necessary in case of reference tracking specifications. As one of the objectives of the full-matrix controller $\boldsymbol{G}_{\alpha}$ is to reduce the loop interaction, the prefilter $\boldsymbol{F}$ can be matrix diagonal. The f_{ii} prefilter element is designed for the equivalent extended plant, $q_{xii} = [p_{ii}^{x*e}]_i^{-1}$, and with the diagonal controller g_{ii}^{β} designed in the previous *Step C.i.1*.

After we finish the *Steps C.i*, $i = 1$ to n, we proceed with *Step D*.

Step D: Final Checks
Once the design of $\boldsymbol{G} = \boldsymbol{G}_{\alpha}\boldsymbol{G}_{\beta}$ has finished, the design has also to fulfill two more stability conditions:[109] (c) no Smith-McMillan pole-zero cancellations have to occur between $\boldsymbol{P}$ and $\boldsymbol{G}$ and (d) no Smith-McMillan pole-zero cancellations have to occur in $|\boldsymbol{P}^* + \boldsymbol{G}|$.

Also, although very remote, the possibility of introducing RHP transmission zeros due to the controller design theoretically exists. This undesirable situation cannot be detected until the whole multivariable system design is completed. To avoid this problem, the multivariable zeros of $\boldsymbol{PG}$ are determined using the Smith-McMillan form over the set of possible plants $\boldsymbol{P}$ due to the uncertainty. If there are no new RHP zeros apart from those that might already be present in $\boldsymbol{P}$, the method concludes. Otherwise, we rectify this undesirable situation by modifying the non-diagonal elements placed in the last column of the matrix G, according to the Smith-McMillan expressions.[115]

Analysis of the 2 × 2 Case

By using Equations 8.90 through 8.92 and 8.94, the expression of the final controller G for a 2 × 2 MIMO plant is

$$G = G_\alpha G_\beta = \begin{bmatrix} g_{11} & g_{12} \\ g_{21} & g_{22} \end{bmatrix} = \begin{bmatrix} \dfrac{\hat{p}_{11}\hat{p}_{22}}{\hat{p}_{11}\hat{p}_{22} - \hat{p}_{12}\hat{p}_{21}} g_{11}^\beta & \dfrac{-\hat{p}_{12}}{\hat{p}_{11} - (\hat{p}_{12}\hat{p}_{21}/\hat{p}_{22})} g_{22}^\beta \\ \dfrac{-\hat{p}_{21}}{\hat{p}_{22} - (\hat{p}_{12}\hat{p}_{21}/\hat{p}_{11})} g_{11}^\beta & \dfrac{\hat{p}_{11}\hat{p}_{22}}{\hat{p}_{11}\hat{p}_{22} - \hat{p}_{12}\hat{p}_{21}} g_{22}^\beta \end{bmatrix} \tag{8.98}$$

For tracking or disturbance rejection at plant output, the expression for 2 × 2 MIMO plants of the previous non-diagonal MIMO QFT methodology (see *Method 1*, Sections 8.5.1 through 8.5.10) was:

$$G = \begin{bmatrix} g_{11} & g_{12} \\ g_{21} & g_{22} \end{bmatrix} = \begin{bmatrix} g_{11} & \dfrac{-\hat{p}_{12}}{\hat{p}_{11}} g_{22} \\ \dfrac{-\hat{p}_{21}}{\hat{p}_{22}} g_{11} & g_{22} \end{bmatrix} \tag{8.99}$$

Method 2 is more general than *Method 1*, as it does not need the hypothesis described in Equations 8.63 and 8.64. The resulting off-diagonal compensators, given by Equation 8.98, are the same as those of *Method 1*, given by Equation 8.99, plus some new terms. Precisely, these new terms in Equation 8.98 disappear and the diagonal compensators reduce to the feedback ones if $\left(\left| \hat{p}_{11}\hat{p}_{22} \right| \gg \left| \hat{p}_{12}\hat{p}_{21} \right| \right)$ is applied, which is the diagonal dominance hypothesis of the *Method 1*—see Equations 8.63 and 8.64. Thus, *Method 2* generalizes and broadens the scope of the preceding *Method 1*. Further discussion on the comparison of both techniques is developed in Section 8.7.

Remark 8.1

Ideal decoupling belongs to the more general group of inverse-based controllers, and has been extensively discussed, especially in the chemical engineering literature and in the context of distillation columns.[239,268,286,292] The works mentioned, which are solely based upon observation of the dual composition control of distillation towers,[314] encountered some drawbacks for this inverse-based approach, such as the sensitivity of decouplers to modeling errors or the possibly complicated expressions for the decouplers.

However, *Method 2* uses ideal decoupling just as a starting expression for Equation 8.94, whose frequency response is approximated in the range of frequencies of interest and for the existing uncertainty. This avoids the use of excessively complicated expressions, potential problems of realization and implementation, and the cancellation of RHP elements—see Section 8.6.2, *Step B*. Moreover, the subsequent G_β feedback controller is designed through MIMO QFT, by taking into account the residual coupling, and not through decentralized control, which has been the approach traditionally used in conjunction with decoupling.

On the other hand, there exists high sensitivity of inverse-based controllers to input uncertainty when the plant has large RGA elements.[260,303] If the input uncertainty is critical for the system under study, the designer should then consider the off-diagonal compensators as a means of fighting against this disturbance at the plant input, and consequently select Case 2 (Section 8.6.3) for their design. Likewise, for extreme cases, one-way or triangular decouplers could be applied instead, since they are much less sensitive to input uncertainty.

Remark 8.2

Notice that if the plant $P(s)$ has no uncertainty and the resulting controller $G_\alpha(s) = P^{-1}(s)$ $P_{diag}(s)$ is stable, minimum phase, and proper, then $P(s)G_\alpha(s) = P_{diag}(s)$ exactly. As a result, the design of $G_\beta(s)$ is very simple. It does not need Equation 8.97, and is carried out as n SISO systems: that is, $g_{11}^\beta(s)$ for $p_{11}(s)$, $g_{22}^\beta(s)$ for $p_{22}(s)$, and so on.

8.6.3 Case 2: Disturbance Rejection at Plant Input

In case the designer wants to use the off-diagonal compensators to deal with disturbance rejection at plant input specifications—see Figure 8.11 with $r' = 0$ and $d_o' = 0$, the controller design methodology is based on the one defined in References 6, 7, 110, 114, 115, 206, 209, and 211, with some additional modifications and remarks. Now the full-matrix controller is given by $G = G_d + G_b$, where

$$G = \begin{bmatrix} g_{11} & g_{12} & \cdots & g_{1n} \\ g_{21} & g_{22} & \cdots & g_{2n} \\ \cdots & \cdots & \cdots & \cdots \\ g_{n1} & g_{n2} & \cdots & g_{nn} \end{bmatrix} = G_d + G_b;$$

$$G_d = \begin{bmatrix} g_{11} & 0 & \cdots & 0 \\ 0 & g_{22} & \cdots & 0 \\ \cdots & \cdots & \cdots & \cdots \\ 0 & 0 & \cdots & g_{nn} \end{bmatrix}; \quad G_b = \begin{bmatrix} 0 & g_{12} & \cdots & g_{1n} \\ g_{21} & 0 & \cdots & g_{2n} \\ \cdots & \cdots & \cdots & \cdots \\ g_{n1} & g_{n2} & \cdots & 0 \end{bmatrix}$$

(8.100)

and the transfer function between the disturbance d_i and the output y is

$$y = [I + PG]^{-1} P d_i = [P^* + G]^{-1} d_i$$

(8.101)

In this particular case, the objectives of the controller are to reject external disturbances at plant input d_i and to reduce the loop interaction. In other words, the disturbance rejection specification can be defined as a diagonal matrix $T_D = [t_{dii}] = y/d_i$, which in Equation 8.101 is

$$T_D = (P^* + G)^{-1}$$

(8.102)

or,

$$G = T_D^{-1} - P^{-1}$$

(8.103)

The methodology for this Case 2 follows the same steps discussed for Case 1, Section 8.6.2, with the changes in *Steps B* and *C.i.1* presented below.

Step B: Design of G_b
Based on Equation 8.103, the off-diagonal elements g_{ij} $(i \neq j)$ of the controller (G_b part) are calculated first, so that

$$g_{ij}(i \neq j) = -\hat{p}^*_{ij} \qquad (8.104)$$

where ($\wedge$) denotes the plant that minimizes the maximum of the nonparametric uncertainty radii comprising the plant templates on the Nichols chart.[6,7] That is to say, the compensator g_{ij} $(i \neq j)$ is shaped following the mean value at every frequency $\omega \in [\omega_{min}, \omega_{max}]$ of the region plotted by the uncertain plants $[-p_{ij}^*]$ in magnitude and phase.

Step C.i.1: Design of G_d
The diagonal elements gii(s) are calculated through standard QFT loop shaping and for the inverse of the equivalent plant $q_{ii} = [p_{ii}^{*e}]_k^{-1}$. As always, they can consider robust stability and robust performance specifications.[6,7,110,114,115,206,209,211] The equivalent plant satisfies the recursive relationship Equation 8.105.[313]

$$\left[p_{ij}^{*e}(s)\right]_k = \left[p_{ij}^{*e}(s)\right]_{k-1} - \frac{\left(\left[p_{i(k-1)}^{*e}(s)\right]_{k-1} + \left[g_{i(k-1)}(s)\right]_{k-1}\right)\left(\left[p_{(k-1)j}^{*e}(s)\right]_{k-1} + \left[g_{(k-1)j}(s)\right]_{k-1}\right)}{\left[p_{(k-1)(k-1)}^{*e}(s)\right]_{k-1} + \left[g_{(k-1)(k-1)}(s)\right]_{k-1}}; \, i,j \geq k; \quad (8.105)$$

$$\left[P^{*e}(s)\right]_{k=1} = P^*(s)$$

Based on Equation 8.103, a starting expression for the diagonal elements g_{ii} in the loop-shaping stage could be: $g_{ii} = t_{dii}^* - \hat{p}_{ii}^*$.

8.6.4 Stability Conditions and Final Implementation

Stability Conditions. The closed-loop stability of the MIMO system with the full-matrix controller $G = [g_{ij}]$ is guaranteed by the following sufficient conditions, previously introduced in Section 8.5.10:[109]

c1. Each $L_i(s) = a_{ii}(s)b_{ii}(s)$, $i = 1, ...,n$, satisfies the Nyquist encirclement condition, where $a_{ii}(s) = g_{ii}^\beta(s)$, $b_{ii}(s) = [p_{ii}^{*e}(s)]_i^{-1}$ for Case 1 (Section 8.6.2), and $a_{ii}(s) = g_{ii}(s)$, $b_{ii}(s) = [p_{ii}^{*e}(s)]_i^{-1}$ for Case 2 (Section 8.6.3),

c2. No RHP pole-zero cancellations occur between $a_{ii}(s)$ and $b_{ii}(s)$, $i = 1,...,n$, $a_{ii}(s)$ and $b_{ii}(s)$ being those defined in the previous point,

c3. No Smith-McMillan pole-zero cancellations occur between $P(s)$ and $G(s)$, and

c4. No Smith-McMillan pole-zero cancellations occur in $|P^*(s) + G(s)|$.

Conditions c1 and c2 are checked loop-by-loop when the compensators $a_{ii}(s)$ are calculated in *Step C*, where $a_{ii}(s) = g_{ii}^\beta(s)$ and $a_{ii}(s) = g_{ii}(s)$, for Case 1 and Case 2, respectively. Conditions c3 and c4 are checked after *Step C* is finished.

Final Implementation
The final implementation of the full-matrix controller G (either $G_\alpha G_\beta$ or $G_d + G_b$) will take the form

$$F_{pd}(G) = \begin{bmatrix} F_{pd}(g_{11}) & F_{pd}(g_{12}) & \cdots & F_{pd}(g_{1n}) \\ F_{pd}(g_{21}) & F_{pd}(g_{22}) & \cdots & F_{pd}(g_{2n}) \\ \cdots & \cdots & \cdots & \cdots \\ F_{pd}(g_{n1}) & F_{pd}(g_{n2}) & \cdots & F_{pd}(g_{nn}) \end{bmatrix} \tag{8.106}$$

where the function $F_{pd}(A)$ means in every case a proper and causal function. If it is necessary to make any modification to get a proper function, it will preserve the dominant poles and zeros (low and medium frequency) of the expression A.

Remark 8.3

Once the design is completed, the Smith-McMillan form of the matrix compensator $G(s)$ must be analyzed to check whether it has introduced additional RHP transmission zeros. If so, a redesign of the non-diagonal elements of the last column of the compensator can typically fix the problem.[115]

8.6.5 Translating Matrix Performance Specifications[119]

One of the main differences between MIMO QFT techniques and the H-infinity MIMO control methods is the way to define the specifications. While the latter are able to define them in terms of matrices (SVD, $||T(j\omega)||_\infty < \delta_1$ or $||S(j\omega)||_\infty < \delta_2$, etc.), QFT usually does the work loop-by-loop.

Once the equilibrium among loops is attained, the interaction between the loops becomes a trade-off, and the burden on any loop cannot be reduced without increasing it on some other loops. In this section an approach to deal with the classical matrix specifications within QFT is introduced, solving the mentioned problem.

8.6.5.1 Case $n \times n$

Consider an $n \times n$ linear MIMO system (Figure 8.11), composed of a set of uncertain plants, $\mathcal{P}(j\omega_i) = \{P(j\omega_i), \omega_i \in \cup \Omega_k\}$, a full-matrix controller G, and a prefilter F, both to be designed. r, n, d_i, d_o and y are vectors representing respectively the reference input, sensor noise input, external disturbances at plant input and output, and plant output.

Tracking
Looking at Figure 8.11, the reference tracking problem gives the following matrix equations:

$$y = T r \tag{8.107}$$

$$y = [P^* + G]^{-1} G F r \tag{8.108}$$

$$[P^* + G]T = G F \tag{8.109}$$

Disturbance Rejection at Plant Output
Analogously, the disturbance rejection at plant output problem gives the following matrix equations:

$$y = S_o \, d_o \tag{8.110}$$

$$y = [P^* + G]^{-1} P^* \, d_o \tag{8.111}$$

$$[P^* + G] S_o = P^* \tag{8.112}$$

Disturbance Rejection at Plant Input

In the same way, the disturbance rejection at plant input problem gives the following matrix equations:

$$y = S_i \, d_i \tag{8.113}$$

$$y = [P^* + G]^{-1} d_i \tag{8.114}$$

$$[P^* + G] S_i = I \tag{8.115}$$

Noise Attenuation

Finally, the noise attenuation problem yields:

$$y = S_n \, n \tag{8.116}$$

$$y = [P^* + G]^{-1} (-G) n \tag{8.117}$$

$$[P^* + G] S_n = -G \tag{8.118}$$

General Expression

The four previous problems (Equations 8.109, 8.112, 8.115, and 8.118) can be written in a compact form, so that,

$$[P^* + G] \alpha = \beta \tag{8.119}$$

where, depending on the case, the matrices α and β are defined so that,

- *Tracking:* $\alpha = T$; $\beta = GF$
- *Disturbance rejection at plant output:* $\alpha = S_o$; $\beta = P^*$
- *Disturbance rejection at plant input:* $\alpha = S_i$; $\beta = I$
- *Noise attenuation:* $\alpha = S_n$; $\beta = -G$

Now, in order to make $(P^* + G)$ in Equation 8.119 a triangular matrix (which allows explicit expressions for each loop problem), and following the Gauss elimination method, the equation is pre-multiplied by

$(M_{n-1}\,M_{n-2}\,M_{n-3}\cdots M_1)$ so that,

$$M_{n-1}\,M_{n-2}\,M_{n-3}\cdots M_1[P^* + G]\,\alpha = M_{n-1}\,M_{n-2}\,M_{n-3}\cdots M_1\beta \qquad (8.120)$$

where

$$M_k = \begin{bmatrix} & & & Column(k) & & & \\ 1 & 0 & \cdots & 0 & & \cdots & 0 \\ 0 & 1 & \cdots & 0 & & \cdots & 0 \\ \cdots & \cdots & \cdots & \cdots & & \cdots & \cdots \\ 0 & 0 & \cdots & 1 & & \cdots & 0 \\ 0 & 0 & \cdots & \dfrac{-\left(p^{*k}_{k+1,k} + g_{k+1,k}\right)}{p^{*k}_{k,k} + g_{k,k}} & & \cdots & 0 & row(k+1) \\ \cdots & \cdots & \cdots & \ddots & & \cdots & \cdots \\ 0 & 0 & \cdots & \dfrac{-\left(p^{*k}_{n,k} + g_{n,k}\right)}{p^{*k}_{k,k} + g_{k,k}} & & \cdots & 1 \end{bmatrix} \qquad (8.121)$$

$$p^{*k}_{ij} = p^{*k-1}_{ij} - \dfrac{\left(p^{*k-1}_{i,k-1} + g_{i,k-1}\right)\left(p^{*k-1}_{k-1,j} + g_{k-1,j}\right)}{p^{*k-1}_{k-1,k-1} + g_{k-1,k-1}}$$

$$\text{for}\quad i = k\cdots n; j = k\cdots n; k = 1\cdots n;\quad\text{and}\quad \left[p^{*1}_{ij}\right] = [P]^{-1} \qquad (8.122)$$

After some calculations, the first part of Equation 8.120 is

$$M_{n-1}\,M_{n-2}\,M_{n-3}\cdots M_1[P^* + G]\,\alpha$$

$$= \begin{bmatrix} \sum\limits_{k=1}^{n}\left(p^{*1}_{1k} + g_{1k}\right)\alpha_{k1} & \sum\limits_{k=1}^{n}\left(p^{*1}_{1k} + g_{1k}\right)\alpha_{k2} & \cdots & \sum\limits_{k=1}^{n}\left(p^{*1}_{1k} + g_{1k}\right)\alpha_{kn} \\ \sum\limits_{k=2}^{n}\left(p^{*2}_{2k} + g_{2k}\right)\alpha_{k1} & \sum\limits_{k=2}^{n}\left(p^{*2}_{2k} + g_{2k}\right)\alpha_{k2} & \cdots & \sum\limits_{k=2}^{n}\left(p^{*2}_{2k} + g_{2k}\right)\alpha_{kn} \\ \cdots & \cdots & \cdots & \cdots \\ \sum\limits_{k=n}^{n}\left(p^{*n}_{nk} + g_{nk}\right)\alpha_{k1} & \sum\limits_{k=n}^{n}\left(p^{*n}_{nk} + g_{nk}\right)\alpha_{k2} & \cdots & \sum\limits_{k=n}^{n}\left(p^{*n}_{nk} + g_{nk}\right)\alpha_{kn} \end{bmatrix} \qquad (8.123)$$

In other words, in terms of the element *ij* of the matrix,

$$M_{n-1}\,M_{n-2}\,M_{n-3}\cdots M_1[P^* + G]\,\alpha = \left[\sum\limits_{k=i}^{n}\left(p^{*i}_{ik} + g_{ik}\right)\alpha_{kj}\right]_{ij} \qquad (8.124)$$

$$\text{for}\quad i = 1\cdots n; j = 1\cdots n$$

with

$$p_{ik}^{*i} = p_{ik}^{*i-1} - \frac{\left(p_{i,i-1}^{*i-1} + g_{i,i-1}\right)\left(p_{i-1,k}^{*i-1} + g_{i-1,k}\right)}{p_{i-1,i-1}^{*i-1} + g_{i-1,i-1}}$$

$$\text{for} \quad i = 1 \cdots n; j = 1 \cdots n; \quad \text{and} \left[p_{ik}^{*1}\right] = [P]^{-1}$$

(8.125)

making

$$q_{ii}^i = \frac{1}{p_{ii}^{*i}}; \quad \text{and} \quad L_i^i = q_{ii}^i \, g_{ii}$$

(8.126)

Equation 8.124 becomes,

$$M_{n-1} M_{n-2} M_{n-3} \cdots M_1 [P^* + G]\alpha = \left[\frac{1+L_i^i}{q_{ii}^i}\alpha_{ij} + \sum_{k=i+1}^{n}\left(p_{ik}^{*i} + g_{ik}\right)\alpha_{kj}\right]_{ij}$$

$$\text{for} \quad i = 1 \cdots n; j = 1 \cdots n$$

(8.127)

$$[\alpha_{ij}] = \begin{cases} T \text{ (for tracking)} \\ S_o \text{ (for disturbance rejection at plant ouput)} \\ S_i \text{ (for disturbance rejection at plant input)} \\ S_n \text{ (for noise attenuation)} \end{cases}$$

(8.128)

Analogously, after some calculations, the second part of Equation 8.120 is

$$M_{n-1} M_{n-2} M_{n-3} \cdots M_1 \beta = \left[\left(\sum_{x=1}^{i-1} a_{ix}\,\beta_{xj}\right) + \beta_{ij}\right]_{ij}$$

(8.129)

where,

$$a_{ij} = 1 \text{ (if } i = j\text{)}; a_{ij} = 0 \text{ (if } i < j\text{)};$$

$$a_{ij} = \sum_{\substack{\text{from } m=1 \\ \text{to } m=i-j}}^{\binom{i-j-1}{m-1}\text{elements}} \prod_{m \text{ elements}} \frac{-\left(p_{ab}^{*b} + g_{ab}\right)}{\left(p_{bb}^{*b} + g_{bb}\right)} \text{ (if } i > j\text{)}$$

(8.130)

and where all the possible combinations of couples *ab* from $a = i$ to $b = j$ exist, without repetition of couples *ab* and with $a \neq b$, with:

$$\binom{i-j-1}{m-1} = \frac{(i-j-1)!}{(m-1)!(i-j-m)!}$$

$$0 \leq (m-1) \leq (i-j-1); 1 \leq m \leq (i-j)$$

and where

$$[\beta_{ij}] = \begin{cases} \boldsymbol{G}\,\boldsymbol{F}\,(\text{for tracking}) \\ \boldsymbol{P}^*\,(\text{for disturbance rejection at plant ouput}) \\ \boldsymbol{I}\,(\text{for disturbance rejection at plant input}) \\ -\boldsymbol{G}\,(\text{for noise attenuation}) \end{cases} \qquad (8.131)$$

Combining Equation 8.127 through 8.131 in Equation 8.120, the explicit expression of α_{ij} is

$$\alpha_{ij} = \frac{q_{ii}^i}{1 + L_i^i}\left[\left(\sum_{x=1}^{i-1} a_{ix}\,\beta_{xj}\right) + \beta_{ij} - \sum_{k=i+1}^{n}\left(p_{ik}^{*i} + g_{ik}\right)\alpha_{kj}\right]_{ij} \qquad (8.132)$$

Now, (1) defining the desired matrix performance specification and its elements $\tau = [\tau_{ij}]$ for the matrices $\alpha = [\alpha_{ij}]$, where $\alpha = T$ or $\alpha = S_o$ or $\alpha = S_i$ or $\alpha = S_n$ (depending on the case); (2) comparing the particular element α_{ij} of Equation 8.132 with its specification, $\alpha_{ij} \leq \tau_{ij}$; and (3) substituting the remaining α_{kj} ($k = i + 1$ to n), which are still unknown, with their correspondent specification τ_{kj}, the general expression for the performance specifications becomes

$$\left|\frac{q_{ii}^i}{1 + L_i^i}\right| \leq \frac{\tau_{ij}}{\left|\left\{\left[\sum_{x=1}^{i-1} a_{ix}\,\beta_{xj}\right] + \beta_{ij} - \sum_{k=i+1}^{n}\left(p_{ik}^{*i} + g_{ik}\right)\tau_{kj}\right\}_{ij}\right|_{\max}} \qquad (8.133)$$

where

$$a_{ix} = 1\,(\text{if } i = x);\ a_{ix} = 0\,(\text{if } i < x);$$

$$a_{ix} = \sum_{\substack{\text{from } m=1 \\ \text{to } m=i-x}}^{\binom{i-x-1}{m-1}\text{elements}} \prod_{m\text{ elements}} \frac{-\left(p_{ab}^{*b} + g_{ab}\right)}{\left(p_{bb}^{*b} + g_{bb}\right)}(\text{if } i > x)$$

and where all the possible combinations of couples ab from $a = i$ to $b = x$ exist, without repetition of couples ab and with $a \neq b$, with:

$$\binom{i-x-1}{m-1} = \frac{(i-x-1)!}{(m-1)!(i-x-m)!}$$
$$0 \leq (m-1) \leq (i-x-1); 1 \leq m \leq (i-x)$$

for the element ij, and where the elements of α_{ij} have been replaced with their specification τ_{ij}, and where:

- *Tracking:* $[\tau_{ij}] = spec(\boldsymbol{T})$; $[\beta_{ij}] = \boldsymbol{G}\,\boldsymbol{F}$
- *Disturbance rejection at plant output:* $[\tau_{ij}] = spec(\boldsymbol{S}_o)$; $[\beta_{ij}] = \boldsymbol{P}^*$

- *Disturbance rejection at plant input:* $[\tau_{ij}] = spec(S_i); \quad [\beta_{ij}] = I$
- *Noise attenuation:* $[\tau_{ij}] = spec(S_n); \quad [\beta_{ij}] = -G$

The notation $|\bullet|_{max}$ in Equation 8.133 means the maximum absolute value within the plant uncertainty. This expression describes the sufficient condition (C_{suf}) to fulfill the specification. It is also possible to define a necessary condition (C_{nec}) to fulfill the specification with the notation $|\bullet|_{min}$, which means the minimum absolute value within the plant uncertainty. In other words:

$$\left| \frac{q_{ii}^i}{1 + L_i^i} \right| \leq C_{suf} \leq C_{nec} \tag{8.134}$$

where the sufficient condition C_{suf} is

$$C_{suf} = \frac{\tau_{ij}}{\left| \left\{ \left(\sum_{x=1}^{i-1} a_{ix} \beta_{xj} \right) \cdot \mid \beta_{ij} - \sum_{k=i+1}^{n} \left(p_{ik}^{*i} + g^{ik} \right) \tau_{kj} \right\}_{ij} \right|_{max}} \tag{8.135}$$

and the necessary condition C_{nec} is

$$C_{nec} = \frac{\tau_{ij}}{\left| \left\{ \left(\sum_{x=1}^{i-1} a_{ix} \beta_{xj} \right) + \beta_{ij} - \sum_{k=i+1}^{n} \left(p_{ik}^{*i} + g^{ik} \right) \tau_{kj} \right\}_{ij} \right|_{min}} \tag{8.136}$$

These last equations generate two different bounds on the Nichols chart for the loop-by-loop controller design method: the *sufficient bound* and the *necessary bound* for every type of matrix specifications and every frequency (see Figure 8.18). The controller design stage (loop shaping) will consider both, depending on the level of conservativeness/robustness needed for the design.

Remark 8.4

It is through the corresponding matrix $[\tau_{ij}]$ for each case (tracking, disturbance rejection at plant output and plant input, and noise attenuation) that global matrix specifications can be taken into account. The designer has at his/her disposal a tool of great versatility. On the one hand, the control engineer can deliberately decide how much burden is allocated to each loop on the basis of the constraints that a particular application presents (restrictions on bandwidth, presence of flexible modes, noise, capacity of actuators, precision needed for a particular output, etc.), which are the real engineering requirements. In this decision on how to share the load among loops in a realistic way (i.e., not demanding simultaneously unachievable loop performance levels), the explicit expressions on interaction among loop specifications developed for the case 2×2 in the next section are used as a guideline. The specification matrix elements τ_{ij} are then accordingly shaped.

At the same time and prior to the design, the designer can make sure that these specifications fulfill a required H-infinity norm (or even the desired frequency response for the maximum singular value of the matrix). This is an additional step which is not typically considered in MIMO QFT techniques, where the H-infinity norm behavior has to be assessed

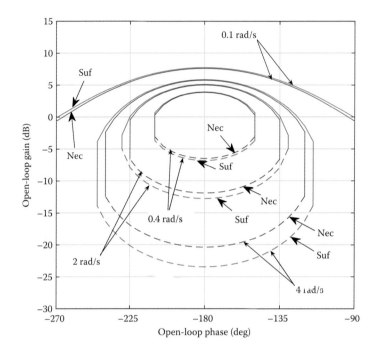

FIGURE 8.18
Example: Necessary and sufficient QFT bounds for matrix specifications on $S(s)$. (Adapted from Garcia-Sanz, M. and Houpis, C.H. 2012. *Wind Energy Systems: Control Engineering Design (Part I: QFT Control, Part II: Wind Turbine Design and Control).* A CRC Press book, Taylor & Francis Group, Boca Raton, FL.)

after the design is performed. The approach introduced in this section allows the QFT designer to make sure that the defined specifications fulfill the matrix requirements as well.

Of course, it leads to choosing one among many possibilities, but this is a degree of freedom explicitly and deliberately used by the control engineer prior to the controller design and on the basis of the rest of the engineering requirements.

On the other hand, if the focus is on optimizing the H-infinity norm instead of predetermining the directionality of the specification matrix $[\tau_{ij}]$, an H-infinity design could be previously performed. The resulting τ_{ij}, optimum from the H-infinity point of view, could then be introduced as specifications for this procedure. In both situations, the methodology assures the translation of constraints on all matrix elements to requirements on the design of each feedback loop. Besides, these translated specifications can be combined with classical QFT loop-by-loop specifications.

8.6.5.2 Case 2 × 2

In this section, Equation 8.133 is particularized for the case 2×2, and for the four classical problems: tracking, disturbance rejection at plant output and input, and noise attenuation.

Tracking
Consider the following specifications:

$$\begin{bmatrix} y_1 \\ y_2 \end{bmatrix} = \begin{bmatrix} a_{11} \leq t_{11} \leq b_{11} & 0 \\ 0 & a_{22} \leq t_{22} \leq b_{22} \end{bmatrix} \begin{bmatrix} r_1 \\ r_2 \end{bmatrix} \quad \text{(for tracking)} \qquad (8.137)$$

$$\begin{bmatrix} y_1 \\ y_2 \end{bmatrix} = \begin{bmatrix} \tau_{c11} & \tau_{c12} \\ \tau_{c21} & \tau_{c22} \end{bmatrix} \begin{bmatrix} r_1 \\ r_2 \end{bmatrix} \quad \text{(for coupling attenuation)} \qquad (8.138)$$

Now, Equation 8.133 yields the following sufficient conditions:

- *Loop 1*

$$a'_{11} \leq \left| \frac{L_1^1}{1 + L_1^1} \right| \leq b'_{11} \quad \text{with } a'_{11} = a_{11} + \tau_{c11}; \ b'_{11} = b_{11} - \tau_{c11} \qquad (8.139)$$

$$\left| \frac{q_{11}^1}{1 + L_1^1} \right| \leq \frac{\tau_{c11}}{\left| \left((1/q_{12}^1) + g_{12} \right) \tau_{c21} \right|_{\max}} \qquad (8.140)$$

$$\left| \frac{q_{11}^1}{1 + L_1^1} \right| \leq \frac{\tau_{c12}}{\left| g_{12} - \left((1/q_{12}^1) + g_{12} \right) b_{22} \right|_{\max}} \qquad (8.141)$$

- *Loop 2*

$$a'_{22} \leq \left| \frac{L_2^2}{1 + L_2^2} \right| \leq b'_{22} \quad \text{with} \quad a'_{22} = a_{22} + \tau_{c22}; \ b'_{22} = b_{22} - \tau_{c22} \qquad (8.142)$$

$$\left| \frac{q_{22}^2}{1 + L_2^2} \right| \leq \frac{\tau_{c22}}{\left| \left(q_{11}^1 / 1 + L_1^1 \right) \left((1/q_{21}^1) + g_{21} \right) g_{12} \right|_{\max}} \qquad (8.143)$$

$$\left| \frac{q_{22}^2}{1 + L_2^2} \right| \leq \frac{\tau_{c21}}{\left| g_{21} - \left(q_{11}^1 / 1 + L_1^1 \right) \left((1/q_{21}^1) + g_{21} \right) g_{11} \right|_{\max}} \qquad (8.144)$$

Note that there is a specific interaction between Equations 8.140 and 8.144, through the definition of the specification τ_{c21}, and between Equations 8.141 and 8.142, through the definition of the specification b_{22}. If τ_{c21} is higher, then Equation 8.140 is more demanding and Equation 8.144 is less so. Analogously, if b_{22} is higher, then Equation 8.141 is more demanding and Equation 8.142 is less. As a consequence, the proper definition of the specifications τ_{c21} and b_{22} establishes a load-sharing between the loops 1 and 2. Observe that this is not the typical trade-off between performance and coupling concerning specifications from the same loop, but a trade-off of design burden among performance specifications from different loops. This load-sharing problem shows that, when it comes to define the specifications, the designer cannot demand too much to a loop without affecting the achievable performance of the rest of the loops, and can therefore be used as a guideline.

Disturbance Rejection at Plant Output

Consider the following specifications:

$$\begin{bmatrix} y_1 \\ y_2 \end{bmatrix} = \begin{bmatrix} \tau_{so11} & \tau_{so12} \\ \tau_{so21} & \tau_{so22} \end{bmatrix} \begin{bmatrix} d_{o1} \\ d_{o2} \end{bmatrix}$$

(8.145)

Now, Equation 8.133 yields the following sufficient conditions,

- *Loop 1*

$$\left| \frac{q_{11}^1}{1+L_1^1} \right| \leq \frac{\tau_{so11}}{\left| \left(1/q_{11}^1 \right) - \left(\left(1/q_{12}^1 \right) + g_{12} \right) \tau_{so21} \right|_{\max}}$$

(8.146)

$$\left| \frac{q_{11}^1}{1+L_1^1} \right| \leq \frac{\tau_{so12}}{\left| \frac{1}{q_{12}^1} - \left(\left(1/q_{12}^1 \right) + g_{12} \right) \bar{\tau}_{so22} \right|_{\max}}$$

(8.147)

- *Loop 2*

$$\left| \frac{q_{22}^2}{1+L_2^2} \right| \leq \frac{\tau_{so22}}{\left| \frac{1}{q_{22}^1} - \frac{q_{11}^1}{1+L_1^1} \left(\left(1/q_{21}^1 \right) + g_{21} \right) \frac{1}{q_{12}^1} \right|_{\max}}$$

(8.148)

$$\left| \frac{q_{22}^2}{1+L_2^2} \right| \leq \frac{\tau_{so21}}{\left| \frac{1}{q_{21}^1} - \frac{1}{1+L_1^1} \left(\left(1/q_{21}^1 \right) + g_{21} \right) \right|_{\max}}$$

(8.149)

As for reference tracking, there exists a load-sharing between loops 1 and 2 through the specifications τ_{so21} and τ_{so22}.

Disturbance Rejection at Plant Input

Consider the following specifications:

$$\begin{bmatrix} y_1 \\ y_2 \end{bmatrix} = \begin{bmatrix} \tau_{si11} & \tau_{si12} \\ \tau_{si21} & \tau_{si22} \end{bmatrix} \begin{bmatrix} d_{i1} \\ d_{i2} \end{bmatrix}$$

(8.150)

Now, Equation 8.133 yields the following sufficient conditions,

- *Loop 1*

$$\left| \frac{q_{11}^1}{1+L_1^1} \right| \leq \frac{\tau_{si11}}{\left| 1 - \left(\left(1/q_{12}^1 \right) + g_{12} \right) \tau_{si21} \right|_{\max}}$$

(8.151)

$$\left| \frac{q_{11}^1}{1 + L_1^1} \right| \leq \frac{\tau_{si12}}{\left| \left((1 / q_{12}^1) + g_{12} \right) \tau_{si22} \right|_{\max}} \tag{8.152}$$

- *Loop 2*

$$\left| \frac{q_{22}^2}{1 + L_2^2} \right| \leq \tau_{si22} \tag{8.153}$$

$$\left| \frac{q_{22}^2}{1 + L_2^2} \right| \leq \frac{\tau_{si21}}{\left| \frac{q_{11}^1}{1 + L_1^1} \left((1 / q_{21}^1) + g_{21} \right) \right|_{\max}} \tag{8.154}$$

Again, a load-sharing appears between loops 1 and 2 through specification elements τ_{si21} and τ_{si22}.

Noise Attenuation
Consider the following specifications:

$$\begin{bmatrix} y_1 \\ y_2 \end{bmatrix} = \begin{bmatrix} b_{n11} & \tau_{sn12} \\ \tau_{sn21} & b_{n22} \end{bmatrix} \begin{bmatrix} n_1 \\ n_2 \end{bmatrix} \quad \text{(for noise attenuation)} \tag{8.155}$$

$$\begin{bmatrix} y_1 \\ y_2 \end{bmatrix} = \begin{bmatrix} \tau_{nc11} & 0 \\ 0 & \tau_{nc22} \end{bmatrix} \begin{bmatrix} n_1 \\ n_2 \end{bmatrix} \quad \text{(for coupling attenuation)} \tag{8.156}$$

Now, Equation 8.133 yields the following sufficient conditions,

- *Loop 1*

$$\left| \frac{L_1^1}{1 + L_1^1} \right| \leq b_{n11}' \quad \text{with} \quad b_{n11}' = b_{n11} - \tau_{nc11} \tag{8.157}$$

$$\left| \frac{q_{11}^1}{1 + L_1^1} \right| \leq \frac{\tau_{nc11}}{\left| \left((1 / q_{12}^1) + g_{12} \right) \tau_{sn21} \right|_{\max}} \tag{8.158}$$

$$\left| \frac{q_{11}^1}{1 + L_1^1} \right| \leq \frac{\tau_{sn12}}{\left| -g_{12} - \left((1 / q_{12}^1) + g_{12} \right) b_{n22} \right|_{\max}} \tag{8.159}$$

- *Loop 2*

$$\left|\frac{L_2^2}{1+L_2^2}\right| \leq b'_{n22} \quad \text{with} \quad b'_{n22} = b_{n22} - \tau_{nc22} \tag{8.160}$$

$$\left|\frac{q_{22}^2}{1+L_2^2}\right| \leq \frac{\tau_{nc22}}{\left|\dfrac{q_{11}^1}{1+L_1^1}\left(\left(1/q_{21}^1\right)+g_{21}\right)g_{12}\right|_{\max}} \tag{8.161}$$

$$\left|\frac{q_{22}^2}{1+L_2^2}\right| \leq \frac{\tau_{sn21}}{\left|-g_{21}-\dfrac{q_{11}^1}{1+L_1^1}\left(\left(1/q_{21}^1\right)+g_{21}\right)g_{11}\right|_{\max}} \tag{8.162}$$

As above, specific interaction appears between loop 1 and 2 specifications through the τ_{sn21} and b_{n22} elements.

8.7 Comparison of Methods 1 and 2[7]

This section makes a comparison between the two methodologies introduced in this chapter: *Method 1* and *Method 2*. Both techniques are based on the same principle: the off-diagonal compensators play the role of feedforward functions that reduce the coupling and facilitate the design of the diagonal compensators, which are feedback functions in charge of fulfilling the robust stability and robust performance specifications. Additionally, for *Method 2*, and particularly for its Case 1 (Section 8.6.2), the diagonal elements are also directly involved in coping with coupling (the pre-compensator G_α is a full-matrix).

The existence of plant uncertainty causes the decoupling (feedforward) to be inexact. This residual coupling, however, is taken into account through a proper design of the diagonal compensators (feedback). Beyond these similarities, the way in which off-diagonal compensators are designed, and the procedure itself, is different for the two methodologies.

Method 1 requires a diagonal dominance hypothesis on the plant, limiting the application of the technique. This hypothesis is a fundamental requirement on the plant itself, and essential to obtain explicit expressions of the coupling matrices, which are defined therein to quantitatively compute the coupling. Besides, it allows two simplifications leading to the explicit optimal expressions for the off-diagonal compensators. By contrast, *Method 2* no longer needs such a hypothesis or any simplification. As a result, this removal extends the application of the method to a larger scope of plants. Furthermore, as is shown in Section 8.6.2, the expressions obtained for the off-diagonal compensators in Equation 8.98 are the same as those of the former method Equation 8.99 plus some new terms which disappear if the abovementioned diagonal dominance hypothesis is applied. So, *Method 2* generalizes *Method 1* and extends its field of application.

Additionally, the procedure has been simplified. In *Method 1*, the approach is performed by columns, designing first the diagonal compensator and then the off-diagonal elements of each column, which is somehow a complicated procedure. In the case of needing a redesign in any diagonal compensator, it implies the redesign of all the elements, diagonal and off-diagonal, from the subsequent loops. However, *Method 2* performs separately the design of the off-diagonal and the diagonal part. Regardless of the case chosen (Case 1, Section 8.6.2, or Case 2, Section 8.6.3), when it comes to designing the feedback diagonal compensators, all the feedforward off-diagonal elements have already been determined. Finally, both methods allow the introduction not only of classical QFT specifications but also of classical matrix specifications, and provide the guideline to perform a load-sharing among loops, distributing the burden as needed.

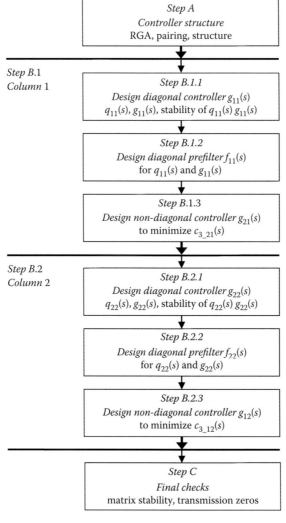

FIGURE 8.19
Flow chart of the MIMO QFT Method 1 for the heat exchanger.

8.8 Heat Exchanger, Example 8.1—MIMO QFT Method 1

To illustrate the MIMO QFT control design *"Method 1"* introduced in Section 8.5, consider again the counter-flow heat exchanger and pasteurization process presented in Example 8.1. The plant definition is given in Equations 8.1 through 8.4. The control specifications are described in Equations 8.6 through 8.8, and the control system block diagram is in Figure 8.3.

Figure 8.19 shows the flow chart to apply the MIMO QFT Method 1 to the design of the control system for this example. The next paragraphs follow this flow chart step by step.

Step A: Controller Structure
Using RGA over the frequencies of interest, and taking into account the requirement of minimum complexity for the controller, we identify the input–output pairing and the most appropriate structure for the matrix compensator. As calculated in Equation 8.5, the RGA matrix for the nominal transfer matrix $P(s)$—Equation 8.4 at $s = 0$—is

$$\boldsymbol{\Lambda} = \boldsymbol{P}_0 \otimes (\boldsymbol{P}_0^{-1})^T = \begin{bmatrix} 7 & 9 \\ 3 & 5 \end{bmatrix} \otimes \left(\begin{bmatrix} 7 & 9 \\ 3 & 5 \end{bmatrix}^{-1} \right)^T = \begin{array}{c} \overbrace{}^{u_1 \qquad u_2} \\ \left[\begin{array}{cc} 4.3750 & -3.3750 \\ -3.3750 & 4.3750 \end{array} \right] \begin{array}{c} y_1 \\ y_2 \end{array} \end{array} \qquad (8.163)$$

We pair the input and output variables according to the elements of the $\boldsymbol{\Lambda}$ matrix that are positive and closer to one. Based on these results, we select $\lambda_{11} = 4.375$ and $\lambda_{22} = 4.375$. This means to pair $[y_1(s) - u_1(s)]$ and $[y_2(s) - u_2(s)]$.

Additionally, we need off-diagonal elements $g_{ij}(s)$ for the controller $G(s)$ in the positions $g_{ij}(s)$ where the loop coupling is significant. As a practical rule, the coupling between i and j is considered important when the absolute value of the corresponding RGA element is greater than 1.2 (or less than 0.8), or $|\lambda_{ij}| \geq 1.2$ (or $|\lambda_{ij}| \leq 0.8$). In this case, as $|\lambda_{12}| = |\lambda_{21}| = 3.3750$, we need one or two of the off-diagonal elements $g_{12}(s)$ and $g_{21}(s)$. For the requirement of the simplest structure, we start with a triangular controller matrix with $g_{21}(s) = 0$. Figure 8.20 shows the two control loops and the matrix controller based on the selected pairing and triangular structure.

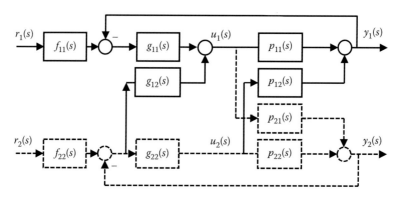

FIGURE 8.20
Proposed control system for the 2 × 2 heat exchanger. Method 1.

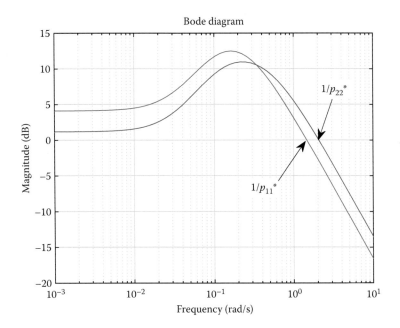

FIGURE 8.21
Bode diagram for the nominal plants $(1/p_{11}{}^*)$ and $(1/p_{22}{}^*)$.

Also, the inverse of the plant matrix is:

$$P(s)^{-1} = P(s)^* = \begin{bmatrix} p_{11}^*(s) & p_{12}^*(s) \\ p_{21}^*(s) & p_{22}^*(s) \end{bmatrix} \tag{8.164}$$

The Bode diagram for the nominal plants of $(1/p_{11}{}^*)$ and $(1/p_{22}{}^*)$ is in Figure 8.21. The controller design is arranged so that we start with the system of smallest bandwidth, which is $(1/p_{11}{}^*)$, and then we continue with $(1/p_{22}{}^*)$.

Step B.1.1: Design of the Diagonal Controller Element $g_{11}(s)$
The element $g_{11}(s)$ is designed by means of the standard SISO QFT loop-shaping technique, for the inverse of the equivalent plant $q_{11}(s)$, and for the control specifications described in Equations 8.6 through 8.8. According to the iterative expression Equation 8.86, the equivalent plant is

$$p_{11}^{*e}(s)\Big|_1 = p_{11}^*(s) \tag{8.165}$$

and the plant to be controlled

$$q_{11}(s) = \frac{1}{p_{11}^{*e}(s)\Big|_1} \tag{8.166}$$

The MATLAB code that calculates the plant $q_{11}(s)$ for all the cases within the parametric uncertainty is included in Appendix 8. This plant and the control specifications defined

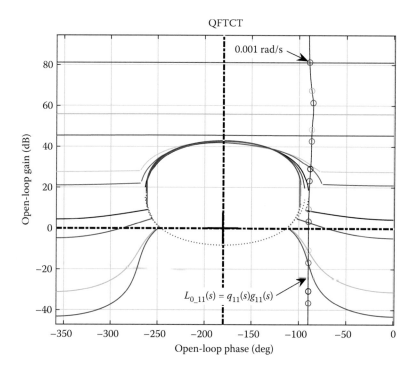

FIGURE 8.22
Loop shaping of controller $g_{11}(s)$. $L_{0\_11}(s) = q_{11}(s)\, g_{11}(s)$.

in Equations 8.6 through 8.8 are introduced in the QFTCT. The QFT bounds and the loop shaping for $g_{11}(s)$ are shown in Figure 8.22.

The expression found for the controller $g_{11}(s)$ is a PID compensator with a first-order low-pass filter, so that

$$g_{11}(s) = 20\frac{((1/0.12)s+1)((1/0.24)s+1)}{s((1/0.014)s+1)} \tag{8.167}$$

The design fulfills the two stability conditions:

1. $L_{0\_11}(s) = q_{11}(s)\, g_{11}(s)$ satisfies the Nyquist encirclement condition and
2. There are no RHP pole-zero cancellations between $q_{11}(s)$ and $g_{11}(s)$.

This is checked following the methodology presented in Chapter 3, Section 3.4, and with the QFTCT, *Controller design* window, *File*, *Check stability* option.

Step B.1.2: Design of the Diagonal Prefilter Element $f_{11}(s)$
The $f_{11}(s)$ prefilter element is designed with the QFTCT, for the equivalent plant $q_{11}(s)$, and for the diagonal controller $g_{11}(s)$ designed in the previous *Step B.1.1*. Figure 8.23 shows the design. The expression found for the prefilter $f_{11}(s)$ is

$$f_{11}(s) = \frac{1}{((1/1.5)s+1)} \tag{8.168}$$

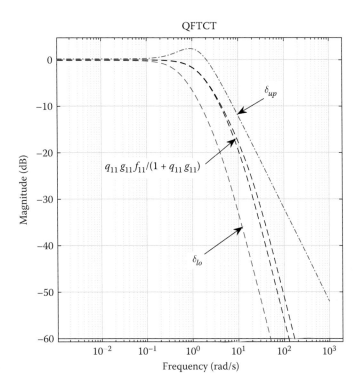

FIGURE 8.23
Design of prefilter $f_{11}(s)$ for $q_{11}(s)$ and $g_{11}(s)$.

Figure 8.24 shows the analysis of the disturbance rejection at the output of the plant and the reference tracking specifications for the equivalent plant $q_{11}(s)$ and with the controller $g_{11}(s)$ and the prefilter $f_{11}(s)$.

Step B.1.3: Design of the Non-Diagonal Controller Element $g_{21}(s)$
Due to the structure for the controller matrix selected in *Step A*, the non-diagonal element $g_{21}(s)$ is initially set to zero, so that

$$g_{21}(s) = 0 \tag{8.169}$$

Step B.2.1: Design of the Diagonal Controller Element $g_{22}(s)$
The element $g_{22}(s)$ is designed by means of the standard SISO QFT loop-shaping technique, for the inverse of the equivalent plant $q_{22}(s)$, and for the control specifications described in Equations 8.6 through 8.8. According to the iterative expression Equation 8.86, the equivalent plant is

$$\left. p_{22}^{*e}(s) \right|_2 = \left. p_{22}^{*}(s) \right|_1 - \frac{\left. p_{21}^{*}(s) \right|_1 + \left. g_{21}(s) \right|_1}{\left. p_{11}^{*}(s) \right|_1 + \left. g_{11}(s) \right|_1} \left. p_{12}^{*}(s) \right|_1 \tag{8.170}$$

and the plant to be controlled

$$q_{22}(s) = \frac{1}{\left. p_{22}^{*e}(s) \right|_2} \tag{8.171}$$

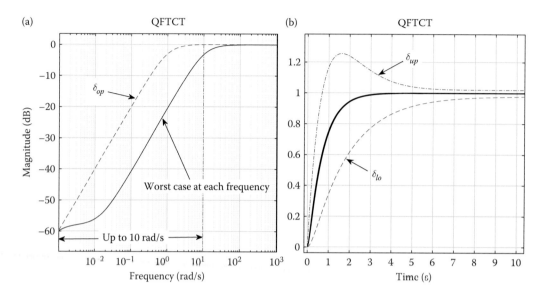

FIGURE 8.24
Analysis of controller $g_{11}(s)$ and prefilter $f_{11}(s)$ for $q_{11}(s)$. (a) Disturbance rejection at plant output: δ_{op}. (b) Reference tracking: δ_{up}, δ_{lo}.

The MATLAB code that calculates the plant $q_{22}(s)$ for all the cases within the parametric uncertainty is also included in Appendix 8. This plant and the control specifications defined in Equations 8.6 through 8.8 are introduced in the QFTCT. The QFT bounds and the loop-shaping for $g_{22}(s)$ are shown in Figure 8.25. The expression found for the controller $g_{22}(s)$ is a PI compensator with a first-order low-pass filter, so that

$$g_{22}(s) = \frac{30((1/0.6)s + 1)}{s((1/300)s + 1)} \tag{8.172}$$

The design fulfills the two stability conditions:

1. $L_{0\_22}(s) = q_{22}(s)\, g_{22}(s)$ satisfies the Nyquist encirclement condition and
2. There are no RHP pole-zero cancellations between $q_{22}(s)$ and $g_{22}(s)$.

This is checked following the methodology presented in Chapter 3, Section 3.4, and with QFTCT, *Controller design* window, *File, Check stability* option.

Step B.2.2: Design of the Diagonal Prefilter Element $f_{22}(s)$
The $f_{22}(s)$ prefilter element is designed with the QFTCT, for the equivalent plant $q_{22}(s)$, and for the diagonal controller $g_{22}(s)$ designed in the previous *Step B.2.1*. Figure 8.26 shows the design. The expression found for the prefilter $f_{22}(s)$ is

$$f_{22}(s) = \frac{1}{((1/0.75)s + 1)} \tag{8.173}$$

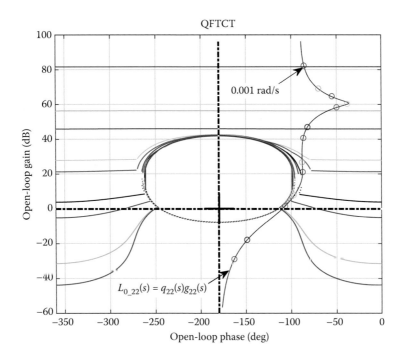

FIGURE 8.25
Loop shaping of controller $g_{22}(s)$. $L_{0\_22}(s) = q_{22}(s)\, g_{22}(s)$.

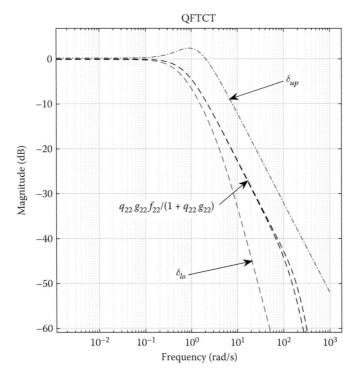

FIGURE 8.26
Design of prefilter $f_{22}(s)$ for $q_{22}(s)$ and $g_{22}(s)$.

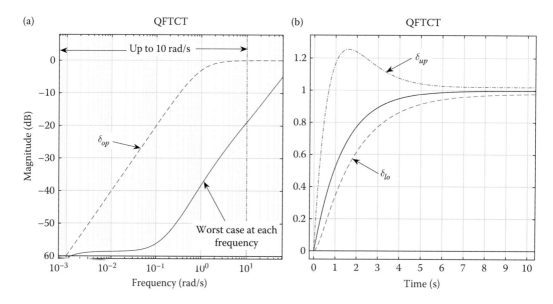

FIGURE 8.27
Analysis of controller $g_{22}(s)$ and prefilter $f_{22}(s)$ for $q_{22}(s)$. (a) Disturbance rejection at plant output: δ_{op}. (b) Reference tracking: δ_{up}, δ_{lo}.

Figure 8.27 shows the analysis of the disturbance rejection at the output of the plant and the reference tracking specifications for the equivalent plant $q_{22}(s)$ and with the controller $g_{22}(s)$ and the prefilter $f_{22}(s)$.

Step B.2.3: Design of the Non-Diagonal Controller Element $g_{12}(s)$
The $g_{12}(s)$ non-diagonal element is designed to minimize the cross-coupling term $c_{3\text{-}12}$ given by Equation 8.73 for disturbance rejection at plant output. The expression for the optimum compensator element Equation 8.77 is utilized in order to achieve this goal. The $g_{12}(s)$ element is designed so that its magnitude and phase over the frequencies of interest pass, respectively, through the middle of the magnitude and phase of the expression $g_{22}(s)\, p_{12}(s)^*/p_{22}(s)^*$ for 256 plants selected within the parametric uncertainty—see Figure 8.28 and Equation 8.174.

$$g_{12}(s) = \text{middle of}\left(g_{22}(s)\frac{p_{12}^{*N}(s)}{p_{22}^{*N}(s)}\right) = \frac{-40\big((1/0.833)s+1\big)}{s\big((1/300)s+1\big)} \tag{8.174}$$

Step C: Final Checks

C.1. Finally, the design also fulfills the other additional two stability conditions:
   No Smith-McMillan pole-zero cancellations occur between $P(s)$ and $G(s)$, and
   No Smith-McMillan pole-zero cancellations occur in $|P^*(s) + G(s)|$.
C.2. The final system $P(s)G(s)$ does not have any additional RHP transmission zero.
C.3. Simulation of the closed-loop MIMO system. Figures 8.29 and 8.30 show the simulation of the two control loops. We use the nominal plant $P = [p_{11}(s)\ p_{12}(s);\ p_{21}(s)\ p_{22}(s)]$ defined in Equation 8.4. The block diagram is the non-diagonal controller

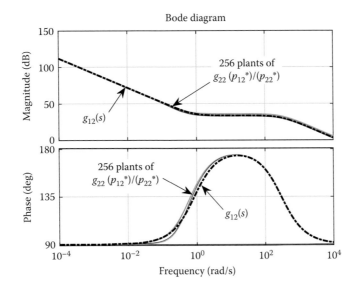

FIGURE 8.28
Design of non-diagonal element $g_{12}(s)$.

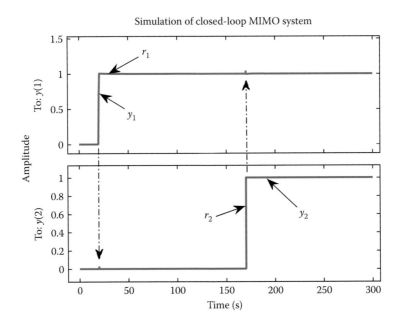

FIGURE 8.29
Closed-loop MIMO system. Figure 8.20. Inputs: $r_1(s)$, $r_2(s)$. Outputs: $y_1(s)$, $y_2(s)$. $P = [p_{11}(s)\ p_{12}(s);\ p_{21}(s)\ p_{22}(s)]$, Equation 8.4, nominal plant. $G = [g_{11}(s)\ g_{12}(s);\ g_{21}(s)\ g_{22}(s)]$, Equations 8.167, 8.169, 8.172, 8.174, and without prefilter: $F = [1\ 0;\ 0\ 1]$. MIMO QFT Method 1.

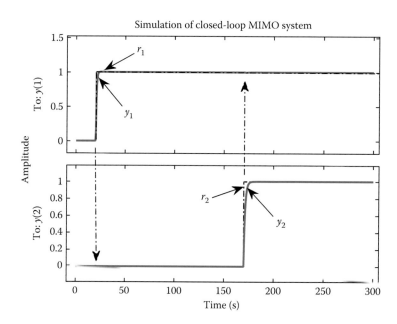

FIGURE 8.30

Closed-loop MIMO system. Figure 8.20. Inputs: $r_1(s)$, $r_2(s)$. Outputs: $y_1(s)$, $y_2(s)$. $P = [p_{11}(s)\ p_{12}(s);\ p_{21}(s)\ p_{22}(s)]$, Equation 8.4, nominal plant. $G = [g_{11}(s)\ g_{12}(s);\ g_{21}(s)\ g_{22}(s)]$, Equations 8.167, 8.169, 8.172, 8.174, and with prefilter: $F = [f_{11}(s)\ 0;\ 0\ f_{22}(s)]$, Equations 8.168, 8.173. MIMO QFT Method 1.

shown in Figure 8.20. The full-matrix controller is $G = [g_{11}(s)\ g_{12}(s);\ g_{21}(s)\ g_{22}(s)]$, Equations 8.167, 8.169, 8.172, 8.174. Figure 8.29 shows the simulation with a unitary diagonal prefilter, $F = [1\ 0;\ 0\ 1]$, and Figure 8.30 with the diagonal prefilter $F = [f_{11}(s)0;\ 0\ f_{22}(s)]$, Equations 8.168, 8.173. The first reference starts at $r_1 = 0$ and changes to $r_1 = 1$ at $t = 20$ s. The second reference starts at $r_2 = 0$ and changes to $r_2 = 1$ at $t = 170$ s. Figure 8.29 shows the results obtained with the triangular controller and unitary diagonal prefilter, which improve significantly the ones presented in Figure 8.9 (independent SISO controllers). Finally, Figure 8.30 shows the results with the triangular controller and diagonal prefilter. They improve the results even more, reducing significantly the effect of the coupling between the control loops.

8.9 Heat Exchanger, Example 8.1—MIMO QFT Method 2

To illustrate the MIMO QFT control design *Method 2* introduced in Section 8.6, consider again the counter-flow heat exchanger and pasteurization process presented in Example 8.1 and Section 8.8. The plant definition is given in Equations 8.1 through 8.4. The control specifications are described in Equations 8.6 through 8.8, and the control system block diagram is in Figure 8.32.

Figure 8.31 shows the flow chart to apply the MIMO QFT Method 2 to the design of the control system for this example. The next paragraphs follow this flow chart step by step.

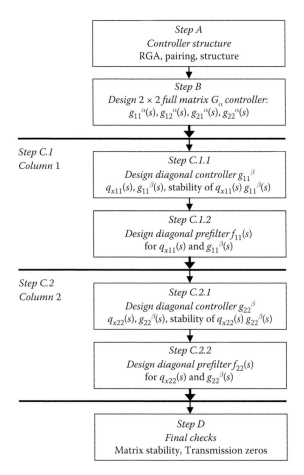

FIGURE 8.31
Flow chart of the MIMO QFT Method 2 for the heat exchanger.

Step A: Controller Structure

Using RGA over the frequencies of interest, and taking into account the requirement of minimum complexity for the controller, we identify the input–output pairing and the appropriate structure for the control system. As calculated in Equation 8.5 or Equation 8.163 for the nominal transfer matrix $P(s)$—Equation 8.4 at $s = 0$, the 2×2 RGA matrix has $\lambda_{11} = 4.375$ and $\lambda_{22} = 4.375$, which means to pair $[y_1(s)-u_1(s)]$ and $[y_2(s)-u_2(s)]$. The controller structure with this *Method 2* requires a 2×2 full matrix $\boldsymbol{G}_\alpha$, a 2×2 diagonal matrix $\boldsymbol{G}_\beta$, and a 2×2 diagonal prefilter matrix $\boldsymbol{F}$. Figure 8.32 shows the two control loops and the controllers based on the selected pairing and control structure.

Step B: Design of 2×2 Full-Matrix G_α Controller

As discussed in Section 8.6.2, the fully populated matrix controller $\boldsymbol{G}$ is composed of two matrices: $\boldsymbol{G} = \boldsymbol{G}_\alpha \boldsymbol{G}_\beta$, so that

$$\boldsymbol{G} = \boldsymbol{G}_\alpha \boldsymbol{G}_\beta = \begin{bmatrix} g_{11}^\alpha(s) & g_{12}^\alpha(s) \\ g_{21}^\alpha(s) & g_{22}^\alpha(s) \end{bmatrix} \begin{bmatrix} g_{11}^\beta(s) & 0 \\ 0 & g_{22}^\beta(s) \end{bmatrix} \tag{8.175}$$

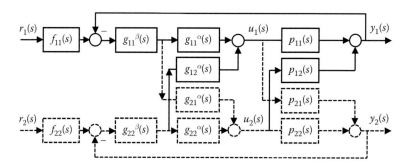

FIGURE 8.32
Proposed control system for the 2 × 2 heat exchanger. Method 2.

The main objective of the pre-compensator G_α is to diagonalize the plant P as much as possible. The expression used to determine G_α is—see also Equation 8.94,

$$G_\alpha(s) = \begin{bmatrix} g_{11}^\alpha(s) & g_{12}^\alpha(s) \\ g_{21}^\alpha(s) & g_{22}^\alpha(s) \end{bmatrix} = \hat{P}^{-1}(s)\hat{P}_{diag}(s) = \begin{bmatrix} \hat{p}_{11}^*(s) & \hat{p}_{12}^*(s) \\ \hat{p}_{21}^*(s) & \hat{p}_{22}^*(s) \end{bmatrix} \begin{bmatrix} \hat{p}_{11}(s) & 0 \\ 0 & \hat{p}_{22}(s) \end{bmatrix} \quad (8.176)$$

where the plant matrix $\hat{P}$, and the corresponding inverse $\hat{P}^{-1}$ and diagonal $\hat{P}_{diag}$ are selected so that the expression of the extended matrix in Equation 8.95 presents the closest form to a diagonal matrix, nulling the off-diagonal terms as much as possible.

The Bode diagrams for the expressions $\hat{p}_{11}^*(s)\hat{p}_{11}(s)$, $\hat{p}_{12}^*(s)\hat{p}_{22}(s)$, $\hat{p}_{21}^*(s)\hat{p}_{11}(s)$, and $\hat{p}_{22}^*(s)\hat{p}_{22}(s)$—see Equation 8.176, including all the model uncertainty are shown in Figures 8.33 through 8.36, respectively. The controller elements $g_{11}^\alpha(s)$, $g_{12}^\alpha(s)$, $g_{21}^\alpha(s)$, and $g_{22}^\alpha(s)$ are calculated as the transfer function that matches the mean value in magnitude and phase of the respective Bode diagram over the frequencies of interest. See also Figures 8.33 through 8.36 and Equations 8.177 through 8.180, respectively.

$$g_{11}^\alpha(s) = \frac{(1.24s + 0.16)}{(s + 0.01)} \quad (8.177)$$

$$g_{12}^\alpha(s) = \frac{-(1.53s + 0.19)}{(s + 0.01)} \quad (8.178)$$

$$g_{21}^\alpha(s) = \frac{-(0.18s + 0.1)}{(s + 0.01)} \quad (8.179)$$

$$g_{22}^\alpha(s) = \frac{(1.24s + 0.16)}{(s + 0.01)} \quad (8.180)$$

Step C.1.1: Design of the Diagonal Controller $g_{11}^\beta(s)$
The element $g_{11}^\beta(s)$ is designed by means of the standard SISO QFT loop-shaping technique, for the inverse of the extended equivalent plant $q_{x11}(s) = [p_{11}^{x*e}]_1^{-1}$, and for the control

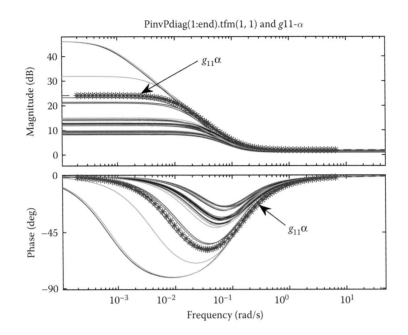

FIGURE 8.33

Design of element g_{11}^{α} as the mean value of $\hat{p}_{11}^{*}(s)\hat{p}_{11}(s)$.

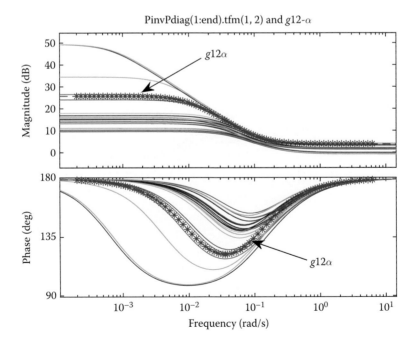

FIGURE 8.34

Design of element $g_{12}^{\alpha}(s)$ as the mean value of $\hat{p}_{12}^{*}(s)\hat{p}_{22}(s)$.

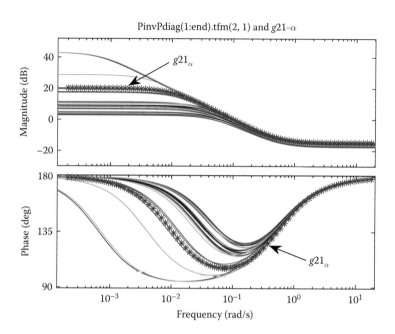

FIGURE 8.35

Design of element $g_{21}{}^{\alpha}(s)$ as the mean value of $\hat{p}_{21}^{*}(s)\hat{p}_{11}(s)$.

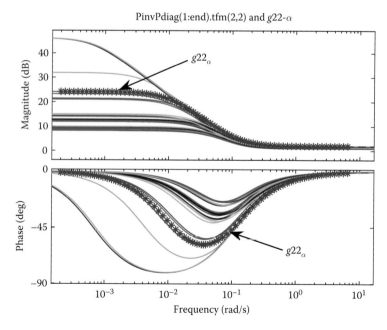

FIGURE 8.36

Design of element $g_{22}{}^{\alpha}(s)$ as the mean value of $\hat{p}_{22}^{*}(s)\hat{p}_{22}(s)$.

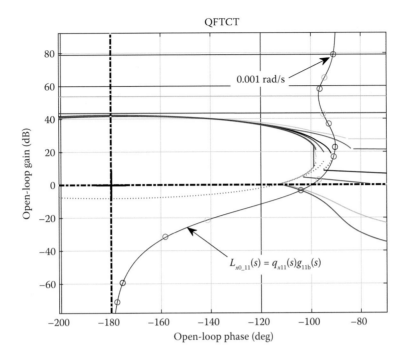

FIGURE 8.37
Loop-shaping of controller $g_{11}^{\beta}(s)$. $L_{x0\_11}(s) = q_{x11}(s)\, g_{11}^{\beta}(s)$.

specifications described in Equations 8.6 through 8.8. According to the iterative expression Equation 8.97, the extended equivalent plant is

$$p_{11}^{x^*e}(s)\Big|_1 = p_{11}^{x^*}(s) \tag{8.181}$$

and the plant to be controlled

$$q_{x11}(s) = \frac{1}{p_{11}^{x^*e}(s)\Big|_1} \tag{8.182}$$

The MATLAB code that calculates the plant $q_{x11}(s)$ for all the cases within the parametric uncertainty is included in Appendix 8. This plant and the control specifications defined in Equations 8.6 through 8.8 are introduced in the QFTCT. The QFT bounds and the loop shaping for $g_{11}^{\beta}(s)$ are shown in Figure 8.37. The expression found for the controller $g_{11}^{\beta}(s)$ is a PI compensator with a first-order low-pass filter, so that

$$g_{11}^{\beta}(s) = \frac{0.9\big((1/0.24)s + 1\big)}{s\big((1/40)s + 1\big)} \tag{8.183}$$

The design fulfills the two stability conditions:

1. $L_{x0\_11}(s) = q_{x11}(s)\, g_{11}^{\beta}(s)$ satisfies the Nyquist encirclement condition and
2. There are no RHP pole-zero cancellations between $q_{x11}(s)$ and $g_{11}^{\beta}(s)$.

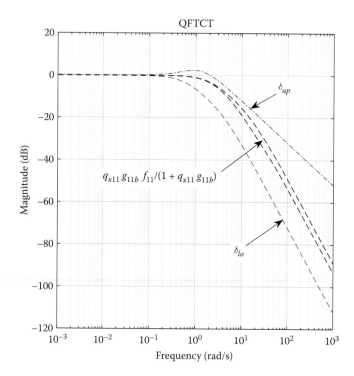

FIGURE 8.38
Design of prefilter $f_{11}(s)$ for $q_{x11}(s)$ and $g_{11}^{\beta}(s)$.

This is checked following the methodology presented in Chapter 3, Section 3.4, and with *QFTCT, Controller design* window, *File, Check stability* option.

Step C.1.2: Design of the Diagonal Prefilter $f_{11}(s)$
The $f_{11}(s)$ prefilter element is designed with the QFTCT, for the equivalent plant $q_{x11}(s)$, and for the diagonal controller $g_{11}^{\beta}(s)$ designed in the previous *Step C.1.1*. Figure 8.38 shows the design. The expression found for the prefilter $f_{11}(s)$ is

$$f_{11}(s) = \frac{\left((1/20)s + 1\right)}{\left((1/2)s + 1\right)} \tag{8.184}$$

Figure 8.39 shows the analysis of the disturbance rejection at the output of the plant and the reference tracking specifications for the equivalent plant $q_{x11}(s)$ and with the controller $g_{11}^{\beta}(s)$ and the prefilter $f_{11}(s)$.

Step C.2.1: Design of the Diagonal Controller $g_{22}^{\beta}(s)$
The element $g_{22}^{\beta}(s)$ is designed by means of the standard SISO QFT loop-shaping technique, for the inverse of the extended equivalent plant $q_{x22}(s) = [p_{22}^{x*e}]_2^{-1}$, and for the control specifications described in Equations 8.6 through 8.8. According to the iterative expression Equation 8.97, the extended equivalent plant is

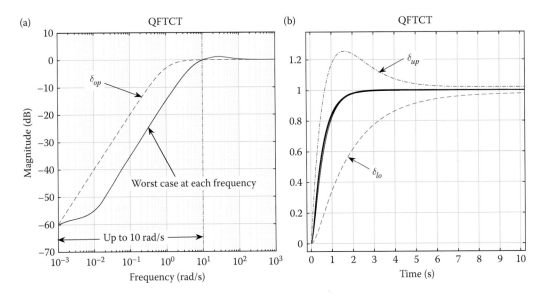

FIGURE 8.39

Analysis of controller $g_{11}{}^{\beta}(s)$ and prefilter $f_{11}(s)$ for $q_{x11}(s)$. (a) Disturbance rejection at plant output: δ_{op}. (b) Reference tracking: δ_{up}, δ_{lo}.

$$p_{22}^{x^*e}(s)\Big|_2 = p_{22}^{x^*}(s)\Big|_1 - \frac{p_{21}^{x^*}(s)\Big|_1 \; p_{12}^{x^*}(s)\Big|_1}{p_{11}^{x^*}(s)\Big|_1 + g_{11}^{\beta}(s)} \tag{8.185}$$

and the plant to be controlled

$$q_{x22}(s) = \frac{1}{p_{22}^{x^*e}(s)\Big|_2} \tag{8.186}$$

The MATLAB code that calculates the plant $q_{x22}(s)$ for all the cases within the parametric uncertainty is included in Appendix 8. This plant and the control specifications defined in Equations 8.6 through 8.8 are introduced in the QFTCT. The QFT bounds and the loop shaping for $g_{22}{}^{\beta}(s)$ are shown in Figure 8.40. The expression found for the controller $g_{22}{}^{\beta}(s)$ is a PI compensator with a first-order low-pass filter, so that

$$g_{22}^{\beta}(s) = \frac{0.4\big((1/0.1)s+1\big)}{s\big((1/40)s+1\big)} \tag{8.187}$$

The design fulfills the two stability conditions:

1. $L_{x0\_22}(s) = q_{x22}(s)\,g_{22}{}^{\beta}(s)$ satisfies the Nyquist encirclement condition and
2. There are no RHP pole-zero cancellations between $q_{x22}(s)$ and $g_{22}{}^{\beta}(s)$.

This is checked following the methodology presented in Chapter 3, Section 3.4, and with *QFTCT, Controller design* window, *File, Check stability* option.

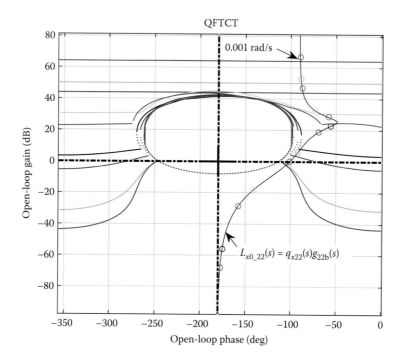

FIGURE 8.40
Loop shaping of controller $g_{22}{}^{\beta}(s)$. $L_{x0\_22}(s) = q_{x22}(s)\,g_{22}{}^{\beta}(s)$.

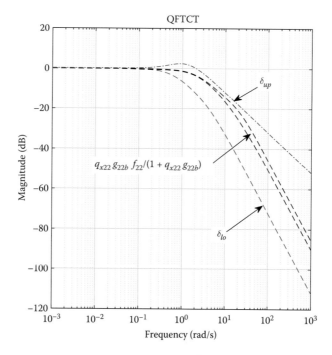

FIGURE 8.41
Design of prefilter $f_{22}(s)$ for $q_{x22}(s)$ and $g_{22}{}^{\beta}(s)$.

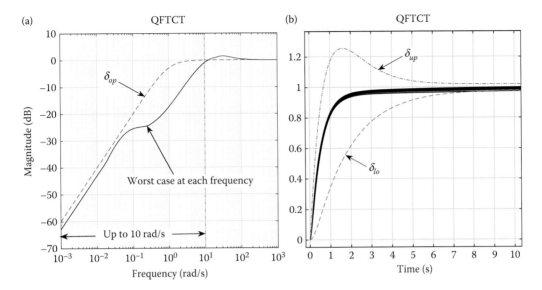

FIGURE 8.42
Analysis of controller g_{22}^{β} (s) and prefilter $f_{22}(s)$ for $q_{x22}(s)$. (a) Disturbance rejection at plant output: δ_{op}. (b) Reference tracking: δ_{up}, δ_{lo}.

Step C.2.2: Design of the Diagonal Prefilter $f_{22}(s)$

The $f_{22}(s)$ prefilter element is designed with the QFTCT, for the equivalent plant $q_{x22}(s)$, and for the diagonal controller $g_{22}^{\beta}(s)$ designed in the previous *Step C.2.1*. Figure 8.41 shows the design. The expression found for the prefilter $f_{22}(s)$ is

$$f_{22}(s) = \frac{((1/20)s+1)}{((1/2)s+1)} \tag{8.188}$$

Figure 8.42 shows the analysis of the disturbance rejection at the output of the plant and the reference tracking specifications for the equivalent plant $q_{x22}(s)$ and with the controller $g_{22}^{\beta}(s)$ and the prefilter $f_{22}(s)$.

Step D: Final Checks

D.1. Finally, the design also fulfills the other additional two stability conditions:

No Smith-McMillan pole-zero cancellations occur between $P(s)$ and $G(s)$, and

No Smith-McMillan pole-zero cancellations occur in $|P^*(s) + G(s)|$.

D.2. The final system $P(s)G(s)$ does not have any additional RHP transmission zero.

D.3. Simulation of the closed-loop MIMO system. Figures 8.43 and 8.44 show the simulation of the two control loops. We use the nominal plant $P = [p_{11}(s)\ p_{12}(s); p_{21}(s)\ p_{22}(s)]$ defined in Equation 8.4. The control block diagram is shown in Figure 8.32. The full-matrix controller is $G = G_{\alpha}G_{\beta}$, with $G_{\alpha} = [g_{11}^{\alpha}(s)\ g_{12}^{\alpha}(s); g_{21}^{\alpha}(s)\ g_{22}^{\alpha}(s)]$, Equations 8.177 through 8.180, and $G_{\beta} = [g_{11}^{\beta}(s)\ 0;\ 0\ g_{22}^{\beta}(s)]$, Equations 8.183 and 8.187. Figure 8.43 shows the simulation with a unitary diagonal prefilter, $F = [1\ 0;\ 0\ 1]$, and Figure 8.44 with the diagonal prefilter $F = [f_{11}(s)\ 0;\ 0\ f_{22}(s)]$, Equations 8.184 and 8.188. The

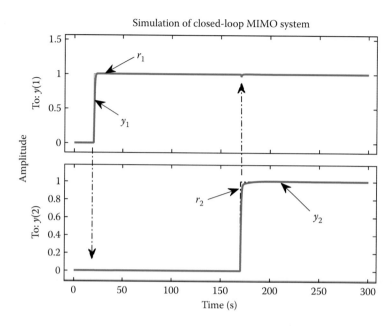

FIGURE 8.43

Closed-loop MIMO system. Figure 8.32. Inputs: $r_1(s)$, $r_2(s)$. Outputs: $y_1(s)$, $y_2(s)$. $P = [p_{11}(s)\ p_{12}(s);\ p_{21}(s)\ p_{22}(s)]$, Equation 8.4, nominal plant. $G = G_\alpha G_\beta$, Equations 8.177 through 8.180, 8.183, 8.187, and without prefilter: $F = [1\ 0;\ 0\ 1]$. MIMO QFT Method 2.

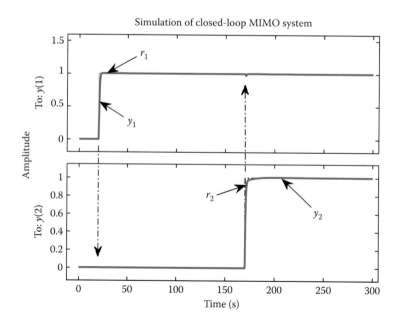

FIGURE 8.44

Closed-loop MIMO system. Figure 8.32. Inputs: $r_1(s)$, $r_2(s)$. Outputs: $y_1(s)$, $y_2(s)$. $P = [p_{11}(s)\ p_{12}(s);\ p_{21}(s)\ p_{22}(s)]$, Equation 8.4, nominal plant. $G = G_\alpha G_\beta$, Equations 8.177 through 8.180, 8.183, 8.187, and with prefilter: $F = [f_{11}(s)\ 0;\ 0\ f_{22}(s)]$, Equations 8.184, 8.188. MIMO QFT Method 2.

first reference starts at $r_1 = 0$ and changes to $r_1 = 1$ at $t = 20$ s. The second reference starts at $r_2 = 0$ and changes to $r_2 = 1$ at $t = 170$ s. Figure 8.43 shows the results obtained with the controller $G = G_\alpha G_\beta$ and the unitary diagonal prefilter, which improve significantly the ones presented in Figure 8.9 (independent SISO controllers). Finally, Figure 8.44 shows the results with the controller $G = G_\alpha G_\beta$ and the diagonal prefilter. They improve even more the results, reducing significantly the effect of the coupling between the control loops.

8.10 Summary

This chapter analyses the main characteristics of MIMO plants and introduces two methodologies (*Method 1* and *Method 2*) to design non-diagonal QFT controllers for MIMO systems with model uncertainty.

Sections 8.1 through 8.4 analyze the special properties that define MIMO systems and propose some tools to understand their dynamics and design appropriate control systems.

Sections 8.5 and 8.8 present the non-diagonal MIMO QFT *Method 1*, including both theory and a detailed example. The method analyses the role played by the non-diagonal compensator elements g_{ij} ($i \neq j$) by means of three coupling matrices (C_1, C_2, and C_3) and a quality function η_{ij}. It quantifies the amount of interaction among the control loops, and proposes a sequential design methodology for the fully populated matrix compensator that yields n equivalent tracking SISO systems and n equivalent disturbance rejection SISO systems. As a result, the off-diagonal elements g_{ij} ($i \neq j$) of the compensator matrix reduce (or cancel if there is no uncertainty) the level of coupling among loops, and the diagonal elements g_{kk} regulate the system with less bandwidth requirements than the diagonal elements of the original Horowitz's methods.[6,7,110,114,115,206,209,211]

Sections 8.6 and 8.9 present the non-diagonal MIMO QFT *Method 2*, including both theory and a detailed example. It is a formal reformulation of *Method 1*. It avoids former hypotheses of diagonal dominance and simplifies the calculations for the off-diagonal elements and the method itself. It also reformulates the classical matrix definition of MIMO specifications by designing a new set of loop-by-loop QFT bounds on the Nichols chart with necessary and sufficient conditions. It gives explicit expressions to share the load among the loops of the MIMO system to achieve the matrix specifications. And all for stability, reference tracking, disturbance rejection at plant input and output, and noise attenuation problems.[119]

The MATLAB code that develops the examples and cases presented in this chapter is also included in Appendix 8.

8.11 Practice

The list shown below summarizes the collection of problems and cases included in this book that apply the control methodologies introduced in this chapter:

- *Example 8.1.* Section 8.1. SISO control of a 2 × 2 heat exchanger system.
- *Example 8.1.* Section 8.8. MIMO QFT control, Method 1, 2 × 2 heat exchanger.
- *Example 8.1.* Section 8.9. MIMO QFT control, Method 2, 2 × 2 heat exchanger.
- Appendix 8. MATLAB code for Example 8.1, all cases.
- *Case study CS3.* A 2 × 2 wastewater treatment plant.
- *Case study CS5.* A 6 × 6 space telescope control.
- *Project P5.* Appendix 1. A 2 × 2 distillation column.
- *Problem Q8.* Appendix 1. Two flow problem.
- *Problem Q9.* Appendix 1. 2 × 2 MIMO system.
- *Problem Q10.* Appendix 1. 2 × 2 MIMO system.
- *Problem Q11.* Appendix 1. Spacecraft flying in formation. Low Earth Orbit.
- *Problem Q12.* Appendix 1. 3 × 3 MIMO system.

9

Control Topologies

9.1 Introduction

Sometimes, the performance specifications required for a system cannot be completely achieved by a simple 2DOF control structure, with a feedback controller $G(s)$ and a prefilter $F(s)$. The presence of significant disturbances, insufficient number of actuators (less actuators than sensors), or the inability to have adequate sensors (less sensors than actuators) are examples of this problem. Fortunately, the design of an appropriate control topology with additional blocks and connections is the solution in many cases. This chapter analyzes some of these common problems and proposes some practical control topologies to achieve the required performance specifications.

The chapter contains nine sections, each one with a specific control topology solution. It includes: cascade control (Section 9.2), feedforward control (Section 9.3), override control (Section 9.4), ratio control (Section 9.5), mid-range control (Section 9.6), load-sharing control (Section 9.7), split-range control (Section 9.8), inferential control (Section 9.9), and auctioneering control (Section 9.10).

Each section starts with a control challenge frequently confronted in industry. We encourage the reader to analyze it carefully and propose some solutions before continuing reading. Then, the second part of each section develops some control topology solutions for the challenge, including additional control blocks, connections, and sometimes supplementary sensors or actuators.

In all cases, once the control topology is chosen and the additional blocks are defined, a QFT robust control methodology can be applied to design the feedback controller $G(s)$, prefilter $F(s)$, and other additional functions. As in previous cases, the design considers the model uncertainty of the plant, the model uncertainty of the additional blocks of the structure, and a variety of simultaneous control specifications.

9.2 Cascade Control Systems

9.2.1 Challenge 9.1

Given the control system shown in Figure 9.1 and its transfer function description in Equation 9.1, modify the diagram topology to significantly reduce the effect of the

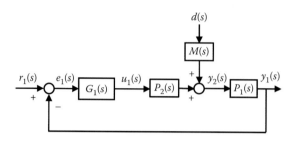

FIGURE 9.1
Challenge 9.1: reduce the effect of the disturbance $d(s)$ on $y_1(s)$.

disturbance $d(s)$ on the output $y_1(s)$. An additional sensor for $y_2(s)$ is available. The dynamics of $P_2(s)$ is faster than the dynamics of $P_1(s)$.

$$y_1(s) = \frac{P_1(s)P_2(s)G_1(s)}{1 + P_1(s)P_2(s)G_1(s)} r_1(s) + \frac{P_1(s)M(s)}{1 + P_1(s)P_2(s)G_1(s)} d(s) \qquad (9.1)$$

9.2.2 Solution 9.1: Cascade Control

Use the additional sensor $y_2(s)$ to close an *inner* (secondary) control loop to regulate $y_2(s)$ and design the corresponding inner controller $G_2(s)$. Then, close an *outer* (primary) control loop to regulate $y_1(s)$ and design the outer controller $G_1(s)$. This is the cascade control topology—see Figure 9.2.

The transfer function description of the cascade control diagram is shown in Equation 9.2. Comparing this expression with the one of Equation 9.1, we can see that the effect of the disturbance $d(s)$ on $y_1(s)$ is attenuated significantly, as $d(s)$ passes through the expression $M(s)/(1 + P_2(s)G_2(s))$ in the cascade control system instead of through $M(s)$ in the original system.

$$y_1(s) = \frac{P_1(s)\left[\dfrac{P_2(s)G_2(s)}{1 + P_2(s)G_2(s)}\right]G_1(s)}{1 + P_1(s)\left[\dfrac{P_2(s)G_2(s)}{1 + P_2(s)G_2(s)}\right]G_1(s)} r_1(s) + \frac{P_1(s)\dfrac{M(s)}{1 + P_2(s)G_2(s)}}{1 + P_1(s)\left[\dfrac{P_2(s)G_2(s)}{1 + P_2(s)G_2(s)}\right]G_1(s)} d(s) \qquad (9.2)$$

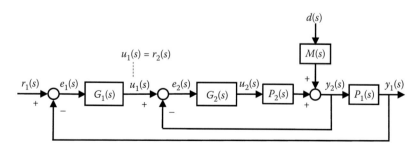

FIGURE 9.2
Cascade control system.

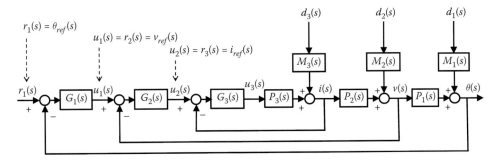

FIGURE 9.3
Three-loop cascade control system for an electrical motor.

A cascade control solution is appropriate when the following five criteria are satisfied:[367]

C1—A single feedback loop does not give a satisfactory disturbance rejection.

C2—A secondary sensor $y_2(s)$ is available for an inner loop.

C3—The sensor $y_2(s)$ measures the effect of the key disturbance $d(s)$.

C4—The sensor $y_2(s)$ is influenced by the actuator (causal relationship u_2 to y_2).

C5—The dynamics of $P_2(s)$ is faster than the dynamics of $P_1(s)$.

The design of the controllers $G_1(s)$ and $G_2(s)$ of the cascade control system has the following sequence:

Step 1: Design the inner (*secondary*) loop controller $G_2(s)$ for disturbance rejection and stability specifications.

Step 2: Design the outer (*primary*) loop controller $G_1(s)$ for disturbance rejection, reference tracking, and stability specifications.

Following these criteria, case study CS4 presents a QFT cascade control solution for a radio telescope servo system.

Finally, a system can include more than two cascade control loops. For instance, Figure 9.3 shows a three-loop cascade control solution for electrical motors. It includes an inner loop for *Current control, i(s)*; a middle loop for *Velocity control, v(s)*; and an outer loop for *Position control, θ(s)*.

9.3 Feedforward Control Systems

9.3.1 Challenge 9.2

Given the control system shown in Figure 9.4 and its transfer function description in Equation 9.3, modify the diagram topology to (1) significantly reduce the effect of the disturbance $d(s)$ on the plant output $y(s)$, and (2) speed-up the variable $y(s)$ tracking the reference $r(s)$. An additional sensor for $d(s)$ is available.

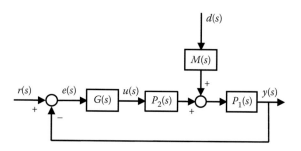

FIGURE 9.4
Challenge 9.2: (1) reduce the effect of the disturbance $d(s)$ on $y(s)$ and (2) speed-up the variable $y(s)$ tracking the reference $r(s)$.

$$y(s) = \frac{P_1(s)P_2(s)G(s)}{1+P_1(s)P_2(s)G(s)}r(s) + \frac{P_1(s)M(s)}{1+P_1(s)P_2(s)G(s)}d(s) \qquad (9.3)$$

9.3.2 Solution 9.2a: For Disturbance Rejection

Use the additional sensor $d(s)$ and a new block $G_f(s)$ to calculate an additional control signal, named feedforward $u_{ff}(s)$. This signal is added to the existing feedback control signal $u_{fb}(s)$. Figure 9.5 shows this feedforward–feedback control topology for disturbance rejection.

The transfer function description of this feedforward–feedback control diagram is shown in Equation 9.4. If we define $G_f(s)$ as in Equation 9.5, with $V(s) = 1$, then the effect of the disturbance $d(s)$ on $y(s)$ is cancelled, according to Equation 9.4. Notice that $\hat{P}_2(s)$ and $\hat{M}(s)$ are the estimated model for $P_2(s)$ and $M(s)$, respectively, and $V(s)$ is typically a low-pass filter.

$$y(s) = \frac{P_1(s)P_2(s)G(s)}{1+P_1(s)P_2(s)G(s)}r(s) + \frac{P_1(s)[M(s)+P_2(s)G_f(s)]}{1+P_1(s)P_2(s)G(s)}d(s) \qquad (9.4)$$

$$G_f(s) = -\hat{P}_2(s)^{-1}\,\hat{M}(s)V(s) \qquad (9.5)$$

$$u(s) = u_{fb}(s) + u_{ff}(s) \qquad (9.6)$$

A feedforward–feedback control solution is appropriate when the following five criteria are satisfied:[367]

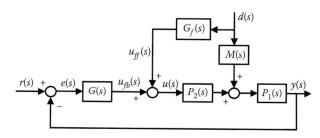

FIGURE 9.5
Feedforward–feedback control system for disturbance rejection.

C1—A single feedback loop does not give a satisfactory disturbance rejection.

C2—A secondary sensor $d(s)$ is available.

C3—The sensor $d(s)$ measures directly the disturbance.

C4—$d(s)$ is not influenced by the actuator (no causal relationship from u to d).

C5—The dynamics of $P_2(s)$ is similar or faster than the dynamics of $M(s)$.

The feedback controller $G(s)$ and the feedforward controller $G_f(s)$ are tuned independently. For $G(s)$ we can use the QFT robust control method and for $G_f(s)$ an expression based on Equation 9.5.

Notice also that a system can deal with more than one feedforward block. For instance, Figure 9.6 shows a two-feedforward feedback control solution. In this case, we have two main disturbances and two additional sensors available, one for each disturbance. The design of the feedforward controllers is based on the expressions given by Equations 9.7 and 9.8.

$$G_{f1}(s) = -\hat{P}_2(s)^{-1}\,\hat{M}_1(s)V_1(s) \qquad (9.7)$$

$$G_{f2}(s) = -\hat{P}_2(s)^{-1}\,\hat{M}_2(s)V_2(s) \qquad (9.8)$$

$$u(s) = u_{fb}(s) + u_{ff1}(s) + u_{ff2}(s) \qquad (9.9)$$

A discussion of the feedforward–feedback control for disturbance rejection in the context of QFT was developed in Section 2.12, including the solution and MATLAB code for the flight control problem presented in Example 2.2.

In some cases the calculation of the $G_f(s)$ controllers could involve unstable poles, non-proper expressions (more poles than zeros), positive exponential functions or even zero-pole cancellations. In these situations, the designers have to find a stable, proper approximation. This is the purpose of the transfer function $V(s)$ included in the expressions Equation 9.5, 9.7, and 9.8. The Example 2.2, Equation 2.76, shows also a solution for that problem.

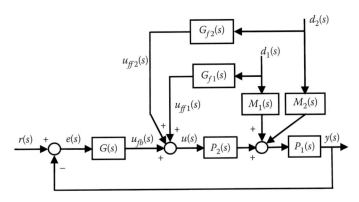

FIGURE 9.6
Feedforward–feedback control system for rejection of two disturbances.

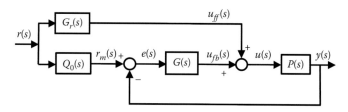

FIGURE 9.7
Feedforward control system for reference tracking. Model matching.

9.3.3 Solution 9.2b: For Reference Tracking. Model Matching

Use an additional block $G_r(s)$ to calculate an additional control signal, named feedforward $u_{ff}(s)$. This signal is added to the existing feedback control signal $u_{fb}(s)$. Figure 9.7 shows this feedforward–feedback control topology for reference tracking. It is also called model matching.

The transfer function description of this feedforward–feedback control diagram for reference tracking is shown in Equation 9.10. The expression $Q_0(s)$, or model to match, is the desired transfer function we want to have from $r(s)$ to $y(s)$. It substitutes the classical pre-filter $F(s)$. The expression $G_r(s)$ is calculated as in Equation 9.11. In this way, the transfer function from $r(s)$ to $y(s)$ is reduced to just the model we want to match, $Q_0(s)$, according to Equation 9.10.

$$y(s) = \frac{P(s)}{1+P(s)G(s)}[G(s)Q_0(s)+G_r(s)]r(s) = \left[Q_0(s)+\frac{P(s)G_r(s)-Q_0(s)}{1+P(s)G(s)}\right]r(s) \qquad (9.10)$$

$$G_r(s) = \hat{P}(s)^{-1}Q_0(s) \qquad (9.11)$$

A discussion of the feedforward–feedback control for model matching in the context of QFT was developed in Section 2.11, including the solution and MATLAB code for the flight control problem presented in Example 2.2.

9.3.4 Solution 9.2c: For Disturbance Rejection and Reference Tracking

We can combine the solutions 9.2a and 9.2b, as shown in Figure 9.8 and Equations 9.12 through 9.15.

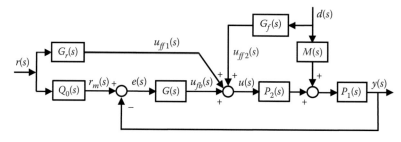

FIGURE 9.8
Feedforward–feedback control for disturbance rejection and reference tracking.

$$y(s) = \frac{P_1(s)P_2(s)}{1+P_1(s)P_2(s)G(s)}[G(s)Q_0(s)+G_r(s)]r(s) + \frac{P_1(s)[M(s)+P_2(s)G_f(s)]}{1+P_1(s)P_2(s)G(s)}d(s)$$

$$= \left[Q_0(s) + \frac{P_1(s)P_2(s)G_r(s)-Q_0(s)}{1+P_1(s)P_2(s)G(s)}\right]r(s) + \frac{P_1(s)[M(s)+P_2(s)G_f(s)]}{1+P_1(s)P_2(s)G(s)}d(s) \tag{9.12}$$

$$G_r(s) = [\hat{P}_1(s)\ \hat{P}_2(s)]^{-1}Q_0(s) \tag{9.13}$$

$$G_f(s) = -\hat{P}_2(s)^{-1}\ \hat{M}(s)V(s) \tag{9.14}$$

$$u(s) = u_{fb}(s) + u_{ff1}(s) + u_{ff2}(s) \tag{9.15}$$

For additional information about QFT methodologies for these two feedforward–feedback solutions concerning disturbance rejection and model matching, see the References 138 and 121 regarding SISO and MIMO cases, respectively.

9.4 Override Control Systems

9.4.1 Challenge 9.3

Given the plant shown in Figure 9.9, with two sensors $y_1(s)$ and $y_2(s)$ and only one actuator $u(s)$, find an appropriate control topology to regulate both output variables $y_1(s)$ and $y_2(s)$ within some limits and with only one actuator $u(s)$.

9.4.2 Solution 9.3: Override Control

As a general rule, we can say that one actuator can control only one output variable at a time. However, it is possible to use the same actuator to control two or more variables depending on which variable is the most critical at a given time. This is the *override control*, also known as *constraint control* or *selector control*. It is composed of a feedback controller $G_k(s)$ and a closed-loop for each output variable, from the sensor $y_k(s)$ to the controller output $u_k(s)$. Additionally, a special block or blocks, named "*Selectors*," receive the controller outputs $u_k(s)$ and apply the most critical one to the actuator $u(s)$ at a given time.

Figure 9.10 shows an override control solution for Figure 9.9 hydraulic system. It requires to regulate the flow rate and simultaneously needs to make sure that the pressure of the

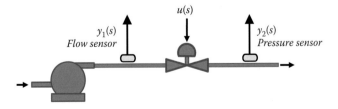

FIGURE 9.9
Challenge 9.3: controlling two outputs with one valve. Two sensors, one actuator.

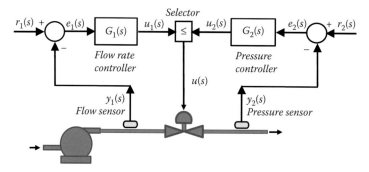

FIGURE 9.10
Override control system. Controlling $y_1(s)$ and keeping $y_2(s) \leq y_{2max}$.

circuit is always below a given limit. The system has two sensors. The first one measures the flow rate $y_1(s)$ and the second one the pressure $y_2(s)$. Also, the system has one valve $u(s)$. The override topology is composed of two controllers: $G_1(s)$ for the flow rate and $G_2(s)$ for the pressure. The control signals of both controllers go to the selector block, which in normal conditions (*pressure < limit*) applies $u(s) = u_1(s)$. However, when the pressure approaches the limit, then the selector applies $u(s) = u_2(s)$ as long as needed. For this, a simple dead-band strategy with two limits ($y_{2maxOff} < y_{2maxOn}$) instead of only one (y_{2max}) can be used in the selector to prevent the override control system oscillating around the limit of y_{2max}.

This one-limit case can be easily extended to confine the second variable between an upper limit and a lower limit. Figure 9.11 shows the block diagram for this new problem. Now, the control objective is to regulate the plant variable $y(s)$ while keeping a second variable $z(s)$ between two limits, $z_{min} \leq z \leq z_{max}$. The override solution is composed of two sensors, $y(s)$ and $z(s)$, one actuator $u(s)$, three controllers, $G_1(s)$, $G_{max}(s)$, and $G_{min}(s)$, and two selector blocks. The reference signals of the three controllers are respectively $y_{ref}(s)$, $z_{max}(s)$, and $z_{min}(s)$. The sensor for the first controller is $y(s)$, and the sensor for both the second and third controllers

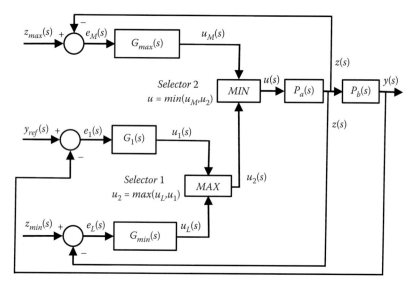

FIGURE 9.11
Override control system. Controlling $y(s)$ and keeping $z_{min} \leq z(s) \leq z_{max}$.

is $z(s)$. The output signals of the three controllers are respectively $u_1(s)$, $u_M(s)$, and $u_L(s)$. The first selector "*MAX*" chooses the largest signal between $u_1(s)$ and $u_L(s)$: $u_2 = max(u_L, u_1)$. Then, the second selector "*MIN*" chooses the smallest signal between $u_M(s)$ and the output of the other selector $u_2(s)$: $u = min(u_M, u_2)$. Then it applies the final signal $u(s)$ to the actuator.

As in the first case, only one control loop is in operation at a time. The three controllers can be tuned in the same way as single-loop controllers, for instance using the QFT robust control methodology. If the controllers have integrators, then it is necessary to track the integral states of the controllers that are not in operation to avoid any bump while switching. Section 10.3.4 describes a bumpless solution that works well in these cases.

9.5 Ratio Control Systems

9.5.1 Challenge 9.4

Given the plant shown in Figure 9.12, with two sensors $y_1(s)$ and $y_2(s)$ and only one actuator $u(s)$, find an appropriate control topology to regulate the ratio between the two variables, $a = y_2/y_1$. Examples for this problem are combustion processes, where the fuel-to-air ratio has to be controlled to improve the combustion efficiency, or blending of chemicals, where the ratio between two flows has to be constant or a function of a third variable.

9.5.2 Solution 9.4: Ratio Control

An intuitive solution to control the ratio $a = y_2/y_1$ is the topology proposed in Figure 9.13a. It measures the signals $y_1(s)$ and $y_2(s)$, divides them, compares the division with the ratio a (reference), and applies the result (error) to the controller $G_r(s)$ and valve $u(s)$. However, although this solution seems to be a natural choice, the structure does not work well due to the division operation. A simple solution is proposed in Figure 9.13b. This ratio control structure avoids the division. It calculates the reference to follow by multiplying the $y_1(s)$ signal by the ratio a. Then, compares it with $y_2(s)$, and applies the resulting error to the controller $G_2(s)$ and valve $u(s)$.

An extension of this correct solution is shown in Figure 9.14a. In this case the rate a is variable and depends on another control loop, which regulates a third variable $y_0(s)$ with a controller $G_0(s)$ that follows a reference $r_0(s)$.

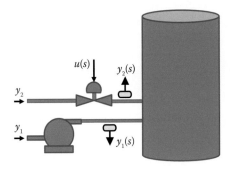

FIGURE 9.12
Challenge 9.4: controlling the ratio between two variables: $a = y_2/y_1$. Two sensors, one actuator.

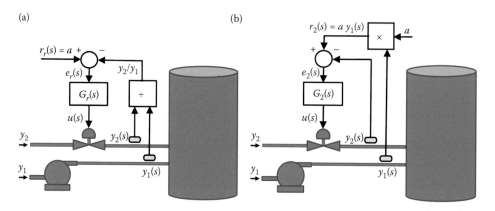

FIGURE 9.13
Ratio control system: (a) wrong topology and (b) correct solution.

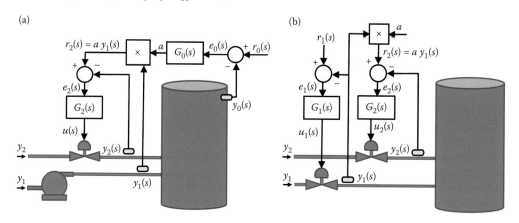

FIGURE 9.14
Ratio control system: (a) with an additional sensor and (b) with two sensors and two actuators.

In case we have two actuators $u_1(s)$ and $u_2(s)$, one for each pipe, the control structure shown in Figure 9.13b can be improved with an additional controller $G_1(s)$, as shown in Figure 9.14b. Now, this main controller $G_1(s)$ regulates the flow, and the second controller $G_2(s)$ follows the first one and controls the ratio.

In this new option, the secondary flow $y_2(s)$ is delayed, as compared to the desired flow $a\, y_1(s)$. To fix this problem, we have to speed up the secondary flow control system. This can be easily done by adding a supplementary feedforward block between the two loops, as discussed in Section 9.3. Figure 9.15a and b shows the solution for this problem, with $0 \leq \gamma \leq 1$.

9.6 Mid-Range Control Systems

9.6.1 Challenge 9.5

Given the plant shown in Figure 9.16, with one sensor $y(s) = y_1 + y_2$ and two actuators, one small and with a very high resolution $u_1(s)$, and another much larger and with less

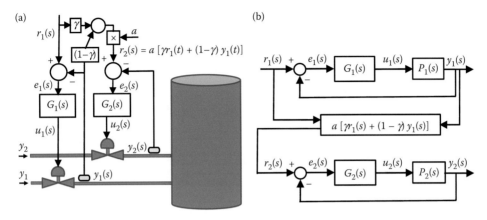

FIGURE 9.15
Ratio control system with two sensors, two actuators and feedforward.

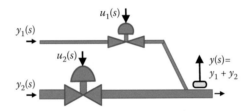

FIGURE 9.16
Challenge 9.5: controlling all range with the high resolution of the small valve, and without saturating it. One sensor, two actuators.

resolution $u_2(s)$, find an appropriate control topology to regulate the output variable $y(s)$ with the high resolution of $u_1(s)$ in all range of operation and without saturating $u_1(s)$.

9.6.2 Solution 9.5: Mid-Range Control

We propose a structure composed of two control loops. The first one measures the sensor signal $y(s)$, compares it with the reference to follow $r_1(s)$, and with the controller $G_1(s)$ moves the small valve $u_1(s)$. The second loop measures the signal $u_1(s)$, compares it with a reference $r_2(s)$, and with the controller $G_2(s)$ moves the large valve $u_2(s)$. The reference $r_2(s)$ of this second loop is the middle operation point of the small valve. Figure 9.17a and b shows this mid-range control topology. In short, the small valve $u_1(s)$ and its controller $G_1(s)$ regulate the flow $y(s)$. As the small valve $u_1(s)$ requires more flow and moves away from the middle point of its range of operation, the controller $G_2(s)$ and the large valve $u_2(s)$ change the flow through the second pipe to keep the small valve $u_1(s)$ in the middle of its range of operation.

The effect of the existing delay between the two control loops can be attenuated by introducing an additional feedforward block $G_f(s)$ from $u_2(s)$ to $u_1(s)$, as shown in Figure 9.18a and b—see also Sections 9.3 and 9.5. This additional block speeds up the reaction of $u_1(s)$ to variations of $u_2(s)$. The design of the feedback and feedforward controllers can be carried out in the context of QFT as discussed in Sections 2.11 and 2.12 and Example 2.2.

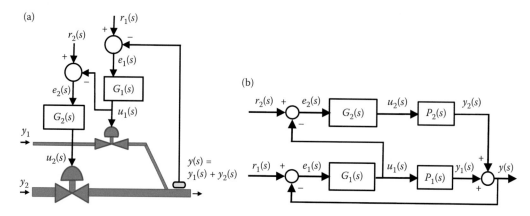

FIGURE 9.17
Mid-range control system.

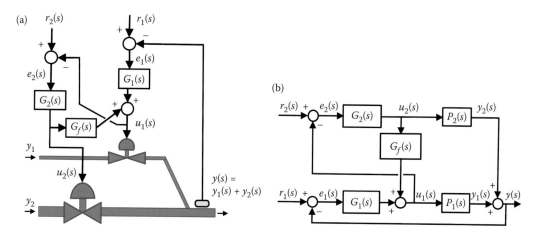

FIGURE 9.18
Mid-range control system with two control loops and feedforward.

9.7 Load-Sharing Control Systems

9.7.1 Challenge 9.6

Given the plant shown in Figure 9.19, with one sensor $\Omega(s)$ and two actuators $u_1(s)$ and $u_2(s)$, find an appropriate control topology to regulate the output variable $\Omega(s)$ coordinating both actuators $u_1(s)$ and $u_2(s)$, and avoiding them to fight each other. The example of Figure 9.19 represents two motors, each one with a pinion, and a central wheel. $T_1(s)$ and $T_2(s)$ are the torques applied by each motor, and $\Omega(s)$ the angular velocity of the central wheel.

9.7.2 Solution 9.6: Load-Sharing Control

An intuitive solution to control the common variable $\Omega(s)$ with the two actuators available, $u_1(s)$ and $u_2(s)$, is to close a control loop around each of them, with two controllers $G_1(s)$ and $G_2(s)$ as shown in Figure 9.20a. However, although this solution seems to be a natural choice, if both controllers have an integral part, the actuators can eventually start fighting

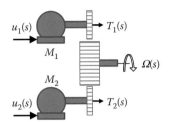

FIGURE 9.19
Challenge 9.6: controlling one variable with several motors. One sensor, several actuators.

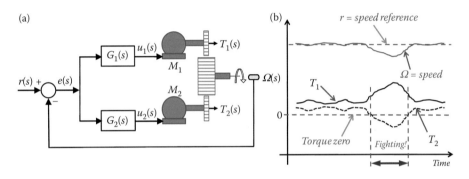

FIGURE 9.20
Potentially wrong solution if channels fight against each other.

each other, providing opposite torques $T_1(s)$ and $T_2(s)$ for a controlled angular speed $\Omega(s)$, as shown in Figure 9.20b.

Three solutions to this problem are proposed here. The first and second one use the block diagram shown in Figure 9.20a, either with only one controller $G_i(s)$ with integral part (first solution), or with no integral parts at all in any controller $G_i(s)$ (second solution). In this second case, simple controllers with high gain can do the job, like the first-order transfer functions proposed by the so-called droop control, commonly used in the electrical grid for frequency control.

A third solution is shown in the block diagram of Figure 9.21a. In this case, there is only one controller $G_1(s)$. Now, it can include an integral part. Its output is divided, or shared, between the two channels according to the γ or $(1 - \gamma)$ weighting factors, $0 \leq \gamma \leq 1$.

The three options can be easily expanded to more than two channels. A MIMO QFT robust control with a load-sharing solution for spacecraft formations can be found in Reference 209. Also, the book *Load-Sharing Control*, written by Prof. Eduard Eitelberg, is an excellent reference for this matter.[147]

9.8 Split-Range Control Systems

9.8.1 Challenge 9.7

Given the plant shown in Figure 9.22, with one sensor $y(s)$ and two actuators $u_1(s)$ and $u_2(s)$, find an appropriate control topology to regulate the output variable $y(s)$ moving the actuators one at a time, depending on the selected regime.

(a)

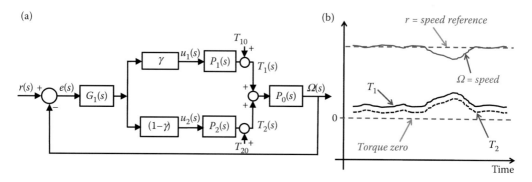

FIGURE 9.21
Load-sharing control system solution. (a) Block diagram and (b) control signals.

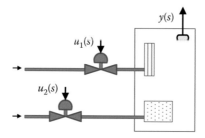

FIGURE 9.22
Challenge: controlling one variable with two valves. One sensor, two actuators.

9.8.2 Solution 9.7 Split-Range Control

Control the common variable $y(s)$ with the two actuators available, $u_1(s)$ and $u_2(s)$, by closing a control loop around each actuator, with two controllers $G_1(s)$ and $G_2(s)$, as shown in
Figure 9.23a and b. Add a selector block, with a dead-band in the middle, as shown in
Figure 9.23c. This selector chooses one controller at a time, depending on the state of the
system. The dead-band prevents the system from oscillating while in the frontier of the
system states. Also, if the controllers have integrators, then it is necessary to track the integral states of the controllers that are not in operation to avoid any bump while switching.
Section 10.3.4 describes a bumpless solution that works well in these cases.

A typical example of this split-range control topology is an air-conditioning system,
with a heating system $u_1(s)$, a cooling system $u_2(s)$, and a common temperature sensor $y(s)$.

9.9 Inferential Control Systems

9.9.1 Challenge 9.8

Let us suppose a plant with the block diagram shown in Figure 9.24. It has one sensor $z(s)$ and one actuator $u(s)$. The variable to be controlled is $y(s)$, for which there
is no sensor available. Find an appropriate control topology to regulate this output

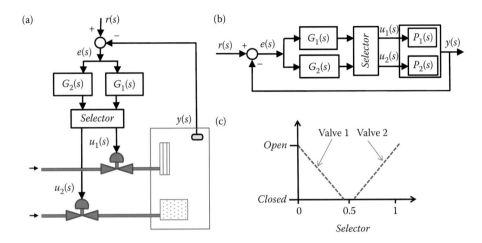

FIGURE 9.23
Split-range control system.

variable $y(s)$. The expressions for the variable to be controlled $y(s)$ and the sensor $z(s)$, from the actuator $u(s)$ and a disturbance $d(s)$, are described by Equations 9.16 and 9.17, respectively.

$$y(s) = P_1(s)u(s) + M_1(s)d(s) \tag{9.16}$$

$$z(s) = P_2(s)u(s) + M_2(s)d(s) \tag{9.17}$$

9.9.2 Solution 9.8: Inferential Control

From Equation 9.17, we can express $d(s)$ in terms of $z(s)$ and $u(s)$—see Equation 9.18. Introducing this result in Equation 9.16, we get the expression of Equation 9.19, where we have substituted the plants $P_1(s)$, $P_2(s)$, $M_1(s)$, and $M_2(s)$ by the corresponding estimated models $\hat{P}_1(s), \hat{P}_2(s), \hat{M}_1(s), \hat{M}_2(s)$.

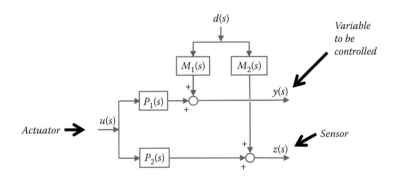

FIGURE 9.24
Challenge 9.8: no sensor available for the variable $y(s)$ to be controlled.

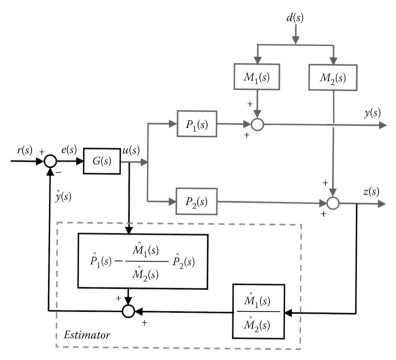

FIGURE 9.25
Inferential control system.

The left term of Equation 9.19 is the estimated value for the signal $y(s)$ we need to control, which has no sensor available. Figure 9.25 shows the implementation of the inferential control topology to regulate $y(s)$. It is based on Equation 9.19 for the estimated signal $\hat{y}(s)$.

$$d(s) = \frac{1}{M_2(s)} z(s) - \frac{P_2(s)}{M_2(s)} u(s) \tag{9.18}$$

$$\hat{y}(s) = \left[\hat{P}_1(s) - \frac{\hat{M}_1(s)}{\hat{M}_2(s)} \hat{P}_2(s) \right] u(s) + \frac{\hat{M}_1(s)}{\hat{M}_2(s)} z(s) \tag{9.19}$$

9.10 Auctioneering Control Systems

9.10.1 Challenge 9.9

Given a long reactor with many sensors $y_1(s)$, $y_2(s)$,...,$y_n(s)$ spatially distributed as shown in Figure 9.26, and one actuator $u(s)$, find an appropriate control topology to maintain all variables $y_k(s)$ below (or above) a given limit.

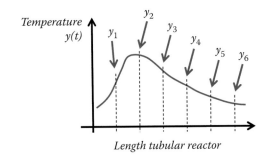

FIGURE 9.26

Challenge 9.9: maintain all variables $y_k(s)$ under a given limit. Many sensors, one actuator.

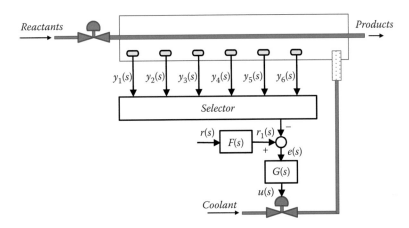

FIGURE 9.27

Auctioneering control system.

9.10.2 Solution 9.9: Auctioneering Control

Figure 9.27 shows the topology for the auctioneering control problem. It is composed of one control loop, with the feedback controller $G(s)$ and the prefilter $F(s)$, a set of sensors $y_1(s)$, $y_2(s) \cdots y_n(s)$ spatially distributed along the reactor, an actuator $u(s)$, and a selector that uses the sensor with the higher (lower) value at any time.

9.11 Summary

This chapter has proposed control topology solutions for nine common problems with a different number of sensors and actuators in the system, including cascade control, feed-forward control, override control, ratio control, mid-range control, load-sharing control, split-range control, inferential control, and auctioneering control.

9.12 Practice

The list shown below summarizes the collection of problems and cases included in this book that apply the control methodologies introduced in this chapter.

- *Example 2.2.* Section 2.11. Model matching. Airplane flight control.
- *Example 2.2.* Section 2.12. Feedforward/feedback. Airplane flight control.
- Case study CS4. Cascade control for a radio telescope servo system.
- *Project P3.* Override control for inverted pendulum.
- *Project P4.* Load-sharing control for interconnected micro-grids.
- *Project P7.* Feedforward and cascade control of a multi-tank hydraulic system.

10

Controller Implementation

10.1 Introduction

This chapter deals with the analog and digital implementation of controllers. Section 10.2 studies the analog implementation with active RC (resistor-capacitor) electrical circuits. Section 10.3 presents a variety of practical issues concerning the digital implementation with microcontrollers. Finally, Section 10.4 studies the resiliency of the implemented controllers with a fragility analysis method based on QFT.

10.2 Analog Implementation

Sensors and actuators used in control systems work habitually with electrical signals as outputs and inputs, respectively. Also, control commands are often generated as electrical signals. In all these cases, the controllers can be easily implemented with *active RC circuits*, which are simple combinations of operational amplifiers, resistors, and capacitors.

Although a digital implementation of controllers is a common and attractive solution nowadays (see Section 10.3), an analog implementation can still be the best option in some cases. In particular, an analog implementation has to be considered when extreme reliability is needed (for instance in spacecraft missions), when the bandwidth of the dynamics of the plant is very large and the available sampling time for the potential digital solution is not fast enough (for instance in power electronics), or when the cost of the implementation is an important factor for the project (for instance in low-cost systems).

The main element of an *active RC circuit* is the operational amplifier (op-amp). The magnitude frequency response of the op-amp is typically constant (about 100–120 dB) for a very large bandwidth (up to 100 kHz or 1 GHz). The electrical behavior of the op-amp within this flat-response range has two main characteristics:

1. The input impedance at the positive and negative ports is infinite, which means that the input currents are null: $i_+ = i_- = 0$.
2. The voltage of the positive and negative ports is the same: $v_+ = v_-$.

Figure 10.1 shows a simple active RC circuit with one op-amp and two resistors, R_1 and R_2. As mentioned, $i_+ = i_- = 0$. Also, the circuit imposes $v_+ = 0$, which also means $v_- = 0$.

As a result, $i_1 = (v_1 - 0)/R_1$, $i_2 = i_1$, and $v_0 = 0 - R_2\, i_2$. Substituting $i_2 = i_1 = v_1/R_1$, the v_0/v_1 transfer function is $-R_2/R_1$, which is Equation 10.1.

$$\frac{v_0(s)}{v_1(s)} = g_1(s) = -\frac{R_2}{R_1} \tag{10.1}$$

A collection of active RC circuits commonly employed in the analog implementation of controllers are shown in Figures 10.1 through 10.15. The transfer functions for these circuits are also shown in Equations 10.1 through 10.14.

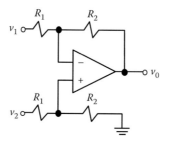

FIGURE 10.1
Gain (negative), Equation 10.1.

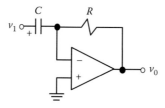

FIGURE 10.2
Summer, Equation 10.2.

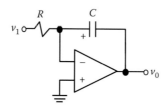

FIGURE 10.3
Differentiator and negative gain, Equation 10.3.

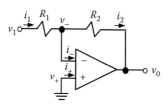

FIGURE 10.4
Integrator and negative gain, Equation 10.4.

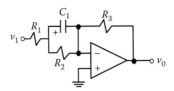

FIGURE 10.5
Lead controller and negative gain, Equation 10.5.

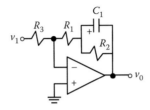

FIGURE 10.6
Lag controller and negative gain, Equation 10.6.

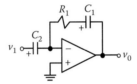

FIGURE 10.7
PD with negative gain, Equation 10.7.

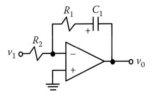

FIGURE 10.8
PI with negative gain, Equation 10.8.

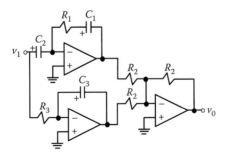

FIGURE 10.9
PID with positive gain, Equation 10.9.

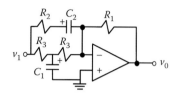

FIGURE 10.10
Complex zeros and negative gain, Equation 10.10.

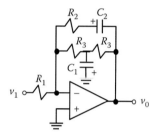

FIGURE 10.11
Complex poles and negative gain, Equation 10.11.

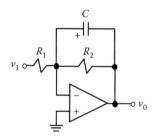

FIGURE 10.12
Real pole and negative gain, Equation 10.12.

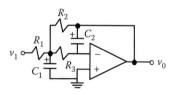

FIGURE 10.13
Complex poles and negative gain, Equation 10.13.

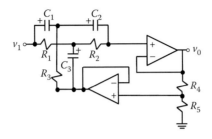

FIGURE 10.14
Notch filter, Equation 10.14.

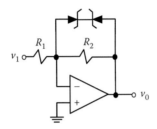

FIGURE 10.15
Nonlinearity (saturation).

These figures and expressions include the circuits needed for the implementation of the controller elements selected in the QFT design process, including gain, integrators, differentiators, real poles and zeros, complex poles and zeros, lead and lag controllers, notch filters, PI, PD, and PID controllers, and nonlinearities. As a result, the QFT feedback controllers and prefilters, feedforward blocks (see Chapter 9), sum of signals, and even nonlinear control components (see Chapters 6 and 7) can be implemented with a combination of these circuits: that is, $G(s) = g_i(s)\, g_j(s)\, g_k(s)\, \ldots$, etc.

$$v_0(s) = \frac{R_2}{R_1}[v_2(s) - v_1(s)] \tag{10.2}$$

$$\frac{v_0(s)}{v_1(s)} = g_3(s) = [-RCs] \tag{10.3}$$

$$\frac{v_0(s)}{v_1(s)} = g_4(s) = \frac{-1}{RCs} \tag{10.4}$$

$$\frac{v_0(s)}{v_1(s)} = g_5(s) = \left(\frac{-R_3}{R_1 + R_2}\right)\frac{R_2C_1s + 1}{((R_1R_2C_1)/(R_1 + R_2))s + 1} \tag{10.5}$$

$$\frac{v_0(s)}{v_1(s)} = g_6(s) = \left(\frac{-(R_1 + R_2)}{R_3}\right)\frac{((R_1R_2C_1)/(R_1 + R_2))s + 1}{R_2C_1s + 1} \tag{10.6}$$

$$\frac{v_0(s)}{v_1(s)} = g_7(s) = (k_p + k_d s) = -\left(\frac{C_2}{C_1} + R_1C_2s\right) \tag{10.7}$$

$$\frac{v_0(s)}{v_1(s)} = g_8(s) = \left(k_p + \frac{k_i}{s}\right) = -\left(\frac{R_1}{R_2} + \frac{1}{R_2 C_1 s}\right) \tag{10.8}$$

$$\frac{v_0(s)}{v_1(s)} = g_9(s) = \left(k_p + \frac{k_i}{s} + k_d s\right) = \left(\frac{C_2}{C_1} + \frac{1}{R_3 C_3 s} + R_1 C_2 s\right) \tag{10.9}$$

$$\frac{v_0(s)}{v_1(s)} = g_{10}(s) = \left(\frac{-R_1}{R_2 R_3}\right) \frac{(R_2 R_3^2 C_1 C_2)s^2 + R_3(2R_2 C_2 + R_3 C_1)s + (2R_3 + R_2)}{R_3 C_1 s + 2} \tag{10.10}$$

$$\frac{v_0(s)}{v_1(s)} = g_{11}(s) = \left(\frac{-R_2 R_3}{R_1}\right) \frac{R_3 C_1 s + 2}{(R_2 R_3^2 C_1 C_2)s^2 + R_3(2R_2 C_2 + R_3 C_1)s + (2R_3 + R_2)} \tag{10.11}$$

$$\frac{v_0(s)}{v_1(s)} = g_{12}(s) = \left(\frac{-R_2}{R_1}\right) \frac{1}{R_2 C s + 1} \tag{10.12}$$

$$\frac{v_0(s)}{v_1(s)} = g_{13}(s) = \left(\frac{-R_2}{R_1}\right) \frac{(R_2 R_3 C_1 C_2)^{-1}}{s^2 + s(R_1^{-1} + R_2^{-1} + R_3^{-1})C_1^{-1} + (R_2 R_3 C_1 C_2)^{-1}} \tag{10.13}$$

$$\frac{v_0(s)}{v_1(s)} = g_{14}(s) = \frac{s^2}{R_1 C_1 s^2 + 4(1 - (R_5/(R_4 + R_5)))s + 1} \tag{10.14}$$

The values for the resistors are typically chosen from 5 kΩ to 2 MΩ, and the values for the capacitors from 1 pF to 100 mF. The impedance between v_- and v_0 (like R_2 in Figure 10.1) should not be too small to avoid a large consumption of power in this impedance. On the other hand, the impedance between v_1 and v_- (like R_1 in Figure 10.1) should not be too large, since it reduces the input signal and increases the thermal noise effect.[256]

Finally, the ground configuration of the circuits needs special attention. In particular, we have to avoid ground paths where the current can flow and produce voltage drops. For instance, when analog and digital systems are present, we must connect the analog and digital grounds at one point only (this is quite often at the A/D converter). See Reference 256 for more details.

10.3 Digital Implementation

A digital control system (Figure 10.16) utilizes a digital computer or microprocessor (μP) to compute the control law or algorithm; an analog-to-digital converter (A/D C) to sample the analog sensor signals and convert them into digits or numbers for the algorithm; and a digital-to-analog converter (D/A C) to convert the result of the control algorithm (numbers) into an analog control signal for the actuators. The A/D and D/A conversions are made once every sampling time (see Section 10.3.1).

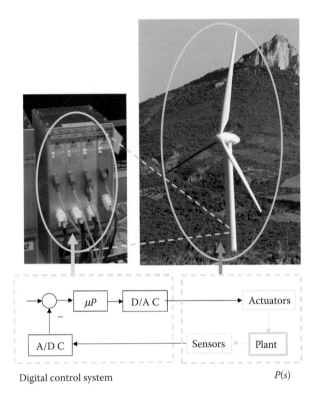

FIGURE 10.16
Digital control system. (Adapted from Garcia-Sanz, M. and Houpis, C.H. 2012. *Wind Energy Systems: Control Engineering Design (Part I: QFT Control, Part II: Wind Turbine Design and Control).* A CRC Press book, Taylor & Francis Group, Boca Raton, FL.)

10.3.1 Sample and Hold

Sampling a continuous-time signal is to replace the signal by its values at a discrete set of points. Sampling instants are equally spaced in time at $t_n = nh$, where h is the sampling period or sampling time (s), and $f_s = 1/h$ (Hz) or $\omega_s = 2\pi/h$ (rad/s) the sampling frequency. Figure 10.17 shows the sampling of a continuous signal. The frequency $f_N = 1/(2h)$ (Hz) or $\omega_N = \pi/h$ (rad/s) is known as the Nyquist frequency.

The sampling process can create new frequency components in the post-processed signal. This effect is known as *aliasing*. In particular, sampling of a signal of frequency ω_{signal} creates additional signal components ω_{added} with frequencies: $\omega_{added} = n\,\omega_s \pm \omega_{signal}$, where $\omega_s = 2\pi/h$ (rad/s) is the sampling frequency and n is an arbitrary integer. For example, in Figure 10.18, the additional component added by the aliasing has a frequency of $f_{added} = n\,f_s \pm f_{signal} = 0.52632 - 0.5 = 0.02632$ Hz, which is $T_{added} = 38$ s (period).

To avoid the aliasing problem, we need $f_{signal} < f_N < f_s$, where $f_N = f_s/2$ (Hz) is the Nyquist frequency. Figure 10.19 shows an example with no aliasing, where the signals are: $f_{signal} = 0.50 < f_N = 1.25 < f_s = 2.50$ Hz.

The Shannon sampling theorem says that: "A continuous-time signal with a Fourier transform that is zero outside the interval $(-\omega_0, \omega_0)$ is given uniquely by its values in equidistant points if the sampling frequency ω_s is higher than $2\omega_0$. That is to say: $\omega_s > 2\omega_0$." Best practices however recommend a much higher sampling frequency, with ω_s larger than 20 times the plant crossover frequency.

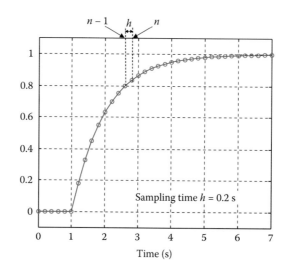

FIGURE 10.17
Sampling a continuous signal.

The analog-to-digital converter (A/D C) samples the analog sensor signals and converts them into digits or numbers for the algorithm (Figure 10.16). The sampling operation has a transfer function of

$$\text{Sampler}(s) = \frac{1}{h} \tag{10.15}$$

On the other hand, the digital-to-analog converter (D/A C) converts the result of the control algorithm (numbers) into an analog control signal for the actuators (Figure 10.16).

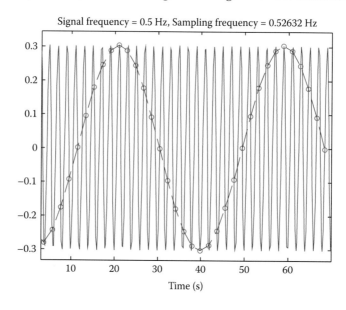

FIGURE 10.18
Aliasing problem when sampling a continuous signal. Original signal (solid line); sampled signal (dashed line with "o").

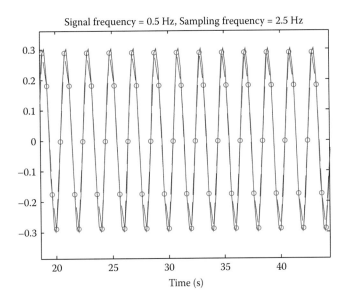

FIGURE 10.19
No aliasing when sampling a continuous signal. Original signal (solid line); sampled signal (dashed line with "o").

The D/A C output signal is piecewise constant, continuous from the left. Standard digital-to-analog converters (D/A C) are designed as a zero-order hold (ZOH). They hold constant the old value until new conversion is ordered (see Figure 10.20).

The ZOH has a function $G_{ZOH}(t) = u_{unitStep}(t) - u_{unitStep}(t - h)$, which is

$$G_{ZOH}(s) = \frac{1 - e^{-sh}}{s} \tag{10.16}$$

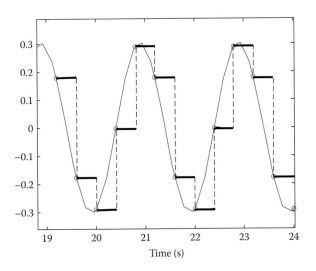

FIGURE 10.20
ZOH process.

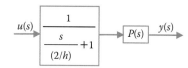

FIGURE 10.21
Sampling and ZOH model and plant.

Using the first Pade approximation for the time delay, which is valid when the value of h is small enough

$$e^{-hs} \approx \frac{1-sh/2}{1+sh/2} \qquad (10.17)$$

the transfer function of the ZOH becomes

$$G_{ZOH}(s) = \frac{1-e^{-sh}}{s} \approx \frac{2h}{hs+2} \qquad (10.18)$$

Now, combining Equations 10.15 and 10.18, sampler and hold ($S\&H$), the transfer function results

$$G_{S\&H}(s) \approx \frac{1}{h}\frac{2h}{hs+2} = \frac{1}{(s/(2/h))+1} \qquad (10.19)$$

This $S\&H$ transfer function has to be included in the block diagram and considered in the controller design process. This means that the plant to be controlled becomes $P_{new}(s) = P(s)$ $G_{S\&H}(s)$, as it is shown in Figure 10.21.

If h is very small (very high sampling frequency in comparison to the bandwidth of the dynamics of the plant), the additional pole at $s = -2/h$ included by $G_{S\&H}(s)$ is at a very high frequency, and then it affects very little the design of the controller (loop shaping). However, if h is not small enough, the effect of this pole can be quite significant in the system stability and design of the controller.

10.3.2 Computer Control Algorithms

Once the controllers [$G(s)$, $F(s)$, etc.] have been designed in the manner described in the previous chapters, their s-domain expressions are transformed into digital control algorithms following three steps:

Step 1: The s-domain transfer functions [$G(s)$, $F(s)$, etc.] are transformed into z-domain expressions [$G(z)$, $F(z)$, etc.] by use of an s-to-z transformation.

Step 2: The z-domain expressions [$G(z)$, $F(z)$, etc.] are rearranged in terms of z^{-m} factors, $m = 0, 1, 2,\ldots$. This is easily done by multiplying the z-domain transfer functions by $z^{-\alpha}$, being α the exponent of the highest order of the polynomials. In this way, the expression for $u(z)$ is obtained.

Step 3: The $u(z)$ expression is translated into algorithms in terms of $u(n)$, $u(n-1)$, $u(n-2),\ldots, e(n), e(n-1), e(n-2),\ldots$, being $u(n-\alpha)=u(z)\,z^{-\alpha}$, $e(n-\alpha)=e(z)\,z^{-\alpha}$, and (n) the current sampling period, $(n-1)$ the previous sampling period, $(n-2)$ two sampling periods ago, and so on.

Step 1 translates the controller s-domain functions into z-domain functions. The exact s-to-z transformation is shown in Equation 10.20. As this expression is irrational, we use rational approximations for the s-to-z transformation. As an illustration, Equation 10.21 shows the *Forward difference* or *Euler* approximation, Equation 10.22 the *Backward difference* approximation, and Equation 10.23 the *Tustin or Bilinear* approximation. Example 10.1 applies these three steps to a simple PI controller.

$$\text{Exact expression}: z = e^{sh} \tag{10.20}$$

$$\text{Forward approximation}: s = \frac{z-1}{h} \tag{10.21}$$

$$\text{Backward approximation}: s = \frac{z-1}{zh} \tag{10.22}$$

$$\text{Tustin approximation}: s = \frac{2}{h}\frac{z-1}{z+1} \tag{10.23}$$

EXAMPLE 10.1

- *Step 0*: The controller as an s-domain transfer function.
  Equation 10.24 is a PI controller, designed according to previous chapters.

$$\frac{u(s)}{e(s)} = G_{PI}(s) = K\left[1+\frac{1}{T_i s}\right] = \left(\frac{K}{T_i}\right)\frac{(T_i s+1)}{s} \tag{10.24}$$

- *Step 1*: The controller as a z-domain transfer function.
  Applying Equation 10.21, *Forward approximation*, through Equation 10.24, we obtain Equation 10.25.

$$\frac{u(z)}{e(z)} = G_{PI}(z) = \left(\frac{K}{T_i}\right)\frac{[T_i((z-1)/h)+1]}{((z-1)/h)} \tag{10.25}$$

which also is

$$(z-1)u(z) = K\left[(z-1)+\frac{h}{T_i}\right]e(z) \tag{10.26}$$

- *Step 2a*: The z-domain expression in terms of z^{-m} factors.
  Multiplying Equation 10.26 by z^{-1}, we obtain Equation 10.27.

$$(1-z^{-1})u(z) = K\left[(1-z^{-1})+\frac{h}{T_i}z^{-1}\right]e(z) \tag{10.27}$$

- *Step 2b*: The $u(z)$ expression in terms of z^{-m} factors.
  Solving Equation 10.27 for $u(z)$, we obtain Equation 10.28.

$$u(z) = u(z)z^{-1} + Ke(z) + \left(\frac{Kh}{T_i} - K \right) e(z) z^{-1} \tag{10.28}$$

- *Step 3*: The algorithm expression in terms of (n), $(n-1)$, $(n-2)$, …
  Applying $u(n-\alpha) = u(z)\,z^{-\alpha}$ and $e(n-\alpha) = e(z)\,z^{-\alpha}$ to Equation 10.28, we obtain
  Equation 10.29.

$$u(n) = u(n-1) + Ke(n) + K\left(\frac{h}{T_i} - 1 \right) e(n-1) \tag{10.29}$$

Finally, Figure 10.22 shows the code implementation of Equation 10.29 in the digital control algorithm. In each iteration (every sampling time h), the algorithm measures $y(n)$ from the A/D converter and $r(n)$ from the desired reference signal, calculates $e(n)$, processes $u(n)$ according to Equation 10.29, sends $u(n)$ to the D/A converter (which is connected to the actuator), and renames $u(n)$ and $e(n)$ as $u(n-1)$ and $e(n-1)$, respectively, for the next iteration.

Note 1: the *s-to-z* transformation calculated in Step 1 can also be done in MATLAB using the command "c2d," which converts continuous-time transfer functions into discrete-time transfer functions. As an illustration, we reconsider Example 10.1 again, now with $K = 12$, $T_i = 6$, and a sampling period $h = 3$ s. Then,

$$\frac{u(s)}{e(s)} = G_{PI}(s) = \left(\frac{K}{T_i} \right) \frac{(T_i s + 1)}{s} = \left(\frac{12}{6} \right) \frac{(6s+1)}{s} \tag{10.30}$$

$$GPIz = c2d(tf(2*[6\ 1],[1\ 0]),3,'Tustin') - Matlab\ command \tag{10.31}$$

$$\frac{u(z)}{e(z)} = G_{PI}(z) = \frac{15z - 9}{z - 1} \tag{10.32}$$

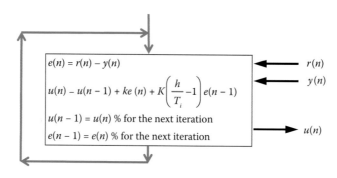

FIGURE 10.22
Algorithm implementation.

$$(z-1)u(z) = (15z-9)e(z) \tag{10.33}$$

$$(1-z^{-1})u(z) = (15-9z^{-1})e(z) \tag{10.34}$$

$$u(n) = u(n-1) + 15e(n) - 9e(n-1) \tag{10.35}$$

Note 2: Frequency warping. The frequency scale can be distorted due to the discretization approximation. This can significantly affect notch filters and low-pass filters, resulting in incorrect applied notch frequencies and bandwidth. To understand the problem, we apply the exact *s-to-z* transformation, shown in Equation 10.20 to the Tustin transformation shown in Equation 10.23, so that

$$s = \frac{2}{h}\frac{z-1}{z+1} = \frac{2}{h}\frac{e^{sh}-1}{e^{sh}+1} = \frac{2}{h}\frac{e^{sh/2}-e^{-sh/2}}{e^{sh/2}+e^{-sh/2}} = \frac{2j}{h}\tan\left(\frac{\omega h}{2}\right) \tag{10.36}$$

which in the plant $P(s)$ is

$$P(s) = P\left(\frac{2}{h}\frac{e^{sh}-1}{e^{sh}+1}\right) = P\left(j\left[\frac{2}{h}\tan\left(\frac{\omega h}{2}\right)\right]\right) = P(j\omega_p) \tag{10.37}$$

with

$$\omega_p = \frac{2}{h}\tan\left(\frac{\omega h}{2}\right) \tag{10.38}$$

The difference between the original frequency (ω) and the digital implemented frequency (ω_p) is shown in Figure 10.23. As we can see, the higher the frequency, the larger the difference.

A solution to eliminate the warping distortion at a given frequency $\omega = \omega_1$ is to modify the Tustin approximation with a pre-warping factor, as shown in Equation 10.39

$$s = \left[\frac{\omega_1 h}{2\tan(\omega_1 h/2)}\right]\frac{2}{h}\frac{z-1}{z+1} = \frac{\omega_1}{\tan(\omega_1 h/2)}\frac{z-1}{z+1} \tag{10.39}$$

Note 3: Ringing. If the discretization method gives a denominator of $G(z)$ with stable negative poles (poles inside the unitary circle with negative real part, Figure 10.24a), then these negative poles of the controller will change the sign of the output $u(t)$ at each sampling time (see Figure 10.24b). This effect is called ringing, and should be avoided by selecting another more appropriate discretization method to obtain stable positive poles (poles inside the unitary circle with positive real part).

Note 4: Quantization. A digital computer has a finite precision in the calculations (number of bits). This limitation affects the terms of the digital controller. As the coefficients of the digital controller are proportional to the sampling time h, when the sampling time h is very small, the controller terms can be rounded

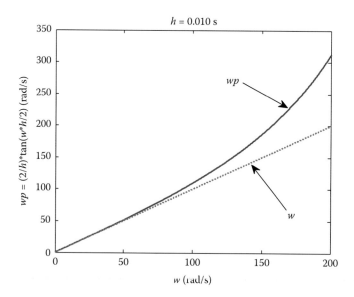

FIGURE 10.23
Frequency warping.

off if the computer has not enough precision. This will produce a steady state error, or integral offset. As an example, consider the digital PI controller shown in Equation 10.29. If the coefficients are $K = 0.1$, $T_i = 1200$, and $h = 0.02$ s, the coefficient $Kh/T_i = 1.7 \times 10^{-6} = 2^{-19.2}$. To represent this number in the computer, we need 20 bits. If, however, $h = 1$ s, then we only need 14 bits.

In summary, the sampling frequency $f_s = 1/h$ has to be fast enough (h small enough) to pick the dynamics of the system, avoid aliasing problems and warping distortion, and put the effect of the sample and hold additional function $G_{S\&H}(s)$ at a frequency much higher than the crossover frequency. On the other hand, the sampling frequency f_s is upper-limited by the finite precision (number of bits) of the calculations. A good sampling frequency selection is $\omega_s = 2\pi/h = 2\pi f_s > 20\omega_1$, ω_1 being the frequency of the fastest pole or zero of the plant, or the crossover frequency (whoever is higher).

10.3.3 Positional and Velocity Algorithms

Positional algorithms calculate the complete expression for the controller output $u(n)$ at every sample. On the contrary, velocity algorithms (or incremental) only calculate the increment $\Delta u(n)$ or difference from the previous controller output $u(n-1)$ at every sample.

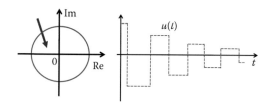

FIGURE 10.24
Ringing problem.

As an example, consider the standard PID controller shown in Equations 2.85 through 2.88, and repeated here as Equation 10.40, with a low-pass first-order filter (at $s = -N/T_d$) and a kick-off solution ($r = 0$) for the derivative part, and with a weight coefficient b for the proportional part.

$$u(s) = K\left\{[br(s) - y(s)] + \frac{1}{T_i s}[r(s) - y(s)] - \frac{T_d s}{1 + (T_d / N)s} y(s)\right\} \tag{10.40}$$

10.3.3.1 Positional Algorithm

Applying the Tustin transformation (Equation 10.23) to Equation 10.40, we find the Equations 10.41 through 10.44 for the *positional algorithm*, with the coefficients shown in Equations 10.45 through 10.48. In this case, the controller output applied at every sample is the complete $u(n)$ signal, calculated with Equation 10.44.

$$P(n) - K[br(n) - y(n)] \tag{10.41}$$

$$I(n) = I(n-1) + b_{i1}e(n) + b_{i2}e(n-1) \tag{10.42}$$

$$D(n) = a_d D(n-1) - b_d[y(n) - y(n-1)] \tag{10.43}$$

$$u(n) = P(n) + I(n) + D(n) \tag{10.44}$$

with

$$b_{i1} = \frac{Kh}{2T_i} \tag{10.45}$$

$$b_{i2} = \frac{Kh}{2T_i} \tag{10.46}$$

$$a_d = \frac{2T_d - Nh}{2T_d + Nh} \tag{10.47}$$

$$b_d = \frac{2KT_dN}{2T_d + Nh} \tag{10.48}$$

10.3.3.2 Velocity Algorithm

On the contrary, if we calculate the increment of every term as shown in Equations 10.49 through 10.51, that is $\Delta P(n)$, $\Delta I(n)$, and $\Delta D(n)$ as the difference between the current term (at instant n) minus the previous term (at instant $n-1$), we can combine all of them to calculate the total increment $\Delta u(n)$—see Equation 10.52, and then add it to the previous controller output $u(n-1)$ to have $u(n)$—see Equation 10.53. This is the *velocity algorithm* which, as mentioned, only calculates the increment $\Delta u(n)$ or difference from the previous controller output $u(n-1)$ at every sample.

$$\Delta P(n) = P(n) - P(n-1) = K[br(n) - y(n) - br(n-1) + y(n-1)] \tag{10.49}$$

$$\Delta I(n) = I(n) - I(n-1) = b_{i1}e(n) + b_{i2}e(n-1) \tag{10.50}$$

$$\Delta D(n) = D(n) - D(n-1) = a_d D(n-1) - b_d[y(n) - 2y(n-1) + y(n-2)] \tag{10.51}$$

$$\Delta u(n) = \Delta P(n) + \Delta I(n) + \Delta D(n) \tag{10.52}$$

$$u(n) = u(n-1) + \Delta u(n) \tag{10.53}$$

10.3.4 Switching and Bumpless Algorithms

Practically speaking, many control solutions switch controllers among several modes: (a) manual/automatic, (b) override solutions—see Section 9.4, (c) split-range solutions—see Section 9.8, (d) gain scheduling—see Chapter 6, or (e) nonlinear dynamic control—see Chapter 7. To avoid bump problems under these switching events, we have to make sure that the state of the system is correct when switching. A practical solution is to implement the controller in the velocity (incremental) form described in the previous section. In the switching operation, the previous element $u(n-1)$ is calculated by the previous mode, and the increment $\Delta u(n)$ is calculated by the new mode. Figure 10.25 and Equations 10.54 and 10.55 show the switching at the time n.

$$\Delta u(n) = \Delta P(n) + \Delta I(n) + \Delta D(n)\ldots\text{in a PID controller} \tag{10.54}$$

$$u(n) = u(n-1) + \Delta u(n) \tag{10.55}$$

10.3.5 Pulse Width Modulation

Occasionally, we have limited actuators that do not accept a continuous signal, but only two values: *On* (1) or *Off* (0)—see Figure 10.26. In this situation, we can translate the continuous output of the controller $u(t)$ into an *On/Off* signal (only 1 or 0) with a variable pulse

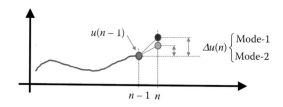

FIGURE 10.25
Switching between velocity algorithms.

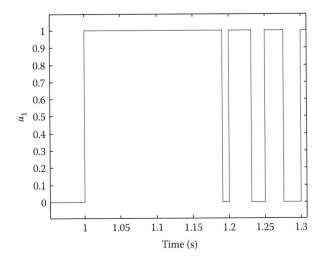

FIGURE 10.26
PWM control signal $u_1(t)$.

time (T_{on} or T_{off} times). This method is known as a *pulse width modulation* (PWM) technique. For a given *On* signal value (u_{max}), *Off* signal value (u_{min}), and cycle time (T_{cycle}), the time of the cycle that the applied *On/Off* signal $u_1(t)$ is at the maximum value (*On*), also known as T_{pulse}, is calculated as shown in Equation 10.56 and Figure 10.26. Example 10.2 shows how to apply this expression.

$$T_{pulse}(t) = \frac{u(t) - u_{min}}{u_{max} - u_{min}} T_{cycle} \qquad (10.56)$$

EXAMPLE 10.2

Consider the plant $P(s)$ shown in Equation 10.57, the calculated PI analog controller $G(s)$ in Equation 10.58, and the corresponding digital controller $G(z)$ in Equation 10.59, calculated with the Tustin approximation and a sampling time $h = 0.01$ s.

$$P(s) = \frac{9}{(2.5s + 1)} \qquad (10.57)$$

$$G(s) = \frac{2s + 1}{s} \qquad (10.58)$$

$$G(z) = \frac{2.005z - 1.995}{z - 1} \qquad (10.59)$$

Consider also that the PWM has a $T_{cycle} = 0.05$ s, $u_{max} = 1$, and $u_{min} = 0$. Figures 10.27 and 10.28 show the analog controller and PWM solution respectively, with a simulation of the closed-loop control system following a unitary step reference change at $t = 1$ s. As we can see, both cases produce similar control results.

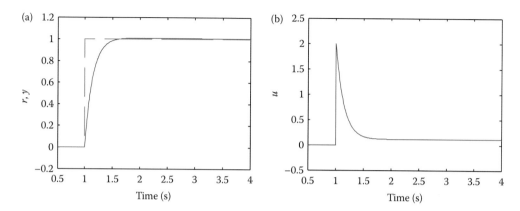

FIGURE 10.27
Analog controller, Equation 10.58. (a) $r(t)$ and $y(t)$; (b) $u(t)$.

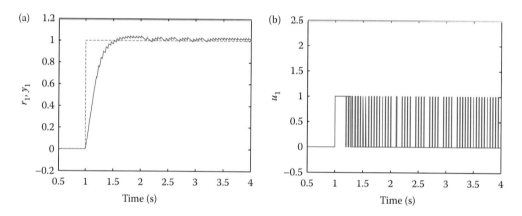

FIGURE 10.28
PWM controller, Equation 10.59. (a) $r(t)$ and $y(t)$; (b) $u(t)$.

10.4 Fragility Analysis with QFT

Once we have designed the controllers and have implemented them with either an analog active RC circuit or a digital control algorithm, we have to study the resiliency/fragility of the implemented solution. A fragility problem arises when the implementation of the controller compromises the system stability. In the digital case, as the number of bits of the microprocessor is limited, the digital implementation requires to round the coefficients of the controller to a given number of decimals. Additionally, some round off errors in numerical computations and inaccuracies in the analog-to-digital or digital-to-analog conversions can appear. In the analog case, imprecisions in the selection of resistors and capacitors, or potential changes of their values with temperature, can produce the same problem. When these small changes in the controller coefficients compromise the system stability, we say that the controller is *fragile*. If these imprecisions do not affect the stability, the controller is *resilient*.[85]

The fragility problem is primarily found in high-order controllers, usually designed by some optimal control synthesis methods, standard model-based techniques, and certain parameterizations. In 1997, Keel and Bhattacharyya wrote a paper entitled *"Robust, Fragile or Optimal?"*[315] which shocked the international community and started off a large set of comments and replies.[316–325] They demonstrated the inherent fragility of some well-known H_∞, H_2, L_2, and μ controllers when small changes are introduced in the controller coefficients.

This section introduces a practical method based on QFT to analyze the controller fragility/resilience to changes in its coefficients.[85] The method simultaneously takes into account the plant model with parametric and nonparametric uncertainty, the robust stability and performance specifications, and the uncertainty in the controller coefficients. It studies the controller fragility at every frequency and for both stability and performance specifications.

The method starts with the plant model $P(s)$ with uncertainty, the stability and performance specifications, and the designed controller $G_0(s)$, which we call the nominal controller. It is verified that, with a perfect controller implementation, the closed-loop system meets the stability and performance specifications for all the plants within the uncertainty.

Now, consider that the uncertainty also affects the controller, which is represented as $G(s)$. To analyze the fragility introduced by the controller, we define an *extended plant* as

$$P_e(s) = P(s)G(s) \tag{10.60}$$

Then, a set of new templates are defined (the *extended templates*). They combine the plant model uncertainty with the controller uncertainty at every frequency of interest. With these extended templates, and taking into account the specifications, a set of new bounds are defined (the *extended bounds*). Now, and based on these extended bounds, the controller fragility is analyzed. The methodology is composed of the following steps:

Step 1. Define the plant model with uncertainty $P(s)$.

Step 2. Select the nominal plant $P_0(s)$ within the set of plants defined by the uncertainty.

Step 3. Define the desired stability and performance specifications for the control system.

Step 4. Design a robust controller $G_0(s)$.

Step 5. Select the interval of variation of the controller coefficients $G(s)$.

Step 6. Define the extended plant as, $P_e(s) = P(s) G(s)$.

Step 7. Choose $P_e^o(s) = P_0(s) G_0(s)$ as the nominal extended plant.

Step 8. Generate the extended templates from the extended plant $P_e(s)$ and for the frequencies of interest.

Step 9. Calculate the extended bounds from the extended templates and the previous stability and performance specifications.

Step 10. Apply a unitary controller to shape $P_e^o(s)$ over the extended bounds, that is $L_0(s) = P_e^o(s) \times 1$.

Step 11. If $L_0(s)$ penetrates the forbidden region delimited by the extended robust stability bounds, then the controller is *stability-fragile* at these frequencies. If $L_0(s)$ penetrates the forbidden region delimited by the extended robust performance bounds, then the controller is *performance-fragile* at these frequencies. Otherwise it is resilient.

EXAMPLE 10.3: A FRAGILE CONTROLLER

Consider the simple plant and the parametric uncertainty defined by Equation 10.61. The nominal plant is chosen with $k = 1$ and $a = 1$—see Equation 10.62.

$$P(s) = \frac{k\,a}{s\,(s+a)}, \quad \text{with} \quad k \in [1,10], \quad a \in [1,10] \tag{10.61}$$

$$P_0(s) = \frac{1}{s\,(s+1)} \tag{10.62}$$

The robust stability specification is selected as $W_s = 1.2$, which is equivalent to a gain and phase margin of GM = 1.8333 dB and PM = 50°. The controller designed for the above plant and stability specification (nominal controller) is:

$$G_0(s) = \frac{k_G((1/z_1)s+1)((1/z_2)s+1)}{((1/p_1)s+1)((1/p_2)s+1)((1/p_3)s+1)} \tag{10.63}$$

with

$$k_G = 1.1 \quad z_2 = 2.7 \quad p_2 = 20$$
$$z_1 = 23 \quad p_1 = 220 \quad p_3 = 18$$

Figure 10.29 shows the stability bounds and the loop shaping of the nominal open-loop transfer function $L_0(s) = P_0(s)\,G_0(s)$.

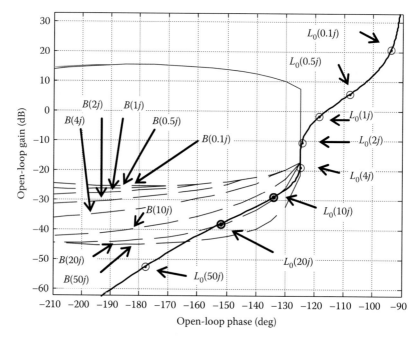

FIGURE 10.29
Nichols chart with the stability bounds of the system $P(s)$ and the nominal open-loop function $L_0(s)$ with the designed controller $G_0(s)$. (Adapted from Garcia-Sanz, M. and Houpis, C.H. 2012. *Wind Energy Systems: Control Engineering Design (Part I: QFT Control, Part II: Wind Turbine Design and Control)*. A CRC Press book, Taylor & Francis Group, Boca Raton, FL.)

The fragility analysis is now undertaken assuming that the uncertainty in the controller parameters is

$$k_G \in [1.1, \ 2] \quad z_2 \in [2.7, \ 3] \quad p_2 \in [20, \ 21]$$
$$z_1 \in [23, \ 24] \quad p_1 \in [220, \ 221] \quad p_3 \in [18, \ 19]$$

These intervals are related to the applied implementation technique. Here for the sake of clarity, they are taken arbitrarily in order to see the effect of the fragility associated with the controller uncertainty. The model of the extended system results in the expression $P_e(s) = P(s) G(s)$, so that

$$P_e(s) = \frac{ka}{s(s+a)} \frac{k_G((1/z_1)s+1)((1/z_2)s+1)}{((1/p_1)s+1)((1/p_2)s+1)((1/p_3)s+1)} \tag{10.64}$$

The parametric uncertainty of the extended plant is

$$k \in [1, \ 10], \quad z_1 \in [23, \ 24], \quad p_2 \in [20, \ 21], \quad k_G \in [1.1, \ 2]$$
$$a \in [1, \ 10], \quad z_2 \in [2.7, \ 3], \quad p_3 \in [18, \ 19], \quad p_1 \in [220, \ 221]$$

Next, the templates of this extended plant are established, that is, the *extended templates*. From these, the new bounds (*extended bounds*) are drawn up for the same robust stability specification $W_s = 1.2$.

The nominal plant of the extended system is $P_e^o(s) = P_0(s)G_0(s)$. Note that the chosen extended nominal plant coincides with the previous nominal plant multiplied by the original controller.

Subsequently, a unitary controller to shape $P_e^o(s)$ over the extended bounds is chosen as shown in Equation 10.65. In this way, the open-loop transfer function remains the same as before and thus it enables the designer to make the comparison and determine the controller resilience. The performance of this new expanded nominal plant is checked out with regard to the extended bounds (stability bounds of the extended system), as is shown in Figure 10.30.

$$L_e(s) = P_e^o(s) \times 1 = P_0(s)G_0(s) = L_0(s) \tag{10.65}$$

When comparing both Figures 10.29 and 10.30, one notices that the extended system violates the stability requirements at the frequencies 10 rad/s and 20 rad/s, that is, $L_0(j10)$ and $L_0(j20)$ are above $B_e(j10)$ and $B_e(j20)$, respectively. In other words, the open-loop transfer function $L_0(s)$ penetrates the forbidden region delimited by the extended robust stability bounds, although the original open-loop transfer function (see Figure 10.29) met these specifications. In conclusion, $G_0(s)$ is a fragile controller for this system.

10.5 Summary

This chapter studied the analog implementation of controllers with active RC electrical circuits and a variety of practical issues concerning the digital implementation with

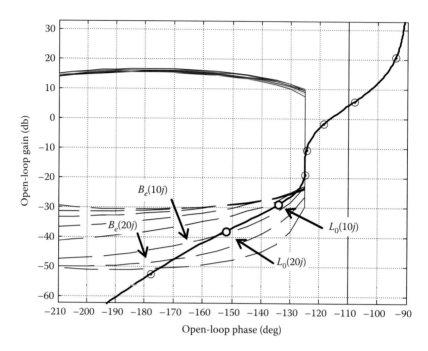

FIGURE 10.30
Nichols chart of the extended stability bounds of the system $P_e(s)$. (Adapted from Garcia-Sanz, M. and Houpis, C.H. 2012. *Wind Energy Systems: Control Engineering Design (Part I: QFT Control, Part II: Wind Turbine Design and Control).* A CRC Press book, Taylor & Francis Group, Boca Raton, FL.)

microcontrollers. The chapter also studied the resiliency of the implemented controllers with a fragility analysis method based on QFT.

10.6 Practice

The list shown below summarizes the collection of problems and cases included in this book that apply the control methodologies introduced in this chapter.

- Example 10.1. Section 10.3.2. Digital PI algorithm
- Example 10.2. Section 10.3.5. Digital PWM-PI algorithm
- Example 10.3. Section 10.4. Fragility analysis of a control system

Case Study 1: Satellite Control

CS1.1 Description

Space technology is one of the main achievements of the twentieth century. It has evolved extraordinarily since the very first satellites, the Soviet Sputnik (October 1957) and the U.S. Explorer (January 1958). Our modern society depends greatly on satellites. They have become the key instruments for telecommunications, airplane navigation, GPS systems, meteorology, scientific research, cellphones, TV, internet, etc.

Satellites can operate from low Earth orbits (LEO) around 300 km altitude, to geostationary Earth orbits (GEO) about 42,000 km, and beyond the Earth gravity, toward the exploration of other planets. Satellites can be very light, from just a few kilograms (nano-satellites), or quite heavy, with more than 1000 kg (large satellites). In most cases, they have solar panels to harvest energy from the Sun—see Figure CS1.1. This is usually possible if they are located not farther than the Asteroid belt, located between the orbits of the planets Mars and Jupiter. The physical laws that describe the dynamics and motion of a satellite are well known. The control system, and in particular the so-called attitude control system (ACS), is one of the key components of every satellite. For additional information about spacecraft dynamics and control, see Reference 363, written by Marcel Sidi, the first PhD student of Professor Isaac Horowitz. Incidentally, notice that the QFT-bounds are also known as the Horowitz–Sidi bounds.

This case study CS1 presents the design of an ACS for a satellite with flexible solar panels. The control system has three simultaneous objectives for every plant within the model uncertainty: (1) to regulate the angle of motion of the satellite, (2) to reject unpredictable disturbances, and (3) to minimize the solar panel vibrations.

We propose a classical 2DOF control system that includes a feedback compensator $G(s)$ and a prefilter $F(s)$, as shown in Figure CS1.3. In the following sections, we model and analyze the dynamics of the satellite, propose a set of stability and performance specifications, design the controllers, and validate them in both the frequency and time domains.

CS1.2 Plant Model

Figure CS1.2 shows a model diagram for a satellite with two solar panels. It includes the variables and parameters that describe the dynamics of the system. We consider the satellite as a rigid central body with two equal and flexible solar panels, each one modeled as a lumped mass at the end of a flexible beam.

The variable we want to control is $\theta(t)$, which is the angle of motion of the central body, in [rad]. The satellite is equipped with reaction wheels that provide a symmetric torque $T_m(t)$

FIGURE CS1.1
Hispasat. A telecommunication satellite with solar panels.

about the axis of rotation, in [Nm]. In addition, $y_p(t)$ is the deformation of the solar panel, in [m]. The main parameters of the system are

- J_0, the moment of inertia of the central body, (kg m^2)
- m, the mass of each solar panel, as a mass-point at the end of the panel, (kg)
- l_p, the length of each solar panel, from center of the mass of the central body, (m)
- $J = J_0 + 2\, m\, l_p^2$, the moment of inertia of the entire system, (kg m^2)
- K_p, the panel translational stiffness coefficient, (N/m)
- B_p, the panel translational damping coefficient, (Ns/m)
- t, the time (s).

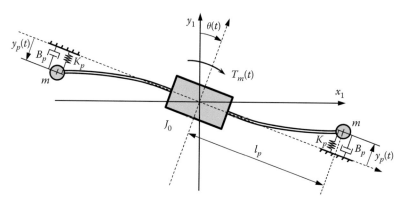

FIGURE CS1.2
Satellite with solar panels. Model description.

We start the modeling with the general equation of motion of a mechanical system, which is

$$\mathcal{M}\ddot{q} + \mathcal{C}\dot{q} + \mathcal{K}q = \mathcal{Q}(\dot{q}, q, u, t) \tag{CS1.1}$$

where $\mathcal{M}$, $\mathcal{C}$, and $\mathcal{K}$ are the *mass*, *damping*, and *stiffness* matrices, $\mathcal{Q}$ is the matrix of inputs, q is the vector of generalized coordinates, and u is the vector of inputs. Also, the general form of the Euler–Lagrange equation is

$$\frac{d}{dt}\left(\frac{\partial L}{\partial \dot{q}_i}\right) - \frac{\partial L}{\partial q_i} + \frac{\partial D_d}{\partial \dot{q}_i} = Q_i, \quad i = 1, 2, \ldots, n \quad \text{and} \quad L = E_k - E_p \tag{CS1.2}$$

where q_i are the generalized coordinates or degrees of freedom of the system, Q_i the generalized forces applied to each subsystem "i," E_k and E_p the kinetic and potential energies, D_d the dissipator co-content, and L the Lagrangian.[236]

The generalized coordinates for the satellite with two solar panels are the angle of motion of the central body and the deformation of each solar panel, which are, respectively: $q_1 = \theta(t)$, and $q_2 = y_p(t)$—see Figure CS1.2. The system contains the kinetic, potential, and dissipative functions described by Equations CS1.3 through CS1.5. Notice that we consider a linear dissipator, which means that the D_d expression or dissipator co-content for the Lagrange equations equals one-half of the power being absorbed by the dissipator.[236]

$$E_k = \frac{1}{2}J_0\dot{\theta}^2 + 2\left[\frac{1}{2}m\left(\dot{y}_p + l_p\dot{\theta}\right)^2\right] \tag{CS1.3}$$

$$E_p = 2\left(\frac{1}{2}K_p y_p^2\right) \tag{CS1.4}$$

$$D_d = 2\left(\frac{1}{2}B_p\dot{y}_p^2\right) \tag{CS1.5}$$

The generalized force applied about the axis of rotation of the central body, $i = 1$, is the torque: $Q_1 = T_m(t)$. Using Equations CS1.3 through CS1.5, the terms for the Euler–Lagrange equation are

$$\frac{d}{dt}\left(\frac{\partial L}{\partial \dot{q}_i}\right) = \frac{d}{dt}\begin{bmatrix}\dfrac{\partial E_k}{\partial \dot{\theta}} \\[2mm] \dfrac{\partial E_k}{\partial \dot{y}_p}\end{bmatrix} = \begin{bmatrix}J_0 + 2ml_p^2 & 2ml_p \\ 2ml_p & 2m\end{bmatrix}\begin{bmatrix}\ddot{\theta} \\ \ddot{y}_p\end{bmatrix} = \mathcal{M}\ddot{q} \tag{CS1.6}$$

$$-\frac{\partial L}{\partial q_i} = \frac{\partial E_p}{\partial q_i} = \begin{bmatrix}\dfrac{\partial E_p}{\partial \theta} \\[2mm] \dfrac{\partial E_p}{\partial y_p}\end{bmatrix} = \begin{bmatrix}0 & 0 \\ 0 & 2K_p\end{bmatrix}\begin{bmatrix}\theta \\ y_p\end{bmatrix} = \mathcal{K}q \tag{CS1.7}$$

$$\frac{\partial D_d}{\partial \dot{q}_i} = \begin{bmatrix} \dfrac{\partial D_d}{\partial \dot{\theta}} \\[2mm] \dfrac{\partial D_d}{\partial \dot{y}_p} \end{bmatrix} = \begin{bmatrix} 0 & 0 \\ 0 & 2B_p \end{bmatrix} \begin{bmatrix} \dot{\theta} \\ \dot{y}_p \end{bmatrix} = \mathcal{C}\dot{q} \qquad (\text{CS1.8})$$

$$\mathcal{Q} = \begin{bmatrix} 1 & 0 \\ 0 & 0 \end{bmatrix} \begin{bmatrix} T_m \\ 0 \end{bmatrix} = \mathcal{R}u \qquad (\text{CS1.9})$$

Rearranging the equation of motion Equation CS1.1 as

$$\ddot{q} = -\mathcal{M}^{-1}\mathcal{C}\dot{q} - \mathcal{M}^{-1}\mathcal{K}q + \mathcal{M}^{-1}\mathcal{R}u \qquad (\text{CS1.10})$$

and using also Equations CS1.6 through CS1.9, we find the state space description of the system ($\dot{x} = Ax + Bu$; $y = Cx$) with the expressions,

$$\dot{x} = \left[\begin{array}{c|c} -\mathcal{M}^{-1}\mathcal{C} & -\mathcal{M}^{-1}\mathcal{K} \\ \hline I & 0 \end{array}\right] x + \left[\begin{array}{c} \mathcal{M}^{-1}\mathcal{R} \\ \hline 0 \end{array}\right] u$$

$$y = [0 \mid I]x \qquad (\text{CS1.11})$$

with $I = [1\ 0;\ 0\ 1]$, $0 = [0\ 0;\ 0\ 0]$, and,

$$A = \begin{bmatrix} 0 & \dfrac{2l_p B_p}{J_0} & 0 & \dfrac{2l_p K_p}{J_0} \\[3mm] 0 & \dfrac{-B_p(J_0 + 2ml_p^2)}{mJ_0} & 0 & \dfrac{-K_p(J_0 + 2ml_p^2)}{mJ_0} \\[3mm] 1 & 0 & 0 & 0 \\[1mm] 0 & 1 & 0 & 0 \end{bmatrix} \qquad (\text{CS1.12})$$

$$B = \begin{bmatrix} \dfrac{1}{J_0} & 0 \\[3mm] \dfrac{-l_p}{J_0} & 0 \\[2mm] 0 & 0 \\[1mm] 0 & 0 \end{bmatrix} \qquad (\text{CS1.13})$$

$$C = \begin{bmatrix} 0 & 0 & 1 & 0 \\ 0 & 0 & 0 & 1 \end{bmatrix} \qquad (\text{CS1.14})$$

$$x = \begin{bmatrix} \dot{q} \\ q \end{bmatrix} = \begin{bmatrix} \dot{\theta} & \dot{y}_p & \theta & y_p \end{bmatrix}^T \qquad (\text{CS1.15})$$

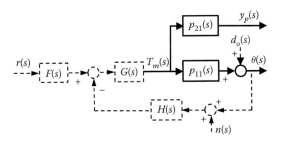

FIGURE CS1.3
ACS for satellite with solar panels.

$$u = \begin{bmatrix} T_m & 0 \end{bmatrix}^T \qquad \text{(CS1.16)}$$

$$y = \begin{bmatrix} \theta & y_p \end{bmatrix}^T \qquad \text{(CS1.17)}$$

Now, applying $P(s) = C(sI-A)^{-1}B$ for $y(s) = P(s)u(s)$, the transfer function $p_{11}(s)$, from the torque T_m to the angular movement θ of the satellite, is

$$\frac{\theta(s)}{T_m(s)} = p_{11}(s) = \left(\frac{1}{J_0}\right) \frac{\left(s^2 + (B_p/m)s + (K_p/m)\right)}{s^2 \left(s^2 + (B_p J/(m J_0))s + (K_p J/(m J_0))\right)} \qquad \text{(CS1.18)}$$

with $\quad J = J_0 + 2m l_p^2$

and the transfer function $p_{21}(s)$, from the torque T_m to the deformation of the solar panel y_p, is—see also Figure CS1.3

$$\frac{y_p(s)}{T_m(s)} = p_{21}(s) = \left(\frac{1}{J_0}\right) \frac{-l_p}{\left(s^2 + (B_p J/(m J_0))s + (K_p J/(m J_0))\right)} \qquad \text{(CS1.19)}$$

with $\quad J = J_0 + 2 m l_p^2$

CS1.2.1 Parametric Uncertainty

For this study, we consider a medium size satellite with the parameters suggested in Reference 363. In particular, the mass of each panel is $m = 20$ kg, the length of each panel, from the center of the mass of the satellite to the end of the panel $l_p = 2$ m, and the panel translational stiffness coefficient $K_p = 320$ N/m. The moment of inertia of the entire system follows the expression: $J = J_0 + 2m l_p^2 = J_0 + 160$.

Many satellites also have a significant mass of fuel to carry out orbit maneuvers with thrusters. This mass of fuel is typically up to a 40% of the mass of the central body. As this mass changes during the lifetime of the satellite, we include the corresponding uncertainty in the moment of inertia J_0 of the system. In addition, we include uncertainty in B_p, which is the panel translational damping coefficient for the flexible panels, as it is quite difficult to measure with enough precision. Thus, the transfer functions $p_{11}(s)$ and $p_{21}(s)$ described in Equations CS1.18 and CS1.19 become

$$\frac{\theta(s)}{T_m(s)} = p_{11}(s) = \left(\frac{1}{J_0}\right)\frac{\left(s^2+(B_p/20)s+16\right)}{s^2\left(s^2+(B_p(J_0+160)/(20J_0))s+(16(J_0+160)/J_0)\right)} \tag{CS1.20}$$

$$\frac{y_p(s)}{T_m(s)} = p_{21}(s) = \left(\frac{1}{J_0}\right)\frac{-2}{\left(s^2+(B_p(J_0+160)/(20J_0))s+(16(J_0+160)/J_0)\right)} \tag{CS1.21}$$

with: $J_0 \in [432, 720]$ kg m², $B_p \in [0.24, 0.48]$ Ns/m.

CS1.2.2 Frequencies

Figure CS1.5 shows the Bode diagram of the plant $p_{11}(s)$, including also the parametric uncertainty. Based on this picture, we select an array of frequencies so that

$$w = [0.001\ 0.005\ 0.01\ 0.02\ 0.05\ 0.1\ 0.2\ 0.3\ 0.4\ 0.5\ 0.6\ 0.7\ 0.8\ 0.9\ 1\ 2\ 3\ 4\ 4.68\ 5\ 10\ 15]\mathrm{rad/s} \tag{CS1.22}$$

CS1.2.3 Nominal Plant

For the nominal plants $p_{11,0}(s)$ and $p_{21,0}(s)$, we select $J_0 = 432$ and $B_p = 0.24$, resulting in the transfer functions

$$\left.\frac{\theta(s)}{T_m(s)}\right|_0 = p_{11,0}(s) = 0.002315\frac{(s^2+0.012s+16)}{s^2(s^2+0.0164s+21.93)} \tag{CS1.23}$$

$$\left.\frac{y_p(s)}{T_m(s)}\right|_0 = p_{21,0}(s) = \frac{-0.00463}{(s^2+0.0164s+21.93)} \tag{CS1.24}$$

CS1.2.4 QFT Templates

Figure CS1.4 shows the QFT templates for this case. They are calculated using the plant model and parametric uncertainty defined with Equation CS1.20, and the array of frequencies of interest defined in Equation CS1.22.

CS1.3 Preliminary Analysis

Before starting the design of the controller, we conduct a brief preliminary analysis of the dynamics of the satellite. Figure CS1.5 shows the Bode diagram and the NC of $p_{11}(s)$ with four cases within the uncertainty:

Case $p_{11\text{-}1}(s)$: $J_0 = 720$, $B_p = 0.48$;   Case $p_{11\text{-}2}(s)$: $J_0 = 720$, $B_p = 0.32$
Case $p_{11\text{-}3}(s)$: $J_0 = 720$, $B_p = 0.24$;   Case $p_{11\text{-}4}(s)$: $J_0 = 432$, $B_p = 0.48$

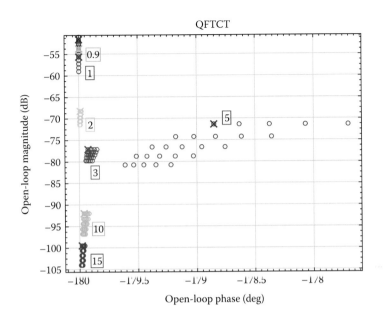

FIGURE CS1.4
QFT templates for $p_{11}(s)$.

As we can see, the system $p_{11}(s)$ has a double integrator. This means that there is initially no damping. The system is at the very limit of stability. If a disturbance enters into the system, it will oscillate forever, as we can expect for any body in orbit without a control system.

In addition, there is a severe resonance mode around $\omega = 4.50$ rad/s, with an extremely low damping—see peak in the Bode diagram or loop in the Nichols chart (Figure CS1.5). This resonance is a consequence of the flexible panels. As a result, the bandwidth of the control system is limited due to this resonance mode.

Moreover, this resonance can also vary with the parametric uncertainty. Thus, the natural frequency and damping are, respectively, $\omega_n \in [4.42, 4.68]$ rad/s and $\zeta \in [0.0017, 0.0037]$. As a consequence, this variation of the parameters could eventually change the natural frequency and increase the peak of the resonance, making the system unstable.

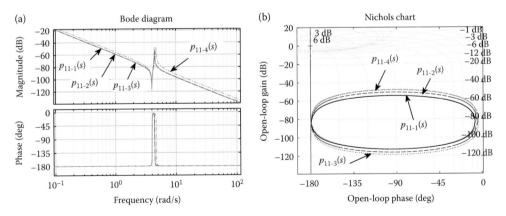

FIGURE CS1.5
Frequency-domain $p_{11-i}(s)$. (a) Bode diagram and (b) Nichols chart.

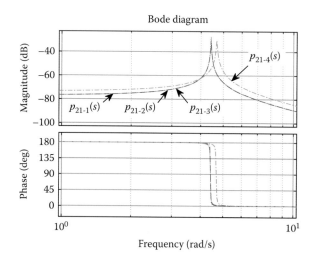

FIGURE CS1.6
Bode diagram of $p_{21\text{-}i}(s)$.

Similarly, Figure CS1.6 shows the Bode diagram of the plant $p_{21\_i}(s)$ for the same four cases. This plant describes the effect of the torque command T_m on the solar panel displacement y_p. Again, the peak at $\omega_n \in [4.42, 4.68]$ rad/s is the resonance mode of the solar panels.

CS1.4 Control Specifications

Taking into account the dynamics of the satellite and the operator objectives and limitations, we define the following robust control specifications—with $H(s) = 1$:

- *Type 1: Stability specification*

$$|T_1(j\omega)| = \left|\frac{p_{11}(j\omega)G(j\omega)}{1 + p_{11}(j\omega)G(j\omega)}\right| \le \delta_1(\omega) = W_s = 1.93$$

$$\omega \in [0.001\ 0.005\ 0.01\ 0.02\ 0.05\ 0.1$$

$$0.2\ 0.3\ 0.4\ 0.5\ 0.6\ 0.7\ 0.8\ 0.9\ 1\ 2\ 3\ 4\ 4.68\ 5\ 10\ 15]\,\text{rad/s}$$

(CS1.25)

  which is equivalent to $PM = 30.03°$, $GM = 3.63$ dB—see Equations 2.30 and 2.31.

- *Type 3: Sensitivity or disturbances at plant output specification*—see Figure CS1.7a

$$|T_3(j\omega)| = \left|\frac{\theta(j\omega)}{d_o(j\omega)}\right| = \left|\frac{1}{1 + p_{11}(j\omega)G(j\omega)}\right| \le \delta_3(\omega) = \frac{(s/a_d)}{(s/a_d)+1}\,;\ a_d = 0.2$$

$$\omega \in [0.001\ 0.005\ 0.01\ 0.02\ 0.05\ 0.1]\,\text{rad/s}$$

(CS1.26)

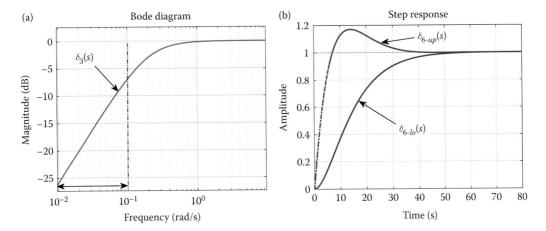

FIGURE CS1.7
Control specifications for $p_{11}(s)$. (a) Sensitivity or disturbance rejection at the output of the plant: δ_3. (b) Reference tracking: $\delta_{6\text{-}up}$, $\delta_{6\text{-}lo}$.

- *Type 6: Reference tracking specification*—see Figure CS1.7b

$$\delta_{6\text{-}lo}(\omega) < |T_6(j\omega)| = \left|F(j\omega)\frac{p_{11}(j\omega)G(j\omega)}{1+p_{11}(j\omega)G(j\omega)}\right| \leq \delta_{6\text{-}up}(\omega)$$

$$\omega \in [0.001\ 0.005\ 0.01\ 0.02\ 0.05\ 0.1$$
$$0.2\ 0.3\ 0.4\ 0.5\ 0.6\ 0.7\ 0.8\ 0.9\ 1]\text{rad/s}$$

(CS1.27)

$$\delta_{6\text{-}lo}(s) = \frac{1}{\left[(s/a_L)+1\right]^2};\quad a_L = 0.13$$

(CS1.28)

$$\delta_{6\text{-}up}(s) = \frac{[(s/a_U)+1]}{\left[(s/\omega_n)^2+(2\zeta s/\omega_n)+1\right]};\quad a_U = 0.1;\quad \zeta = 0.8;\quad \omega_n = \frac{1.25\,a_U}{\zeta}$$

(CS1.29)

- *Type k1: External disturbances $d_o(s)$ over $y_p(s)$ via $p_{21}(s) \times G(s)$ and control loop*

$$|T_{k1}(j\omega)| = \left|\frac{y_p(j\omega)}{d_o(j\omega)}\right| = \left|\frac{p_{21}(j\omega)G(j\omega)}{1+p_{11}(j\omega)G(j\omega)}\right| \leq \delta_{k1}(\omega) = 0.20$$

$$\omega \in [4\ 4.68\ 5]\ \text{rad/s}$$

(CS1.30)

CS1.5 Controller Design

CS1.5.1 QFT-Bounds

The QFT-bounds integrate the dynamics of the plant, the model uncertainty, and the control specifications at each frequency of interest. Figure CS1.8a–d shows the bounds for stability,

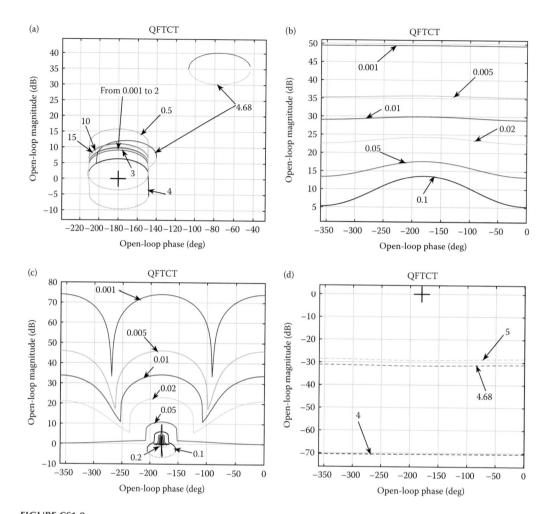

FIGURE CS1.8
(a) Stability bounds, (b) sensitivity bounds, (c) reference tracking bounds, and (d) solar panel vibration attenuation bounds.

sensitivity, reference tracking, and solar panel vibration attenuation, respectively. Figure CS1.9 presents the intersection of all these bounds at each frequency. As we have a bound solution for each frequency, the selected control specifications for the satellite are compatible.

CS1.5.2 Loop-Shaping—$G(s)$

The design of the feedback controller $G(s)$ is carried out on the Nichols chart (NC). It is done by adding poles and zeros until the nominal loop, defined as $L_0(j\omega) = p_{11,0}(j\omega)G(j\omega)$, meets the QFT-bounds presented in Figure CS1.9.

We start the design of $G(s)$ by adding a low frequency zero ($z_1 = 0.1$) to move $L_0(s)$ from $-180°$ line to the right, and pass the stability bounds through the right. Then, we increase the controller gain to $k = 20$. Afterward, to meet the dashed-line bound at 4.68 rad/s, we add a notch filter $(s^2 + 2\,\zeta_a\,\omega_n\,s + \omega_n^2)/(s^2 + 2\,\zeta_b\,\omega_n\,s + \omega_n^2)$, with a natural frequency of $\omega_n = 4.68$ rad/s, a damping numerator coefficient of $\zeta_a = 0.07$, and a damping denominator coefficient of $\zeta_b = 1$.

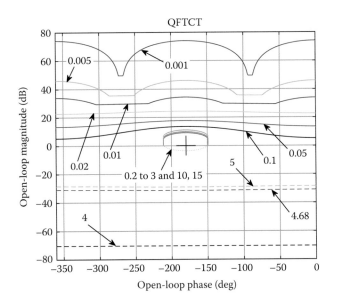

FIGURE CS1.9
QFT-bounds. Intersection of bounds.

This gives a depth of $20 \times \log_{10}(\zeta_a/\zeta_b)$ dB $= -23.1$ dB, which is enough to place $L_0(4.68)$ below the $B(4.68)$ dashed-line bound. Finally, we add two additional poles at a higher frequency ($p_1 = p_2 = 1$), to filter out high frequency noise. The final controller $G(s)$ is shown in Equation CS1.31 and the loop-shaping in Figure CS1.10.

$$G(s) = \frac{20((s/0.1)+1)}{(s+1)^2} \frac{(s^2 + 3.066s + 21.9)}{(s^2 + 43.80s + 21.9)} \tag{CS1.31}$$

CS1.5.3 Prefilter Design—*F*(s)

Taking into account the controller $G(s)$ defined in Equation CS1.31, the reference tracking specifications presented in Equations CS1.27 through CS1.29, and the plant model described in Equation CS1.20, we design a prefilter $F(s)$ to assure that all the input/output functions $p_{11}(s)G(s)F(s)/[1 + p_{11}(s)G(s)]$ from low frequencies to 1 rad/s (which is before the flexible mode) are inside the band defined by the limits $\delta_{6-up}(\omega)$ and $\delta_{6-lo}(\omega)$. The prefilter is shown in Equation CS1.32, and the input/output functions and limits in Figure CS1.11.

$$F(s) = \frac{((s/0.5)+1)}{((s/0.1)+1)} \tag{CS1.32}$$

CS1.6 Analysis and Validation

The analysis of the closed-loop stability in the frequency domain is shown in Figure CS1.12a. The dashed-line is the stability specification W_s, defined in Equation CS1.25.

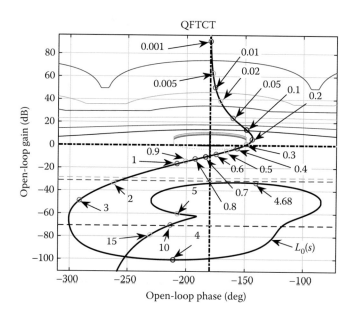

FIGURE CS1.10
QFT-bounds and $G(s)$ design—loop-shaping.

The solid-line represents the worst case of all the possible functions $p_{11}G/(1 + p_{11}G)$ at each frequency and due to the model uncertainty. The control system meets the stability specification (the solid-line is below the dashed-line W_s) for all the plants within the uncertainty.

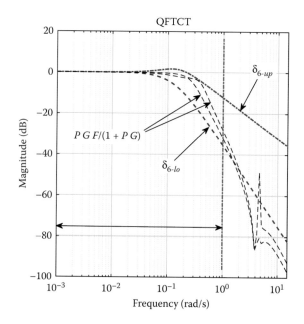

FIGURE CS1.11
Prefilter $F(s)$ for reference tracking specs: $\delta_{6\text{-}lo}(\omega) \leq |T_6| \leq \delta_{6\text{-}up}(\omega)$.

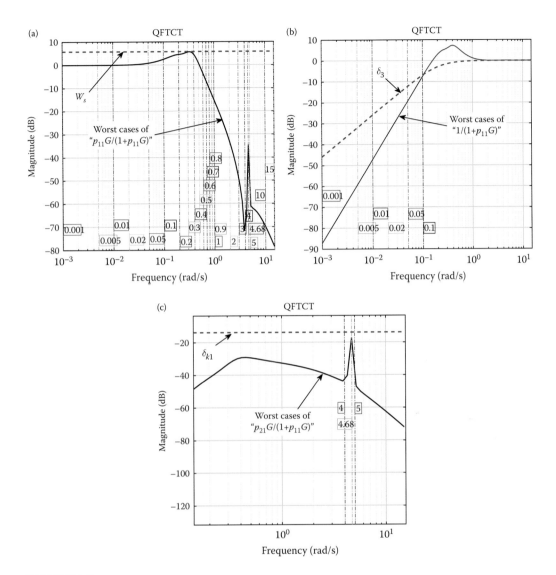

FIGURE CS1.12
Frequency-domain analysis: (a) stability, (b) sensitivity or disturbance rejection at plant output, and (c) panel vibration attenuation.

The analysis of the robust sensitivity specification, or rejection of disturbances at the output of the plant, is shown in Figure CS1.12b. The dashed-line is the sensitivity specification $\delta_3(\omega)$, defined in Equation CS1.26. The solid-line represents the worst case of all the possible functions $1/(1 + p_{11}G)$ at each frequency and due to the model uncertainty. The control system meets the sensitivity specification in all cases (the solid-line is below the dashed-line δ_3) from 0 to 0.1 rad/s.

The panel vibration analysis in the frequency domain is shown in Figure CS1.12c. It represents the attenuation of the effect of disturbances $d_o(s)$ over the panel movement $y_p(s)$ through $p_{21}(s) \times G(s)$ and the control loop—see also Figure CS1.3. The dashed-line is the specification $\delta_{k1}(\omega)$ defined in Equation CS1.30. The solid-line represents the worst case of all the possible functions $p_{21}G/(1 + p_{11}G)$ at each frequency due to the model uncertainty.

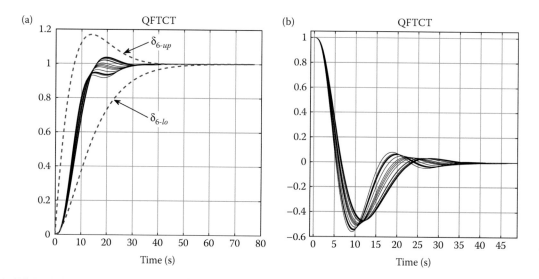

FIGURE CS1.13

Time-domain analysis. (a) Reference tracking: $\delta_{6\text{-}lo}$ and $\delta_{6\text{-}up}$ specs, and $\theta(t)$ of 200 cases of $p_{11}GF/(1 + p_{11}G)$ to a unitary step reference $r(t)$. (b) Sensitivity: $\theta(t)$ of 200 cases of $1/(1 + p_{11}G)$ to a unitary step disturbance $d_o(t)$.

Once more, the control system meets the specification in all cases (solid-line is below dashed-line δ_{k1}).

The time-domain analysis of the reference tracking specification is shown in Figure CS1.13a. The figure shows the limits $\delta_{6\text{-}up}(\omega)$ and $\delta_{6\text{-}lo}(\omega)$—see Equations CS1.29 and CS1.28, and the time responses of the angle $\theta(t)$ to a unitary step reference $r(t)$. We simulate 200 cases of the closed-loop transfer function $p_{11}GF/(1 + p_{11}G)$—see Figure CS1.3 with $H(s) = 1$. The control system meets the specification (is between the upper and lower limits) in all cases.

In addition, the time-domain analysis of the sensitivity is shown in Figure CS1.13b. The figure presents the time responses of the angle $\theta(t)$ to a unitary step disturbance $d_o(t)$—see also Figure CS1.3 with $H(s) = 1$. We simulate 200 cases of $1/(1 + p_{11}G)$. The control system achieves a good disturbance rejection in all cases.

In addition, Figure CS1.14 shows the solar panels oscillation when a unitary step disturbance input $d_o(s)$ is introduced into the system. The disturbance $d_o(s)$ enters at the output of the plant $p_{11}(s)$, travels the control loop until $T_m(s)$, and finally goes through $p_{21}(s)$ to $y_p(s)$—see the block diagram in Figure CS1.3. We study two cases. The first one, represented in Figure CS1.14a, uses the complete controller $G(s)$ described in Equation CS1.31. The second case, represented in Figure CS1.14b, uses this controller without the notch filter.

The solar panels oscillation is significantly reduced when we add this notch filter in the feedback controller $G(s)$, that is, the complete Equation CS1.31. Figure CS1.14a shows the $y_p(t)$ results with the complete $G(s)$, and Figure CS1.14b without the notch filter. Both analysis are performed for the same cases within the plant uncertainty. The figures have the same scale.

Finally, Figure CS1.15 zooms in one case of the simulations of Figure CS1.14a and b. As we can see, the frequency of the oscillation in the figure (15 peaks in 20 s) is the natural frequency of the solar panel oscillation: 0.75 Hz (4.68 rad/s). The results of the complete controller $G(s)$ with the notch filter are represented by the solid-line—Equation CS1.31, and the results without the notch filter by the dashed-line. As anticipated, the complete

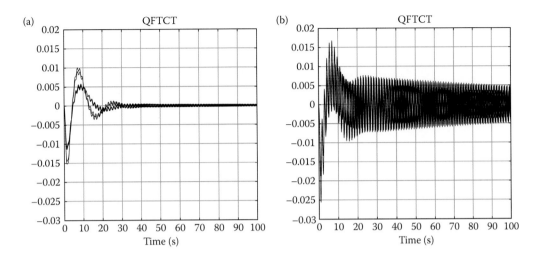

FIGURE CS1.14
Solar panels oscillation $y_b(t)$. Simulation of 25 plants. (a) $G(s)$ with the notch filter—see Equation CS1.31. (b) $G(s)$ without the notch filter.

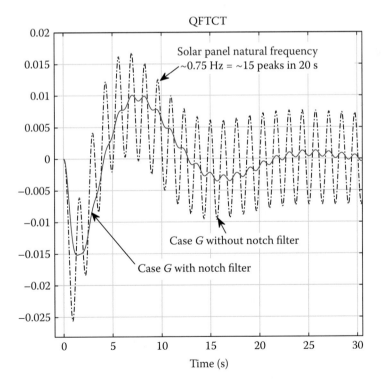

FIGURE CS1.15
Zooming in. Solar panel oscillation $y_b(t)$. $G(s)$ with notch filter (solid-line), and without notch filter (dashed-line).

controller $G(s)$ with the notch filter obtains a much better peak to peak reduction of the solar panels vibration.

CS1.7 Summary

This case study has successfully designed a robust QFT Attitude Control System for a satellite with flexible solar panels. The design considers the model uncertainty introduced by the fuel consumption and the imprecise knowledge of the damping coefficients. It accomplishes four simultaneous control objectives:

- Stability
- Reference tracking and regulation of the satellite angle
- Rejection of unpredictable disturbances
- Minimization of the solar panel vibrations

Case Study 2: Wind Turbine Control

CS2.1 Description

Wind turbines are complex systems with large and flexible structures that work under unpredictable environmental conditions and a variable electrical grid. The efficiency and reliability of a wind turbine depend on how the control system deals with the nonlinear characteristics, high model uncertainty, stability limitations, energy capture maximization, and load and mechanical fatigue reduction. These objectives require the control system to coordinate numerous variables in real time, including the blade pitch angles, torques, active and reactive powers, rotor speed, yaw orientation, temperatures, currents, voltages, and power factors.

This case study presents the design of a pitch control system to regulate the rotor speed of a variable-speed pitch-controlled wind turbine in *Region 3* (above-rated region, $12 <$ wind speed < 17 m/s, see Figure CS2.19), while attenuating simultaneously the tower and blades vibration. The machine is composed of a gearless drive-train, a multipole synchronous generator, and a full-power converter. It is also called a *Type-4* wind turbine. As an example, Figure CS2.1 shows a picture of the first prototype of the TWT-1.65, a 1.65 MW *Type-4* wind turbine designed by the author and the MTOI team in the period 1998–2011.

The four objectives pursued in this case study are: (1) to control the rotor speed Ω_r by pitching simultaneously the angles β of the turbine blades—also called collective pitch, (2) to reject the effect of unpredictable wind disturbances, (3) to minimize the tower fore-aft oscillations, and (4) to attenuate the blades flap-wise vibrations.

For these objectives, we propose a classical 2DOF control system that includes a feedback compensator $G(s)$ and a prefilter $F(s)$, as shown in Figure CS2.5. In the following sections we model and analyze the dynamics of the wind turbine, propose a set of stability and performance specifications, design the controllers, and validate them in both frequency and time domains.

CS2.2 Plant Model

Figure CS2.2 shows the model diagram for the *Type-4* wind turbine. The model considers four degrees of freedom: the rotor rotation $\theta_r(t)$, the generator rotation $\theta_g(t)$, the flap-wise blade bending $y_b(t)$, and the fore-aft tower bending $y_t(t)$. The main variables and parameters of the system are the following:

FIGURE CS2.1
TWT-1.65 variable-speed pitch-controlled wind turbine, MTOI, 2001.

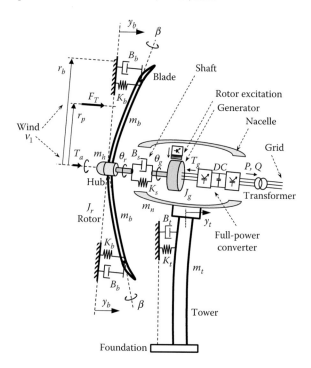

FIGURE CS2.2
Variable-speed pitch-controlled gearless wind turbine with full-power converter (*Type-4* wind turbine). Model description.

Inputs

- v_1 undisturbed upstream wind speed, [m/s]
- F_T thrust force applied by the wind on the rotor, [N]
- T_a aerodynamic torque applied by the wind on the rotor, [Nm]
- T_g antagonistic electrical torque applied by the rectifier/generator, [Nm]
- β pitch angle of the blades, [rad]

- T_{gd} desired electrical torque, output of torque controller, [Nm]
- β_d desired pitch angle, output of pitch controller, [rad]

Outputs

- y_t fore-aft displacement of the tower at the nacelle level, [m]
- y_b flap-wise displacement of the tip of the blade, [m]
- θ_r rotor angular position, [rad]
- θ_g generator angular position, [rad]
- $\Omega_r = \dot{\theta}_r$ rotor angular speed, [rad/s]
- $\Omega_g = \dot{\theta}_g$ generator angular speed, [rad/s]
- P active power at the output of the inverter, [W]
- Q reactive power at the output of the inverter, [VA]

Parameters

- ρ density of the air, [kg/m³]
- N number of blades, [−]
- r_b radius of the blade, [m]
- r_p distance from the center of the rotor to the center of pressure, or point where the equivalent lumped force F_T is applied. $r_p = (2/3)\, r_b$, [m]
- J_g moment of inertia of the generator, at Ω_g, [kg m²]
- J_h moment of inertia of the hub, at Ω_r, [kg m²]
- J_b moment of inertia of one blade, about root, at Ω_r, [kg m²]
- $J_r = J_h + N\, J_b$, moment of inertia of the rotor, at Ω_r, [kg m²]
- m_b mass of one blade, [kg]
- m_h mass of the hub, [kg]
- m_n mass of the nacelle, [kg]
- m_t mass of the tower, [kg]
- K_t tower translational fore-aft stiffness coefficient, [N/m]
- B_t tower translational fore-aft damping coefficient, [Ns/m]
- K_b blade translational flap-wise stiffness coefficient, [N/m]
- B_b blade translational flap-wise damping coefficient, [Ns/m]
- K_s shaft torsional stiffness coefficient, [Nm]
- B_s shaft torsional damping coefficient, [Nms]
- $m_1 = m_n + m_h + (1/3)m_t + Nm_b$. Equivalent mass at the tip of the tower, for fore-aft y_t movement, [kg]
- $m_2 = J_b/r_b^2$. Equivalent mass at the tip of the blade, for flap-wise y_b movement, [kg]

As in case study CS1, we start with the general equation of motion for the mechanical system, which is

$$\mathcal{M}\ddot{q} + \mathcal{C}\dot{q} + \mathcal{K}q = \mathcal{Q}(\dot{q}, q, u, t) \tag{CS2.1}$$

where $\mathcal{M}$, $\mathcal{C}$, and $\mathcal{K}$ are the *Mass, Damping,* and *Stiffness* matrices, $\mathcal{Q}$ the matrix of inputs, q the vector of generalized coordinates, and u the vector of inputs. Also, we consider the general form of the Euler–Lagrange equation,

$$\frac{d}{dt}\left(\frac{\partial L}{\partial \dot{q}_i}\right) - \frac{\partial L}{\partial q_i} + \frac{\partial D_d}{\partial \dot{q}_i} = Q_i,\, i = 1, 2, \ldots, n \text{ and } L = E_k - E_p \tag{CS2.2}$$

where q_i are the generalized coordinates or degrees of freedom of the system, Q_i the generalized forces applied to each subsystem "i," E_k the kinetic energy, E_p the potential energy, D_d the dissipator co-content, and L the Lagrangian.[236]

The generalized coordinates for the wind turbine are the linear displacement of the tower $y_t(t)$, the linear displacement of the blades $y_b(t)$, the rotor angle $\theta_r(t)$ and the generator angle $\theta_g(t)$, which are respectively: $q_1 = y_t(t)$, $q_2 = y_b(t)$, $q_3 = \theta_r(t)$, and $q_4 = \theta_g(t)$—see Figure CS2.2. The system contains the kinetic, potential, and dissipative functions described by Equations CS2.3 through CS2.5. Notice that we consider linear dissipators, which means that the D_d expression or dissipator co-content equals one-half of the power being absorbed by the dissipators.[236]

$$E_k = \frac{1}{2} m_1 \dot{y}_t^2 + \frac{1}{2} N m_2 (\dot{y}_t + \dot{y}_b)^2 + \frac{1}{2} J_r \dot{\theta}_r^2 + \frac{1}{2} J_g \dot{\theta}_g^2 \tag{CS2.3}$$

$$E_p = \frac{1}{2} K_t\, y_t^2 + \frac{1}{2} N K_b\, y_b^2 + \frac{1}{2} K_s (\theta_r - \theta_g)^2 \tag{CS2.4}$$

$$D_d = \frac{1}{2} B_t\, \dot{y}_t^2 + \frac{1}{2} N B_b\, \dot{y}_b^2 + \frac{1}{2} B_s (\dot{\theta}_r - \dot{\theta}_g)^2 \tag{CS2.5}$$

The generalized forces applied to the wind turbine are the thrust force $F_T(t)$ and aerodynamic torque $T_a(t)$, both applied by the wind to the rotor, and the antagonistic electrical torque $T_g(t)$, applied by the power electronics and electrical generator. Now, using Equations CS2.3 through CS2.5, the terms for the Euler–Lagrange equation are

$$\frac{d}{dt}\left(\frac{\partial L}{\partial \dot{q}_i}\right) = \frac{d}{dt}\begin{bmatrix} \dfrac{\partial E_k}{\partial \dot{y}_t} \\[6pt] \dfrac{\partial E_k}{\partial \dot{y}_b} \\[6pt] \dfrac{\partial E_k}{\partial \dot{\theta}_r} \\[6pt] \dfrac{\partial E_k}{\partial \dot{\theta}_g} \end{bmatrix} = \begin{bmatrix} m_1 + N m_2 & N m_2 & 0 & 0 \\ N m_2 & N m_2 & 0 & 0 \\ 0 & 0 & J_r & 0 \\ 0 & 0 & 0 & J_g \end{bmatrix}\begin{bmatrix} \ddot{y}_t \\ \ddot{y}_b \\ \ddot{\theta}_r \\ \ddot{\theta}_g \end{bmatrix} = \mathcal{M}\ddot{q} \tag{CS2.6}$$

$$-\frac{\partial L}{\partial q_i} = \frac{\partial E_p}{\partial q_i} = \begin{bmatrix} \dfrac{\partial E_p}{\partial y_t} \\[6pt] \dfrac{\partial E_p}{\partial y_b} \\[6pt] \dfrac{\partial E_p}{\partial \theta_r} \\[6pt] \dfrac{\partial E_p}{\partial \theta_g} \end{bmatrix} = \begin{bmatrix} K_t & 0 & 0 & 0 \\ 0 & N K_b & 0 & 0 \\ 0 & 0 & K_s & -K_s \\ 0 & 0 & -K_s & K_s \end{bmatrix}\begin{bmatrix} y_t \\ y_b \\ \theta_r \\ \theta_g \end{bmatrix} = \mathcal{K}q \tag{CS2.7}$$

$$\frac{\partial D_d}{\partial \dot{q}_i} = \begin{bmatrix} \dfrac{\partial D_d}{\partial \dot{y}_t} \\[6pt] \dfrac{\partial D_d}{\partial \dot{y}_b} \\[6pt] \dfrac{\partial D_d}{\partial \dot{\theta}_r} \\[6pt] \dfrac{\partial D_d}{\partial \dot{\theta}_g} \end{bmatrix} = \begin{bmatrix} B_t & 0 & 0 & 0 \\ 0 & NB_b & 0 & 0 \\ 0 & 0 & B_s & -B_s \\ 0 & 0 & -B_s & B_s \end{bmatrix} \begin{bmatrix} \dot{y}_t \\ \dot{y}_b \\ \dot{\theta}_r \\ \dot{\theta}_g \end{bmatrix} = \mathcal{C}\dot{q} \tag{CS2.8}$$

$$\mathcal{Q} = \begin{bmatrix} 1 & 0 & 0 \\ 1 & 0 & 0 \\ 0 & 1 & 0 \\ 0 & 0 & -1 \end{bmatrix} \begin{bmatrix} F_T \\ T_a \\ T_g \end{bmatrix} = \mathcal{R}u \tag{CS2.9}$$

Rearranging the equation of motion Equation CS2.1 as

$$\ddot{q} = -\mathcal{M}^{-1}\mathcal{C}\dot{q} - \mathcal{M}^{-1}\mathcal{K}q + \mathcal{M}^{-1}\mathcal{R}u \tag{CS2.10}$$

and applying Equations CS2.6 through CS2.9, we find a state space description of the system ($\dot{x} = Ax + Bu$; $y = Cx$) with the expressions,

$$\dot{x} = \left[\begin{array}{c|c} -\mathcal{M}^{-1}\mathcal{C} & -\mathcal{M}^{-1}\mathcal{K} \\ \hline I & 0 \end{array}\right] x + \left[\begin{array}{c} \mathcal{M}^{-1}\mathcal{R} \\ \hline 0 \end{array}\right] u \tag{CS2.11}$$

$$y = [I \mid 0]x$$

with $I = [1\,0\,0\,0;\, 0\,1\,0\,0;\, 0\,0\,1\,0;\, 0\,0\,0\,1]$, $0 = [0\,0\,0\,0;\, 0\,0\,0\,0;\, 0\,0\,0\,;\, 0\,0\,0\,0]$, and,

$$A = \begin{bmatrix} -\dfrac{B_t}{m_1} & \dfrac{NB_b}{m_1} & 0 & 0 & -\dfrac{K_t}{m_1} & \dfrac{NK_b}{m_1} & 0 & 0 \\[8pt] \dfrac{B_t}{m_1} & \dfrac{-B_b(m_1 + Nm_2)}{m_1 m_2} & 0 & 0 & \dfrac{K_t}{m_1} & \dfrac{-K_b(m_1 + Nm_2)}{m_1 m_2} & 0 & 0 \\[8pt] 0 & 0 & \dfrac{-B_s}{J_r} & \dfrac{B_s}{J_r} & 0 & 0 & \dfrac{-K_s}{J_r} & \dfrac{K_s}{J_r} \\[8pt] 0 & 0 & \dfrac{B_s}{J_g} & \dfrac{-B_s}{J_g} & 0 & 0 & \dfrac{K_s}{J_g} & \dfrac{-K_s}{J_g} \\[8pt] 1 & 0 & 0 & 0 & 0 & 0 & 0 & 0 \\ 0 & 1 & 0 & 0 & 0 & 0 & 0 & 0 \\ 0 & 0 & 1 & 0 & 0 & 0 & 0 & 0 \\ 0 & 0 & 0 & 1 & 0 & 0 & 0 & 0 \end{bmatrix} \tag{CS2.12}$$

$$B = \begin{bmatrix} 0 & 0 & 0 \\ \dfrac{m_1 + N\,m_2}{N\,m_1\,m_2} & -\dfrac{1}{m_1} & 0 & 0 \\ 0 & \dfrac{1}{J_r} & 0 \\ 0 & 0 & \dfrac{-1}{J_g} \\ 0 & 0 & 0 \\ 0 & 0 & 0 \\ 0 & 0 & 0 \\ 0 & 0 & 0 \end{bmatrix} \qquad (CS2.13)$$

$$C = \begin{bmatrix} 1 & 0 & 0 & 0 & 0 & 0 & 0 & 0 \\ 0 & 1 & 0 & 0 & 0 & 0 & 0 & 0 \\ 0 & 0 & 1 & 0 & 0 & 0 & 0 & 0 \\ 0 & 0 & 0 & 1 & 0 & 0 & 0 & 0 \end{bmatrix} \qquad (CS2.14)$$

$$x = \begin{bmatrix} \dot{q} \\ q \end{bmatrix} = [\dot{y}_t \quad \dot{y}_b \quad \dot{\theta}_r \quad \dot{\theta}_g \quad y_t \quad y_b \quad \theta_r \quad \theta_g]^T \qquad (CS2.15)$$

$$u = [F_T \quad T_a \quad T_g]^T \qquad (CS2.16)$$

$$y = [\dot{y}_t \quad \dot{y}_b \quad \dot{\theta}_r \quad \dot{\theta}_g]^T \qquad (CS2.17)$$

The thrust force F_T and aerodynamic torque T_a applied by the wind to the rotor are proportional to the rotor area (πr_b^2), the density of the air ρ, and the square and cubic of the wind velocity v_1, respectively, as described by Equations CS2.18 and CS2.19, so that

$$F_T = \frac{1}{2}\rho\pi r_b^2 C_T(\lambda,\beta)v_1^2 \qquad (CS2.18)$$

$$T_a = \frac{\rho\pi r_b^2 C_P(\lambda,\beta)v_1^3}{2\Omega_r} \qquad (CS2.19)$$

where C_p is the power coefficient and C_T the torque coefficient. Both coefficients are a nonlinear function of the tip-speed ratio $\lambda = \Omega_r r_p/v_1$ and the pitch angle β, as shown in Figure CS2.3. For more information see the author's book, *Wind Energy Systems: Control Engineering Design*, Reference 7.

According to these expressions and the characteristics of Cp and C_T, the inputs F_T and T_a of Equation CS2.16 depend on v_1, β, and Ω_r in a nonlinear way. Notice that $\Omega_r = \dot{\theta}_r$. Now, linearizing these equations around a working point $(v_{10}, \beta_0, \Omega_{r0})$, and ignoring the bias

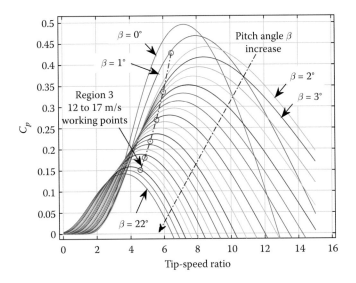

FIGURE CS2.3
C_p/λ curves for pitch angles $0° \leq \beta \leq 22°$ and working points.

components, the inputs F_T and T_a are described by a transfer matrix whose elements are just gains, so that

$$\begin{bmatrix} F_T(s) \\ T_a(s) \end{bmatrix} = \begin{bmatrix} K_{F\Omega} & K_{FV} & K_{F\beta} \\ K_{T\Omega} & K_{TV} & K_{T\beta} \end{bmatrix} \begin{bmatrix} \Omega_r(s) \\ v_1(s) \\ \beta(s) \end{bmatrix} \tag{CS2.20}$$

where the gains are calculated by using the C_T and C_p curves (Figure CS2.3) and Equations CS2.18 and CS2.19, so that

$$K_{F\Omega} = \left.\frac{\partial F_T(t)}{\partial \Omega_r(t)}\right|_0 = \frac{1}{2}\rho\pi r_b^2 \left.\frac{\partial C_T}{\partial \Omega_r}\right|_0 v_{10}^2 \tag{CS2.21}$$

$$K_{FV} = \left.\frac{\partial F_T(t)}{\partial v_1(t)}\right|_0 = \frac{1}{2}\rho\pi r_b^2 \left(\left.\frac{\partial C_T}{\partial v_1}\right|_0 v_{10}^2 + 2v_{10}C_{T0} \right) \tag{CS2.22}$$

$$K_{F\beta} = \left.\frac{\partial F_T(t)}{\partial \beta(t)}\right|_0 = \frac{1}{2}\rho\pi r_b^2 \left.\frac{\partial C_T}{\partial \beta}\right|_0 v_{10}^2 \tag{CS2.23}$$

$$K_{T\Omega} = \left.\frac{\partial T_a(t)}{\partial \Omega_r(t)}\right|_0 = \frac{1}{2}\rho\pi r_b^2 \left(\left.\frac{\partial C_p}{\partial \Omega_r}\right|_0 \frac{1}{\Omega_{r0}} - C_{p0}\frac{1}{\Omega_{r0}^2} \right) v_{10}^3 \tag{CS2.24}$$

$$K_{TV} = \left.\frac{\partial T_a(t)}{\partial v_1(t)}\right|_0 = \frac{1}{2}\rho\pi r_b^2 \frac{1}{\Omega_{r0}} \left(\left.\frac{\partial C_p}{\partial v_1}\right|_0 v_{10}^3 + 3C_{p0}v_{10}^2 \right) \tag{CS2.25}$$

$$K_{T\beta} = \left.\frac{\partial T_a(t)}{\partial \beta(t)}\right|_0 = \left.\frac{1}{2}\rho\pi r_b^2 \frac{1}{\Omega_{r0}}\frac{\partial C_p}{\partial \beta}\right|_0 v_{10}^3 \tag{CS2.26}$$

Then, adding the electrical torque, we have

$$\begin{bmatrix} M^{-1}\mathcal{R} \\ 0 \end{bmatrix} u = B \begin{bmatrix} F_T(s) \\ T_a(s) \\ T_g(s) \end{bmatrix} = B \begin{bmatrix} K_{F\Omega} & K_{FV} & K_{F\beta} & 0 \\ K_{T\Omega} & K_{TV} & K_{T\beta} & 0 \\ 0 & 0 & 0 & 1 \end{bmatrix} \begin{bmatrix} \Omega_r(s) \\ v_1(s) \\ \beta(s) \\ T_g(s) \end{bmatrix} \tag{CS2.27}$$

On the other hand, the transfer functions of the actuators are

$$\beta(s) = A_\beta(s)\beta_d(s) \tag{CS2.28}$$

$$T_g(s) = A_T(s)T_{gd}(s) \tag{CS2.29}$$

where β_d is the demanded blade pitch angle and T_{gd} the demanded electrical torque, both calculated by the control system, and where $A_\beta(s)$ and $A_T(s)$ are the transfer functions from the control signals (β_d, T_{gd}) to the actual value of the actuators (β, T_g). Using these expressions with Equation CS2.27, we have

$$\begin{bmatrix} M^{-1}\mathcal{R} \\ 0 \end{bmatrix} u = B \begin{bmatrix} K_{F\Omega} & K_{FV} & K_{F\beta}A_\beta(s) & 0 \\ K_{T\Omega} & K_{TV} & K_{T\beta}A_\beta(s) & 0 \\ 0 & 0 & 0 & A_T(s) \end{bmatrix} \begin{bmatrix} \Omega_r(s) \\ v_1(s) \\ \beta_d(s) \\ T_{gd}(s) \end{bmatrix} = B_K \begin{bmatrix} \Omega_r(s) \\ v_1(s) \\ \beta_d(s) \\ T_{gd}(s) \end{bmatrix} \tag{CS2.30}$$

Now, putting all the $\Omega_r = \dot{\theta}_r$ variables together, that is, adding the first column of B_K to the third column of A {i.e., $A(1:8,3) = A(1:8,3) + B_K(1:8,1)$}, and removing it from B_K, {i.e., $B_K(:,1) = []$}, the new vector of inputs is

$$u = [v_1 \quad \beta_d \quad T_{gd}]^T \tag{CS2.31}$$

and applying $P(s) = C(sI - A)^{-1}B_K$ for $y(s) = P(s)u(s)$, the transfer matrix from the three independent inputs to the four outputs is

$$\begin{bmatrix} \dot{y}_t(s) \\ \dot{y}_b(s) \\ \Omega_r(s) \\ \Omega_g(s) \end{bmatrix} = P_{4\times3}(s) \begin{bmatrix} v_1(s) \\ \beta_d(s) \\ T_{gd}(s) \end{bmatrix} \tag{CS2.32}$$

The third row of $P_{4\times3}(s)$ contains the transfer functions of the rotor speed $\Omega_r(s)$ from the wind speed $v_1(s)$, the demanded blade pitch angle $\beta_d(s)$, and the demanded electrical torque $T_{gd}(s)$, respectively, so that

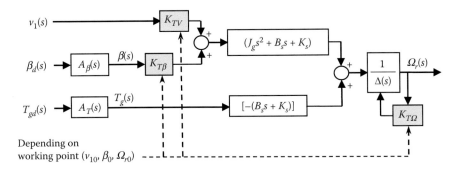

FIGURE CS2.4
Rotor speed linear transfer functions, $p_{31}(s)$, $p_{32}(s)$, $p_{33}(s)$, from the wind speed, demanded pitch angle, and electrical torque, respectively. The parameters K_{TV}, $K_{T\beta}$ and $K_{T\Omega}$ have uncertainty to absorb the nonlinear characteristics.

$$\Omega_r(s) = p_{31}(s)v_1(s) + p_{32}(s)\beta_d(s) + p_{33}(s)T_{gd}(s) \tag{CS2.33}$$

with

$$p_{31}(s) = \frac{\left(J_g s^2 + B_s s + K_s\right)K_{TV}}{\Delta(s)} \tag{CS2.34}$$

$$p_{32}(s) = \frac{\left(J_g s^2 + B_s s + K_s\right)K_{T\beta}A_\beta(s)}{\Delta(s)} \tag{CS2.35}$$

$$p_{33}(s) = \frac{-(B_s s + K_s)A_T(s)}{\Delta(s)} \tag{CS2.36}$$

$$\Delta(s) = J_g J_r s^3 + (B_s J_g + B_s J_r - J_g K_{T\Omega})s^2 + (J_g K_s - B_s K_{T\Omega} + J_r K_s)s - K_{T\Omega}K_s \tag{CS2.37}$$

Figure CS2.4 shows graphically the rotor speed signal $\Omega_r(s)$ as a function of the wind speed $v_1(s)$, the demanded blade pitch angle $\beta_d(s)$, and the demanded electrical torque $T_{gd}(s)$ according to Equations CS2.33 through CS2.37. In addition, Figure CS2.5 shows the complete block diagram representation of Equation CS2.32 and the rotor speed/pitch control system, with the feedback controller $G(s)$ and the prefilter $F(s)$ to be designed in the next sections.

CS2.2.1 Parametric Uncertainty

The parameters for the model of the wind turbine considered in this case study are shown in Table CS2.1. They are taken from the NREL 5-MW baseline wind turbine described in Reference 364, and from the author's book, Reference 7.

NOTE: the parameters $K_{T\Omega}$, K_{TV} and $K_{T\beta}$ have been calculated from Figure CS2.3 and according to Equations CS2.20 through CS2.26, and for a scenario of wind velocity between 12 and 17 m/s. See Figure CS2.19 and Reference 7 for more details.

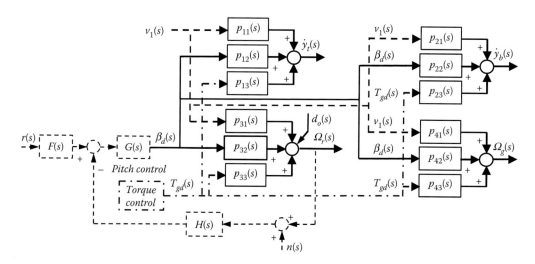

FIGURE CS2.5
Complete block diagram representation of Equation CS2.32 and wind turbine rotor speed/pitch control system.

TABLE CS2.1

Wind Turbine Parameters

Power to the grid [W]—rated	5.0×10^6	ρ (air) [kg/m³]	1.225
Efficiency shaft and generator	94.5%	*Efficiency converter*	98.5%
$T_{g.max}$ to generator [Nm]—rated	400,6111	h [m]—tower height	87.6
Ω_{r_nom} [rad/s]	1.2671	N (number of blades)	3
r_b (blade radius) [m]	61.5	r_p [m]	41
J_g [kg m²]—generator	5,025,497	J_h [kg m²]—hub	115,926
J_b [kg m²]—one blade	11,776,047	J_r [kg m²]—rotor	35,444,067
m_b [kg]—one blade	17,740	m_h [kg]—hub	56,780
m_n [kg]—nacelle	240,000	m_t [kg]—tower	347,460
m_1[kg] $= m_n + m_h + (1/3)m_t + Nm_b$	465,820	m_2[kg] $= J_b/r_p^2$	3113.5
K_s [Nm]—shaft	76,387,512	B_s [Nms]—shaft	6,215,000
K_t [N/m]—tower	1,930,491	B_t [Ns/m]—tower	18,965
K_b [N/m]—one blade	54,586	B_b [Ns/m]—one blade	124.5
$K_{T\beta}$ min [Nm/deg]	-4.1439×10^5	$K_{T\beta}$ max [Nm/deg]	-2.7894×10^4
$K_{T\Omega}$ min [Nm/(rad/s)]	-3.3095×10^6	$K_{T\Omega}$ max [Nm/(rad/s)]	-8.8646×10^5
K_{TV} min [Nm/(m/s)]	5.0115×10^5	K_{TV} max [Nm/(m/s)]	6.1698×10^5
$A_\beta(s)$	1.0	$A_T(s)$	1.0

The wind turbine studied here presents three flexible modes: one for the shaft torsion, another for the tower fore-aft movement, and another for the blade flap-wise movement. These displacements are described in Figure CS2.2. Using the parameters of Table CS2.1, the natural frequency and damping coefficients of these three movements are

$$\omega_{n.shaft} = \sqrt{\frac{K_s}{J_g}} = 3.8987 \text{ rad/s (0.6205 Hz)}, \ \zeta_{shaft} = \frac{B_s}{2\sqrt{K_s J_g}} = 0.1586$$

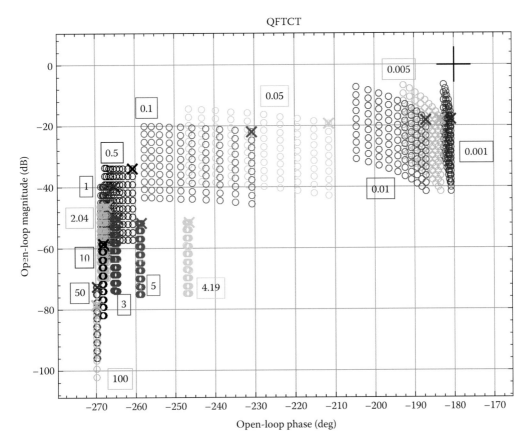

FIGURE CS2.6
QFT templates for $p_{32}(s)$.

$$\omega_{n.tower.fore.aft} = \sqrt{\frac{K_t}{m_1}} = 2.0358 \, \text{rad/s} \, (0.3240 \, \text{Hz}), \quad \zeta_{tower.fore.aft} = \frac{B_t}{2\sqrt{K_t m_1}} = 0.01$$

$$\omega_{n.blade.flapwise} = \sqrt{\frac{K_b}{m_2}} = 4.1871 \, \text{rad/s} \, (0.6664 \, \text{Hz}), \quad \zeta_{blade.flapwise} = \frac{B_b}{2\sqrt{K_b m_2}} = 0.004775$$

This leads to the following expressions:

$$\frac{\Omega_r(s)}{v_1(s)} = p_{31}(s) = \frac{(5.025497 \times 10^6 s^2 + 6,215,000 s + 7.63875 \times 10^7) K_{TV}}{1.78124 \times 10^{14} s^3 + (2.51518 \times 10^{14} - 5.025497 \times 10^6 K_{T\Omega}) s^2 + \cdots} \quad \text{(CS2.38)}$$
$$+ (3.09137 \times 10^{15} - 6,215,000 K_{T\Omega}) s - (7.63875 \times 10^7 \, K_{T\Omega})$$

$$\frac{\Omega_r(s)}{\beta_d(s)} = p_{32}(s) = \frac{(5.025497 \times 10^6 s^2 + 6,215,000 s + 7.63875 \times 10^7) K_{T\beta}}{1.78124 \times 10^{14} s^3 + (2.51518 \times 10^{14} - 5.025497 \times 10^6 K_{T\Omega}) s^2 + \cdots} \quad \text{(CS2.39)}$$
$$+ (3.09137 \times 10^{15} - 6,215,000 K_{T\Omega}) s - (7.63875 \times 10^7 \, K_{T\Omega})$$

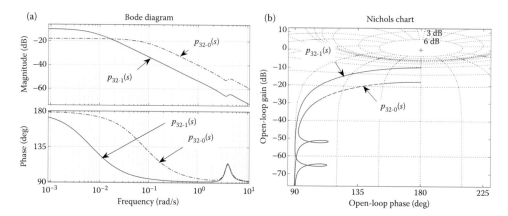

FIGURE CS2.7

$p_{32\text{-}0}(s)$ and $p_{32\text{-}1}(s)$. (a) Bode diagram and (b) Nichols chart. See shaft resonant mode at $\omega_{n.shaft} = 3.90$ rad/s.

with the parametric uncertainty in the aerodynamic coefficients,

$$K_{T\beta} \in [-41.439, -2.7894] \times 10^4 \text{ Nm/deg},$$
$$K_{T\Omega} \in [-33.095, -8.8646] \times 10^5 \text{ Nm/(rad/s)},$$
$$K_{TV} \in [5.0115, 6.1698] \times 10^5 \text{ Nm/(m/s)}.$$

CS2.2.2 Frequencies

Figures CS2.7 and CS2.8 show the Bode diagrams for the transfer functions $p_{32}(s)$, $p_{12}(s)$, and $p_{22}(s)$. We select the array of frequencies of interest based on this information—see Equation CS2.40. Notice that we include frequencies that represent the resonances (peaks) that appear in the diagrams, and which represent the flexible modes of the wind turbine for shaft torsion, and tower fore-aft and blade flap-wise displacements.

$$\omega = [0.001 \quad 0.005 \quad 0.01 \quad 0.05 \quad 0.1 \quad 0.5 \quad 1 \quad 2.04 \quad 3 \quad 4.19 \quad 5 \quad 10 \quad 50 \quad 100] \text{ rad/s}$$

$$(CS2.40)$$

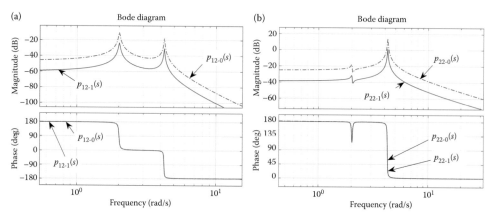

FIGURE CS2.8

Bode diagrams of (a) $p_{12\text{-}0}(s)$ and $p_{12\text{-}1}(s)$, (b) $p_{22\text{-}0}(s)$, and $p_{22\text{-}1}(s)$. See tower and blade modes at $\omega_{n.tower} = 2.04$ rad/s and $\omega_{n.blade} = 4.19$ rad/s.

CS2.2.3 Nominal Plant

Equations CS2.41 and CS2.42 represent the nominal plants $p_{32\text{-}0}(s)$ and $p_{31\text{-}0}(s)$, respectively. They are based on the following selection of parameters within the uncertainty:

$$K_{T\beta 0} = -41.439 \times 10^4 \text{ Nm/deg},$$
$$K_{T\Omega 0} = -33.095 \times 10^5 \text{ Nm/(rad/s), and}$$
$$K_{TV0} = 5.0115 \times 10^5 \text{ Nm/(m/s), resulting the next transfer functions}$$

$$\left.\frac{\Omega_r(s)}{v_1(s)}\right|_0 = p_{31.0}(s) = \frac{0.01414s^2 + 0.01749s + 0.2149}{s^3 + 1.505s^2 + 17.47s + 1.419} \tag{CS2.41}$$

$$\left.\frac{\Omega_r(s)}{\beta_d(s)}\right|_0 = p_{32.0}(s) = \frac{-0.01169s^2 - 0.01446s - 0.1777}{s^3 + 1.505s^2 + 17.47s + 1.419} \tag{CS2.42}$$

CS2.2.4 QFT Templates

Figure CS2.6 shows the QFT templates for $p_{32}(s)$, which is the main plant of the control loop. They are calculated using the model and uncertainty defined after Equation CS2.39, and the array of frequencies of interest defined in Equation CS2.40.

CS2.3 Preliminary Analysis

Before starting the design of the controller, we conduct a brief preliminary analysis of the dynamics of the wind turbine. Figure CS2.7 shows the Bode diagram and the Nichols chart of $p_{32}(s)$ with two extreme cases within the uncertainty:

Case 0: $p_{32,0}(s)$. $K_{T\Omega} = -33.095 \times 10^5$, $K_{T\beta} = -41.439 \times 10^4$, $K_{TV} = 5.0115 \times 10^5$
Case 1: $p_{32,1}(s)$. $K_{T\Omega} = -2.7894 \times 10^5$, $K_{T\beta} = -8.8646 \times 10^4$, $K_{TV} = 6.1698 \times 10^5$

The plant $p_{32}(s)$ describes the effect of the pitch command β_d on the rotor velocity Ω_r. As we can see, the system has no integrators (system type 0) and has a negative gain. The uncertainty affects significantly the magnitude of the gain. It increases by a factor of 2.5 as we go from case 0 ($v_1 = 12$ m/s) to case 1 ($v_1 = 17$ m/s). Numerically, $dcgain \in [-0.125, -0.318]$. This variation is due to the nonlinear changes of the aerodynamics between 12 and 17 m/s, respectively.

In addition, there is a resonance mode at $\omega = 3.90$ rad/s. This is represented by the peak in the Bode diagram and the loop in the Nichols chart. The damping coefficient is about $\zeta = 0.16$. This resonance mode is a consequence of the shaft torsion. As a result, the bandwidth of the control system will have to be below this resonance frequency to reduce the excitation of this shaft mode.

Furthermore, Figure CS2.8a shows the Bode diagram of the plant $p_{12}(s)$ for both cases, 0 and 1. This plant describes the effect of the pitch command β_d on the fore-aft tower movement y_t. The first peak in the Bode diagram, at $\omega = 2.04$ rad/s, is the resonance mode related to the fore-aft tower movement, which is particularly undamped, with a damping coefficient of about $\zeta = 0.01$. The second peak, at $\omega = 4.19$ rad/s, is the resonance mode related to the flap-wise blade movement. It is also extremely undamped, with a damping coefficient of about $\zeta = 0.004775$.

Similarly, Figure CS2.8b shows the Bode diagram of the plant $p_{22}(s)$ for both cases, 0 and 1. This plant describes the effect of the pitch command β_d on the flap-wise blade displacement y_b. Again, the first peak, at $\omega = 2.04$ rad/s, is the resonance mode of the tower, and the second peak, at $\omega = 4.19$ rad/s, the resonance mode of the blade.

As we can see, both movements tower fore-aft and blade flap-wise are coupled. The two resonance modes appear in both transfer functions, $p_{12}(s)$ and $p_{22}(s)$, from the pitch command β_d to the tower and blade movements, y_t and y_b, respectively. They are also much more severe than the shaft torsion mode. The bandwidth of the control system will have to be below these two resonance modes to reduce the movement of the tower and blades at these frequencies.

CS2.4 Control Specifications

Taking into account the dynamics of the wind turbine, the operator objectives and limitations, we define the following robust control specifications—with $H(s) = 1$:

- *Type 1: Stability specification*

$$|T_1(j\omega)| = \left| \frac{p_{32}(j\omega)G(j\omega)}{1 + p_{32}(j\omega)G(j\omega)} \right| \leq \delta_1(\omega) = W_s = 1.46$$

$$\omega \in [0.001 \quad 0.005 \quad 0.01 \quad 0.05 \quad 0.1 \quad 0.5 \quad 1 \quad 2.04 \quad 3 \quad 4.19 \quad 5 \quad 10 \quad 50 \quad 100] \, \text{rad/s}$$

$$(\text{CS2.43})$$

which is equivalent to $PM = 40.05°$, $GM = 4.53$ dB—see Equations 2.30 and 2.31.

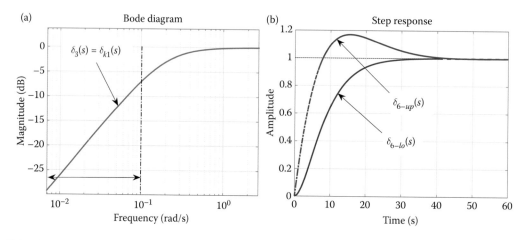

FIGURE CS2.9

Control specifications. (a) Disturbance rejection specifications: $\delta_3 = \delta_{k1}$ and (b) reference tracking: δ_{6-up}, δ_{6-lo}.

- *Type 3: Sensitivity or disturbances at plant output specification—see Figure CS2.9a*

$$|T_3(j\omega)| = \left| \frac{\Omega_r(j\omega)}{d_o(j\omega)} \right| = \left| \frac{1}{1 + p_{32}(j\omega)G(j\omega)} \right| \le \delta_3(\omega) = \frac{\left(\dfrac{s}{a_d} \right)}{\left(\dfrac{s}{a_d} \right) + 1}; \, a_d = 0.2 \tag{CS2.44}$$

$\omega \in [0.001 \quad 0.005 \quad 0.01 \quad 0.05 \quad 0.1] \, \text{rad/s}$

- *Type k1: Wind disturbances $v_1(s)$ over $\Omega_r(s)$ via $p_{31}(s)$ and loop—also Figure CS2.9a*

$$|T_{k1}(j\omega)| = \left| \frac{\Omega_r(j\omega)}{v_1(j\omega)} \right| = \left| \frac{p_{31}(j\omega)}{1 + p_{32}(j\omega)G(j\omega)} \right| \le \delta_{k1}(\omega) = \frac{\left(\dfrac{s}{a_{k1}} \right)}{\left(\dfrac{s}{a_{k1}} \right) + 1}; \, a_{k1} = 0.2 \tag{CS2.45}$$

$\omega \in [0.001 \quad 0.005 \quad 0.01 \quad 0.05 \quad 0.1] \, \text{rad/s}$

- *Type 6: Reference tracking specification—see Figure CS2.9b*

$$\delta_{6\text{-}lo}(\omega) < |T_6(j\omega)| = \left| F(j\omega) \frac{p_{32}(j\omega)G(j\omega)}{1 + p_{32}(j\omega)G(j\omega)} \right| \le \delta_{6\text{-}up}(\omega) \tag{CS2.46}$$

$\omega \in [0.001 \quad 0.005 \quad 0.01 \quad 0.05 \quad 0.1 \quad 0.5] \, \text{rad/s}$

$$\delta_{6\text{-}lo}(s) = \frac{1}{\left[\left(\dfrac{s}{a_L} \right) + 1 \right]^2}; \, a_L = 0.22 \tag{CS2.47}$$

$$\delta_{6\text{-}up}(s) = \frac{\left[\left(\dfrac{s}{a_U} \right) + 1 \right]}{\left[\left(\dfrac{s}{\omega_n} \right)^2 + \left(\dfrac{2\zeta s}{\omega_n} \right) + 1 \right]}; \quad a_U = 0.09; \, \zeta = 0.8; \, \omega_n = \frac{1.25a_U}{\zeta} \tag{CS2.48}$$

- *Type k2: Wind disturbances $v_1(s)$ over $y_t(s)$ via $p_{12}(s) \times G(s) \times p_{31}(s)$—Figure CS2.10*

$$|T_{k2}(j\omega)| = \left| \frac{y_t(j\omega)}{v_1(j\omega)} \right| = \left| \frac{p_{12}(j\omega)p_{31}(j\omega)G(j\omega)}{1 + p_{32}(j\omega)G(j\omega)} \right| \le \delta_{k2}(\omega) = \frac{1}{5s + 1} \tag{CS2.49}$$

$\omega \in [2.04 \quad 4.19] \, \text{rad/s}$

- *Type k3: Wind disturbances $v_1(s)$ over $y_b(s)$ via $p_{22}(s) \times G(s) \times p_{31}(s)$ and loop*

$$|T_{k3}(j\omega)| = \left| \frac{y_b(j\omega)}{v_1(j\omega)} \right| = \left| \frac{p_{22}(j\omega)p_{31}(j\omega)G(j\omega)}{1 + p_{32}(j\omega)G(j\omega)} \right| \le \delta_{k3}(\omega) = 0.85 \tag{CS2.50}$$

$\omega \in [2.04 \quad 4.19] \, \text{rad/s}$

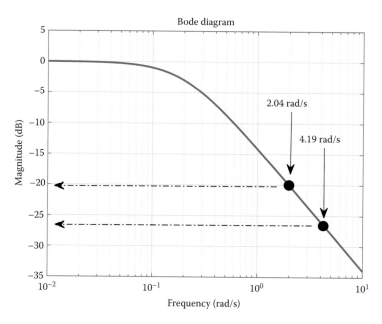

FIGURE CS2.10

Control specifications. *Type k2*: attenuation of the effect of wind disturbances $v_1(s)$ over the tower $y_t(s)$ through $p_{12}(s) \times G(s) \times p_{31}(s)$ and control loop.

CS2.5 Controller Design

CS2.5.1 QFT Bounds

As mentioned, the QFT bounds take into account the plant dynamics, model, uncertainty, and control specifications. Figure CS2.11a–f, show the bounds for stability, sensitivity, reference tracking, wind disturbance rejection, tower fore-aft oscillation attenuation, and blade flap-wise vibration reduction, respectively. Figure CS2.12 presents the intersection of all these six bounds at each frequency. As we have a bound solution for each frequency, the selected control specifications for the wind turbine are compatible.

CS2.5.2 Loop Shaping—*G(s)*

The design of the feedback controller $G(s)$ is carried out on the Nichols chart. It is done by adding poles and zeros until the nominal loop, defined as $L_0(j\omega) = p_{32,0}(j\omega)G(j\omega)$, meets the QFT bounds presented in Figure CS2.12.

We start the design of $G(s)$ by adding a negative sign, as the plant has a negative DC gain. Then, we add an integrator to obtain zero steady-state error for step reference inputs, and also to place $L_0(s)$ at the same phase of the indentation of the low-frequency bound $B(0.001)$. After this, we increase the magnitude of the controller gain to $k = 35$, to put $L_0(0.001)$ above the $B(0.001)$ bound. Next, we add a low-frequency zero ($z_1 = 0.11$), to move $L_0(s)$ to the right, and to pass the stability bounds ("circles") through the right.

Afterward, to meet the dashed-line bound at 2.04 rad/s, we add a notch filter $\left(s^2 + 2\zeta_a\omega_n s + \omega_n^2\right) / \left(s^2 + 2\zeta_b\omega_n s + \omega_n^2\right)$, with a natural frequency of $\omega_n = 2.04$ rad/s, a damping numerator coefficient of $\zeta_a = 0.2$, and a damping denominator coefficient of $\zeta_b = 1$.

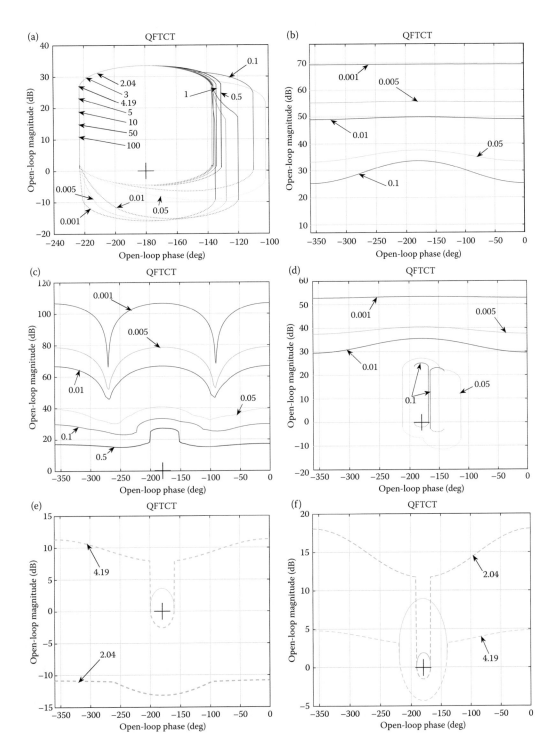

FIGURE CS2.11
QFT bounds. (a) Stability bounds, (b) sensitivity bounds, (c) reference tracking bounds, (d) wind disturbance rejection bounds, (e) fore-aft tower vibration attenuation bounds, and (f) blade flap-wise vibration reduction bounds.

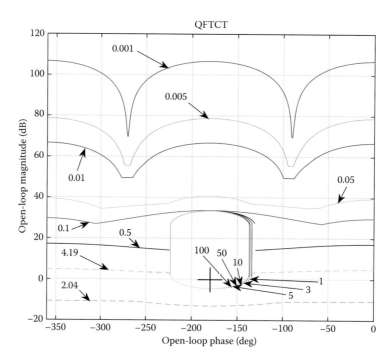

FIGURE CS2.12
QFT bounds. Intersection of bounds.

This gives a depth of $20 \times \log_{10}(\zeta_a/\zeta_b)\mathrm{dB} = -13.98$ dB, which is enough to place $L_0(2.04)$ below the $B(2.04)$ dashed-line bound.

Finally, we add one more additional pole at high frequency ($p_1 = 40$), to filter out high-frequency noise. The final controller $G(s)$ is shown in Equation CS2.51 and the loop shaping in Figure CS2.13.

$$G(s) = \frac{-35\left(\left(\dfrac{s}{0.11}\right)+1\right)}{s\left(\left(\dfrac{s}{40}\right)+1\right)} \frac{(s^2 + 0.816s + 4.1616)}{(s^2 + 4.08s + 4.1616)} \qquad \text{(CS2.51)}$$

CS2.5.3 Prefilter Design—$F(s)$

With the controller $G(s)$ just designed—see Equation CS2.51, the reference tracking specifications presented in Equations CS2.46 through CS2.48, and the plant model $p_{32}(s)$ described in Equation CS2.39, we design a prefilter $F(s)$ to assure that all the input/output functions $p_{32}(s)G(s)F(s)/[1 + p_{32}(s)G(s)]$, from low frequencies up to 0.5 rad/s are inside the band defined by the limits $\delta_{6\text{-}up}(\omega)$ and $\delta_{6\text{-}lo}(\omega)$. The designed prefilter is shown in Equation CS2.52, and the input/output functions and limits in Figure CS2.14.

$$F(s) = \frac{\left(\left(\dfrac{s}{2.5}\right)+1\right)}{\left(\left(\dfrac{s}{0.2}\right)+1\right)} \qquad \text{(CS2.52)}$$

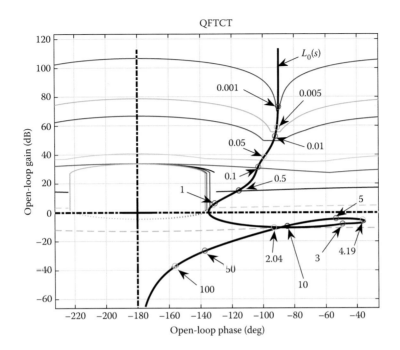

FIGURE CS2.13
QFT bounds and $G(s)$ design—loop shaping.

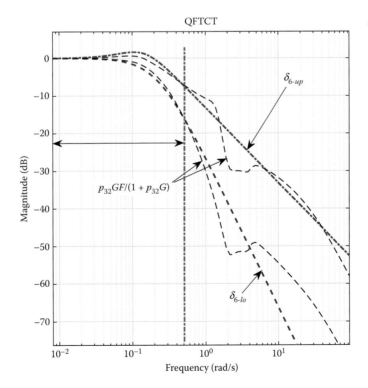

FIGURE CS2.14
Prefilter $F(s)$ for reference tracking specs: $\delta_{6-lo}(\omega) \leq |T_6| \leq \delta_{6-up}(\omega)$.

CS2.6 Analysis and Validation

The analysis of the closed-loop stability in the frequency domain is shown in Figure CS2.15a. The dashed line is the stability specification W_s defined in Equation CS2.43. The solid line represents the worst case of all the possible functions $p_{32}G/(1 + p_{32}G)$ at each frequency due to the model uncertainty. The control system meets the stability specification (the solid line is below the dashed line W_s) for all the plants within the uncertainty.

The analysis of the robust sensitivity specification, or rejection of disturbances at the output of the plant, is shown in Figure CS2.15b. The dashed line is the sensitivity specification $\delta_3(\omega)$ defined in Equation CS2.44. The solid line represents the worst case of all the possible functions $1/(1 + p_{32}G)$ at each frequency due to the model uncertainty. The control system meets the sensitivity specification in all cases (the solid line is below the dashed line δ_3) from 0 to 0.1 rad/s.

The analysis of the robust wind disturbance rejection in the frequency domain is shown in Figure CS2.15c. The dashed line is the specification $\delta_{k1}(\omega)$ defined in Equation CS2.45. The solid line represents the worst case of all the possible functions $p_{31}/(1 + p_{32}G)$ at each frequency due to the model uncertainty. Again, the control system meets the specification for all the plants within the uncertainty (the solid line is below the dashed line δ_{k1}).

The fore-aft tower vibration analysis in the frequency domain is shown in Figure CS2.15d. It represents the attenuation of the effect of wind disturbances $v_1(s)$ over the tower movement $y_t(s)$ through $p_{12}(s) \times G(s) \times p_{31}(s)$ and the control loop—see also Figure CS2.5. The dashed line is the specification $\delta_{k2}(\omega)$ defined in Equation CS2.49. The solid line represents the worst case of all possible functions $p_{12}p_{31}G/(1 + p_{32}G)$ at each frequency due to the model uncertainty. Once more, the control system meets the specification in all cases (solid line is below dashed line δ_{k2}).

The flap-wise blade vibration analysis in the frequency domain is shown in Figure CS2.15e. It represents the attenuation of the effect of wind disturbances $v_1(s)$ over the blades movement $y_b(s)$ through $p_{22}(s) \times G(s) \times p_{31}(s)$ and the control loop—see Figure CS2.5. The dashed line is the specification $\delta_{k3}(\omega)$ defined in Equation CS2.50. The solid line represents the worst case of all possible functions $p_{22}p_{31}G/(1 + p_{32}G)$ at each frequency due to the model uncertainty. The control system meets the specification in all cases (solid line is below dashed line δ_{k3}).

The time-domain analysis of the reference tracking specification is presented in Figure CS2.16a. The figure shows the limits $\delta_{6\text{-}up}(\omega)$ and $\delta_{6\text{-}lo}(\omega)$—see Equations CS2.48 and CS2.47, and the time responses of the rotor velocity $\Omega_r(t)$ to a unitary step reference $r(t)$. We simulate 100 cases of the closed-loop transfer function $p_{32}GF/(1 + p_{32}G)$—see Figure CS2.5, with $H(s) = 1$ and the electrical torque constant at $T_{gd} = T_{gdmax} = 4{,}006{,}111$ Nm, and Figure CS2.19, *Region 3, above-rated*. The results are given in per unit, the unit being $\Omega_{r\_nom} = 1.2671$ rad/s $= 12.1$ rpm. The control system meets the specification (is between the upper and lower limits) in all cases.

Additionally, the time-domain analysis of the wind disturbance rejection is shown in Figure CS2.16b. The figure presents the time responses of the rotor velocity $\Omega_r(t)$ to a unitary step wind speed disturbance $v_1(t)$. We simulate 100 cases within the uncertainty of $p_{31}/(1 + p_{32}G)$, also with $H(s) = 1$ and $T_{gd} = T_{gdmax} = 4{,}006{,}111$ Nm. Now, the zero level is the non-perturbed $\Omega_{r\_nom} = 1.2671$ rad/s $= 12.1$ rpm. The control system achieves a good disturbance rejection in all cases.

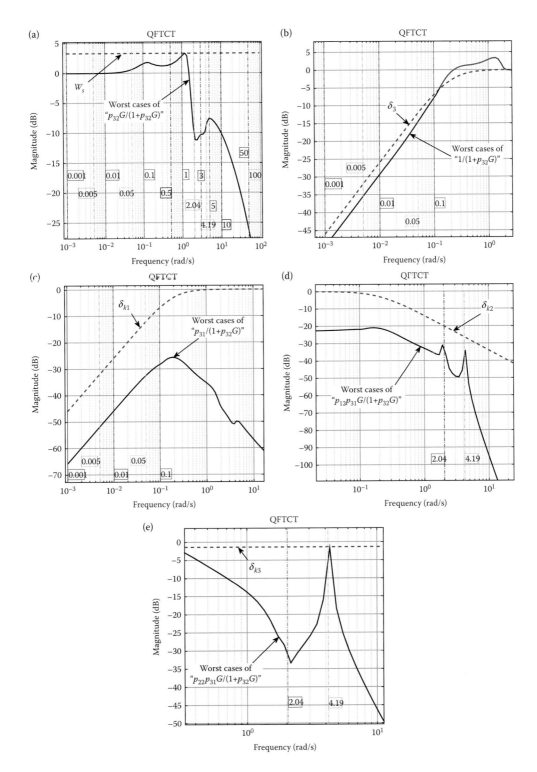

FIGURE CS2.15
Frequency-domain analysis: (a) stability, (b) sensitivity, (c) wind disturbance rejection, (d) fore-aft tower vibration, and (e) flap-wise blades vibration.

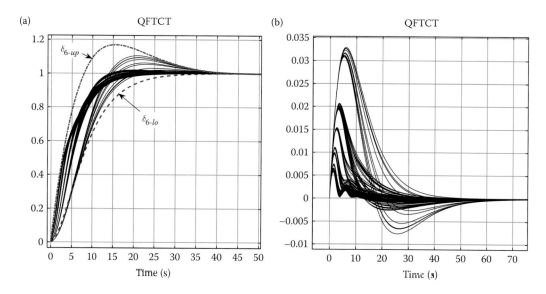

FIGURE CS2.16
Time domain. (a) Reference tracking $\Omega_r(s) = p_{32}GF/(1 + p_{32}G)r(s)$ and (b) wind disturbance rejection $\Omega_r(s) = p_{31}/(1 + p_{32}G)v_1(s)$.

Additionally, Figure CS2.17 shows the tower fore-aft oscillation when a unitary step wind velocity input $v_1(s)$ is introduced into the system. The disturbance $v_1(s)$ enters through $p_{31}(s)$, travels the pitch control loop until $\beta_d(s)$, and finally goes through $p_{12}(s)$ to $y_t(s)$—see the block diagram in Figure CS2.5. We study two cases. The first one, represented in Figure CS2.17a, uses the complete controller described in Equation CS2.51. The second case, represented in Figure CS2.17b, uses this controller without the notch filter.

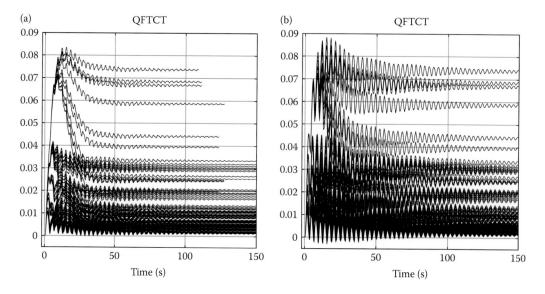

FIGURE CS2.17
Tower fore-aft oscillation, $y_t(t)$. Simulation of 100 plants. (a) $G(s)$ with the notch filter—see Equation CS2.51 and (b) $G(s)$ without the notch filter.

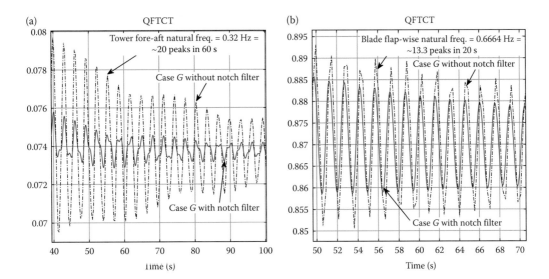

FIGURE CS2.18

Zooming in. (a) Tower fore-aft oscillation y_t and (b) blade flap-wise oscillation y_b. $G(s)$ with notch filter (solid line) and without (dashed line).

The tower oscillation is significantly reduced when we add the notch filter in the feedback controller $G(s)$, that is, the complete Equation CS2.51. Figure CS2.17a shows the $y_t(t)$ results with the complete $G(s)$, and Figure CS2.17b without the notch filter. Both analyses are performed for the same 100 cases within the plant uncertainty. The figures have the same scale.

Figure CS2.18a zooms in one case of the simulations of Figure CS2.17a and b. As we can see, the frequency of the oscillation in the figure (20 peaks in 60 s) is the natural frequency of the tower fore-aft oscillation: 0.320 Hz. The results of $y_t(t)$ using the complete controller $G(s)$ are represented by the solid line—Equation CS2.51, and the results of $G(s)$ without the notch filter by the dashed line. As anticipated, the complete controller $G(s)$ with the notch filter obtains a much better peak to peak reduction of the tower fore-aft oscillation.

Finally, Figure CS2.18b presents the attenuation of the blade flap-wise vibration $y_b(t)$ at its resonance frequency of 0.6664 Hz (13.3 peaks in 20 s), also with the controller $G(s)$ with and without the notch filter. In this case, the step input disturbance $v_1(s)$ enters through $p_{31}(s)$, travels the pitch control loop until $\beta_d(s)$, and goes through $p_{22}(s)$ to $y_b(s)$—see also the block diagram in Figure CS2.5. As in the tower vibration case, the complete controller $G(s)$ with the notch filter obtains a better peak to peak reduction of the blade flap-wise oscillation.

CS2.7 Extension to Higher Wind Velocities

This case study has designed the wind turbine pitch control system for the first part of the above-rated region, also called *Region 3*—see Figure CS2.19, range of wind speed v_1 from $v_1 = v_r \sim 12$ m/s to $v_1 = 17$ m/s.

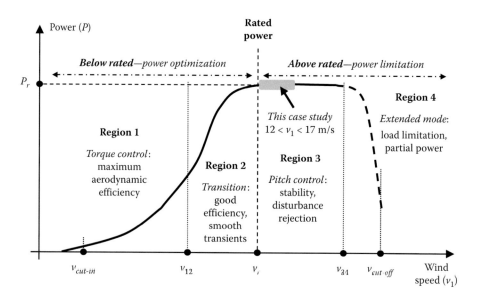

FIGURE CS2.19
Wind turbine power curve: Active power P versus wind speed v_1.

The proposed controller $G(s)$ can be easily extended to the entire *Region 3*, from v_r to v_{34} in Figure CS2.19. As the aerodynamics changes significantly with the wind velocity, the parameters $K_{T\Omega}$, K_{TV}, and $K_{T\beta}$ also change considerably—see Figure CS2.3 and Equations CS2.20 through CS2.26. As a result, the gains and one pole of the denominators of $p_{31}(s)$, $p_{32}(s)$, and $p_{33}(s)$ change with the wind speed—see Equations CS2.38 and CS2.39. A quantitative analysis of these variations is in the author's book, *Wind Energy Systems: Control Engineering Design*, Chapter 12, Reference 7.

To deal with this problem, the controller $G(s)$, Equation CS2.51, had to modify its gain and zero in real time, as a function of the desired pitch angle β_d, so that

$$k = -35 \times f_1(\beta_d) \tag{CS2.53}$$

$$z_1 = 0.11 \times f_2(\beta_d) \tag{CS2.54}$$

As the magnitude of the gain of the plant $\Omega_r(s)/\beta_d(s) = p_{32}(s)$ increases with the wind speed increase, the correcting function $f_1(\beta_d)$ typically decreases with the pitch angle β_d to compensate this effect. Also, as one pole of the plant $p_{32}(s)$ becomes faster with the wind speed increase, the correcting function $f_2(\beta_d)$ typically increases with the pitch angle to make the controller zero faster as well.

The design, stability analysis, and validation of this extended control strategy can be easily done according to the methodology presented in Chapters 6 and 7.

CS2.8 Summary

This case study has successfully designed a robust QFT control system for a variable-speed pitch-controlled gearless wind turbine. The design considers the model nonlinearities and uncertainty introduced by the aerodynamics. It accomplishes five simultaneous control objectives:

- Stability
- Tracking and regulation of the rotor speed by pitching the angle of the blades
- Rejection of unpredictable wind disturbances
- Minimization of the tower fore-aft oscillations and
- Attenuation of the blades flap-wise vibrations

Case Study 3: Wastewater Treatment Plant Control

CS3.1 Description

Human activity is seriously increasing the amount of nitrogen and phosphorus in the environment. When the concentration of nitrogen or phosphorus is too high, the water becomes polluted. This affects streams, rivers, lakes, and ground water, often resulting in human health issues, environmental destruction, and economic impact. Fortunately, the environmental policies and standards on water pollution have become increasingly stringent during the last few decades. Wastewater treatment plants (WWTP) with activated sludge processes (ASP) play a vital role in removing harmful organic matter, nitrogen, and phosphorus from domestic and industrial wastewater.

The high cost of WWTPs and the stringent water quality standards justify the effort to design advanced control strategies to manage ASP with maximum efficiency and minimum cost. The large variety of biological mechanisms and processes involved, their nonlinearity and multi-input multi-output (MIMO) characteristics, and the uncertainties in the composition and flow of the influent challenge the design of control solutions.

This Case Study CS3 presents the design of a MIMO QFT robust control system for an ASP-WWTP that simultaneously reduces the concentration of ammonia (NH_4) and nitrates (NO_3) in the plant effluent to meet the environmental standards. As an example, Figure CS3.1 shows a picture of the Crispijana-Vitoria plant, a municipal water treatment plant for nitrogen removal, able to deal with a 5000 m^3/h inflow, for which the author with the CEIT Research Centre designed an advanced robust control system in the 90s.[192,208,212]

In the following sections, we model and analyze the dynamics of an ASP-WWTP, analyze the MIMO characteristics, propose a set of robust stability and performance specifications, design SISO and MIMO control solutions, and validate them in both the frequency and time domains.

CS3.2 Plant Model

Figure CS3.2 shows the plant layout of a municipal ASP-WWTP that reduces the concentration of nitrogen (NH_4 and NO_3) and phosphorus (PO_4) in water. In this case study, we propose the design of a MIMO controller for simultaneous removal of NH_4 and NO_3.

The influent flow (Q_{IN}) is the wastewater coming from the city, typically with a high concentration of nitrogen, that is NH_4 and NO_3. The plant objective is to reduce these concentrations in the effluent flow (Q_{OUT}). The WWTP hydraulic configuration is composed of

FIGURE CS3.1
WWTP, Crispijana, Spain. Inflow 5000 m³/h.

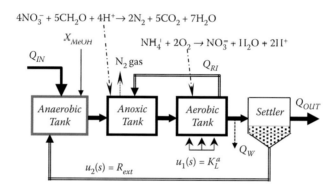

FIGURE CS3.2
ASP-WWTP configuration.

three biological tanks in series (*Anaerobic, Anoxic,* and *Aerobic*) and one settler for effluent clarification and sludge thickening—see Figure CS3.2.

The purpose of the *anaerobic tank* (first tank) is to promote phosphorous removal through growth of PAO bacteria, which accumulate phosphorus in approximately 35% of their weight, compared with regular bacteria which accumulate only 2%. Additionally, an external feed of ferric hydroxide (X_{MeOH}) is provided to ensure efficient phosphorus removal. In this way, the phosphates not biologically removed by the PAO bacteria are coagulated through a chemical reaction.

In the *anoxic tank* (second tank), heterotrophic bacteria are responsible for the denitrification. These bacteria biodegrade the organic matter, which at the same time turns NO_3 recycled from the aerobic tank (Q_{RI}) into nitrogen (N_2) gas. This reaction takes place in an anaerobic environment where the bacteria responsible for denitrification respire with nitrate instead of oxygen. The process, called *denitrification*, can be summarized by the formula:

$$4NO_3^- + 5CH_2O + 4H^+ \rightarrow 2N_2 + 5CO_2 + 7H_2O \qquad \text{(CS3.1)}$$

NO_3 that enters the previous expression is both coming from the influent Q_{RI} and also the product of a nitrification process taking place in the *aerobic tank* (third tank) by means of nitrifying bacteria and the injection of air ($u_1 = K_L a$). In other words, NH_4 is oxidized to NO_3 with an aeration system in the *aerobic tank*. The process, called *nitrification*, can be summarized by the formula:

$$NH_4^+ + 2O_2 \rightarrow NO_3^- + H_2O + 2H^+ \tag{CS3.2}$$

Additionally, in the *settler* the activated sludge is thickened, so that the clarified supernatant overflows into the effluent. At the same time, the activated sludge is recycled to the anaerobic tank ($u_2 = R_{ext}$) to maintain a high biomass concentration in the reactors.

As a result, NH_4 and NO_3 that enter into the WWTP are removed from the water using two biological processes: *nitrification* in the 3rd tank and *denitrification* in the 2nd tank, converting them finally into N_2 gas. The case is a 2×2 MIMO plant, being the plant outputs (sensors) and actuators respectively:

- $y_1(s) = NH_4$, or concentration of ammonia in the effluent Q_{OUT}
- $y_2(s) = NO_3$, or concentration of nitrates in the effluent Q_{OUT}
- $u_1(s) = K_La$, or air flow injected in the aerobic tank
- $u_2(s) = R_{ext}$, or external sludge recirculation flow, from settler to anaerobic tank

Other variables of the plant are the internal recirculation flow Q_{RI}, the waste sludge flow Q_W and the ferric hydroxide added in anaerobic tank X_{MeOH}. In this problem, we consider these three variables working at a fixed rate.

We use the *International Water Association* (IWA) ASM2d nonlinear model for our ASP-WWTP. We linearize the system around the working points, take the effect of the nonlinearities as parametric uncertainty, and identify and calibrate a resulting 2×2 MIMO model, as shown in Equation CS3.3,[212]

$$\begin{bmatrix} NH_4(s) \\ NO_3(s) \end{bmatrix} = P_w(s) = \begin{bmatrix} p_{w11}(s) & p_{w12}(s) \\ p_{w21}(s) & p_{w22}(s) \end{bmatrix} \begin{bmatrix} K_La(s) \\ R_{ext}(s) \end{bmatrix} \tag{CS3.3}$$

where

$$p_{w11}(s) = \frac{k_{11}}{\left(\dfrac{s}{a_{11}}+1\right)}; \quad p_{w12}(s) = \frac{k_{12}\left((s/z_{12.1})+1\right)\left((s/z_{12.2})+1\right)}{\left((s/a_{12})+1\right)\left[(s/\omega_{n12})^2 + (2\zeta_{12}/\omega_{n12})s + 1\right]}$$

$$p_{w21}(s) = \frac{k_{21}}{\left(\dfrac{s}{a_{21}}+1\right)}; \quad p_{w22}(s) = \frac{k_{22}}{\left(\dfrac{s}{a_{22}}+1\right)}$$

and where the parameters are

- $k_{11} \in [-0.045, -0.035]$; $a_{11} \in [6.25 \times 10^{-5}, 8.25 \times 10^{-5}]$.
- $k_{12} = -6.239 \times 10^{-6}$; $z_{12.1} = 7.534 \times 10^{-4}$; $z_{12.2} = -3.17 \times 10^{-5}$; $a_{12} = 8.04 \times 10^{-5}$; $\omega_{n12} = 4.58 \times 10^{-4}$; $\zeta_{12} = 0.8493$.
- $k_{21} = 0.0464$; $a_{21} = 1.008 \times 10^{-4}$.
- $k_{22} \in [-2.2 \times 10^{-5}, -1.8 \times 10^{-5}]$; $a_{22} \in [1.57 \times 10^{-4}, 1.77 \times 10^{-4}]$.

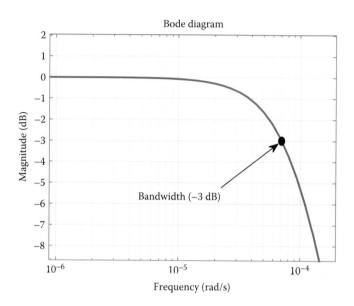

FIGURE CS3.3
Bode diagram for daily-average low-pass filter $f_{LP}(s)$, Equation CS3.4.

Additionally, environmental regulations and operation procedures require the control of WWTPs following daily-average measurements of the output concentrations. For this reason, we multiply each $p_{wij}(s)$ plant by a low-pass filter $f_{LP}(s)$ with a unitary dc-gain and a $2\pi/(24 \times 60 \times 60) = 7.27 \times 10^{-5}$ rad/s bandwidth (as a day is 24 h, with 60 min/h and 60 s/min)—see Figure CS3.3. In this way, the daily-averaged plant models, now called $p_{ij}(s)$, are

$$P(s) = P_w(s)\begin{bmatrix} f_{LP}(s) \\ f_{LP}(s) \end{bmatrix},$$

with $p_{ij}(s) = p_{wij}(s)\,f_{LP}(s)$; $i = 1,2$; $j = 1,2$

with $f_{LP}(s) = \dfrac{1}{\left[\left(\dfrac{s}{1.1\times10^{-4}}\right)^2 + \left(\dfrac{2}{1.1\times10^{-4}}\right)s + 1\right]}$

(CS3.4)

CS3.2.1 Frequencies

Figure CS3.4 shows the Bode diagram of the 2×2 MIMO plant $P(s) = [p_{ij}(s)]$, $i = 1,2, j = 1,2$. Based on this picture, we select the array of frequencies of interest, so that

$$\omega = [1\times10^{-7}\ 5\times10^{-7}\ 1\times10^{-6}\ 5\times10^{-6}\ 1\times10^{-5}\ 2\times10^{-5}\ 5\times10^{-5}\ 1\times10^{-4}\ 2\times10^{-4}$$
$$5\times10^{-4}\ 1\times10^{-3}\ 2\times10^{-3}\ 5\times10^{-3}]\,\text{rad/s}$$

(CS3.5)

CS3.2.2 Nominal Plant

The nominal plant is selected at the operating point of the WWTP, which is described by the mean values in the intervals of uncertainty in Equations CS3.3.

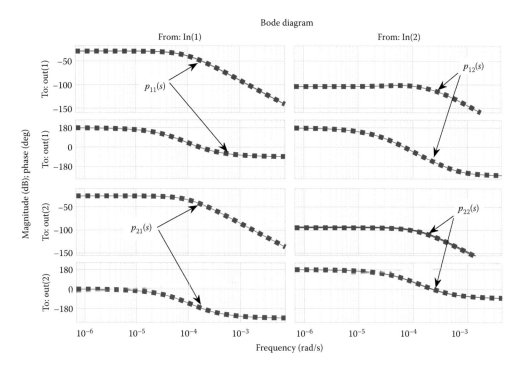

FIGURE CS3.4

Bode diagram of $P(s) = [p_{ij}(s)]$, $i = 1,2$, $j = 1,2$.

CS3.3 Preliminary Analysis

Before starting the design of the controllers, we conduct a brief preliminary analysis of the dynamics of the WWTP. As just mentioned, Figure CS3.4 shows the Bode diagram of the 2×2 MIMO plant $P(s) = [p_{ij}(s)]$, $i = 1, 2$, $j = 1, 2$, including also the parametric uncertainty.

As expected, the plant elements $p_{ij}(s)$ have no integrators (system type 0) and include negative (k_{11}, k_{12}, and k_{22}) and positive (k_{21}) dc gains. Also, the phases indicate that there is a non-minimum phase zero component. This is analyzed in-depth with the relative gain array.

Figure CS3.5 presents the relative gain analysis (RGA) over the frequencies of interest of the nominal 2×2 MIMO plant $P_0(s)$. These graphical results are also presented numerically for very low frequency or $\omega = 0$ rad/s in Equation CS3.6, and for very high frequency or $\omega = \infty$ rad/s in Equation CS3.7.

$$\Lambda_{(\omega=0)} = \begin{matrix} & u_1 & u_2 \\ & \begin{bmatrix} 0.7414 & 0.2586 \\ 0.2586 & 0.7414 \end{bmatrix} & \begin{matrix} y_1 \\ y_2 \end{matrix} \end{matrix} \tag{CS3.6}$$

$$\Lambda_{(\omega=\infty)} = \begin{matrix} & u_1 & u_2 \\ & \begin{bmatrix} -0.8898 & 1.8898 \\ 1.8898 & -0.8898 \end{bmatrix} & \begin{matrix} y_1 \\ y_2 \end{matrix} \end{matrix} \tag{CS3.7}$$

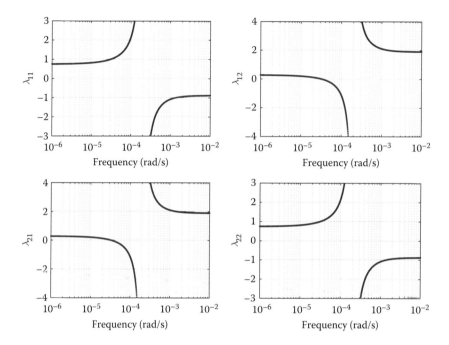

FIGURE CS3.5
RGA of $P_0(s)$ with frequency: λ_{ij}, $i = 1,2$ and $j = 1,2$.

By looking at the matrix $\Lambda_{(\omega=0)}$, Equation CS3.6, we select the input–output pairing for the 2×2 MIMO system as: $(K_L a, NH_4)$ and (R_{ext}, NO_3), that is, (u_1, y_1) and (u_2, y_2).

Also, comparing the two matrices $\Lambda_{(\omega=0)}$ and $\Lambda_{(\omega=\infty)}$, Equations CS3.6 and CS3.7, we see that the signs of λ_{11}, λ_{12}, λ_{21}, and λ_{22} change. As we discussed in Chapter 8, Section 8.3.3, this means that the MIMO system has a non-minimum phase (nmp) zero. By inspection of the transfer matrix, we find the nmp zero at: $z_1 = -2 \times 10^{-4}$ rad/s, which is the frequency of the discontinuity in the diagrams of Figure CS3.5. As we will see next, this nmp zero will limit the bandwidth of the control system.

CS3.4 Control Specifications

Taking into account the dynamics of the WWTP and the operator objectives and limitations, we define the following robust control specifications. They are the same for both channels, and include stability, output disturbance rejection, and reference tracking objectives as presented below:

- *Type 1: Stability specification*

$$\left| T_1(j\omega) \right| = \left| \frac{p_{ii}(j\omega)g_{ii}(j\omega)}{1 + p_{ii}(j\omega)g_{ii}(j\omega)} \right| \leq \delta_1(\omega) = W_s = 1.66, \, i = 1,2$$

$$\omega \in [1 \times 10^{-7} \; 5 \times 10^{-7} \; 1 \times 10^{-6} \; 5 \times 10^{-6} \; 1 \times 10^{-5} \; 2 \times 10^{-5} \; 5 \times 10^{-5}$$

$$1 \times 10^{-4} \; 2 \times 10^{-4} \; 5 \times 10^{-4} \; 1 \times 10^{-3} \; 2 \times 10^{-3} \; 5 \times 10^{-3}] \; \text{rad/s}$$

(CS3.8)

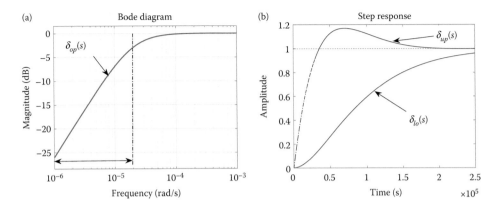

FIGURE CS3.6
Control specifications for all channels. (a) Disturbance rejection at the output of the plant: δ_{op}. (b) Reference tracking: δ_{up}, δ_{lo}.

which is equivalent to $PM = 35.06°$, $GM = 4.09$ dB—see Equations 2.30 and 2.31.

- *Type 3: Sensitivity or disturbances at plant output specification*—see Figure CS3.6a

$$|T_3(j\omega)| = \left|\frac{y(j\omega)}{d_o(j\omega)}\right| = \left|\frac{1}{1 + p_{ii}(j\omega)g_{ii}(j\omega)}\right| \le \delta_{op}(\omega) = \frac{\left(\dfrac{s}{a_d}\right)}{\left(\dfrac{s}{a_d}\right)+1}; a_d = 2\times10^{-5}, \tag{CS3.9}$$

$$i = 1,2. \quad \omega \in [1\times10^{-7}\ 5\times10^{-7}\ 1\times10^{-6}\ 5\times10^{-6}\ 1\times10^{-5}\ 2\times10^{-5}] \text{ rad/s}$$

- *Type 6: Reference tracking specification*—see Figure CS3.6b

$$\delta_{lo}(\omega) \le \left|\frac{p_{ii}(j\omega)g_{ii}(j\omega)}{1 + p_{ii}(j\omega)g_{ii}(j\omega)} f_{ii}(j\omega)\right| \le \delta_{up}(\omega), i = 1,2 \tag{CS3.10}$$

$$\delta_{lo}(s) = \frac{1}{\left[\left(\dfrac{s}{a_L}\right)+1\right]^2}; a_L = 2\times10^{-5} \tag{CS3.11}$$

$$\delta_{up}(s) = \frac{\left[\left(\dfrac{s}{a_U}\right)+1\right]}{\left[\left(\dfrac{s}{\omega_n}\right)^2 + \left(\dfrac{2\zeta s}{\omega_n}\right)+1\right]}; \quad a_U = 2\times10^{-5}; \zeta = 0.8; \omega_n = \frac{1.25a_U}{\zeta} \tag{CS3.12}$$

$$\omega \in [1\times10^{-7}\ 5\times10^{-7}\ 1\times10^{-6}\ 5\times10^{-6}\ 1\times10^{-5}\ 2\times10^{-5}\ 5\times10^{-5}\ 1\times10^{-4}] \text{ rad/s}$$

CS3.5 Controller Design

This section presents two control solutions for the WWTP. The first one is composed of two independent SISO PI controllers, and is implemented according to the structure in

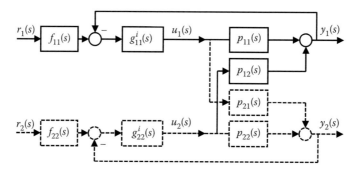

FIGURE CS3.7
SISO QFT control system for the WWTP.

Figure CS3.7. The second solution applies the MIMO QFT Method 2 presented in Chapter 8, Sections 8.6 and 8.9, and is implemented according to the diagram of Figure CS3.14.

CS3.5.1 Independent SISO QFT Control

Controller Structure
Figure CS3.7 shows the *SISO QFT* controller structure for the WWTP. It is composed of a 2×2 diagonal matrix $G^i(s) = \left[g^i_{11}(s)\, 0;\, 0\, g^i_{22}(s) \right]$ and a 2×2 diagonal prefilter matrix $F(s) = [f_{11}(s)\, 0;\, 0\, f_{22}(s)]$.

Design of 2×2 Diagonal Matrix $G^i(s)$ Controller and Prefilter $F(s)$
The SISO controller, or 2×2 diagonal controller G^i, is composed of two independent controllers, $g^i_{11}(s)$ and $g^i_{22}(s)$. The element $g^i_{11}(s)$ is designed by means of the standard SISO QFT loop-shaping technique, for the plant $p_{11}(s)$, and for the control specifications described in Equations CS3.8 through CS3.12. The QFT bounds and the loop shaping for $g^i_{11}(s)$ are shown in Figure CS3.8. The expression found for the controller $g^i_{11}(s)$ is a PI so that

$$g^i_{11}(s) = \frac{-0.0006 \left(\dfrac{1}{3 \times 10^{-5}} s + 1 \right)}{s} \tag{CS3.13}$$

The $f_{11}(s)$ prefilter element is designed for the plant $p_{11}(s)$ and the controller $g^i_{11}(s)$. Figure CS3.9 shows the design. The expression found for the prefilter $f_{11}(s)$ is

$$f_{11}(s) = \frac{1}{\left(\dfrac{1}{3.2 \times 10^{-5}} s + 1 \right)} \tag{CS3.14}$$

Figure CS3.10 shows the analysis of the disturbance rejection at the output of the plant and the reference tracking specifications. The analysis is made using the plant $p_{11}(s)$ with uncertainty, the controller $g^i_{11}(s)$ and the prefilter $f_{11}(s)$. When considering no coupling ($p_{12} = p_{21} = 0$), the first channel of the SISO control system meets the specifications for all the $p_{11}(s)$ plants within the uncertainty.

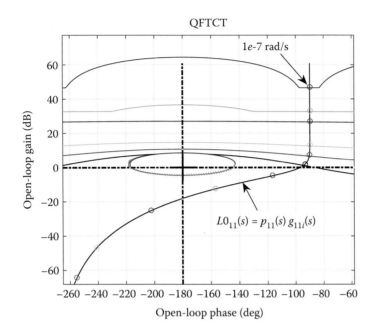

FIGURE CS3.8
Loop shaping of controller $g_{11}^i(s)$. $L_{0\_11}(s) = p_{11}(s)\, g_{11}^i(s)$.

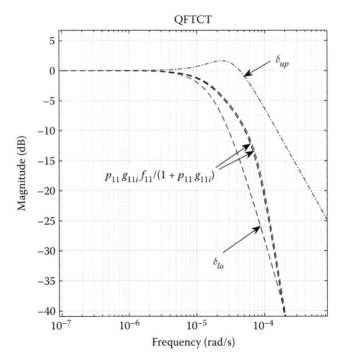

FIGURE CS3.9
Design of prefilter $f_{11}(s)$ for $p_{11}(s)$ and $g_{11}^i(s)$.

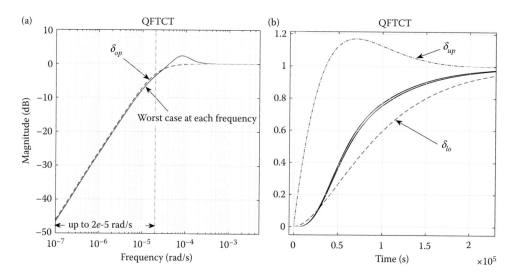

FIGURE CS3.10
Analysis of controller $g_{11}^i(s)$ and prefilter $f_{11}(s)$ for $p_{11}(s)$. (a) Disturbance rejection at plant output: δ_{op}. (b) Reference tracking: δ_{up}, δ_{lo}.

The element $g_{22}^i(s)$ is also designed by means of the standard SISO QFT loop-shaping technique, for the plant $p_{22}(s)$, and for the control specifications described in Equations CS3.8 through CS3.12. The QFT bounds and the loop shaping for $g_{22}^i(s)$ are shown in Figure CS3.11. The expression found for the controller $g_{22}^i(s)$ is also a PI, so that

$$g_{22}^i(s) = \frac{-1.5\left(\dfrac{1}{4.5\times10^{-5}}s+1\right)}{s} \tag{CS3.15}$$

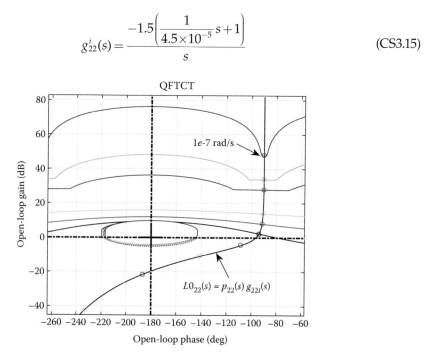

FIGURE CS3.11
Loop shaping of controller $g_{22}^i(s)$. $L_{0\_22}(s) = p_{22}(s)\, g_{22}^i(s)$.

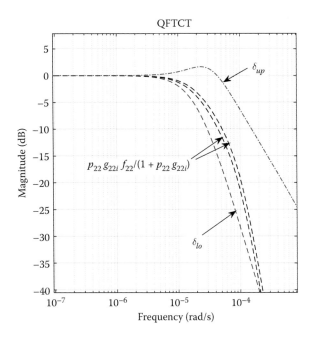

FIGURE CS3.12
Design of prefilter $f_{22}(s)$ for $p_{22}(s)$ and $g_{22}^i(s)$.

The $f_{22}(s)$ prefilter element is designed for the plant $p_{22}(s)$ and the controller $g_{22}^i(s)$. Figure CS3.12 shows the design. The expression for the prefilter $f_{22}(s)$ is

$$f_{22}(s) = \frac{1}{\left(\dfrac{1}{3.2 \times 10^{-5}} s + 1 \right)} \tag{CS3.16}$$

Figure CS3.13 shows the analysis of the disturbance rejection at the output of the plant and the reference tracking specifications, both for the plant $p_{22}(s)$ with the controller $g_{22}^i(s)$ and the prefilter $f_{22}(s)$. When considering no coupling ($p_{12} = p_{21} = 0$), the second channel of the SISO control system meets the specifications for all the $p_{22}(s)$ plants within the uncertainty.

CS3.5.2 MIMO QFT Control, Method 2

In the next subsections, we follow the flow chart and steps presented in Figure 8.31 for the design of a MIMO QFT controller according to Method 2. See also Sections 8.6 and 8.9 for more details.

Step A: Controller Structure
Figure CS3.14 shows the *MIMO QFT Method-2* controller structure for the WWTP. It is composed of a 2 × 2 full matrix $\boldsymbol{G}_\alpha$, a 2 × 2 diagonal matrix $\boldsymbol{G}_\beta$ and a 2 × 2 diagonal prefilter matrix $\boldsymbol{F}$.

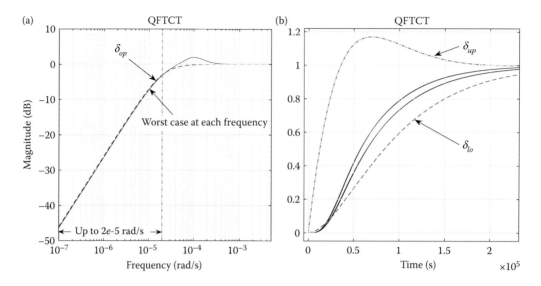

FIGURE CS3.13
Analysis of controller $g_{22}^i(s)$ and prefilter $f_{22}(s)$ for $p_{22}(s)$. (a) Disturbance rejection at plant output: δ_{op}. (b) Reference tracking: δ_{up}, δ_{lo}.

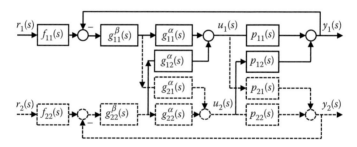

FIGURE CS3.14
MIMO QFT control system for the WWTP. Method 2.

Step B: Design of 2 × 2 Full Matrix G_α Controller
As discussed in Section 8.6.2, the fully populated matrix controller G is composed of two matrices: $G = G_\alpha G_\beta$, so that

$$G = G_\alpha G_\beta = \begin{bmatrix} g_{11}^\alpha(s) & g_{12}^\alpha(s) \\ g_{21}^\alpha(s) & g_{22}^\alpha(s) \end{bmatrix} \begin{bmatrix} g_{11}^\beta(s) & 0 \\ 0 & g_{22}^\beta(s) \end{bmatrix} \qquad \text{(CS3.17)}$$

The main objective of the pre-compensator G_α is to diagonalize the plant P as much as possible. As discussed in Sections 8.6 and 8.9, the expression used to calculate G_α is based on

$$G_\alpha(s) = \begin{bmatrix} g_{11}^\alpha(s) & g_{12}^\alpha(s) \\ g_{21}^\alpha(s) & g_{22}^\alpha(s) \end{bmatrix} = P^{-1}(s)\, P_{\text{diag}}(s) = \begin{bmatrix} p_{11}^*(s) & p_{12}^*(s) \\ p_{21}^*(s) & p_{22}^*(s) \end{bmatrix} \begin{bmatrix} p_{11}(s) & 0 \\ 0 & p_{22}(s) \end{bmatrix} \qquad \text{(CS3.18)}$$

where the plant matrix P, the corresponding inverse P^{-1}, and the diagonal P_{diag} are selected so that the expression of the extended matrix $P^x = P\, G_\alpha$ presents the closest form to a diagonal matrix, nulling the off-diagonal terms as much as possible.

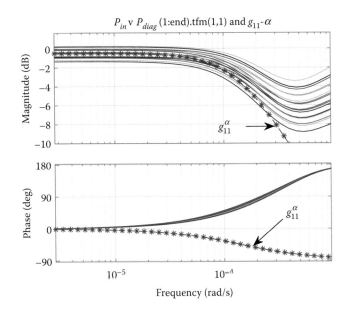

FIGURE CS3.15
Controller element $g_{11}^{\alpha}(s)$ and $p_{11}^{*}(s) \times p_{11}(s)$.

The Bode diagrams for the four expressions $p_{11}^{*}(s) \times p_{11}(s)$, $p_{12}^{*}(s) \times p_{22}(s)$, $p_{21}^{*}(s) \times p_{11}(s)$, and $p_{22}^{*}(s) \times p_{22}(s)$, including all the model uncertainty, are shown in Figures CS3.15 through CS3.18, respectively. The controller elements $g_{11}^{\alpha}(s)$, $g_{12}^{\alpha}(s)$, $g_{21}^{\alpha}(s)$, and $g_{22}^{\alpha}(s)$ are calculated as the transfer function that matches the mean value of the respective Bode diagram at low frequencies and then filters out the dynamics before the nmp zero at -2×10^{-4} rad/s—see also Equations CS3.19 through CS3.22, respectively.

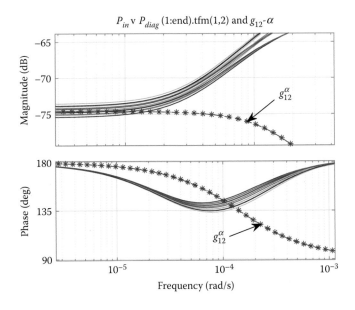

FIGURE CS3.16
Controller element $g_{12}^{\alpha}(s)$ and $p_{12}^{*}(s) \times p_{22}(s)$.

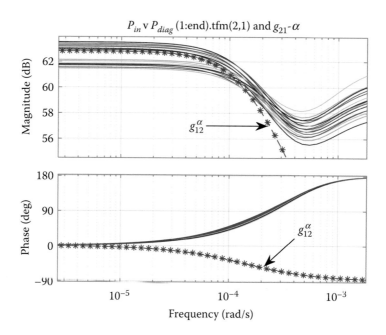

FIGURE CS3.17
Controller element $g_{21}^{\alpha}(s)$ and $p_{21}^{*}(s) \times p_{11}(s)$.

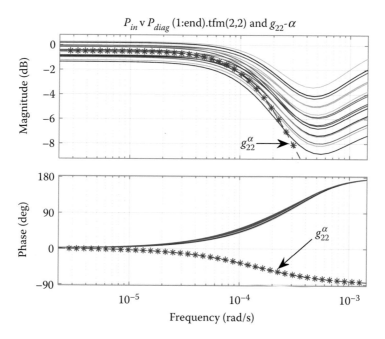

FIGURE CS3.18
Controller element $g_{22}^{\alpha}(s)$ and $p_{22}^{*}(s) \times p_{22}(s)$.

$$g_{11}^{\alpha}(s) = \frac{0.9439}{7143s + 1} \tag{CS3.19}$$

$$g_{12}^{\alpha}(s) = \frac{-0.0001858}{7143s + 1} \tag{CS3.20}$$

$$g_{21}^{\alpha}(s) = \frac{1382}{7143s + 1} \tag{CS3.21}$$

$$g_{22}^{\alpha}(s) = \frac{0.9439}{7143s + 1} \tag{CS3.22}$$

Step C.1.1: Design of the Diagonal Controller $g_{11}^{\beta}(s)$
The element $g_{11}^{\beta}(s)$ is designed by means of the standard SISO QFT loop-shaping technique, for the inverse of the extended equivalent plant $q_{x11}(s) = [p_{11}^{x^*e}]_1^{-1}$, and for the control specifications described in Equations (CS3.8 through CS3.12). According to the iterative expression Equation 8.97, the extended equivalent plant is

$$p_{11}^{x^*e}(s)\Big|_1 = p_{11}^{x^*}(s) \tag{CS3.23}$$

and the plant to be controlled by $g_{11}^{\beta}(s)$

$$q_{x11}(s) = \frac{1}{p_{11}^{x^*e}(s)\Big|_1} \tag{CS3.24}$$

The MATLAB code that calculates the plant $q_{x11}(s)$ for all the cases within the parametric uncertainty is similar to the one included in Appendix 8, Example 8.1, and Section 8.9 (QFT MIMO Method 2). The QFT bounds and the loop shaping for $g_{11}^{\beta}(s)$ are shown in Figure CS3.19. The expression found for the controller $g_{11}^{\beta}(s)$ is

$$g_{11}^{\beta}(s) = \frac{-0.0008\left(\dfrac{1}{5 \times 10^{-5}}s + 1\right)\left(\dfrac{1}{9 \times 10^{-5}}s + 1\right)\left(\dfrac{1}{0.00014}s + 1\right)\left(\dfrac{1}{0.0003}s + 1\right)}{s\left(\dfrac{1}{0.0009}s + 1\right)\left(\dfrac{1}{0.04}s + 1\right)^2\left(\dfrac{1}{0.2}s + 1\right)} \tag{CS3.25}$$

The design also fulfils the two stability conditions:
a. $L_{x0\_11}(s) = q_{x11}(s)\, g_{11}^{\beta}(s)$ satisfies the Nyquist encirclement condition and
b. There are no RHP pole-zero cancellations between $q_{x11}(s)$ and $g_{11}^{\beta}(s)$.

This is checked following to the methodology presented in Chapter 3, Section 3.4, with the *QFT Control Toolbox* (QFTCT), *Controller design* window, *File, Check stability* option.

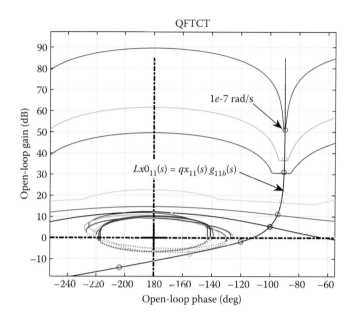

FIGURE CS3.19
Loop shaping of controller $g_{11}^{\beta}(s)$. $L_{x0\_11}(s) = q_{x11}(s)\, g_{11}^{\beta}(s)$.

Step C.1.2: Design of the Diagonal Prefilter $f_{11}(s)$
The $f_{11}(s)$ prefilter element is designed for the equivalent plant $q_{x11}(s)$ and the diagonal controller $g_{11}^{\beta}(s)$ designed in the previous Step C.1.1. Figure CS3.20 shows the design. The expression found for the prefilter $f_{11}(s)$ is

$$f_{11}(s) = \frac{1}{\left(\dfrac{1}{3.2\times10^{-5}}s+1\right)} \tag{CS3.26}$$

Figure CS3.21 shows the analysis of the disturbance rejection at the output of the plant and the reference tracking specifications for the equivalent plant $q_{x11}(s)$ and with the controller $g_{11}^{\beta}(s)$ and the prefilter $f_{11}(s)$. The control system meets the specifications with $q_{x11}(s)$ in all cases within the uncertainty.

Step C.2.1: Design of the Diagonal Controller $g_{22}^{\beta}(s)$
The element $g_{22}^{\beta}(s)$ is designed by means of the standard SISO QFT loop-shaping technique, for the inverse of the extended equivalent plant $q_{x22}(s) = [p_{22}^{x*e}]_2^{-1}$, and for the control specifications described in Equations CS3.8 through CS3.12. According to the iterative expression Equation 8.97, the extended equivalent plant is

$$\left. p_{22}^{x*e}(s)\right|_2 = \left. p_{22}^{x*}(s)\right|_1 - \frac{\left. p_{21}^{x*}(s)\right|_1 \left. p_{12}^{x*}(s)\right|_1}{\left. p_{11}^{x*}(s)\right|_1 + g_{11}^{\beta}(s)} \tag{CS3.27}$$

and the plant to be controlled by $g_{22}^{\beta}(s)$

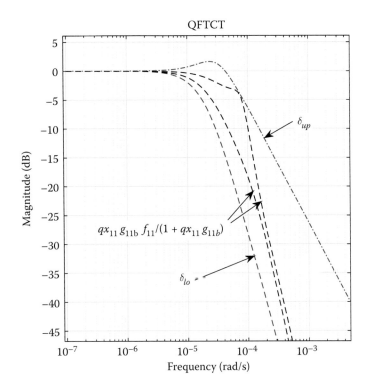

FIGURE CS3.20
Design of prefilter $f_{11}(s)$ for $q_{x11}(s)$ and $g_{11}^{\beta}(s)$.

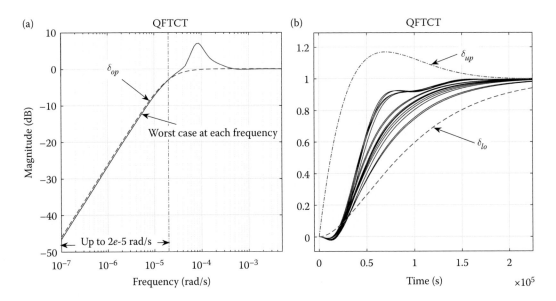

FIGURE CS3.21
Analysis of controller $g_{11}^{\beta}(s)$ and prefilter $f_{11}(s)$ for $q_{x11}(s)$. (a) Disturbance rejection at plant output: δ_{op}. (b) Reference tracking: δ_{up}, δ_{lo}.

$$q_{x22}(s) = \frac{1}{p_{22}^{x^*e}(s)\big|_2} \qquad \text{(CS3.28)}$$

The MATLAB code that calculates the plant $q_{x22}(s)$ for all the cases within the parametric uncertainty is similar to the one included in Appendix 8, Example 8.1, and Section 8.9 (QFT MIMO Method 2). The QFT bounds and the loop shaping for $g_{22}^{\beta}(s)$ are shown in Figure CS3.22. The expression for the controller $g_{22}^{\beta}(s)$ is

$$g_{22}^{\beta}(s) = \frac{-0.52\left(\dfrac{1}{5\times10^{-5}}s+1\right)^3}{s\left(\dfrac{1}{0.001}s+1\right)\left(\dfrac{1}{0.0013}s+1\right)^2} \qquad \text{(CS3.29)}$$

The design also fulfils the two stability conditions:

a. $L_{x0\_22}(s) = q_{x22}(s)\,g_{22}^{\beta}(s)$ satisfies the Nyquist encirclement condition, and
b. There are no RHP pole-zero cancellations between $q_{x22}(s)$ and $g_{22}^{\beta}(s)$.

This is checked following to the methodology presented in Chapter 3, Section 3.4, with the *QFT Control Toolbox* (QFTCT), *Controller design* window, *File, Check stability* option.

Step C.2.2: Design of the Diagonal Prefilter $f_{22}(s)$
The $f_{22}(s)$ prefilter element is designed for the equivalent plant $q_{x22}(s)$ and the diagonal controller $g_{22}^{\beta}(s)$ designed in the previous Step C.2.1. Figure CS3.23 shows the design. The expression found for the prefilter $f_{22}(s)$ is

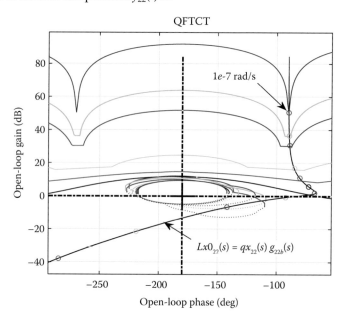

FIGURE CS3.22
Loop shaping of controller $g_{22}^{\beta}(s)$. $L_{x0\_22}(s) = q_{x22}(s)\,g_{22}^{\beta}(s)$.

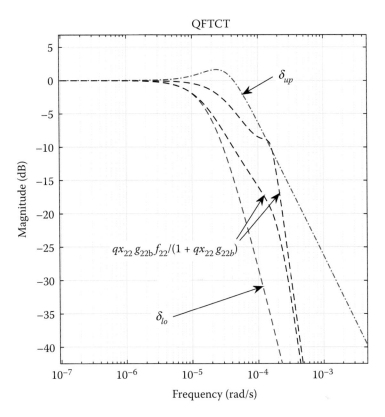

FIGURE CS3.23
Design of prefilter $f_{22}(s)$ for $q_{x22}(s)$ and $g_{22}^{\beta}(s)$.

$$f_{22}(s) = \cfrac{1}{\left(\cfrac{1}{3.2 \times 10^{-5}}s + 1\right)} \tag{CS3.30}$$

Figure CS3.24 shows the analysis of the disturbance rejection at the output of the plant and the reference tracking specifications for the equivalent plant $q_{x22}(s)$ and with the controller $g_{22}^{\beta}(s)$ and the prefilter $f_{22}(s)$. The control system meets the specifications with $q_{x22}(s)$ in all cases within the uncertainty.

(a) Disturbance rejection at plant output: δ_{op}. (b) Reference tracking: δ_{up}, δ_{lo}.

Step D: Final Checks
D.1. The design also fulfils the other additional two stability conditions:

- No Smith-McMillan pole-zero cancellations occur between $P(s)$ and $G(s)$, and
- No Smith-McMillan pole-zero cancellations occur in $|P^*(s) + G(s)|$.

D.2. The final system $P(s)G(s)$ does not have any additional RHP transmission zero.

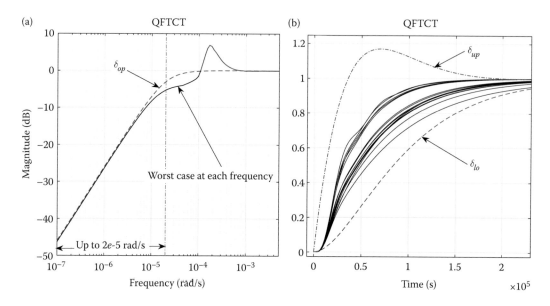

FIGURE CS3.24

Analysis of controller $g_{22}^{\beta}(s)$ and prefilter $f_{22}(s)$ for $q_{x22}(s)$. (a) Disturbance rejection at plant output: δ_{op}. (b) Reference tracking: δ_{up}, δ_{lo}.

CS3.6 Analysis and Validation

This section presents the time-domain analysis of the controllers designed in this case study. The analysis includes the simulation of the complete MIMO ASP-WWTP with both the SISO solution (Section CS3.5.1) and the MIMO QFT solution (Section CS3.5.2).

Figure CS3.25 shows the time-domain simulation of the two outputs, $y_1(s) = NH_4$ and $y_2(s) = NO_3$, of the MIMO nominal plant $P_0(s) = [p_{11}(s)\ p_{12}(s); p_{21}(s)\ p_{22}(s)]$, Equations CS3.3 and CS3.4.

We introduce two reference inputs, $r_1(s)$ and $r_2(s)$, and two external disturbances, $d_1(s)$ and $d_2(s)$ in the system. The reference inputs start at a polluted water level and change to the level required by the standards, that is NH_4 $r_1(s)$ from 2 g/m³ to 1 g/m³ at 2×10^6 s and NO_3 $r_2(s)$ from 8 g/m³ to 4 g/m³ at 3×10^6 s—see dashed lines in the figure.

The external disturbances introduce a significant step increase of the concentration of NH_4 and NO_3, respectively, in the water, with $d_1(s)$ from 0 to 0.5 g/m³ at 4×10^6 s and $d_2(s)$ from 0 to 2 g/m³ at 5×10^6 s.

The results with the SISO controllers are represented by the dotted lines. They use the control structure shown in Figure CS3.7, and the expressions for $g_{11}^i(s)$, $g_{22}^i(s)$, $f_{11}(s)$, and $f_{22}(s)$ in Equations CS3.13, CS3.15, CS3.14, and CS3.16, respectively.

The results with the MIMO QFT controller are represented by the solid lines. They use the control structure shown in Figure CS3.14, and the expressions for $G = G_\alpha\,G_\beta$ in Equations CS3.19 through CS3.22, CS3.25, and CS3.29, and for $F = [f_{11}(s)\ 0; 0\ f_{22}(s)]$ in Equations CS3.26 and CS3.30.

The independent SISO controllers achieve a good performance. The MIMO controller improves the results, reducing significantly the effect of the coupling between the control loops.

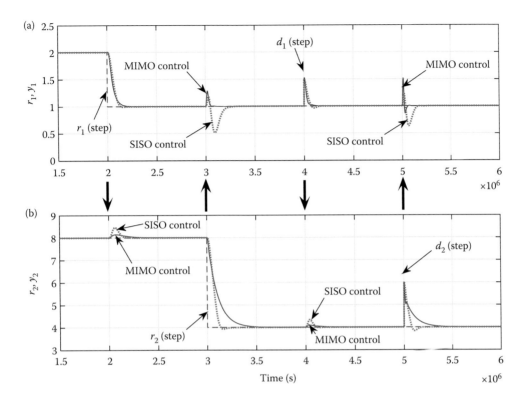

FIGURE CS3.25
MIMO WWTP. SISO controllers (dotted line), MIMO controller (solid line), references (dashed line). (a) 1st channel NH_4, (b) 2nd channel NO_3.

CS3.7 Summary

This case study has designed a robust 2×2 MIMO QFT control system for a wastewater treatment plant (WWTP) with an activated sludge process (ASP) that simultaneously reduces the concentration of ammonia (NH_4) and nitrates (NO_3) in the plant effluent. The design considers the model uncertainty and MIMO loop interaction. It accomplishes four simultaneous control objectives:

- Stability
- Reference tracking
- Rejection of unpredictable disturbances and
- Minimization of loop interaction

Case Study 4: Radio-Telescope Control

CS4.1 Description

Radio astronomy is a young scientific field. It was born by serendipity in 1931, when Karl Jansky, a physicist and radio engineer at Bell Telephone Laboratories, built an antenna to study the properties of 20.5 MHz radio waves for use in transatlantic telephone services. While recording signals from all directions, Jansky found a unique radiation repeated every 23 h and 56 min, which is the period of the Earth's rotation relative to the stars (sidereal day), instead of 24 h (solar day). Based on this unexpected observation, Jansky concluded that the radio signals came from the Milky Way, and radio astronomy was born.[366]

Since then, many radio telescopes have been built, and thousands of new astronomical discoveries have changed our understanding of the universe. One of the most prominent facilities is the Robert C. Byrd Green Bank Telescope, or GBT (WV). It is the world's largest single-dish fully steerable radio telescope, operating at meter to millimeter wavelengths (0.1–116 GHz)—see Figure CS4.1. It has an enormous 100×110 m elliptical collecting area, an unblocked aperture, a 148 m structure, and an excellent surface accuracy with over 2200 actuators.

The telescope is driven in the azimuth axis by 16 motors, each of 30 hp, coupled together in four trucks. In the elevation axis, there are eight motors, each of 40 hp. All motors are preloaded to remove backlash. The telescope has an off-axis geometric focus, with a feed arm and a sub-reflector, as shown in Figure CS4.2.

This case study presents the design of the velocity and position cascade control loops for the azimuth axis of a radio telescope similar to GBT. Coincidentally, the author has worked with National Radio Astronomy Observatory (NRAO) engineers on the design of advanced control solutions for the azimuth, elevation, and sub-reflector servo systems of GBT for a number of years.

In the following sections, we model and analyze the dynamics of the radio telescope, propose a set of stability and performance specifications, design the velocity and position cascade control systems, improve the results with a nonlinear dynamic control (NDC) strategy, and validate them in both frequency and time domains.

CS4.2 Plant Model

Figure CS4.2 shows the diagram for the azimuth axis of the radio telescope. The model considers four degrees of freedom: the angular displacement of the dish $\theta_d(t)$, the angular displacement of the central base of the structure $\theta_b(t)$, the angular displacement of each

FIGURE CS4.1
Green Bank Telescope (GBT), NRAO, West Virginia.

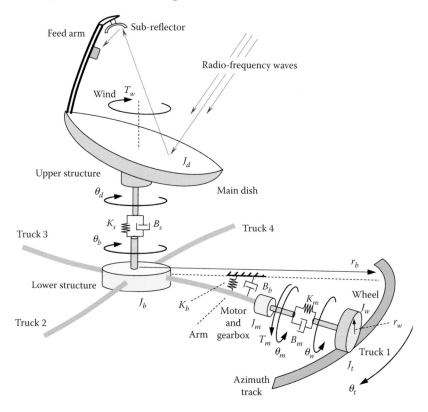

FIGURE CS4.2
Radio-telescope diagram. Azimuth model description.

truck system $\theta_t(t)$, and the angular displacement of each motor $\theta_m(t)$. The main variables and parameters of the system are the following:

Inputs

- T_w: aerodynamic torque applied by the wind on the dish and structure, [Nm]
- T_m equivalent torque applied by the four motors of each truck, [Nm]

Outputs

- θ_d: dish angular position, [rad]
- θ_b: base (lower structure) angular position, [rad]
- θ_t: truck angular position, [rad]
- θ_w: truck wheel angular position, $\theta_w = \theta_t\,(r_b/r_w)$ [rad]
- θ_m: motor shaft angular position at the gearbox output, [rad]

Parameters

- N: number of truck systems. Each truck is composed of one arm and a drive-train system (with one equivalent motor, gearbox, brake, shaft, and pinion), [-]
- r_b: radius of the arm between the lower structure and each truck, [m]
- r_w: radius of the equivalent truck wheel, [m]
- R: radius rate, $R = r_b/r_w$, [-]
- J_d: moment of inertia of dish, feed arm, and upper structure, at θ_d, [kg m^2]
- J_b: moment of inertia of lower structure, at θ_b, [kg m^2]
- J_t: moment of inertia of each truck, at θ_t, [kg m^2]
- J_w: moment of inertia of each equivalent truck wheel, at θ_w, [kg m^2]
- J_m: moment of inertia of each equivalent motor with its brake, gearbox, and shaft, at θ_m, [kg m^2]
- K_s: upper structure torsional stiffness coefficient, [Nm]
- B_s: upper structure torsional damping coefficient, [Nms]
- K_b: one arm torsional stiffness coefficient, [Nm]
- B_b: one arm torsional damping coefficient, [Nms]
- K_m: one motor shaft torsional stiffness coefficient, [Nm]
- B_m: one motor shaft torsional damping coefficient, [Nms]

As in previous cases, we start with the general equation of motion for the mechanical system, which is

$$\mathcal{M}\ddot{q} + \mathcal{C}\dot{q} + \mathcal{K}q = \mathcal{Q}(\dot{q}, q, u, t) \tag{CS4.1}$$

where $\mathcal{M}$, $\mathcal{C}$, and $\mathcal{K}$ are the *Mass*, *Damping*, and *Stiffness* matrices, $\mathcal{Q}$ the matrix of inputs, q the vector of generalized coordinates, and u the vector of inputs. Also, we consider the general form of the Euler–Lagrange equation,

$$\frac{d}{dt}\left(\frac{\partial L}{\partial \dot{q}_i}\right) - \frac{\partial L}{\partial q_i} + \frac{\partial D_d}{\partial \dot{q}_i} = Q_i\,,\ i = 1, 2, ..., n \quad \text{and} \quad L = E_k - E_p \tag{CS4.2}$$

where q_i are the generalized coordinates or degrees of freedom of the system, Q_i the generalized forces applied to each subsystem "i", E_k and E_p the kinetic and potential energies, D_d the dissipator co-content, and L the Lagrangian.[236]

The generalized coordinates for the radio telescope are the angular displacement of the dish $\theta_d(t)$, the angular displacement of the central base of the structure $\theta_b(t)$, the angular displacement of each truck system $\theta_t(t)$, and the angular displacement of each motor $\theta_m(t)$, which are, respectively: $q_1 = \theta_d(t)$, $q_2 = \theta_b(t)$, $q_3 = \theta_t(t)$, and $q_4 = \theta_m(t)$—see Figure CS4.2. The azimuth-axis displacement contains the kinetic, potential, and dissipative functions described by Equations CS4.3 through CS4.5. Notice that we consider linear dissipators, which means that the D_d expression or dissipator co-content equals one-half of the power being absorbed by the dissipators.[236]

$$E_k = \frac{1}{2} J_d \dot{\theta}_d^2 + \frac{1}{2} J_b \dot{\theta}_b^2 + \frac{1}{2} N J_t \dot{\theta}_t^2 + \frac{1}{2} N J_w (R\dot{\theta}_t)^2 + \frac{1}{2} N J_m \dot{\theta}_m^2 \tag{CS4.3}$$

$$E_p = \frac{1}{2} K_s (\theta_d - \theta_b)^2 + \frac{1}{2} N K_b (\theta_b - \theta_t)^2 + \frac{1}{2} N K_m (R\theta_t \quad \theta_m)^2 \tag{CS4.4}$$

$$D_d = \frac{1}{2} B_s (\dot{\theta}_d - \dot{\theta}_b)^2 + \frac{1}{2} N B_b (\dot{\theta}_b - \dot{\theta}_t)^2 + \frac{1}{2} N B_m (R\dot{\theta}_t - \dot{\theta}_m)^2 \tag{CS4.5}$$

The generalized forces applied to the radio telescope are the aerodynamic torque applied by the wind on the dish and upper structure $T_w(t)$, and the torque applied by the N trucks, $N\,T_m(t)$. Now, using Equations CS4.3 through CS4.5, the terms for the Euler–Lagrange equation are

$$\frac{d}{dt}\left(\frac{\partial L}{\partial \dot{q}_i}\right) = \frac{d}{dt}\begin{bmatrix}\dfrac{\partial E_k}{\partial \dot{\theta}_d}\\[4pt]\dfrac{\partial E_k}{\partial \dot{\theta}_b}\\[4pt]\dfrac{\partial E_k}{\partial \dot{\theta}_t}\\[4pt]\dfrac{\partial E_k}{\partial \dot{\theta}_m}\end{bmatrix} = \begin{bmatrix} J_d & 0 & 0 & 0 \\ 0 & J_b & 0 & 0 \\ 0 & 0 & N(J_t + R^2 J_w) & 0 \\ 0 & 0 & 0 & N J_m \end{bmatrix}\begin{bmatrix}\ddot{\theta}_d\\\ddot{\theta}_b\\\ddot{\theta}_t\\\ddot{\theta}_m\end{bmatrix} = \mathcal{M}\,\ddot{q} \tag{CS4.6}$$

$$-\frac{\partial L}{\partial q_i} = \frac{\partial E_p}{\partial q_i} = \begin{bmatrix}\dfrac{\partial E_p}{\partial \theta_d}\\[4pt]\dfrac{\partial E_p}{\partial \theta_b}\\[4pt]\dfrac{\partial E_p}{\partial \theta_t}\\[4pt]\dfrac{\partial E_p}{\partial \theta_m}\end{bmatrix} = \begin{bmatrix} K_s & -K_s & 0 & 0 \\ -K_s & K_s + N K_b & -N K_b & 0 \\ 0 & -N K_b & N(K_b + R^2 K_m) & -N R K_m \\ 0 & 0 & -N R K_m & N K_m \end{bmatrix}\begin{bmatrix}\theta_d\\\theta_b\\\theta_t\\\theta_m\end{bmatrix} = \mathcal{K}\,q \tag{CS4.7}$$

$$\frac{\partial D_d}{\partial \dot{q}_i} = \begin{bmatrix} \dfrac{\partial D_d}{\partial \dot{\theta}_d} \\[6pt] \dfrac{\partial D_d}{\partial \dot{\theta}_b} \\[6pt] \dfrac{\partial D_d}{\partial \dot{\theta}_t} \\[6pt] \dfrac{\partial D_d}{\partial \dot{\theta}_m} \end{bmatrix} = \begin{bmatrix} B_s & -B_s & 0 & 0 \\ -B_s & B_s + N B_b & -N B_b & 0 \\ 0 & -N B_b & N(B_b + R^2 B_m) & -N R B_m \\ 0 & 0 & -N R B_m & N B_m \end{bmatrix} \begin{bmatrix} \dot{\theta}_d \\ \dot{\theta}_b \\ \dot{\theta}_t \\ \dot{\theta}_m \end{bmatrix} = \mathcal{C}\,\dot{q} \qquad \text{(CS4.8)}$$

$$\mathcal{Q} = \begin{bmatrix} 1 & 0 \\ 0 & 0 \\ 0 & 0 \\ 0 & N \end{bmatrix} \begin{bmatrix} T_w \\ T_m \end{bmatrix} = \mathcal{R}\,u \qquad \text{(CS4.9)}$$

Rearranging the equation of motion Equation CS4.1 as

$$\ddot{q} = -\mathcal{M}^{-1}\mathcal{C}\,\dot{q} - \mathcal{M}^{-1}\mathcal{K}\,q + \mathcal{M}^{-1}\mathcal{R}\,u \qquad \text{(CS4.10)}$$

and using Equations CS4.6 through CS4.9, we find a state space description of the system ($\dot{x} = Ax + Bu; y = Cx$) with the expressions

$$\dot{x} = \left[\begin{array}{c|c} -\mathcal{M}^{-1}\mathcal{C} & -\mathcal{M}^{-1}\mathcal{K} \\ \hline I & 0 \end{array} \right] x + \left[\begin{array}{c} \mathcal{M}^{-1}\mathcal{R} \\ \hline 0 \end{array} \right] u \qquad \text{(CS4.11)}$$

$$y = Cx$$

with $I = [1\,0\,0\,0; 0\,1\,0\,0; 0\,0\,1\,0; 0\,0\,0\,1]$, $0 = [0\,0\,0\,0; 0\,0\,0\,0; 0\,0\,0\,; 0\,0\,0\,0]$, and,

$$A = \begin{bmatrix} -\dfrac{B_s}{J_d} & \dfrac{B_s}{J_d} & 0 & 0 & -\dfrac{K_s}{J_d} & \dfrac{K_s}{J_d} & 0 & 0 \\[10pt] \dfrac{B_s}{J_b} & \dfrac{-(B_s + N B_b)}{J_b} & \dfrac{N B_b}{J_b} & 0 & \dfrac{K_s}{J_b} & \dfrac{-(K_s + N K_b)}{J_b} & \dfrac{N K_b}{J_b} & 0 \\[10pt] 0 & \dfrac{B_b}{J_w R^2 + J_t} & \dfrac{-(B_b + R^2 B_m)}{J_w R^2 + J_t} & \dfrac{R B_m}{J_w R^2 + J_t} & 0 & \dfrac{K_b}{J_w R^2 + J_t} & \dfrac{-(K_b + R^2 K_m)}{J_w R^2 + J_t} & \dfrac{R K_m}{J_w R^2 + J_t} \\[10pt] 0 & 0 & \dfrac{R B_m}{J_m} & \dfrac{-B_m}{J_m} & 0 & 0 & \dfrac{R K_m}{J_m} & \dfrac{-K_m}{J_m} \\[6pt] 1 & 0 & 0 & 0 & 0 & 0 & 0 & 0 \\ 0 & 1 & 0 & 0 & 0 & 0 & 0 & 0 \\ 0 & 0 & 1 & 0 & 0 & 0 & 0 & 0 \\ 0 & 0 & 0 & 1 & 0 & 0 & 0 & 0 \end{bmatrix}$$

$$\text{(CS4.12)}$$

$$B = \begin{bmatrix} \dfrac{1}{J_d} & 0 \\ 0 & 0 \\ 0 & 0 \\ 0 & \dfrac{1}{J_m} \\ 0 & 0 \\ 0 & 0 \\ 0 & 0 \\ 0 & 0 \end{bmatrix} \qquad \text{(CS4.13)}$$

$$C = \begin{bmatrix} 0 & 0 & 0 & 1 & 0 & 0 & 0 & 0 \\ 0 & 0 & 0 & 0 & 1 & 0 & 0 & 0 \end{bmatrix} \qquad \text{(CS4.14)}$$

$$x = \begin{bmatrix} \dot{q} \\ q \end{bmatrix} = [\dot{\theta}_d \quad \dot{\theta}_b \quad \dot{\theta}_t \quad \dot{\theta}_m \quad \theta_d \quad \theta_b \quad \theta_t \quad \theta_m]^T \qquad \text{(CS4.15)}$$

$$u = [T_w \quad T_m]^T \qquad \text{(CS4.16)}$$

$$y = [\dot{\theta}_m \quad \theta_d]^T \qquad \text{(CS4.17)}$$

Now, applying $P(s) = C(sI - A)^{-1}B$ for $y(s) = P(s)u(s)$, the transfer matrix from the two independent inputs to the two outputs of interest is

$$\begin{bmatrix} \dot{\theta}_m(s) \\ \theta_d(s) \end{bmatrix} = P_{2\times 2}(s) \begin{bmatrix} T_w(s) \\ T_m(s) \end{bmatrix} \qquad \text{(CS4.18)}$$

where the velocity of the motors $d\theta_m(s)/dt = \theta'_m(s)$ and the dish angular position $\theta_d(s)$, from both the wind torque $T_w(s)$ and the motor torque $T_m(s)$, are

$$\dot{\theta}_m(s) = p_{11}(s)T_w(s) + p_{12}(s)T_m(s) \qquad \text{(CS4.19)}$$

$$\theta_d(s) = p_{21}(s)T_w(s) + p_{22}(s)T_m(s) \qquad \text{(CS4.20)}$$

Figure CS4.3 graphically shows the motor speed $\theta'_m(s)$, position $\theta_m(s)$, and dish displacement $\theta_d(s)$ as a function of the motor and wind torques, $T_m(s)$ and $T_w(s)$, according to Equations CS4.19 and CS4.20. In addition, the figure also shows the cascade control servo system, with the velocity controller $G_v(s)$ and the position controllers, $G_p(s)$ and $F_p(s)$, all to be designed in the next sections.

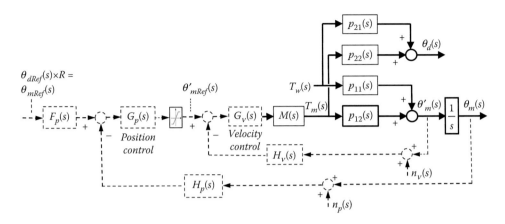

FIGURE CS4.3
Block diagram representation of the radio-telescope dynamics and velocity/position control servo systems.

Parameters and Parametric Uncertainty

The parameters for the model of the radio telescope considered in this case study are shown in Table CS4.1. They are based on the results published in Reference 365 about the NRAO-GBT. The parameters with uncertainty are the inertia of the dish and upper structure, which changes with the variation of the elevation-axis angle, and the three damping coefficients, so that

$$J_d \in [86.7, \ 117.3] \times 10^6 \ \text{kg m}^2 \tag{CS4.21}$$

$$B_s \in [54.8, 74.2] \times 10^6 \ \text{Ns/m}$$

$$B_b \in [2.66, \ 3.59] \times 10^5 \ \text{Ns / m}$$

$$B_m \in [233.67, \ 316.14] \ \text{Ns/m}$$

TABLE CS4.1

Radio Telescope Parameters (Nominal Case)

Radio-telescope weight [kg]	3.85×10^6	L_d (dish dimensions) [m]	100×110
Sensitivity radio freq. [GHz]	0.1 to 116	h (total height) [m]	148
N—number of azimuth tracks	4	R—radius ratio r_b / r_w	50
J_d [kg m²]—dish, f.a., up.struct.	102×10^6	J_b [kg m²]—lower structure	34.3×10^6
J_w [kg m²]—equiv wheel, truck	426	J_m [kg m²]—equiv. motor, truck	517.56
J_t [kg m²]—equiv truck system	3.43×10^6	$M(s)$—motor dynamics	1
K_s [N/m]—upper structure	1.9731×10^9	B_s [Ns/m]—upper structure	4.4862×10^6
K_b [N/m]—lower structure	94.9×10^6	B_b [Ns/m]—lower structure	3.1249×10^5
K_m [N/m]—motor shaft	1.6851×10^6	B_m [Ns/m]—motor shaft	274.89
H_v—feedback dynamics	1	H_p—feedback dynamics	1

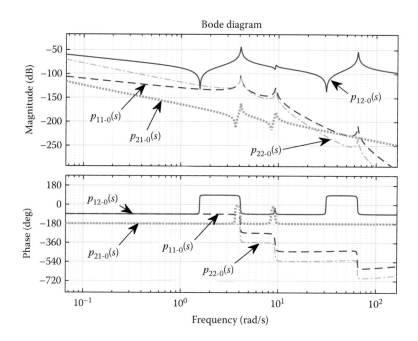

FIGURE CS4.4
Bode diagram, nominal plants: $p_{11\text{-}0}(s)$, $p_{12\text{-}0}(s)$, $p_{21\text{-}0}(s)$, and $p_{22\text{-}0}(s)$.

Frequencies

Figures CS4.4 through CS4.6 show the Bode diagrams for the transfer functions $p_{11}(s)$, $p_{12}(s)$, $p_{21}(s)$, and $p_{22}(s)$. We select an array of frequencies of interest based on this information—see Equation CS4.22. Notice that we include frequencies that represent the resonances (peaks) that appear in the diagrams, and which represent the flexible modes of the radio telescope for upper structure and feed-arm torsion, lower structure and truck-arm torsion, and motor shaft torsion.

$$\omega = [0.01\ 0.05\ 0.1\ 0.5\ 1\ 1.5\ 2\ 3\ 4\ 5\ 10\ 50\ 60\ 65\ 70\ 100\ 500]\,\text{rad/s} \qquad \text{(CS4.22)}$$

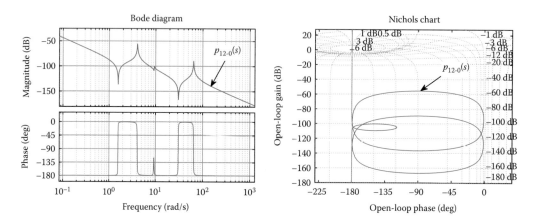

FIGURE CS4.5
Bode diagram and Nichols diagram of $p_{12\text{-}0}(s)$.

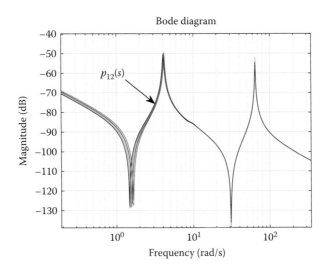

FIGURE CS4.6
Bode diagram (magnitude) of $p_{12}(s)$ with $\pm 15\%$ of uncertainty on J_d, B_s, B_b, and B_m.

Nominal Plant

The nominal plant for all, $p_{11\text{-}0}(s)$, $p_{12\text{-}0}(s)$, $p_{21\text{-}0}(s)$, and $p_{22\text{-}0}(s)$, is selected according to the following parameters within the uncertainty:

$$J_d = 102 \times 10^6 \ \text{kg m}^2 \tag{CS4.23}$$

$$B_s = 4.4862 \times 10^6 \ \text{Ns/m}$$

$$B_b = 3.1249 \times 10^5 \ \text{Ns/m}$$

$$B_m = 274.89 \ \text{Ns/m}$$

CS4.3 Preliminary Analysis

Before starting the design of the controllers, we conduct a preliminary analysis of the dynamics of the radio telescope. Figure CS4.4 shows the Bode diagram of the nominal four plants: $p_{11\text{-}0}(s)$, $p_{12\text{-}0}(s)$, $p_{21\text{-}0}(s)$, and $p_{22\text{-}0}(s)$.

The plants $p_{11}(s)$ and $p_{12}(s)$ describe, respectively, the effect of torque applied by the wind $T_w(s)$ and the torque applied by the motor $T_m(s)$ on the motor speed $\theta'_m(s)$. Both transfer functions are system-type 1, that is, they have one integrator.

The plants $p_{21}(s)$ and $p_{22}(s)$ describe, respectively, the effect of torque applied by the wind $T_w(s)$ and the torque applied by the motor $T_m(s)$ on the dish position $\theta_d(s)$. Both transfer functions are system-type 2, that is, they have two integrators.

All the plants contain three resonance modes—see Figures CS4.4. The first one, at $\omega = 4.1$ rad/s (0.65 Hz), represents the vibration of the dish, feed arm, and upper structure in the azimuth direction (θ_d). The second mode, at $\omega = 9.35$ rad/s (1.49 Hz), represents the vibration of the lower structure and arms in the azimuth direction (θ_b), and the third mode, at $\omega = 64.8$ rad/s (10.31 Hz), represents the vibration of the shaft of the motors (θ_m).

The main plant to be controlled is $p_{12}(s)$, which is the transfer function from the truck equivalent motor torque to the motor velocity $\theta'_m(s)/T_m(s)$—see Figure CS4.3. The resonance modes of $p_{12}(s)$ limit the potential bandwidth of the control system, which will be below these modes to reduce the vibration of the dish and feed arm at these frequencies—see Figures CS4.5 and CS4.6.

CS4.4 Azimuth Axis. Velocity Control

Figure CS4.3 shows the cascade control diagram for the telescope azimuth-axis servo system. It is composed of two control loops: an inner loop, which is the velocity (rate) control system, and an outer loop, which is the position control system. In this section, we design the inner controller: $G_v(s)$.

CS4.4.1 QFT Templates

Figure CS4.7 shows the QFT templates for $p_{12}(s)$, which is the main plant of the velocity control loop. They are calculated using the model and uncertainty defined in Equations CS4.18 and CS4.21, and the array of frequencies defined in Equation CS4.22.

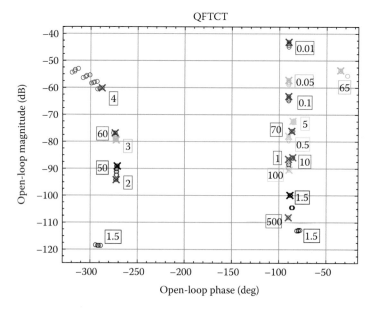

FIGURE CS4.7
QFT templates for $p_{12}(s)$.

CS4.4.2 Control Specifications

Taking into account the dynamics of the radio telescope, the astronomers' objectives and the system limitations, we define the following robust control specifications—with $H_v(s) = 1$:

Type 1: Stability specification

$$|T_1(j\omega)| = \left| \frac{p_{12}(j\omega)G_v(j\omega)}{1 + p_{12}(j\omega)G_v(j\omega)} \right| \leq \delta_1(\omega) = W_s = 1.46 \tag{CS4.24}$$

$$\omega \in [0.01 \ 0.05 \ 0.1 \ 0.5 \ 1 \ 1.5 \ 2 \ 3 \ 4 \ 5 \ 10 \ 50 \ 60 \ 65 \ 70 \ 100 \ 500] \text{ rad/s}$$

which is equivalent to $PM = 40.05°$, $GM = 4.53$ dB—see Equations 2.30 and 2.31.

Type k1: Wind disturbances $T_w(s)$ over $\theta'_m(s)$ via $p_{11}(s)$—Figures CS4.3 and CS4.8.

$$|T_{k1}(j\omega)| = \left| \frac{\dot{\theta}_m(j\omega)}{T_w(j\omega)} \right| = \left| \frac{p_{11}(j\omega)}{1 + p_{12}(j\omega)G_v(j\omega)} \right| \leq \delta_{k1}(\omega) \tag{CS4.25}$$

$$\delta_{k1}(\omega) = \frac{(s/a_{k1})}{(s/a_{k1}) + 1}; a_{k1} = 3 \times 10^5; \omega \in [0.01 \ 0.05 \ 0.1 \ 0.5] \text{ rad/s}$$

CS4.4.3 Controller Design

QFT Bounds

As mentioned, the QFT bounds take into account the plant dynamics, model, uncertainty, and control specifications. Figures CS4.9a and b show the bounds for stability and wind disturbance rejection calculated from Equations CS4.24 and CS4.25, respectively. Figure CS4.10 presents the intersection of the two bounds at each frequency. As we have

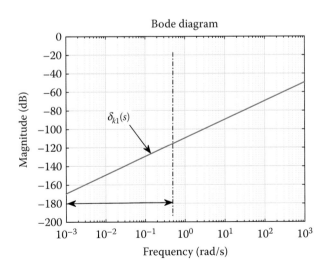

FIGURE CS4.8
Attenuation of wind disturbances $T_w(s)$ over $\theta'_m(s)$—*Type k1*.

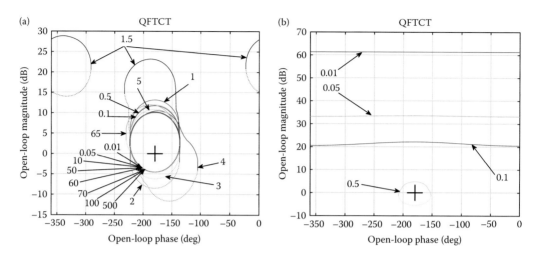

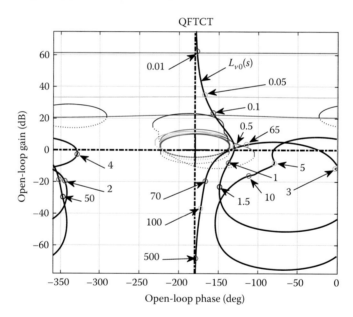

FIGURE CS4.9

QFT bounds. (a) Stability and (b) wind disturbance rejection.

FIGURE CS4.10

QFT bounds and $G_v(s)$ design—loop shaping.

a bound solution for each frequency, these control specifications for the radio telescope are compatible.

Loop Shaping—$G_v(s)$

The design of the feedback controller $G_v(s)$ is carried out on the Nichols chart. It is done by adding poles and zeros until the nominal loop, defined as $L_{v0}(j\omega) = p_{12,0}(j\omega)G_v(j\omega)$, meets the QFT bounds presented in Figure CS4.10.

We start the design of $G_v(s)$ by adding an integrator to obtain zero steady-state error for ramp reference inputs. After this, we increase the magnitude of the controller gain to $k = 1900$, to put $L_{v0}(0.01)$ above the $B(0.01)$ bound. Then, we add a low-frequency zero ($z_1 = 0.2$), to move $L_{v0}(s)$ to the right, and to pass the stability bounds ("circles") through the right.

Afterward, to attenuate the effect of the first flexible mode at 4 rad/s, we add a notch filter $(s^2 + 2\,\zeta_a\,w_n s + w_n^2)/(s^2 + 2\,\zeta_b\,w_n\,s + w_n^2)$, with a natural frequency of $w_n = 4$ rad/s, a damping numerator coefficient of $\zeta_a = 0.1$, and a damping denominator coefficient of $\zeta_b = 1$. This gives a notch depth of $20 \times \log_{10}(\zeta_a/\zeta_b)$ dB $= -20$ dB.

Finally, we add one more additional pole at high frequency ($p_1 = 5$), to filter out high-frequency noise. The final expression for $G_v(s)$ is the PI with a low-pass filter and a notch filter controller shown in Equation CS4.26. The loop shaping is shown in Figure CS4.10.

$$G_v(s) = \frac{1900\left((s/0.2)+1\right)}{s\left((s/5)+1\right)} \frac{(s^2+0.8s+16)}{(s^2+8s+16)} \tag{CS4.26}$$

CS4.4.4 Analysis and Validation of $G_v(s)$

The analysis of the closed-loop stability in the frequency domain is shown in Figure CS4.11a. The dashed line is the stability specification W_s defined in Equation CS4.24. The solid line represents the worst case of all the possible functions $p_{12}G_v/(1 + p_{12}G_v)$ at each frequency due to the model uncertainty. The control system meets the stability specification (the solid line is below the dashed line W_s) for all the plants within the uncertainty, and including the flexible modes at 4.1, 9.35, and 64.8 rad/s.

The analysis of the robust wind disturbance rejection in the frequency domain is shown in Figure CS4.11b. The dashed line is the specification $\delta_{k1}(\omega)$ defined in Equation CS4.25. The solid line represents the worst case of all the possible functions $p_{11}/(1 + p_{12}G_v)$ at each frequency due to the model uncertainty. The control system meets the specification for all the plants within the uncertainty (the solid line is below the dashed line δ_{k1}).

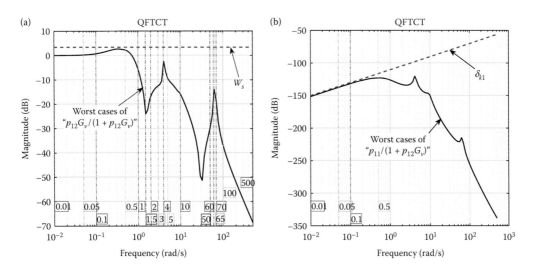

FIGURE CS4.11
Frequency-domain analysis. (a) Stability and (b) wind disturbance rejection.

Additionally, Figures CS4.20b and CS4.32a show the time-domain simulation of the velocity loop under different circumstances. In all cases, the controller $G_v(s)$ achieves a good performance according to the required specifications.

CS4.5 Azimuth Axis—Position Control

This section presents the design of the position control system of the telescope azimuth-axis servo system. The following sections propose a solution for the 2DOF controller, $G_p(s)$ and $F_p(s)$, of the outer loop. See again Figure CS4.3 with the cascade control diagram. We use the velocity controller $G_v(s)$ designed in the previous section for the inner loop.

CS4.5.1 Plant

The plant $p_p(s)$ to be controlled by the position controller $G_p(s)$ is composed of the inner veloc-ity loop and the additional integrator, as shown in Figure CS4.3. In this way, the plant $p_p(s)$ is

$$p_p(s) = \frac{p_{12}(s)\,M(s)\,G_v(s)}{1 + p_{12}(s)\,M(s)\,G_v(s)\,H_v(s)}\left(\frac{1}{s}\right) \tag{CS4.27}$$

CS4.5.2 QFT Templates

Figure CS4.12 shows the QFT templates for $p_p(s)$—Equation CS4.27. They are calculated using the $p_p(s)$ plant model, the uncertainty defined in Equation CS4.21, and the array of frequencies of interest selected in Equation CS4.22.

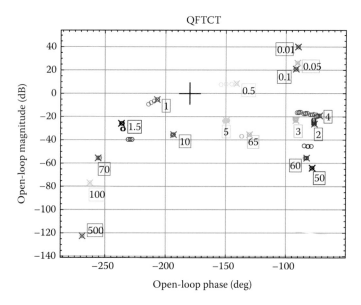

FIGURE CS4.12
QFT templates for $p_p(s)$.

CS4.5.3 Control Specifications

Taking into account the dynamics of the radio telescope, the astronomers' objectives, and the system limitations, we define the following robust control specifications for the position control system—with $H_p(s) = 1$:

Type 1: Stability specification

$$|T_1(j\omega)| = \left|\frac{p_p(j\omega)G_p(j\omega)}{1 + p_p(j\omega)G_p(j\omega)}\right| \leq \delta_1(\omega) = W_s = 1.46 \tag{CS4.28}$$

$$\omega \in [0.01\ \ 0.05\ \ 0.1\ \ 0.5\ \ 1\ \ 1.5\ \ 2\ \ 3\ \ 4\ \ 5\ \ 10\ \ 50\ \ 60\ \ 65\ \ 70\ \ 100\ \ 500]\ \text{rad/s}$$

which is equivalent to $PM = 40.05°$, $GM = 4.53$ dB—see Equations 2.30 and 2.31.

Type 3: Sensitivity or disturbances at plant output specification—see Figure CS4.13a.

$$|T_3(j\omega)| = \left|\frac{\theta_m(j\omega)}{d_o(j\omega)}\right| = \left|\frac{1}{1 + p_p(j\omega)G_p(j\omega)}\right| \leq \delta_3(\omega) = \frac{(s/a_d)}{(s/a_d) + 1}; \ a_d = 0.5 \tag{CS4.29}$$

$$\omega \in [0.01\ \ 0.05\ \ 0.1\ \ 0.5]\ \text{rad/s}$$

Type 6: Reference tracking specification—see Figure CS4.13b.

$$\delta_{6\_lo}(\omega) < |T_6(j\omega)| = \left|F(j\omega)\frac{p_p(j\omega)G_p(j\omega)}{1 + p_p(j\omega)G_p(j\omega)}\right| \leq \delta_{6\_up}(\omega) \tag{CS4.30}$$

$$\omega \in [0.01\ \ 0.05\ \ 0.1\ \ 0.5\ \ 1]\ \text{rad/s}$$

$$\delta_{6\_lo}(s) = \frac{1}{\left[(s/a_L) + 1\right]^2}; \ a_L = 0.292 \tag{CS4.31}$$

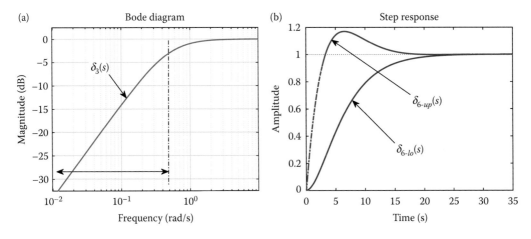

FIGURE CS4.13
Control specifications. (a) Disturbance rejection specifications: δ_3 and (b) reference tracking: δ_{6-up}, δ_{6-lo}.

$$\delta_{6\_up}(s) = \frac{[(s/a_U)+1]}{[(s/\omega_n)^2+(2\zeta s/\omega_n)+1]}; \quad a_U = 0.215; \ \zeta = 0.8; \ \omega_n = \frac{1.25\,a_U}{\zeta} \quad \text{(CS4.32)}$$

CS4.5.4 Controller Design

QFT Bounds

As mentioned, the QFT bounds take into account the plant dynamics, model, uncertainty, and control specifications. Figures CS4.14a through c, show the bounds for stability, sensitivity, and reference tracking for the position loop, see Equations CS4.28 through CS4.32. Figure CS4.14d presents the intersection of the three bounds at each frequency of interest. As we have a bound solution for each frequency, the selected control specifications for the position loop are compatible.

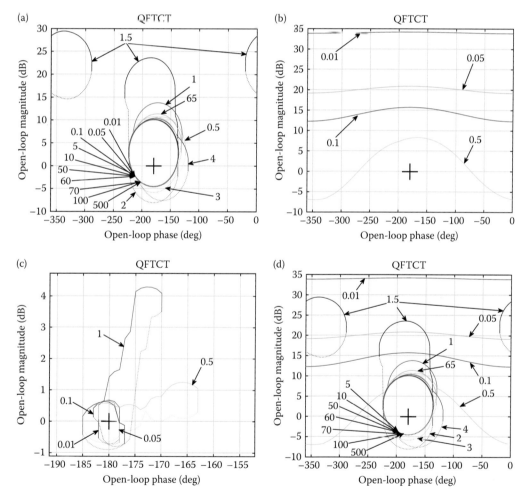

FIGURE CS4.14

QFT bounds. (a) Stability, (b) sensitivity, (c) reference tracking, and (d) intersection of bounds.

Loop Shaping—$G_p(s)$

The design of the feedback controller $G_p(s)$ is carried out on the Nichols chart. It is done by adding poles and zeros until the nominal loop, defined as $L_{p0}(j\omega) = p_{p,0}(j\omega)G_p(j\omega)$, meets the QFT bounds presented in Figure CS4.14d.

We start the design of $G_p(s)$ by increasing the magnitude of the controller gain to $k = 0.5$, to put $L_{p0}(0.01)$ above the $B(0.01)$ bound. Next, we add a low-frequency zero ($z_1 = 0.6$), to move $L_{p0}(s)$ to the right, and to pass the stability bounds ("circles") through the right.

Finally, we add one more additional pole at high frequency ($p_1 = 25$), to filter out high-frequency noise. The final expression for $G_p(s)$ is the lead controller shown in Equation CS4.33. The loop shaping is shown in Figure CS4.15.

$$G_p(s) = \frac{0.5\left((s/0.6)+1\right)}{\left((s/25)+1\right)} \tag{CS4.33}$$

Prefilter Design—$F_p(s)$

With the controller $G_p(s)$ just designed—see Equation CS4.33, the reference tracking specifications presented in Equations CS4.30 through CS4.32, and the plant model $p_p(s)$ described in Equation CS4.27, we design a prefilter $F_p(s)$ to assure that all the input/output functions $p_p(s)G_p(s)F_p(s)/[1 + p_p(s)G_p(s)]$, from low frequencies up to 1 rad/s are inside the band defined by the limits $\delta_{6-up}(\omega)$ and $\delta_{6-lo}(\omega)$. The designed prefilter is shown in Equation CS4.34, and the input/output functions and limits in Figure CS4.16.

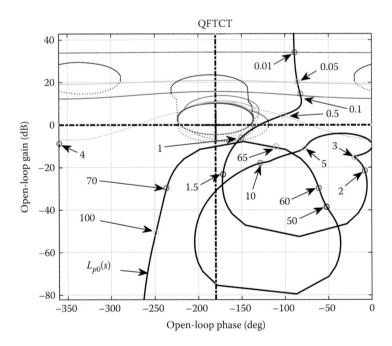

FIGURE CS4.15
QFT bounds and $G_p(s)$ design—loop shaping.

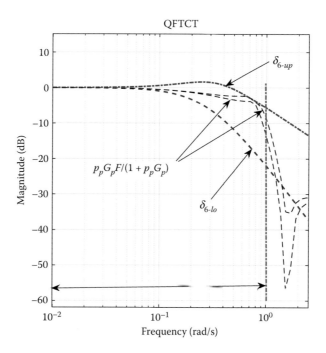

FIGURE CS4.16

Prefilter $F(s)$ for reference tracking specs: $\delta_{6\text{-}lo}(\omega) \leq |T_6| \leq \delta_{6\text{-}up}(\omega)$.

$$F_p(s) = \frac{\big((s/0.3)+1\big)}{\big((s/0.38)+1\big)^2} \tag{CS4.34}$$

CS4.5.5 Analysis and Validation of $G_p(s)$ and $F_p(s)$

The analysis of the closed-loop stability of the position loop in the frequency domain is shown in Figure CS4.17a. The dashed line is the stability specification W_s defined in Equation CS4.28. The solid line represents the worst case of all the possible functions $p_pG_p/(1 + p_pG_p)$ at each frequency due to the model uncertainty. The control system meets the stability specification (the solid line is below the dashed line W_s) for all the plants within the uncertainty.

The analysis of the robust output disturbance rejection of the position loop in the frequency domain is shown in Figure CS4.17b. The dashed line is the specification $\delta_3(\omega)$ defined in Equation CS4.29. The solid line represents the worst case of all the possible functions $1/(1 + p_pG_p)$ at each frequency due to the model uncertainty. The control system meets the specification for all the plants within the uncertainty (the solid line is below the dashed line δ_3 from 0 to 0.1 rad/s).

The time-domain analysis of the reference tracking specification of the position loop is presented in Figure CS4.18a. The figure shows the limits $\delta_{6\text{-}up}(\omega)$ and $\delta_{6\text{-}lo}(\omega)$—see Equations CS4.31 and CS4.32, and the time responses of the motor position $\theta_m(t)$ to a unitary step reference $\theta_{mRef}(t)$. We simulate multiple cases of the closed-loop transfer function $p_pG_pF_p/(1 + p_pG_p)$ within the uncertainty—see Figure CS4.3, with $H_p(s) = 1$. The results are given per unit. The control system meets the specification (is between the upper and lower limits) in all cases.

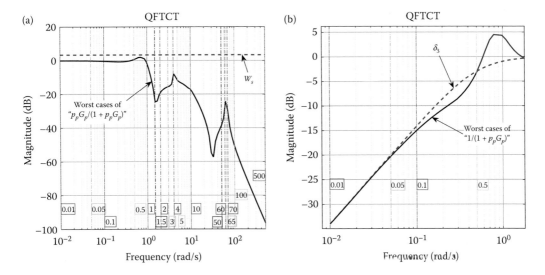

FIGURE CS4.17
Frequency-domain analysis. (a) Stability and (b) sensitivity.

Additionally, the time-domain analysis of the output disturbance rejection is shown in Figure CS4.18b. The figure presents the time responses of the motor position $\theta_m(t)$ to a unitary disturbance at the output of the plant. We simulate multiple cases of $1/(1 + p_pG_p)$ within the uncertainty, also with $H_p(s) = 1$. The zero level is the non-perturbed case. The control system achieves a good disturbance rejection in all cases.

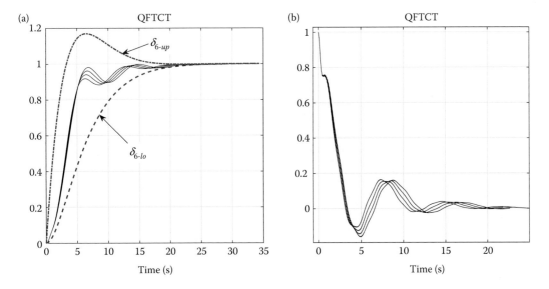

FIGURE CS4.18
Time domain. (a) Reference tracking $\theta_m = p_pG_pF_p/(1 + p_pG_p) \theta_{mRef}$ and (b) output disturbance rejection $\theta_m = 1/(1 + p_pG_p) d_o$.

CS4.6 Simulation: Position and Velocity Loops

This section presents the time-domain simulation of the complete cascade control system presented in Figure CS4.3, including the velocity controller $G_v(s)$ in the inner loop—Equations CS4.26, and the position controller $G_p(s)$ and $F_p(s)$ in the outer loop—Equations CS4.33 and CS4.34. We simulate the nominal plants—see Equation CS4.23, and include the velocity limitation of the azimuth axis, which is 0.67 deg/s or 0.0117 rad/s in terms of the dish movement $d\theta_d/dt$, or 33.52 deg/s or 0.585 rad/s in terms of the motor movement $d\theta_m/dt$—see Figure CS4.19.

The motor reference input θ_{mRef} changes as a step at $t = 1$ s, from 0° to 750°, which in terms of the dish (θ_{dRef}) means from 0° to 15°.

Also, at $t = 60$ s we introduce a step wind disturbance T_W from 0 to 3×10^5 Nm, and keep this value until the end of the simulation. This torque T_W is equivalent to a wind gust of about $v = 6.12$ m/s, and considers the worst case scenario where the wind affects only half of the surface of the dish (taken as a circular dish), and where the aerodynamic thrust coefficient is $C_T = 0.1$. That is to say

$$T_w = F_T\, r_p = \frac{1}{2}\rho_{air}\left(\frac{\pi\, r_{dish}^2}{2}\right)C_T\, v^2\left(\frac{2}{3}r_{dish}\right) \tag{CS4.35}$$

where F_T is the thrust force applied by the wind on the dish, and r_p the distance from the center of the dish to the center of pressure, or point where the equivalent lumped force F_T is applied, $r_p = (2/3)\, r_{dish}$. Also, the density of the air is $\rho_{air} = 1.225$ kg/m^3, and the radius of the dish $r_{dish} = 50$ m.

Figure CS4.20a shows the results of the position loop controlling the motor angle $\theta_m(t)$ and Figure CS4.20b the corresponding velocity loop controlling the motor velocity $d\theta_m(t)/dt$ at the same time. The two figures are in degrees of the motor, and confirm a good performance in both reference tracking and wind disturbance rejection.

Additionally Figures CS4.21a and b show, respectively, the complete movement and the details of the dish position in the same simulation. The peak of the excursion of the dish angle $\theta_d(t)$ under the wind gust of 6.12 m/s is 0.35°. Also, notice that although the steady-state error of the motor "$\theta_{dRef} - \theta_m/R$" is zero, the dish $\theta_d(t)$ ends up with a permanent error of 0.06° due to the constant force applied by the wind.

The effect of the wind in the dish angle is a limiting factor that compromises the quality of astronomical observations. In order to improve the results of the wind disturbance rejection achieved here, the next section proposes a new design based on NDC methodology introduced in Chapter 7.

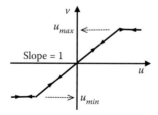

FIGURE CS4.19
Velocity saturation. In dish: 0.6704 deg/s = 0.0117 rad/s. In motor: 33.52 deg/s = 0.585 rad/s. Slope = 1. $u_{max} = +0.585$, $u_{min} = -0.585$.

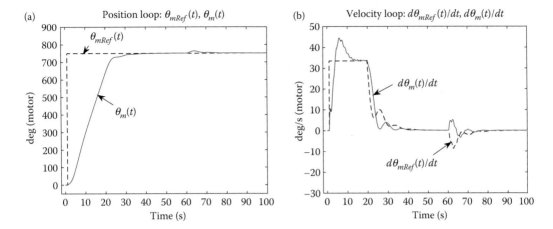

FIGURE CS4.20
Simulation with $G_p(s)$, $F_p(s)$, and $G_v(s)$. (a) Position loop, $\theta_m(t)$. (b) Velocity loop, $d\theta_m(t)/dt$.

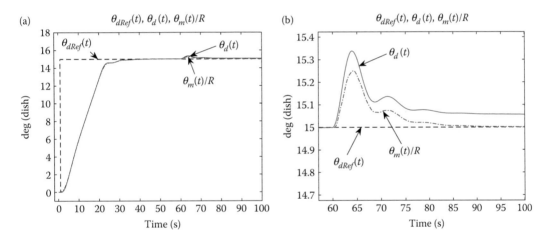

FIGURE CS4.21
Simulation with $G_p(s)$, $F_p(s)$, and $G_v(s)$. Dish position $\theta_d(t)$ and motor $\theta_m(t)/R$. (a) Full signals and (b) zoom at $t = 60$–100 s.

CS4.7 Improving with NDC

This section presents a nonlinear dynamic control (*NDC*) solution for the position loop of the azimuth axis of the radio telescope. The design is based on the theory and examples developed in Chapter 7. In the next sub-sections, we follow the 5-step methodology discussed for the *NDC*—see Section 7.5.

Step A: Controller Structure

The *NDC* structure proposed for the position control system is shown in Figure CS4.22. It consists of two parallel channels, $G_{p1}(s) N_1$ and $G_{p2}(s) N_2$. The first channel $G_{p1}(s) N_1$ is the aggressive one, and the second channel $G_{p2}(s) N_2$ the moderate one. We take $G_{p2}(s) = G_p(s)$, which is the position controller already designed in the previous section. Also, we keep

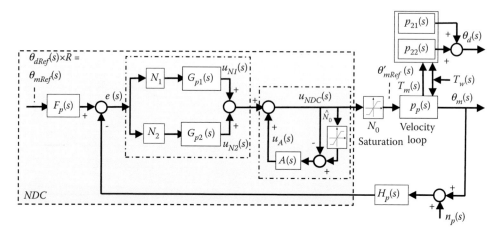

FIGURE CS4.22
NDC block diagram $H_p(s) = 1$.

the same prefilter $F_p(s)$ designed previously. The structure also includes an anti-wind-up nonlinear inner loop for the first channel, with a linear transfer function $A(s)$ and a saturation model $\hat{N}_0$. Notice that it is not necessary any anti-wind-up loop for the second channel, as G_{p2} does not include any integrator—see Equation CS4.33.

The nonlinear functions N_1 and N_2 are defined in Figure CS4.23—see also Figures 7.19 and 7.20, respectively, where $N_1 = N_{01}$ is defined as in Equations 7.17 and 7.18, and $N_2 = N_{02}$ as in Equations 7.19 and 7.20.

The parameters selected for these two nonlinearities are: $\delta_h = 0.75$ and $\delta_l = 0.60$. Notice that $N_{01} + N_{02} = 1$. As a summary, the system has four nonlinearities: the actuator saturation N_0, the actuator saturation model $\hat{N}_0$ and the two control channels N_1 and N_2.

Step B.1: Control Elements for First Channel, $G_{p1}(s)$—Aggressive
The control specifications for the position controller $G_{p1}(s)$ are shown in Equations CS4.36 and CS4.37. They are

Type 1: Stability specification

$$|T_1(j\omega)| = \left|\frac{p_p(j\omega)G_{p1}(j\omega)}{1 + p_p(j\omega)G_{p1}(j\omega)}\right| \le \delta_1(\omega) = W_s = 4.0 \qquad \text{(CS4.36)}$$
$$\omega \in [0.01\ \ 0.05\ \ 0.1\ \ 0.5\ \ 1\ \ 1.5\ \ 2\ \ 3\ \ 4\ \ 5\ \ 10\ \ 50\ \ 60\ \ 65\ \ 70\ \ 100\ \ 500]\ \text{rad/s}$$

which is equivalent to $PM = 14.36°$, $GM = 1.94$ dB—see Equations 2.30 and 2.31.

Type 3: Sensitivity or disturbances at plant output specification

$$|T_3(j\omega)| = \left|\frac{\theta_m(j\omega)}{d_o(j\omega)}\right| = \left|\frac{1}{1 + p_p(j\omega)G_{p1}(j\omega)}\right| \le \delta_3(\omega) = \frac{(s/a_d)}{(s/a_d) + 1}; \ a_d = 3.3333 \qquad \text{(CS4.37)}$$
$$\omega \in [0.01\ \ 0.05\ \ 0.1\ \ 0.5]\ \text{rad/s}$$

As the main objective for the NDC is to achieve a much better wind disturbance rejection, these two specifications for $G_{p1}(s)$ are more aggressive than the ones defined for

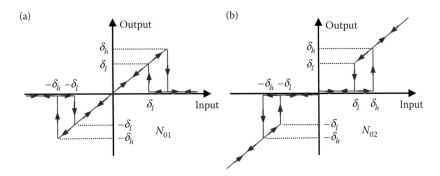

FIGURE CS4.23
Nonlinear elements for the *NDC*: (a) $N_1 = N_{01}$. (b) $N_2 = N_{02}$.

$G_{p2}(s) = G_p(s)$ in the previous section—see Equation CS4.29. Figure CS4.24 compares the disturbance rejection specifications required for $G_{p1}(s)$ and $G_{p2}(s)$.

The QFT bounds are calculated taking into account the $p_p(s)$ transfer function—Equation CS4.27, the parameter uncertainty given by Equation CS4.21, the array of frequencies of interest—Equation CS4.22, and the stability and output disturbance rejection specifications given by Equations CS4.36 and CS4.37. Figures CS4.25a and b show the QFT bounds of each specification, and Figure CS4.26 the intersection of bounds, which is compatible at each frequency.

To meet the specifications, we design a sixth order structure for the $G_{p1}(s)$ controller, as shown in Equation CS4.38. The QFT bounds and the loop shaping of $L_{p0}(s) = p_{p0}(s)G_{p1}(s)$ are shown in Figure CS4.26.

$$G_{p1}(s) = \frac{0.04\,\big((s/0.134)+1\big)\big((s/0.25)+1\big)\big((s/0.3)+1\big)\big((s/0.33)+1\big)}{s^2\,\big((s/50)+1\big)\big((s/70)+1\big)}\,\frac{(s^2+0.13s+4225)}{(s^2+130s+4225)}$$

(CS4.38)

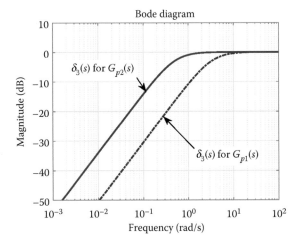

FIGURE CS4.24
Comparison disturbance rejection specifications for $G_{p1}(s)$ and $G_{p2}(s)$.

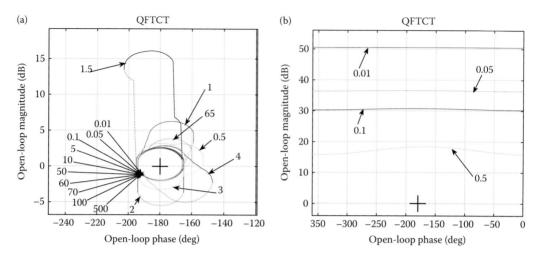

FIGURE CS4.25
QFT bounds for $G_{p1}(s)$. (a) Stability and (b) sensitivity.

The analysis of the closed-loop stability of the position loop in the frequency domain is shown in Figure CS4.27a. The dashed line is the stability specification W_s defined in Equation CS4.36. The solid line represents the worst case of all the possible functions $p_p G_{p1}/(1 + p_p G_{p1})$ at each frequency due to the model uncertainty. The control system meets the stability specification (the solid line is below the dashed line W_s) for all the plants within the uncertainty.

The analysis of the robust output disturbance rejection of the position loop in the frequency domain is shown in Figure CS4.27b. The dashed line is the specification $\delta_3(\omega)$

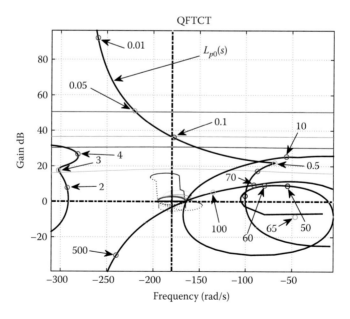

FIGURE CS4.26
QFT bounds and $G_{p1}(s)$ design—loop shaping.

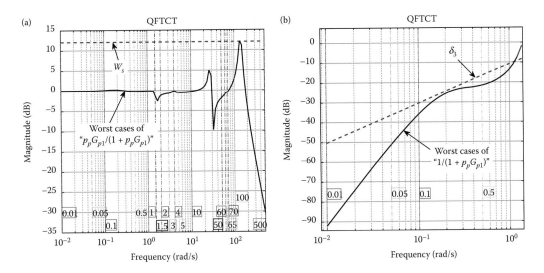

FIGURE CS4.27
Frequency-domain analysis, $G_{p1}(s)$. (a) Stability and (b) sensitivity.

defined in Equation CS4.37. The solid line represents the worst case of all the possible functions $1/(1 + p_p G_{p1})$ at each frequency due to the model uncertainty. The control system meets the specification for all the plants within the uncertainty (the solid line is below the dashed line δ_3).

Step B.2: Control Elements for Second Channel, $G_{p2}(s)$ & Prefilter $F_p(s)$—Moderate

The controller proposed for the second channel (moderate) is the same expression we used in Section CS4.6, Equation CS4.33,

$$G_{p2}(s) = G_p(s) = \frac{0.5\left((s/0.6)+1\right)}{\left((s/25)+1\right)} \tag{CS4.39}$$

The purpose of implementing two channels, $G_{p1}(s)$ and $G_{p2}(s)$, is to improve the system performance beyond the linear limitations. The element $G_{p1}(s)$ has been designed as an active/aggressive controller to work with small errors (see $N_1 = N_{01}$) and the element $G_{p2}(s)$ as a moderate controller to work with large errors (see $N_2 = N_{02}$). The Bode diagram shown in Figure CS4.28 compares the two controllers, the aggressive $G_{p1}(s)$ and the moderate $G_{p2}(s)$. Notice that the magnitude of $G_{p1}(s)$ is much larger for almost all frequencies, except at $\omega = 65$ rad/s (motor shaft resonance), where a notch filter reduces $G_{p1}(s)$ to the $G_{p2}(s)$ level. Also, at low frequency, two integrators make $G_{p1}(s)$ much more active rejecting external disturbances.

Finally, we keep the same prefilter designed in Section CS4.6, Equation CS4.34, for the NDC structure. Again, the expression for this prefilter is

$$F_p(s) = \frac{\left((s/0.3)+1\right)}{\left((s/0.38)+1\right)^2} \tag{CS4.40}$$

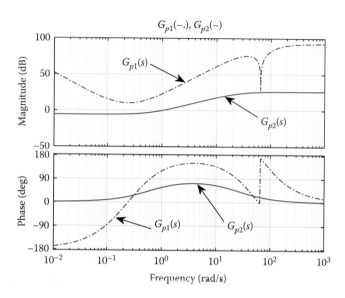

FIGURE CS4.28
Controllers: aggressive $G_{p1}(s)$, moderate $G_{p2}(s)$.

Step C: Inner Loop and Control Element A(s)

The anti-wind-up inner loop is necessary for the first channel, as $G_{p1}(s)$ includes two integrators. At the same time, the second channel does not need this anti-wind-up inner loop, as $G_{p2}(s)$ does not have any integrator. Then, the expression for $A(s)$ is

$$A(s) = \frac{k_a}{s}, \text{ being:}$$
$$\rightarrow \text{ case only } G_{p1} : k_a = 1$$
$$\rightarrow \text{ case only } G_{p2} : k_a = 0 \qquad\qquad\qquad (CS4.41)$$
$$\rightarrow \text{ case } NDC : \begin{cases} \text{for } G_{p1}, k_a = 0.020 \\ \quad \text{for } G_{p2}, k_a = 0 \end{cases}$$

Step D: Isolines—Stability Analysis for the Complete System

The describing functions *DFs* for each nonlinearity, both in the controller and in the plant, are given by the following expressions:

- $N_0 = N_{05}$, actuator saturation—Equations 7.25 and 7.26.
- $\hat{N}_0 = N_0 = N_{05}$, controller actuator saturation model—Equations 7.25 and 7.26.
- $N_1 = N_{01}$, controller first channel nonlinearity—Equations 7.17 and 7.18.
- $N_2 = N_{02}$, controller second channel nonlinearity—Equations 7.19 and 7.20.

The isolines are calculated cascading the linear and nonlinear elements. They include the *DF* of each nonlinearity at their specific position in the loop, from the input (right of the expression) to the output (left of the expression), as expressed by Equation CS4.42

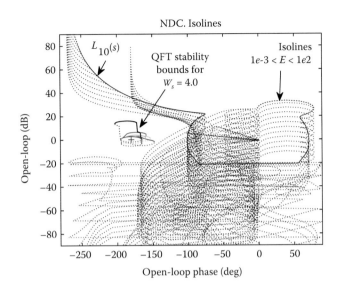

FIGURE CS4.29
Isolines in the Nichols chart, for error E from 10^{-3} to 10^2, ω from 10^{-4} to 10^2 rad/s, and QFT stability bounds for $W_s = 4.0$ (or $PM = 14.3$ deg).

$$L = p_p(s)DF_{N0}\left[\left(\frac{1}{1 + A(s)DF_{\hat{N}0}}\right)G_{p1}(s)DF_{N1} + G_{p2}(s)DF_{N2}\right] \qquad \text{(CS4.42)}$$

The algorithm that calculates the isolines for every amplitude E of the input and frequency ω is similar to the one included in Appendix 7. Figure CS4.29 represents in the Nichols chart the isolines L—Equation CS4.42, for the input errors $10^{-3} \leq E \leq 10^2$ and for the frequencies $10^{-4} \leq \omega \leq 10^2$ rad/s. The figure is calculated for the nominal plant—Equation CS4.23. It also shows the QFT stability bounds found for this nominal plant and $W_s = 4.0$ (or $PM = 14.3°$). As there is no RHP poles and the isolines do not enter into the QFT bounds, the *NDC* system is stable for all the cases within the model uncertainty. The method gives *sufficient conditions* for stability—see also Chapter 3.

Step E: Simulations and Discussion—Position and Velocity Loops

Finally, Figures CS4.30 through CS4.33 show the results of the time-domain simulation of the complete cascade control system with the *NDC* presented in Figure CS4.22, including the inner velocity loop shown in Figure CS4.3. We simulate the nominal plants—see Equation CS4.23, and include the velocity limitation of the azimuth axis, which is ± 0.585 rad/s in terms of the motor movement $d\theta_m/dt$—see Figure CS4.19. The expressions for the controllers are

- Inner loop.
  Velocity controller $G_v(s)$: Equation CS4.26
- Outer loop.
  Prefilter $F_p(s)$: Equation CS4.40
  Channel 1. $G_{p1}(s)$: Equation CS4.38, N_1: Equations 7.17 and 7.18, $\delta_h = 0.75$, $\delta_l = 0.60$

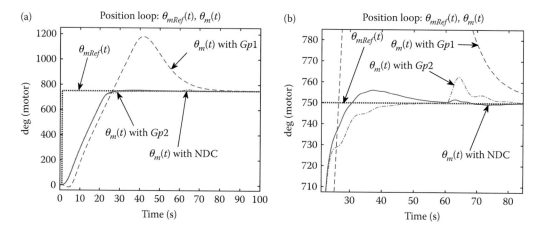

FIGURE CS4.30

Simulation, motor angle $\theta_m(t)$. Position control loop. Strategies: aggressive G_{p1}, moderate G_{p2}, and NDC solutions. (a) Complete simulation and (b) zoom of the disturbance rejection part.

Channel 2. $G_{p2}(s)$: Equation CS4.39, N_2: Equations 7.19 and 7.20, $\delta_h = 0.75$, $\delta_1 = 0.60$

Anti-wind-up. $A(s)$: Equation CS4.41, $\hat{N}_0$: Equations 7.25 and 7.26

We repeat the same simulation conditions of Section CS4.7. The motor reference input θ_{mRef} changes as a step at $t = 1$ s, from 0° to 750°, which in terms of the dish (θ_{dRef}) means from 0° to 15°. Also, at $t = 60$ s we introduce a step wind disturbance T_W from 0 to 3×10^5 Nm, and keep it until the end of the simulation.

Figure CS4.30 shows the simulation of the position loop—motor angle $\theta_m(t)$, including the results with only the aggressive controller $G_{p1}(s)$—dashed line, with only the moderate controller $G_{p2}(s)$—dashed-dotted line, and with the NDC solution—solid line. The controllers $G_{p2}(s)$ and NDC perform better than $G_{p1}(s)$ on reference tracking—see Figure CS4.30a. Also, as we intended, the NDC performs better than the original $G_{p2}(s)$ on disturbance rejection, as shown in Figure CS4.30b.

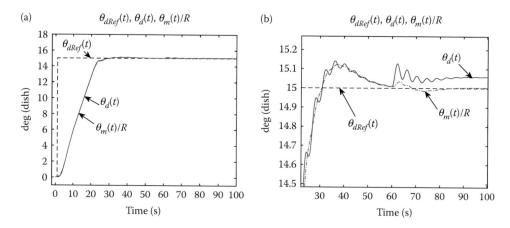

FIGURE CS4.31

Simulation, dish angle $\theta_d(t)$. NDC solution. (a) Complete simulation and (b) zoom of the disturbance rejection part.

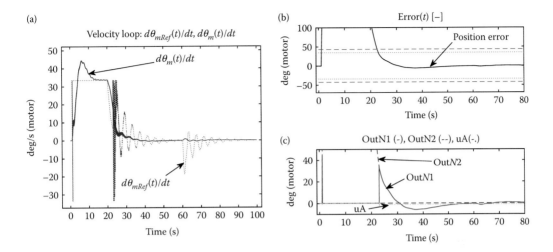

FIGURE CS4.32

Simulation *NDC* combining G_{p1} and G_{p2}. (a) Motor velocity loop $d\theta_m(t)/dt$, (b) motor angle error, and (c) switching signals $uN1$, $uN2$, and uA.

Additionally, Figures CS4.31a and b show, respectively, the complete movement and the details of the dish position with the *NDC* solution. Also with the *NDC* solution, Figure CS4.32a presents the corresponding velocity loop controlling the motor velocity $d\theta_m(t)/dt$ simultaneously, and Figure CS4.32b the motor angle error $e(t)$ and switching signals $uN1(t)$, $uN2(t)$, $uA(t)$—see signals in Figure CS4.22.

Finally, Figure CS4.33 shows the details of the dish angle $\theta_d(t)$ and motor angle $\theta_m(t)/R$ in dish degrees when we use the original moderate controller $G_{p2}(s)$—dashed-dotted line, and the *NDC* solution—solid line. The peak of the excursion of the dish angle $\theta_d(t)$ under the wind gust of 6.12 m/s is reduced from the 0.35° with $G_{p2}(s)$ to 0.13° with *NDC*. Also, notice that although the steady-state error of the motor "$\theta_{dRef}-\theta_m/R$" is always zero, the

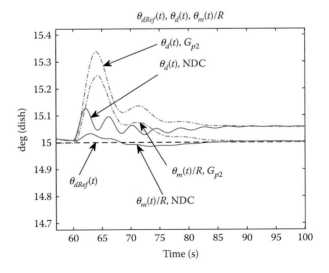

FIGURE CS4.33

Disturbance rejection, dish angle $\theta_d(t)$ and motor angle $\theta_m(t)/R$. Strategies: G_{p2} (dashed-dotted line) and *NDC* (solid line).

dish $\theta_d(t)$ ends up with a permanent error of 0.06° due to the constant force applied by the wind in both cases, and also because the feedback sensor used in this case study is $\theta_m(t)$ and not $\theta_d(t)$.

CS4.8 Summary

This case study has designed two robust QFT control solutions for the velocity and position loops of a radio-telescope servo system. The first solution is based on the classical QFT methodology and the second one proposes an NDC strategy. Both designs accomplish five simultaneous control objectives:

- Stability
- Tracking of the azimuth-axis telescope position
- Regulation of the azimuth-axis telescope velocity
- Rejection of unpredictable wind disturbances
- Reduction of dish and feed-arm vibration

Case Study 5: Attitude and Position Control of Spacecraft Telescopes with Flexible Appendages

CS5.1 Introduction

With high priority science objectives to break the current barriers of our understanding of the universe, and dealing with weight limitations of launch vehicles for cost-effective access to space, new NASA and ESA missions involve both formation flying technology and satellites with large flexible structures. See for instance the Terrestrial Planet Finder, Stellar and Planet Imager, Life Finder, Darwin, Lisa, and Proba-3 missions, etc.

Control of spacecraft with large flexible structures and very demanding astronomical specifications involves significant difficulties due to the combination of flexible modes with small damping, model uncertainty, and coupling among the inputs and outputs.

This case study deals with the design of a MIMO QFT robust control strategy to simultaneously regulate the position and attitude of a spacecraft telescope with large flexible appendages. The spacecraft is part of a multiple formation flying constellation of ESA cornerstone project: the Darwin mission (Figure CS5.1).[211,216]

The scientific objectives of this mission require high-performance control specifications, including micrometer accuracy for position and milli-arc-second precision for attitude, high disturbance rejection properties, loop-coupling attenuation, and low controller complexity and order. The dynamics of the spacecraft presents a 6-inputs × 6-outputs MIMO plant, with 36 transfer functions of 50th order, significant model uncertainty, and high-loop interactions introduced by the flexible modes of the low-stiffness appendages.

The results presented in this case study are part of a project developed by the author for the international benchmark proposed by ESA for the design of the control system of the Darwin mission. The design introduced here is based on the MIMO QFT methodology (Chapter 8), and got the best performance in that international competition.

CS5.2 System Description and Modeling

The Darwin mission consists of one master satellite (central hub) and three to six telescopes flying in formation (Figure CS5.1). They will operate together to analyze the atmosphere of remote planets through appropriate spectroscopy techniques. The mission will employ nulling interferometry to detect dim planets close to bright stars. The infrared light collected by the free-flying telescopes will be recombined inside the hub-satellite in such a way that the light from the central star will be cancelled out by destructive interference, allowing the much fainter planets to stand out. That interferometry technology requires very accurate and stable positioning of the spacecraft in the constellation, which puts high

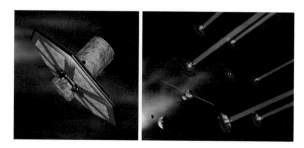

FIGURE CS5.1
Spacecraft telescopes with large flexible appendages flying in formation. (Courtesy ESA.)

demands on the attitude and position control system. Instead of an orbit around the Earth, the mission will be placed further away, at a distance of 1.5 million kilometers from Earth, in the opposite direction from the Sun (Earth–Sun Lagrangian Point L2).

Each telescope flyer is cylindrically shaped (2 m diameter, 2 m height) and weighs 500 kg. In order to protect the instrument from sunlight, it is equipped with a sunshield modeled with six large flexible beams (4 m long and 7 kg) attached to the rigid structure (see Figure CS5.2; beam end-point coordinates in brackets).

For every beam, two different frequencies for the first modes along Y and Z beam axes are considered. Their frequency can vary from 0.05 to 0.5 Hz, with a nominal value of 0.1 Hz, and their damping can vary from 0.1% to 1%, with a nominal value of 0.5%. As regards spacecraft mass and inertia, the corresponding uncertainty around their nominal value is of 5%.

Based on this description, and using a mechanical modeling formulation for multiple flexible appendages of a rigid body spacecraft, the open-loop transfer function matrix representation of the flyer is given in Equation CS5.1 and Figure CS5.3, where x, y, z are the

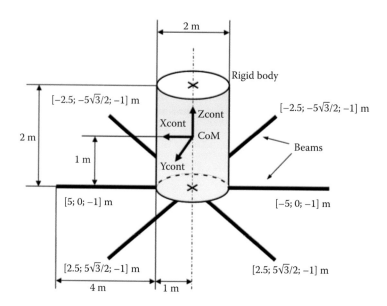

FIGURE CS5.2
Spacecraft description. (Adapted from Garcia-Sanz, M. et al. 2008. *J. ASME J. Dyn. Syst. Meas. Control*, 130, 011006-1–011006-15.)

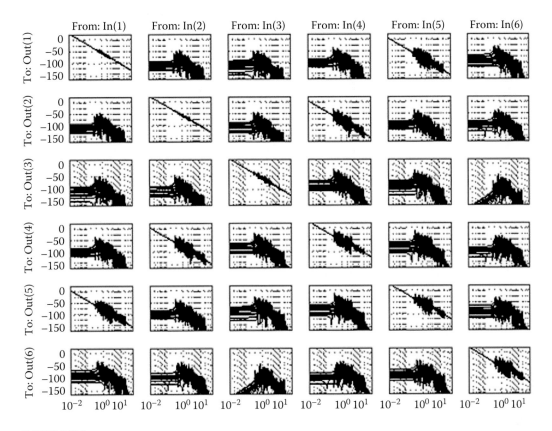

FIGURE CS5.3
Spacecraft model dynamics. (Adapted from Garcia-Sanz, M. et al. 2008. *J. ASME J. Dyn. Syst. Meas. Control*, 130, 011006-1–011006-15.)

position coordinates; φ, θ, ψ are the corresponding attitude angles; F_x, F_y, F_z are the force inputs; T_φ, T_θ, T_ψ are the torque inputs; and where each $p_{ij}(s)$, $i, j = 1,...,6$, is a 50th order Laplace transfer function with uncertainty.

$$\begin{bmatrix} x(s) \\ y(s) \\ z(s) \\ \phi(s) \\ \theta(s) \\ \psi(s) \end{bmatrix} = \begin{bmatrix} p_{11}(s) & p_{12}(s) & p_{13}(s) & p_{14}(s) & p_{15}(s) & p_{16}(s) \\ p_{21}(s) & p_{22}(s) & p_{23}(s) & p_{24}(s) & p_{25}(s) & p_{26}(s) \\ p_{31}(s) & p_{32}(s) & p_{33}(s) & p_{34}(s) & p_{35}(s) & p_{36}(s) \\ p_{41}(s) & p_{42}(s) & p_{43}(s) & p_{44}(s) & p_{45}(s) & p_{46}(s) \\ p_{51}(s) & p_{52}(s) & p_{53}(s) & p_{54}(s) & p_{55}(s) & p_{56}(s) \\ p_{61}(s) & p_{62}(s) & p_{63}(s) & p_{64}(s) & p_{65}(s) & p_{66}(s) \end{bmatrix} \begin{bmatrix} F_x(s) \\ F_y(s) \\ F_z(s) \\ T_\phi(s) \\ T_\theta(s) \\ T_\psi(s) \end{bmatrix} \quad \text{(CS5.1)}$$

The Bode diagram of the plant (Figure CS5.3) shows the dynamics of the 36 matrix elements. Each of them and the MIMO system (matrix) itself are minimum phase. The flexible modes introduced by the appendages (second-order dipoles) affect all the elements around the frequencies $\omega = [0.19, 10]$ rad/s. The diagonal elements $p_{ii}(s)$, $i = 1,...,6$, and the elements $p_{15}(s)$, $p_{51}(s)$, $p_{24}(s)$, and $p_{42}(s)$ are double integrators with additional flexible modes.

CS5.3 Control Specifications

The main objective of the spacecraft control system is to meet a set of astronomical require-ments, which need to keep the flying telescope pointing at both the observed space target and the central hub-satellite. This set of specifications leads to some additional engineer-ing requirements (bandwidth, saturation limits, noise rejection, etc.) and also needs some complementary control requirements (stability, low loop interaction, low controller com-plexity and order, etc.). Specifically, the requirements are

1. Astronomical specifications:
    a. *Position accuracy*: maximum absolute error ≤ 1 μm, and standard deviation ≤ 0.33 μm for x, y, z axes.
    b. *Pointing accuracy*: maximum absolute error ≤ 25 mas, and standard deviation ≤ 8.5 mas for φ, θ, ψ axes.
2. Engineering specifications:
    a. *Bandwidth*: ∼0.01 Hz for all axes.
    b. *Saturation limits*: F_x, F_y, F_z maximum force $= 150$ μN, T_φ, T_θ, T_ψ maximum torque $= 150$ μNm.
    c. *High frequency noise rejection*: high roll-off after the bandwidth.
3. Control specifications:
    a. Loop interaction: minimum.
    b. Rejection of flexible modes: maximum.
    c. Controller complexity and order: minimum.

To achieve these goals, the astronomical, engineering, and control specifications are translated into frequency-domain requirements. This translation into the frequency domain is based on control-ratio models,[259] and takes into account the expected exter-nal disturbances on the spacecraft flexible modes and the coupling among loops. As a result, four specifications are defined to calculate the QFT bounds: Spec.1: Robust stability; Spec.2: Robust sensitivity; Spec.3: Robust disturbance rejection at plant input; and Spec.4: Robust control effort attenuation. The set of frequencies of interest is within the range: $\omega = [6.28 \cdot 10^{-4}, 62.8]$ rad/s.

Spec.1: Robust Stability
This specification, shown in Equation CS5.2, is stated to guarantee a robust stable control. All the required values, displayed loop-by-loop in Equations CS5.3 and CS5.4, imply at least 1.54 (3.75 dB) gain margin and at least 49.25° phase margin. The specifica-tion corresponds not only to the closed-loop transfer function $y_i(s)/r_i(s)$, but also to trans-fer functions $y_i(s)/n_i(s)$ and $u_i(s)/d_i(s)$—see Equations 2.35 and 2.36. Hence, this condition additionally imposes the requirements on sensor noise attenuation, disturbance rejection at plant input, and flexible modes attenuation.

$$\left| \frac{\left[p_{ii}^{*e}(s) \right]_i^{-1} g_{ii}(s)}{1 + \left[p_{ii}^{*e}(s) \right]_i^{-1} g_{ii}(s)} \right| \leq \delta_1(\omega) \qquad \text{(CS5.2)}$$

Note that $\left[p_{ii}^{*e}(s)\right]_{ji}^{-1}$ is the inverse of the equivalent plant—see Equation 8.86, which corresponds to the diagonal plants $p_{ii}(s)$.

$$\text{Loops } 1, 2, \text{ and } 3: \quad \delta_1(\omega) = W_s = 1.85; \quad \forall \omega \tag{CS5.3}$$

$$\text{Loops } 4, 5, \text{ and } 6: \quad \delta_1(\omega) = \left| \frac{0.1687}{s^2 + 0.4s + 0.0912} \right|; \quad \forall \omega \tag{CS5.4}$$

Spec.2: Sensitivity Reduction
The main objective of this specification, Equations CS5.5 and CS5.6, is sensor noise attenuation and reduction of the effect of the parameter uncertainty on the closed-loop transfer function—see Equation 2.37. It corresponds to the $e_i(s)/n_i(s)$ and "$[dt_{ii}(s)/t_{ii}(s)]/[dp_{ii}(s)/p_{ii}(s)]$" transfer functions.

$$\left| \frac{1}{1 + \left[p_{ii}^{*e}(s)\right]_i^{-1} g_{ii}(s)} \right| \leq \delta_2(\omega) \tag{CS5.5}$$

$$\text{All loops}: \quad \delta_2(\omega) = 2; \quad \forall \omega \tag{CS5.6}$$

Spec.3: Rejection of Disturbances at Plant Input
Solar pressure perturbation and gravity gradient are considered to affect at plant input in the form of both force and torque. The purpose of this specification, Equation CS5.7, which corresponds to $e_i(s)/d_i(s)$ and $y_i(s)/d_i(s)$ transfer functions—see Equation 2.38, is to attenuate the effect of plant input disturbances on the control error and the output signal. Thus, a high gain is required in the low-frequency band, Equations CS5.8 through CS5.10. Besides, since $d_i(s)$ also represents the flexible modes, special attention is paid to their frequency range to accomplish the attitude requirements.

$$\left| \frac{\left[p_{ii}^{*e}(s)\right]_i^{-1}}{1 + \left[p_{ii}^{*e}(s)\right]_i^{-1} g_{ii}(s)} \right| \leq \delta_3(\omega) \tag{CS5.7}$$

$$\text{Loops } 1 \text{ and } 2: \quad \delta_3(\omega) = \left| \frac{0.21553(s + 0.385)}{(s + 0.307)(s + 6.18)(s^2 + 0.4s + 0.0912)} \right|; \quad \forall \omega \tag{CS5.8}$$

$$\text{Loop } 3: \delta_3(\omega) = \left| \frac{0.313 \, (s - 0.01705)(s^2 + 0.009974s + 5.104 \cdot 10^{-5})}{(s - 0.01813)(s^2 + 0.02554s + 0.0004754)} \right|; \quad \forall \omega \tag{CS5.9}$$

$$\text{Loops } 4, 5, \text{ and } 6: \quad \delta_3(\omega)$$

$$= \left| \frac{(s + 0.2)(s + 0.186)(s + 0.2044)(s + 0.003892)(s^2 + 0.06014s + 0.02736)}{(s + 0.007333)(s + 0.445)(s^2 + 0.07904s + 0.00326)(s^2 + 0.2352s + 0.0981)} \right|; \quad \forall \omega$$

$$\tag{CS5.10}$$

Spec.4: Control Effort Reduction
Because of saturation limits, control signal movements should be kept reasonably small despite disturbances—see Equation 2.39. This specification, Equation CS5.11, corresponds to $u_i(s)/n_i(s)$ transfer function and is depicted in Equations CS5.12 through CS5.14.

$$\left| \frac{g_{ii}(s)}{1 + \left[p_{ii}^{*e}(s) \right]_i^{-1} g_{ii}(s)} \right| \leq \delta_4(\omega) \tag{CS5.11}$$

$$\text{Loops 1 and 2}: \quad \delta_4(\omega) = \left| \frac{557.1(s+5)}{(s^2 + 3.23 s + 6.5)} \right|; \quad \forall \omega \tag{CS5.12}$$

$$\text{Loop 3}: \quad \delta_4(\omega) = \left| \frac{106.9210(s+0.55)(s^2 + 0.04s + 0.13)}{(s+1.4)^2(s^2 + 0.1227s + 0.097)} \right|; \quad \forall \omega \tag{CS5.13}$$

$$\text{Loops 4, 5, and 6}: \quad \delta_4(\omega) = \left| \frac{4.026(s^2 - 0.1854s + 0.203)(s^2 + 0.04s + 0.54)}{(s^2 + 0.305s + 0.056)(s^2 + 0.115s + 0.095)} \right|; \quad \forall \omega \tag{CS5.14}$$

Reducing Coupling Effects as much as Possible
The coupling effects from other axes are considered as part of the disturbances acting at the input of the equivalent SISO plant. The design of the non-diagonal elements of the matrix compensator intends also to minimize the off-diagonal elements of the coupling matrices—see Equations 8.71 through 8.73.

CS5.4 Control System Design

The MIMO QFT method 1, explained in Sections 8.5 and 8.8, is applied here to design the 6×6 robust control system for the spacecraft telescope described in Section CS5.2, and with the performance specifications defined in Section CS5.3.

Step A: Input–Output Pairing and Loop Ordering
An example of RGA, calculated from low frequency (steady state) up to 0.19 rad/s, and for all cases within the uncertainty, is shown in Equation CS5.15. According to it, the pairing is selected through the main diagonal of the matrix, which contains positive RGA elements. In addition, due to the very demanding specifications, the RGA elements λ_{15}, λ_{24}, λ_{42}, and λ_{51} are considered relevant as well.

$$RGA_{(\omega = 6.28 \cdot 10^{-4} \text{ rad/s})} = \begin{bmatrix} 1.0064 & 0 & 0 & 0 & -0.0064 & 0 \\ 0 & 1.0064 & 0 & -0.0064 & 0 & 0 \\ 0 & 0 & 1 & 0 & 0 & 0 \\ 0 & -0.0064 & 0 & 1.0064 & 0 & 0 \\ -0.0064 & 0 & 0 & 0 & 1.0064 & 0 \\ 0 & 0 & 0 & 0 & 0 & 1 \end{bmatrix} \tag{CS5.15}$$

Also, in accordance with this RGA results and taking into account the requirement of minimum controller complexity and order (Section CS5.3, Specification 3c), we select a controller structure with only six diagonal elements and two additional off-diagonal elements—see Equation CS5.16.

$$G(s) = \begin{bmatrix} g_{11}(s) & 0 & 0 & 0 & 0 & 0 \\ 0 & g_{22}(s) & 0 & 0 & 0 & 0 \\ 0 & 0 & g_{33}(s) & 0 & 0 & 0 \\ 0 & g_{42}(s) & 0 & g_{44}(s) & 0 & 0 \\ g_{51}(s) & 0 & 0 & 0 & g_{55}(s) & 0 \\ 0 & 0 & 0 & 0 & 0 & g_{66}(s) \end{bmatrix} \quad \text{(CS5.16)}$$

From this, four independent control design problems are adopted, two SISO controllers, $[g_{33}(s)]$ and $[g_{66}(s)]$, and two 2×2 MIMO controllers, $[g_{11}(s)\ 0; g_{51}(s)\ g_{55}(s)]$ and $[g_{22}(s)\ 0; g_{42}(s)\ g_{44}(s)]$. The SISO problems are considered as a classical SISO QFT problem, while the two 2×2 MIMO subsystems are studied through the non-diagonal MIMO QFT method 1—see Sections 8.5 and 8.8.

Step B0: Design of Diagonal Controllers $g_{kk}(s)$, $k = 3, 6$—SISO Cases
The controllers $g_{33}(s)$ and $g_{66}(s)$ are independently designed by using classical SISO QFT to satisfy the performance specifications stated in Section CS5.3 for every plant within the uncertainty. The corresponding QFT bounds and the nominal open-loop transfer functions $L_{ii}(s) = p_{ii}(s)\, g_{ii}(s)$, $i = 3, 6$, are plotted on the Nichols charts shown in Figure CS5.4. The expressions for $g_{33}(s)$ and $g_{66}(s)$ are shown in Equations CS5.24 and CS5.27, respectively.

For the two 2×2 MIMO problems, we follow the flow chart presented in Figure 8.19, MIMO QFT Method 1. As we do not have reference tracking specifications, we do not need prefilters and just follow the Steps B.1.1, B.1.3, B.2.1, and B.2.3.

Step B.1.1 (1st): Design of Diagonal Controller $g_{11}(s)$—First MIMO Problem
The compensator $g_{11}(s)$ is designed according to the non-diagonal MIMO QFT method 1, explained in Section 8.5, for the inverse of the equivalent plant $[p_{11}^{*e}(s)]_1 = p_{11}^{*}(s)$.

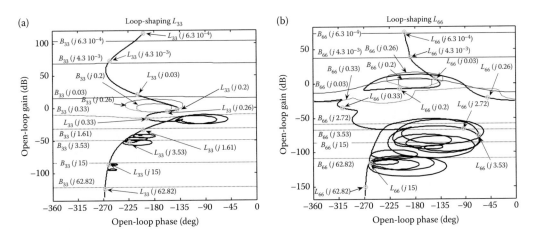

FIGURE CS5.4
Loop shaping: (a) $L_{33}(s) = p_{33}(s)\, g_{33}(s)$ and (b) $L_{66}(s) = p_{66}(s)\, g_{66}(s)$. (Adapted from Garcia-Sanz, M. et al. 2008. *J. ASME J. Dyn. Syst. Meas. Control*, 130, 011006-1–011006-15.)

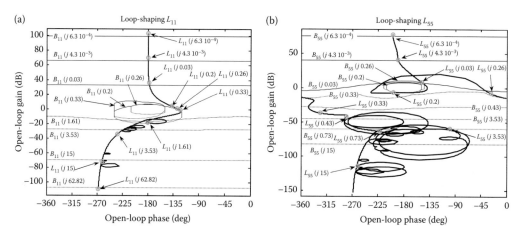

FIGURE CS5.5

Loop shaping: (a) $L_{11}(s) = \left[p_{11}^{*e}(s) \right]_1^{-1} g_{11}(s)$ and (b) $L_{55}(s) = \left[p_{55}^{*e}(s) \right]_2^{-1} g_{55}(s)$. (Adapted from Garcia-Sanz, M. et al. 2008. *J. ASME J. Dyn. Syst. Meas. Control, 130, 011006-1–011006-15.*)

Figure CS5.5a presents the bounds and achieved loop shaping. The expression for $g_{11}(s)$ is shown in Equation CS5.21.

Step B.1.3 (1st): Design of Non-Diagonal Controller $g_{51}(s)$—First MIMO Problem
The non-diagonal compensator $g_{51}(s)$ is designed to minimize the (5,1) element of the coupling matrix in the case of disturbance rejection at plant input, which gives the following expression:

$$g_{51}^{opt}(s) = -p_{51}^{*N}(s) \tag{CS5.17}$$

where N denotes the middle plant that interpolates the expression $[-p_{51}^*(s)]$ from 0 to 10^{-1} rad/s, as shown in Figure CS5.6a. The expression for $g_{51}(s)$ is shown in Equation CS5.22.

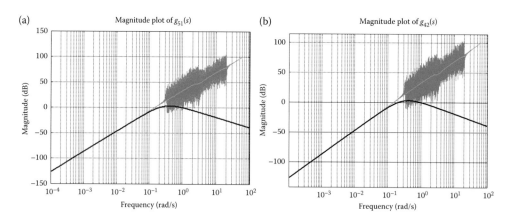

FIGURE CS5.6

(a) Magnitude plot of $[-p_{51}^*(s)]$ with uncertainty and $g_{51}(s)$—solid line and (b) magnitude plot of $[-p_{42}^*(s)]$ with uncertainty and $g_{42}(s)$—solid line. (Adapted from Garcia-Sanz, M. et al. 2008. *J. ASME J. Dyn. Syst. Meas. Control, 130, 011006-1–011006-15.*)

Step B.2.1 (1st): Design of Diagonal Controller $g_{55}(s)$—First MIMO Problem
The compensator $g_{55}(s)$ is designed according to the non-diagonal MIMO QFT method 1, explained in Section 8.5, for the inverse of the equivalent plant $[p_{55}^{*e}(s)]_2$, which is

$$p_{55}^{*e}(s)_2 = \frac{\left[p_{55}^{*e}(s)\right]_1 - \left(\left[p_{51}^{*e}(s)\right]_1 + \left[g_{51}(s)\right]_1\right)\left(\left[p_{15}^{*e}(s)\right]_1\right)}{\left(\left[p_{11}^{*e}(s)\right]_1 + \left[g_{11}(s)\right]_1\right)}$$

Figure CS5.5b presents the bounds and achieved loop shaping. The expression for $g_{55}(s)$ is shown in Equation CS5.26.

Step B.2.3 (1st): Design of Non-Diagonal Controller $g_{15}(s)$—First MIMO Problem
As mentioned, the non-diagonal compensator $g_{15}(s)$ is chosen as null to simplify the controller complexity, according to the specification 3c in Section CS5.3 and the RGA results:

$$g_{15}^{opt}(s) = 0 \qquad\qquad (CS5.18)$$

The second MIMO problem is presented in the following steps. It consists of the design of the elements $g_{22}(s)$, $g_{42}(s)$, $g_{44}(s)$, and $g_{24}(s)$, which are equivalently performed as in the previous Steps B.1.1, B.1.3, B.2.1, and B.2.3, respectively.

Step B.1.1 (2nd): Design of Diagonal Controller $g_{22}(s)$—Second MIMO Problem
The compensator $g_{22}(s)$ is designed according to the non-diagonal MIMO QFT method 1, explained in Section 8.5, for the inverse of the equivalent plant $\left[p_{22}^{*e}(s)\right]_1 = p_{22}^{*}(s)$.
Figure CS5.7a presents the bounds and achieved loop shaping. The expression for $g_{22}(s)$ is shown in Equation CS5.21.

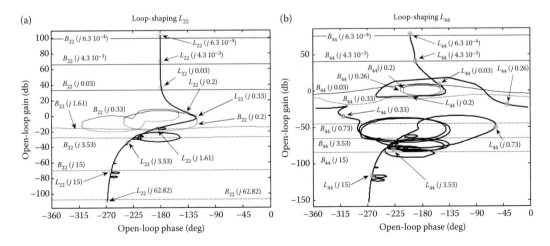

FIGURE CS5.7
Loop shaping (a) $L_{22}(s) = \left[p_{22}^{*e}(s)\right]_1^{-1} g_{22}(s)$ and (b) $L_{44}(s) = \left[p_{44}^{*e}(s)\right]_2^{-1} g_{44}(s)$. (Adapted from Garcia-Sanz, M. et al. 2008. *J. ASME J. Dyn. Syst. Meas. Control*, 130, 011006-1–011006-15.)

Step B.1.3 (2nd): Design of Non-Diagonal Controller $g_{42}(s)$—Second MIMO Problem
The non-diagonal compensator $g_{42}(s)$ is designed to minimize the (4,2) element of
the coupling matrix in the case of disturbance rejection at plant input, which gives
the following expression:

$$g_{42}^{opt}(s) = -p_{42}^{*N}(s) \qquad\qquad\qquad (CS5.19)$$

where N denotes the middle plant that interpolates the expression $[-p_{42}^{*}(s)]$ from 0 to
10^{-1} rad/s, as shown in Figure CS5.6b. The expression for $g_{42}(s)$ is shown in Equation CS5.22.

Step B.2.1 (2nd): Design of Diagonal Controller $g_{44}(s)$—Second MIMO Problem
The compensator $g_{44}(s)$ is designed according to the non-diagonal MIMO QFT method 1,
explained in Section 8.5, for the inverse of the equivalent plant $\left[p_{44}^{*e}(s)\right]_2$, which is

$$p_{44}^{*e}(s)_2 = \frac{\left[p_{44}^{*e}(s)\right]_1 - \left(\left[p_{42}^{*e}(s)\right]_1 + \left[g_{42}(s)\right]_1\right)\left(\left[p_{24}^{*e}(s)\right]_1\right)}{\left(\left[p_{22}^{*e}(s)\right]_1 + \left[g_{22}(s)\right]_1\right)}$$

Figure CS5.7b presents the bounds and achieved loop shaping. The expression for $g_{44}(s)$
is shown in Equation CS5.25.

Step B.2.3 (2nd): Design of Non-Diagonal Controller $g_{24}(s)$—Second MIMO Problem
As in Step B.2.3 (1st), the non-diagonal compensator $g_{24}(s)$ is chosen as null to simplify
the controller complexity, according to the specification 3c in Section CS5.3 and the RGA
results:

$$g_{24}^{opt}(s) = 0 \qquad\qquad\qquad (CS5.20)$$

Steps B.k.2: Design of Prefilters $f_{kk}(s)$, $k = 1, 2,...,6$
There is not prefilter required in this example, because we do not have reference tracking
specifications. Then, $f_{kk}(s) = 1$, $k = 1, 2,...,6$.

Controller Expressions
The notation adopted below for the transfer function expressions of the controllers denotes
the steady state gain as a constant without parenthesis; simple poles and zeros as (ω), which
corresponds to $(s/\omega + 1)$ in the denominator and numerator respectively; poles and zeros
at the origin as (0); conjugate poles and zeros as $[\zeta; \omega_n]$, with $((s/\omega_n)^2 + (2\zeta/\omega_n)s + 1)$ in the
denominator and numerator respectively; n-multiplicity of poles and zeros as an exponent
$()^n$. Following this notation, the MIMO QFT controller designed in this section consists of
the following elements:

$$g_{11}(s) = g_{22}(s) = \frac{\{31.5(0.6194)(0.2138)(0.1663)(0.1649)\}}{\{(0.666)(0.4982)(0.07526)[0.676;1.479]\}} \qquad (CS5.21)$$

$$g_{51}(s) = -g_{42}(s) = \{42.4(0)^2\} / \{(0.3)^3\} \tag{CS5.22}$$

$$g_{15}(s) = g_{24}(s) = 0 \tag{CS5.23}$$

$$g_{33}(s) = \frac{\{125(0.13)(0.057)[0.07019;\ 0.3565][1;\ 0.02]\}}{\{(1.48)(0.7875)(0.2)(0.004)(0.00246)[0.18;\ 0.314]\}} \tag{CS5.24}$$

$$g_{44}(s) = \frac{\{2.242(0.03412)[0.08644;\ 0.7114][0.1131;\ 0.3414][0.1145;\ 0.2604][0.008792;\ 0.2593]}{\{(0.9776)(0.8)(0.0005)^2[0.2451;\ 0.2708][0.297;\ 0.2673][-0.0005;\ 0.254]}$$
$$\frac{[0.7;\ 0.0052][1;\ 0.0007]\}}{[0.14;\ 0.252][0.7;\ 0.0045]\}} \tag{CS5.25}$$

$$g_{55}(s) = \frac{\{2.242(0.13)(0.03)[0.1079;\ 0.7099][0.07069;\ 0.341][0.03;\ 0.2593][0.7;\ 0.0052][1;\ 0.0007]\}}{\{(0.9776)(0.8)(0.12)(0.0005)^2[0.2451;\ 0.2708][-0.0008;\ 0.254][0.3;\ 0.25][0.7;\ 0.0045]\}} \tag{CS5.26}$$

$$g_{66}(s) = \frac{\{2.242\,(0.02584)[0.08644;\ 0.7114][0.1131;\ 0.3414][0.1145;\ 0.2604][0.008792;\ 0.2593]}{\{(0.9776)(0.8)(0.0005)^2[0.2451;\ 0.2708][0.297;\ 0.2673][-0.0007;\ 0.254]}$$
$$\frac{[0.7;\ 0.0052][1;\ 0.0007]\}}{[0.1687;\ 0.241][0.7;\ 0.0045]\}} \tag{CS5.27}$$

CS5.5 Simulation and Validation

Time domain simulations were performed for 300 plants in the ESA spacecraft telescope benchmark simulator—see Figure CS5.8. The plant models were selected randomly within the space of uncertainty using a Monte Carlo technique. The same plants were used for all the participants in the benchmark.

The performance achieved for the position and attitude control by the non-diagonal MIMO QFT was excellent, meeting all the required specifications—see Section CS5.3. Moreover, the MIMO QFT obtained the best results in the benchmark—see maximum and standard deviation results in Table CS5.1, improving by two orders of magnitude the results obtained by the second classified, that used a H-infinity control technique.

At the same time, while the second classified required controller structures with full-matrices of 36 elements of 42nd order, the non-diagonal MIMO QFT design consisted of only eight compensators with simple structures, all from 3rd to 14th order—see Equations CS5.21 through CS5.27, dividing by more than 20 the number of operations per second needed for the real-time control algorithm—see Table CS5.2.

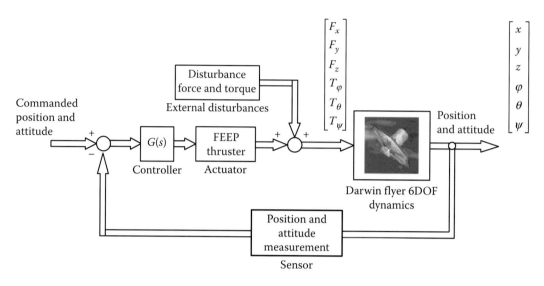

FIGURE CS5.8

ESA telescope-type spacecraft simulator. (Adapted from Garcia-Sanz, M. et al. 2008. *J. ASME J. Dyn. Syst. Meas. Control*, 130, 011006-1–011006-15.)

TABLE CS5.1

Time Simulation Performance with Controllers

	Spec.	Non-Diagonal MIMO QFT Controller	H-Infinity Controller
Maximum Position Error X	<1 μm	**0.0131**	0.0293
		0.0816	0.511
Maximum Position Error Y	<1 μm	**0.0120**	0.0299
		0.0120	0.0299
Maximum Position Error Z	<1 μm	**0.0288**	0.0292
		0.0288	0.0292
Maximum Attitude Error X	<25.5 mas	**25.27**	25.95
		25.27	25.95
Maximum Attitude Error Y	<25.5 mas	**22.91**	23.21
		22.55	28.91
Maximum Attitude Error Z	<25.5 mas	**21.15**	22.84
		21.15	22.84
Standard Deviation of Position Error X	<0.33 μm	**0.00275**	0.00686
		0.0511	0.341
Standard Deviation of Position Error Y	<0.33 μm	**0.00265**	0.00722
		0.00265	0.00722
Standard Deviation of Position Error Z	<0.33 μm	**0.00668**	0.00691
		0.00668	0.00691
Standard Deviation of Attitude Error X	<8.5 mas	**5.57**	5.68
		5.57	5.68
Standard Deviation of Attitude Error Y	<8.5 mas	**5.76**	6.01
		5.80	8.23
Standard Deviation of Attitude Error Z	<8.5 mas	**4.83**	5.00
		4.83	5.00

Note: At each specification (each row), the best result (QFT or H-infinity) is in bold.

TABLE CS5.2

Number of Operations per Second Required by Controllers

Controller	# Multips/s	# Sums/s
Non-diagonal MIMO QFT	130	124
H-infinity	2994	2988

CS5.6 Summary

This case study demonstrated the feasibility of the sequential non-diagonal MIMO robust QFT control strategies to simultaneously regulate the position and attitude of a 6×6 spacecraft telescope with large flexible appendages. The spacecraft was part of a multiple formation flying constellation of a ESA cornerstone mission. The controller satisfactory met the astronomical, engineering, and control requirements.

Appendix 1: Projects and Problems

A1.1 Projects

Project P1. *Vehicle Active Suspension Control*

1. *Chapters and Sections Needed*: Chapter 2; Appendix 2; QFTCT
2. *Project Description*: The marketing department of a major car manufacturer wants to sell a car that offers the smoothest possible ride. For this, the control system must minimize the accelerations felt by the passengers in the vehicle while it passes over uneven surfaces. To achieve this goal, the engineering department has designed an active suspension system that couples a force actuator to a preexisting suspension.

   A schematic of the vehicle suspension system appears in Figure P1.1. The total vehicle chassis has mass $4m_2$ (m_2 affects each wheel). It is suspended on a passive suspension with spring constant k_s and damping coefficient b per wheel. The tire system is modeled as mass m_1 and a stiffness k_t. An accelerometer provides the acceleration of the vehicle body for feedback control, $z(t) = d^2y(t)/dt^2$, $y(t)$ being the height of the vehicle body with respect to the road reference, and $x(t)$ the height of the tire system with respect to the road reference.

   The acceleration of the car chassis $z(s)$, measured by the accelerometer, is the plant output. The plant inputs are the control signal $u(s)$ and the road height $d(s)$ with respect to the road reference. The variation of the road height is considered as the disturbance $d(s)$ on the system to be rejected.

3. *Modeling*: The generalized coordinates are the height of the vehicle body and wheel system, which are, respectively, $q_1 = y(t)$ and $q_2 = x(t)$—see Figure P1.1. The kinetic, potential, and dissipation functions of the system are described by Equations P1.1 through P1.3

$$E_k = \frac{1}{2}m_2\,\dot{y}^2 + \frac{1}{2}m_1\,\dot{x}^2 \tag{P1.1}$$

$$E_p = \frac{1}{2}k_s(y-x)^2 + \frac{1}{2}k_t(x-0)^2 \tag{P1.2}$$

$$D_d = \frac{1}{2}b(\dot{y}-\dot{x})^2 \tag{P1.3}$$

The generalized force applied by the active damper between the masses m_1 and m_2 is $Q_1 = u(t)$. The generalized force applied by the road to the wheel is $Q_2 = d(t)$. Using Equations P1.1 through P1.3, the terms for the Euler–Lagrange expression are

$$\frac{d}{dt}\left(\frac{\partial L}{\partial \dot{q}_i}\right) - \frac{\partial L}{\partial q_i} + \frac{\partial D_d}{\partial \dot{q}_i} = Q_i\,,\ i = 1, 2, \ldots, n \tag{P1.4}$$

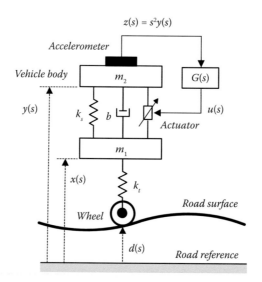

FIGURE P1.1
Vehicle active suspension system.

$$\frac{d}{dt}\left(\frac{\partial L}{\partial \dot{q}_i}\right) = \frac{d}{dt}\begin{bmatrix}\dfrac{\partial E_k}{\partial \dot{y}} \\[2mm] \dfrac{\partial E_k}{\partial \dot{x}}\end{bmatrix} = \begin{bmatrix} m_2 & 0 \\ 0 & m_1 \end{bmatrix}\begin{bmatrix}\ddot{y} \\ \ddot{x}\end{bmatrix} = \mathcal{M}\ddot{q} \tag{P1.5}$$

$$-\frac{\partial L}{\partial q_i} = \frac{\partial E_p}{\partial q_i} = \begin{bmatrix}\dfrac{\partial E_p}{\partial y} \\[2mm] \dfrac{\partial E_p}{\partial x}\end{bmatrix} = \begin{bmatrix} k_s & -k_s \\ -k_s & k_s + k_t \end{bmatrix}\begin{bmatrix}y \\ x\end{bmatrix} = \mathcal{K}q \tag{P1.6}$$

$$\frac{\partial D_d}{\partial \dot{q}_i} = \begin{bmatrix}\dfrac{\partial D_d}{\partial \dot{y}} \\[2mm] \dfrac{\partial D_d}{\partial \dot{x}}\end{bmatrix} = \begin{bmatrix} b & -b \\ -b & b \end{bmatrix}\begin{bmatrix}\dot{y} \\ \dot{x}\end{bmatrix} = \mathcal{C}\dot{q} \tag{P1.7}$$

$$\mathcal{Q} = \begin{bmatrix} 1 & 0 \\ -1 & 1 \end{bmatrix}\begin{bmatrix}u \\ d\end{bmatrix} = \mathcal{R}u \tag{P1.8}$$

As described in the case studies, substituting Equations P1.5 through P1.8 in the Euler–Lagrange equation—Equation P1.4, we write the system as the state space description in Equations P1.9 through P1.12,

$$\dot{x} = \left[\begin{array}{c|c} -\mathcal{M}^{-1}\mathcal{C} & -\mathcal{M}^{-1}\mathcal{K} \\ \hline I & 0 \end{array}\right]x + \left[\begin{array}{c}\mathcal{M}^{-1}\mathcal{R} \\ \hline 0\end{array}\right]u = Ax + Bu \tag{P1.9}$$

$$y = \begin{bmatrix} 0 & | & I \end{bmatrix}x = Cx$$

$$A = \begin{vmatrix} \dfrac{-b}{m_2} & \dfrac{b}{m_2} & \dfrac{-k_s}{m_2} & \dfrac{k_s}{m_2} \\ \dfrac{b}{m_1} & \dfrac{-b}{m_1} & \dfrac{k_s}{m_1} & \dfrac{-(k_s+k_t)}{m_1} \\ 1 & 0 & 0 & 0 \\ 0 & 1 & 0 & 0 \end{vmatrix}$$

(P1.10)

$$B = \begin{vmatrix} \dfrac{1}{m_2} & 0 \\ \dfrac{-1}{m_1} & \dfrac{1}{m_1} \\ 0 & 0 \\ 0 & 0 \end{vmatrix}; \quad C = \begin{bmatrix} 0 & 0 & 1 & 0 \\ 0 & 0 & 0 & 1 \end{bmatrix}$$

(P1.11)

$$x = \begin{bmatrix} \dot{q} \\ q \end{bmatrix} = \begin{bmatrix} \dot{y} & \dot{x} & y & x \end{bmatrix}^T; \quad u = \begin{bmatrix} u & d \end{bmatrix}^T; \quad y = \begin{bmatrix} y & x \end{bmatrix}^T$$

(P1.12)

Now, applying the transformation in Equation P1.13, we obtain the transfer function $p_{11}(s)$ from the controller output $u(s)$ to the vehicle body position $y(s)$, and the transfer function $p_{12}(s)$ from the road height disturbance $d(s)$ to the vehicle body position $y(s)$, as shown in Equations P1.14 and P1.15, respectively.

$$P(s) = C(sI - A)^{-1}B, \quad y(s) = P(s)u(s)$$

(P1.13)

$$\frac{y(s)}{u(s)} = p_{11}(s) = \frac{(m_1 s^2 + k_t)}{m_1 m_2 s^4 + b(m_1 + m_2)s^3 + [k_t m_2 + k_s(m_1 + m_2)]s^2 + bk_t s + k_t k_s}$$

(P1.14)

$$\frac{y(s)}{d(s)} = p_{12}(s) = \frac{(bs + k_s)}{m_1 m_2 s^4 + b(m_1 + m_2)s^3 + [k_t m_2 + k_s(m_1 + m_2)]s^2 + bk_t s + k_t k_s}$$

(P1.15)

The variable we want to control is the vehicle acceleration $z(s)$, which is defined in Equation P1.16 from Equations P1.14 and P1.15—see also Figure P1.2.

$$z(s) = s^2 [p_{11}(s)u(s) + p_{12}(s)d(s)]$$

(P1.16)

The parameters for the system are $m_1 = 10$ kg, $m_2 \in [280, 400]$ kg, $k_s = 27{,}000$ N/m, $k_t \in [70{,}600, 90{,}600]$ N/m, and $b \in [650, 750]$ Ns/m.

4. *Preliminary Analysis*: A quick inspection of the plants $[s^2 p_{11}(s)]$ and $[s^2 p_{12}(s)]$ shows that the system contains a double derivative (due to the accelerometer) and two significant resonances with a low damping around 8 rad/s and 90 rad/s—see

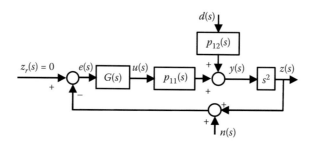

FIGURE P1.2
Control diagram for active suspension system.

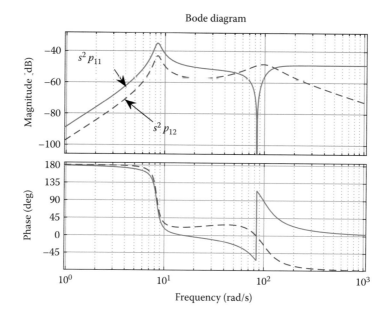

FIGURE P1.3
Bode diagram for nominal plants $z(s)/u(s) = [s^2\, p_{110}(s)]$ and $z(s)/d(s) = [s^2\, p_{120}(s)]$.

Figure P1.3. The controller will have to provide enough damping to reduce the potential oscillations of the system at these frequencies.

5. *Control Specifications*: Taking into account the dynamics of the active suspension system and the vehicle objectives, we define the following robust specifications to control the plant $z(s)/u(s) = s^2\, p_{11}(s)$:

- *Type 1: Stability specification*

$$|T_1(s)| = \left|\frac{s^2 p_{11}(s)G(s)}{1+s^2 p_{11}(s)G(s)}\right| \leq \delta_1(\omega) = W_s = 1.3 \tag{P1.17}$$

$\omega \in [0.001\ 0.005\ 0.01\ 0.05\ 0.1\ 0.5\ 1\ 5\ 7\ 8\ 9\ 10\ 50\ 80\ 90\ 100\ 500]\,\text{rad/s}$

which is equivalent to $PM = 45.24°$, $GM = 4.96$ dB—see Equations 2.30 and 2.31.

- *Type 3: Sensitivity or disturbances at plant output specification*

$$|T_3(s)| = \left|\frac{y(s)}{d(s)}\right| = \left|\frac{1}{1+s^2 p_{11}(s)G(j\omega)}\right| \leq \delta_3(\omega) = \frac{(s/a_d)}{(s/a_d)+1}; \ a_d = 1 \quad \text{(P1.18)}$$

$$\omega \in [0.001 \ 0.005 \ 0.01 \ 0.05 \ 0.1 \ 0.5] \ \text{rad/s}$$

6. *Questions*

   a. Attenuate the road disturbances as much as possible, reducing the vehicle body acceleration $z(s)$. For that, design a suitable controller $G(s)$ to regulate the plant $[s^2 \, p_{11}(s)]$—Equations P1.14 and P1.16, considering all the plants within the model uncertainty, and guaranteeing the stability and disturbance rejection specifications defined in Equations P1.17 and P1.18.

   b. Plot and compare the Bode diagrams of the sensitivity transfer function $z(s)/d(s)$ in both cases, open-loop and closed-loop with the new controller $G(s)$.

   c. Simulate the time response of the vehicle acceleration $z(t)$ when the road introduces a unitary impulse $d(t)$, and in both cases, open loop and closed loop with the new controller $G(s)$.

Project P2. *DVD-Disk Head Control*

1. *Chapters and Sections Needed*: Chapter 2; Appendix 2; QFTCT.

2. *Project Description*: Consider the position control system for the read/write head of a DVD device shown in Figure P2.1. The mechanism consists of a primary motor (under the disk) that rotates the DVD disk at a constant speed, and an arm that holds the read/write head at a constant and small distance above the disk surface. At the same time, a second motor with a control system is used to move the read/write head horizontally, following the reference $x_r(s)$ received from a computer.

3. *Modeling*: The dynamics of the horizontal movement (plane of the paper) of the read/write head is described by Equation P2.1, where $P(s)$ is the transfer function between the motor input $u(s)$ and the horizontal position of the read/write head $x(s)$. The control system for the DVD device is shown in Figure P2.2.

$$\frac{x(s)}{u(s)} = P(s) = \frac{K(bs+1)}{s^2\left((s^2/\omega_n^2)+(2\zeta/\omega_n)s+1\right)} \quad \text{(P2.1)}$$

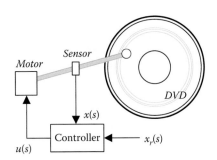

FIGURE P2.1
Position control system for the read/write head of a DVD device (top-down view).

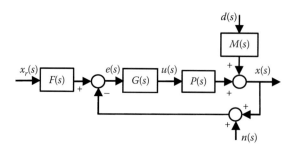

FIGURE P2.2
2DOF control system for the DVD device.

The parameters for the flexible arm are $\omega_n = 50$ rad/s, $\zeta \in [8, 12] \times 10^{-4}$, and for the motor and mechanism $K \in [1.5, 2.5]$, $b \in [3, 5] \times 10^{-5}$, $M(s) = 1$.

4. *Preliminary Analysis*: A quick inspection of the plant $P(s)$ shows that the system contains a double integrator with a very low damping second-order dynamics at $\omega_n = 50$ rad/s due to the flexible arm—see also Figure P2.3. These facts will require the controller to provide a significant damping to reduce the potential oscillations of the read/write head.

5. *Control Specifications*: Taking into account the dynamics of the DVD mechanism and the operation objectives, we define the following robust control specifications:

- *Type 1: Stability specification*

$$|T_1(j\omega)| = \left| \frac{P(j\omega)G(j\omega)}{1 + P(j\omega)G(j\omega)} \right| \le \delta_1(\omega) = W_s = 1.3 \tag{P2.2}$$

$$\omega \in [0.001\ 0.005\ 0.01\ 0.05\ 0.1\ 0.5\ 1\ 5\ 10\ 50\ 100\ 500]\ \text{rad/s}$$

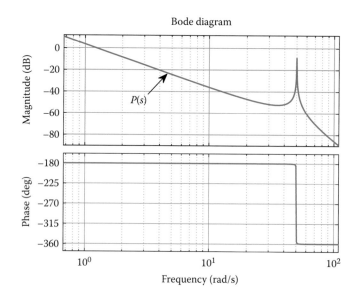

FIGURE P2.3
Bode diagram for nominal plant $P_0(s)$.

which is equivalent to $PM = 45.24°$, $GM = 4.96$ dB—see Equations 2.30 and 2.31.

- *Type 3: Sensitivity or disturbances at plant output specification*

$$|T_3(j\omega)| = \left|\frac{x(j\omega)}{d_o(j\omega)}\right| = \left|\frac{1}{1 + P(j\omega)G(j\omega)}\right| \le \delta_3(\omega) = \frac{(s/a_d)}{(s/a_d) + 1}; \ a_d = 10$$

(P2.3)

$$\omega \in [0.001 \ 0.005 \ 0.01 \ 0.05 \ 0.1 \ 0.5 \ 1 \ 5 \ 10] \text{ rad/s}$$

- *Type 6: Reference tracking specification*

$$\delta_{6\_lo}(\omega) < |T_6(j\omega)| = \left|F(j\omega)\frac{P(j\omega)G(j\omega)}{1 + P(j\omega)G(j\omega)}\right| \le \delta_{6\_up}(\omega)$$

(P2.4)

$$\omega \in [0.001 \ 0.005 \ 0.01 \ 0.05 \ 0.1 \ 0.5 \ 1 \ 5] \text{ rad/s}$$

$$\delta_{6\_lo}(s) = \frac{(1 - \varepsilon_{lo})}{\left[(s/a_L) + 1\right]^2}; \ a_L = 6; \ \varepsilon_{lo} = 0$$

(P2.5)

$$\delta_{6\_up}(s) = \frac{[(s/a_U) + 1] \ (1 + \varepsilon_{up})}{[(s/\omega_n)^2 + (2\zeta s/\omega_n) + 1]}; \ a_U = 2; \ \zeta = 0.8; \ \omega_n = \frac{1.25 \, a_U}{\zeta}; \varepsilon_{up} = 0$$

(P2.6)

6. *Questions*
   a. Design a feedback controller $G(s)$ and a prefilter $F(s)$ to control the DVD head position—plant $P(s)$, Equation P2.1, meeting the robust stability and performance specifications defined above—Equations P2.2 through P2.6.
   b. Simulate the output $x(s)$ of the closed-loop control system when the reference $x_r(s)$ changes as a step input, from 0 to 1, at $t = 1$ s, and when a unitary step disturbance $d(s)$ hits the system at $t = 5$ s—see Figure P2.2.

Project P3. *Inverted Pendulum Control*

1. *Chapters and Sections Needed*: Chapters 2, 3, 9; Appendices 2, 3; QFTCT.
2. *Project Description*: Consider an inverted pendulum mounted on a cart, as shown in Figure P3.1. The cart is moved horizontally by a motor connected to a belt-pulley system. The pendulum moves in the vertical plane and turns about a perpendicular axis fixed to the cart.

   The system has two encoders that measure the two main variables: the angle of the pendulum with the vertical axis $\theta(t)$ and the horizontal position of the cart $x(t)$. The motor applies a horizontal force $F(t)$. The pendulum has a mass m at the end of a rod. The rod is assumed to have zero mass and a length h. The cart mass is M. There is no friction in the system.
3. *Modeling*: The generalized coordinates are the horizontal position of the cart and the angle of the pendulum with the vertical axis, which are respectively $q_1 = x(t)$ and $q_2 = \theta(t)$—see Figure P3.1. The cart and pendulum have no friction. This means

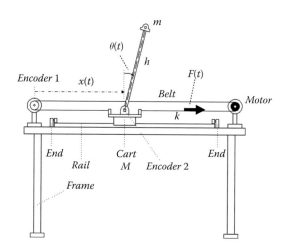

FIGURE P3.1
Inverted pendulum.

that the dissipation function is zero ($D_d = 0$). The kinetic and potential functions of the system are described by Equations P3.1 and P3.2

$$E_k = \frac{1}{2} M \dot{x}^2 + \frac{1}{2} m \left\{ \left[\frac{d}{dt}(x + h\sin\theta) \right]^2 + \left[\frac{d}{dt}(h\cos\theta) \right]^2 \right\} \tag{P3.1}$$

$$E_p = E_{p0} + m g h \cos\theta \tag{P3.2}$$

being E_{p0} the potential energy of the mass m for $\theta = 90°$. The generalized force applied horizontally by the motor to the cart is $u = F(t)$. The expression for the Lagrangian ($L = E_k - E_p$) is

$$L = \frac{1}{2} M \dot{x}^2 + \frac{1}{2} m [(\dot{x} + h\dot{\theta}\cos\theta)^2 + (h\dot{\theta}\sin\theta)^2] - E_{p0} - m g h \cos\theta \tag{P3.3}$$

Using Equation P3.3 in the Euler–Lagrange equation

$$\frac{d}{dt}\left(\frac{\partial L}{\partial \dot{q}_i} \right) - \frac{\partial L}{\partial q_i} + \frac{\partial D_d}{\partial \dot{q}_i} = Q_i, \; i = 1, 2, \ldots, n \tag{P3.4}$$

we obtain the equations related to x and θ, so that

$$(M + m)\ddot{x} + m h \ddot{\theta} \cos\theta - m h \dot{\theta}^2 \sin\theta = F \tag{P3.5}$$

$$\ddot{x} \cos\theta + h\ddot{\theta} - g \sin\theta = 0 \tag{P3.6}$$

which can be linearized for small θ angles as

$$(M+m)\ddot{x}+mh\ddot{\theta}=F \tag{P3.7}$$

$$\ddot{x}+h\ddot{\theta}-g\theta=0 \tag{P3.8}$$

Using these expressions, and considering a state vector $\mathbf{x}=[x \quad \dot{x} \quad \theta \quad \dot{\theta}]^T$, the state-space description of the system ($\dot{\mathbf{x}}=A\mathbf{x}+B u; y=C\mathbf{x}$) becomes

$$\begin{bmatrix} \dot{x} \\ \ddot{x} \\ \dot{\theta} \\ \ddot{\theta} \end{bmatrix} = \begin{bmatrix} 0 & 1 & 0 & 0 \\ 0 & 0 & \dfrac{-mg}{M} & 0 \\ 0 & 0 & 0 & 1 \\ 0 & 0 & \dfrac{(M+m)g}{Mh} & 0 \end{bmatrix} \begin{bmatrix} x \\ \dot{x} \\ \theta \\ \dot{\theta} \end{bmatrix} + \begin{bmatrix} 0 \\ \dfrac{1}{M} \\ 0 \\ \dfrac{-1}{Mh} \end{bmatrix} F \tag{P3.9}$$

$$\begin{bmatrix} x \\ \theta \end{bmatrix} = \begin{bmatrix} 1 & 0 & 0 & 0 \\ 0 & 0 & 1 & 0 \end{bmatrix} \begin{bmatrix} x \\ \dot{x} \\ \theta \\ \dot{\theta} \end{bmatrix} \tag{P3.10}$$

Now, applying $P(s)=C(sI-A)^{-1}B$ for $y(s)=P(s)u(s)$, the transfer function $p_{21}(s)$, from the force $F(s)$ to the angle of the pendulum $\theta(s)$, is

$$\frac{\theta(s)}{F(s)}=p_{21}(s)=\frac{1}{[-Mhs^2+(M+m)g]} \tag{P3.11}$$

Also, for the cases where the mass of the cart is much larger than the mass at the tip of the pendulum, $M\gg m$, the transfer function $p_{11}(s)$ from the force $F(s)$ to the cart position $x(s)$ is

$$\frac{x(s)}{F(s)}=p_{11}(s)=\frac{-hs^2+g}{s^2[-Mhs^2+(M+m)g]}\approx\frac{1}{Ms^2} \tag{P3.12}$$

Figure P3.2 shows the block diagram with the transfer functions $p_{11}(s)$ and $p_{21}(s)$ and the angle control system.

The fixed parameters of the pendulum are $h=0.5$ m, and $g=9.8$ m/s². The parameters with uncertainty are $m \in [0.05, 0.09]$ kg and $M \in [1.5, 2.5]$ kg.

4. *Preliminary Analysis*: A quick inspection of the plant $p_{21}(s)$ shows that the system is unstable. It has two real and symmetric poles, one of them in the left-half plane

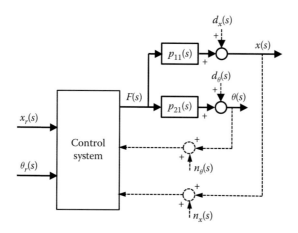

FIGURE P3.2
Control system for inverted pendulum.

(stable) and the other one in the right-half plane (unstable)—see Equation P3.13. The controller will have to stabilize the system.

$$s = \pm\sqrt{\frac{(M+m)\,g}{M\,h}} \tag{P3.13}$$

On the other hand, the plant $p_{11}(s)$—cart position control is a double integrator. The controller will have to provide enough damping to reduce the oscillations of the cart position.

5. *Control Specifications*

- *For the pendulum angle control,* $G_\theta(s)$. Taking into account the dynamics of the inverted pendulum, we define the following robust control specifications— with $H(s) = 1$:

  – *Type 1: Stability specification*

  $$|T_1(j\omega)| = \left|\frac{p_{21}(j\omega)G_\theta(j\omega)}{1 + p_{21}(j\omega)G_\theta(j\omega)}\right| \le \delta_1(\omega) = W_s = 1.08 \tag{P3.14}$$
  $$\omega \in [0.01 \ \ 0.05 \ \ 0.1 \ \ 0.5 \ \ 1 \ \ 5 \ \ 10 \ \ 50 \ \ 100 \ \ 500] \ \text{rad/s}$$

  which is equivalent to $PM = 55.16°$, $GM = 5.69$ dB—see Equations 2.30 and 2.31.

  – *Type 3: Sensitivity or disturbances at plant output specification*

  $$|T_3(j\omega)| = \left|\frac{\theta(j\omega)}{d_o(j\omega)}\right| = \left|\frac{1}{1 + p_{21}(j\omega)G_\theta(j\omega)}\right| \le \delta_3(\omega) = \frac{0.025\,s^2 + 0.2\,s + 0.018}{0.025\,s^2 + 10\,s + 1} \tag{P3.15}$$
  $$\omega \in [0.1 \ \ 0.5 \ \ 1 \ \ 5 \ \ 10 \ \ 50] \ \text{rad/s}$$

- *For the cart position control, $G_x(s)$ and $F_x(s)$. Taking into account the dynamics of the inverted pendulum, we define the following robust control specifications—also with $H(s) = 1$:*
  - *Type 1: Stability specification*

$$\left|T_1(j\omega)\right| = \left|\frac{p_{11}(j\omega)G_x(j\omega)}{1 + p_{11}(j\omega)G_x(j\omega)}\right| \le \delta_1(\omega) = W_s = 1.08 \tag{P3.16}$$

$$\omega \in [0.01 \ 0.05 \ 0.1 \ 0.5 \ 1 \ 5 \ 10 \ 50 \ 100 \ 500] \ \text{rad/s}$$

which is equivalent to $PM = 55.16°$, $GM = 5.69$ dB—see Equations 2.30 and 2.31.
  - *Type 6: Reference tracking specification*

$$\delta_{6\_lo}(\omega) < \left|T_6(j\omega)\right| = \left|F_x(j\omega)\frac{p_{11}(j\omega)G_x(j\omega)}{1 + p_{11}(j\omega)G_x(j\omega)}\right| \le \delta_{6\_up}(\omega) \tag{P3.17}$$

$$\omega \in [0.01 \ 0.05 \ 0.1 \ 0.5 \ 1] \ \text{rad/s}$$

$$\delta_{6\_lo}(s) = \frac{(1 - \varepsilon_{lo})}{[(s/a_L) + 1]^2}; \ a_L = 0.25; \ \varepsilon_{lo} = 0 \tag{P3.18}$$

$$\delta_{6\_up}(s) = \frac{[(s/a_U) + 1](1 + \varepsilon_{up})}{[(s/\omega_n)^2 + (2\zeta s/\omega_n) + 1]}; \ a_U = 0.1; \ \zeta = 0.8; \ \omega_n = \frac{1.25 \, a_U}{\zeta}; \ \varepsilon_{up} = 0.05 \tag{P3.19}$$

6. *Questions*
   a. Design a feedback controller $G_\theta(s)$ to stabilize the angle of the pendulum—plant $p_{21}(s)$, Equation P3.11, keeping it in the upright position ($\theta_r = \theta = 0°$) and meeting the stability and performance specifications defined above—Equations P3.14 and P3.15.
   b. Design a feedback controller $G_x(s)$ and a prefilter $F_x(s)$ to control the position of the cart—plant $p_{11}(s)$, Equation P3.12, meeting the stability and performance specifications defined above—Equation P3.16 through P3.19.
   c. As the system has two sensors, $\theta(s)$ and $x(s)$, and only one actuator, $F(s)$, design an *override control topology* to deal with the two objectives: stabilize and keep upright the angle of the pendulum (primary objective, $\theta_r = 0$), and keep the cart position close to the middle of the rail (secondary objective, $x_r = 0$). Design the selector based on the information of the sensors.
   d. Simulate the complete system: override structure, with the pendulum angle control system and the cart position control system. Being at the initial conditions ($t = 0$) at $\theta_r = \theta = 0°$ (upright position) and $x_r = x = 0$ (middle of the rail), introduce positive and negative small pulse disturbances $d\theta$ at the pendulum angle $\theta(s)$ and analyze the results—see Figure P3.2.

Project P4. *Interconnected Micro-Grids Control*
   1. *Chapters and Sections Needed*: Chapters 2, 9; Appendix 2; QFTCT.
   2. *Project Description*: Consider the two interconnected micro-grids shown in Figure P4.1. The first micro-grid "a" is composed of one conventional generator

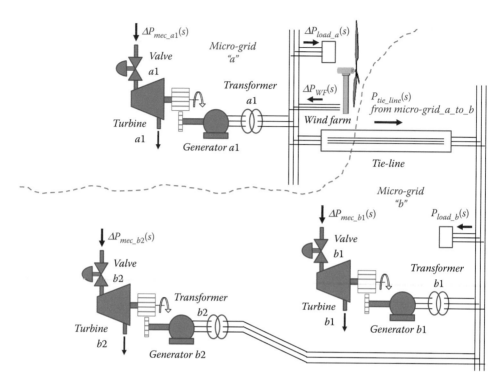

FIGURE P4.1
Interconnected micro-grids.

and a wind farm. The second micro-grid "*b*" is composed of two conventional generators.

Each conventional generator includes a valve, a turbine, a gearbox, an electrical generator, and a transformer. The valve can regulate the mechanical power ΔP_{mec} that enters in the turbine, and as a result the electrical power that goes to the micro-grid. Each micro-grid has an electrical load ΔP_{load} that consumes the power. The wind farm injects a random power ΔP_{WF} into the first micro-grid. The micro-grids are connected through a tie-line. The power that goes from micro-grid "*a*" to micro-grid "*b*" through the tie-line is denoted as $\Delta P_{tie\text{-}line}$. The frequency variation in the first micro-grid is expressed as $\Delta\omega_a = \omega_a - \omega_{ref}$, and the frequency variation in the second micro-grid as $\Delta\omega_b = \omega_b - \omega_{ref}$, being the frequency base $\omega_{ref} = 377$ rad/s (= 60 Hz).

3. *Modeling*: The Laplace model for the two micro-grids, with the electrical generators, turbines valves and the tie line, is shown in Figure P4.2. The details of the control system are shown in Figures P4.3 through P4.6.

The parameters of the system are:

Micro-grid "*a*"

Generator unit a1

- Turbine and generator: $M_{a1} = 4.5$ pu MW/pu freq/s.
- Valve: $T_{va1} = 2$ s. Valve saturation: upper limit = +1, lower limit = 0.

Change in load due to change in frequency in micro-grid "*a*": $D_a = 0.9$.

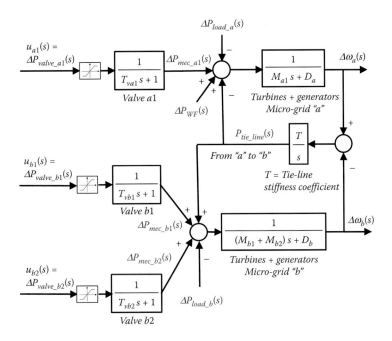

FIGURE P4.2
Model of micro-grids "*a*" and "*b*" and tie-line.

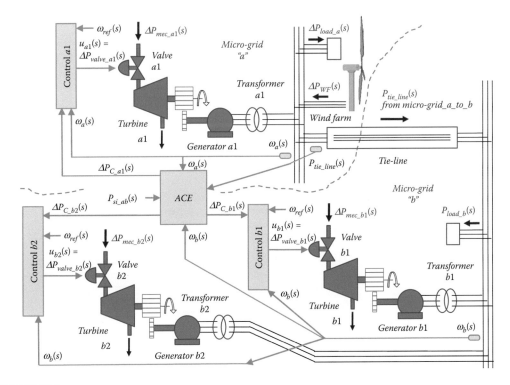

FIGURE P4.3
Frequency control systems (*a1*, *b1*, and *b2*) and area control error system (ACE) for the two interconnected micro-grids.

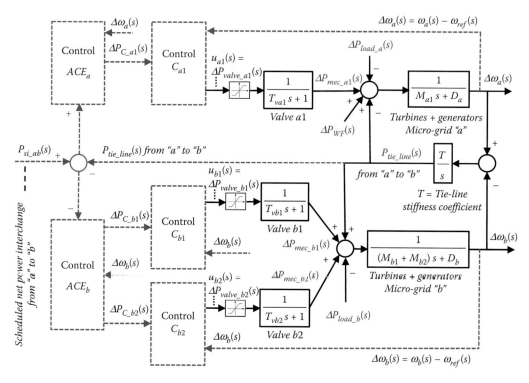

FIGURE P4.4
Frequency control systems (C_{a1}, C_{b1}, and C_{b2}) and area control error system (*ACEa* and *ACEb*) for the two inter-connected micro-grids.

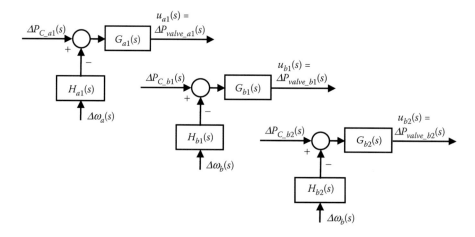

FIGURE P4.5
Frequency control blocks: C_{a1}, C_{b1}, and C_{b2}.

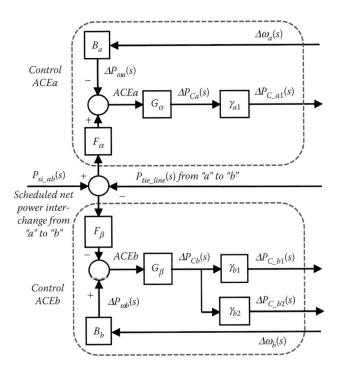

FIGURE P4.6
Area control error blocks: ACE_a and ACE_b.

Participation factors: $\gamma_{a1} = 1$.
Frequency bias factor: $B_a = D_a + H_{a1}$.

Micro-grid "b"
Generator unit b1
- Turbine and generator: $M_{b1} = 2.5$ pu MW/pu freq/s.
- Valve: $T_{vb1} = 5$ s. Valve saturation: upper limit $= +1$, lower limit $= 0$.
Generator unit b2
- Turbine and generator: $M_{b2} = 3$ pu MW/pu freq/s.
- Valve: $T_{vb2} = 4.5$ s. Valve saturation: upper limit $= +1$, lower limit $= 0$.
Change in load due to change in frequency in micro-grid "b": $D_b = 0.8$.
Participation factors: $\gamma_{b1} = 0.7$; $\gamma_{b2} = 0.3$.
Frequency bias factor: $B_b = D_b + H_{b1} + H_{b2}$.

Tie line
Tie-line stiffness coefficient: $T = 377 \times 0.02$ pu $= 7.5$ pu.

Bases
Frequency base $= 377$ rad/s ($= 60$ Hz).

Micro-grid "*a*". MVA base = 2.

Micro-grid "*b*". MVA base = 10.

4. *Control Specifications*

- *For the frequency control loops, $G_{a1}(s)$, $G_{b1}(s)$, and $G_{b2}(s)$.*

  – *Type 1: Stability specification*

$$|T_1(j\omega)| = \left|\frac{p(j\omega)G(j\omega)}{1+p(j\omega)G(j\omega)}\right| \le \delta_1(\omega) = W_s = 2.62 \tag{P4.1}$$

$$\omega \in [0.01\ \ 0.05\ \ 0.1\ \ 0.5\ \ 1\ \ 5\ \ 10\ \ 50\ \ 100\ \ 500]\ \text{rad/s}$$

which is equivalent to $PM = 22.0°$, $GM = 2.8$ dB—see Equations 2.30 and 2.31.

- *For the area control error loops, $G_\alpha(s)$, $F_\alpha(s)$, $G_\beta(s)$, and $F_\beta(s)$.*

  – *Type 1: Stability specification*

$$|T_1(j\omega)| = \left|\frac{p(j\omega)G(j\omega)}{1+p(j\omega)G(j\omega)}\right| \le \delta_1(\omega) = W_s = 1.001 \tag{P4.2}$$

$$\omega \in [0.01\ \ 0.05\ \ 0.1\ \ 0.5\ \ 1\ \ 5\ \ 10\ \ 50\ \ 100\ \ 500]\ \text{rad/s}$$

which is equivalent to $PM = 59.93°$, $GM = 6.01$ dB—see Equations 2.30 and 2.31.

- *Type 6: Reference tracking specification*—see Figure P4.7

$$\delta_{6\_lo}(\omega) < |T_6(j\omega)| = \left|F(j\omega)\,\frac{p(j\omega)G(j\omega)}{1+p(j\omega)G(j\omega)}\right| \le \delta_{6\_up}(\omega) \tag{P4.3}$$

$$\omega \in [0.01\ \ 0.05\ \ 0.1\ \ 0.5\ \ 1]\ \text{rad/s}$$

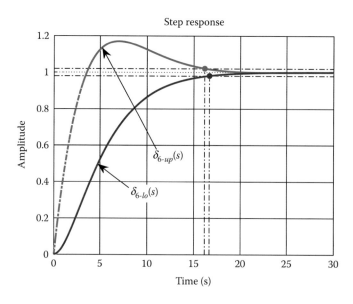

FIGURE P4.7

Reference tracking specifications for area control error.

$$\delta_{6\_lo}(s) = \frac{(1-\varepsilon_{lo})}{[(s/a_L)+1]^2}; \quad a_L = 0.35; \quad \varepsilon_{lo} = 0 \tag{P4.4}$$

$$\delta_{6\_up}(s) = \frac{[(s/a_U)+1](1+\varepsilon_{up})}{[(s/\omega_n)^2+(2\zeta s/\omega_n)+1]}; \quad a_U = 0.2; \quad \zeta = 0.8; \quad \omega_n = \frac{1.25\,a_U}{\zeta}; \quad \varepsilon_{up} = 0 \tag{P4.5}$$

5. *Controllers' Design*: We propose two control solutions for this project. The first one is based on the popular Droop control + PI control, which is very common in power systems. The second one is an open design that allows the designer to use any structure for a higher performance.

- *Case 1. Classical droop control and PI control.*
  - *Frequency loops with droop control*: Due to the load-sharing problem, the blocks $G_{a1}(s)$, $G_{b1}(s)$, and $G_{b2}(s)$ are first-order systems and the blocks $H_{a1}(s)$, $H_{b1}(s)$, and $H_{b2}(s)$ pure gains, as shown in Equations P4.6 through P4.8.
  - *Area control error with PI control*: The blocks $G_\alpha(s)$, $G_\beta(s)$ are PI controllers and the blocks $F_\alpha(s)$, $F_\beta(s)$ unitary gains, as shown in Equations P4.9 and P4.10.

$$G_{a1}(s) = \frac{1}{(s/(k_{ia1}\,R_{a1}))+1}; \quad H_{a1}(s) = \frac{1}{R_{a1}} \tag{P4.6}$$

$$G_{b1}(s) = \frac{1}{(s/(k_{ib1}\,R_{b1}))+1}; \quad H_{b1}(s) = \frac{1}{R_{b1}} \tag{P4.7}$$

$$G_{b2}(s) = \frac{1}{(s/(k_{ib2}\,R_{b2}))+1}; \quad H_{b2}(s) = \frac{1}{R_{b2}} \tag{P4.8}$$

$$G_\alpha(s) = \left(\frac{k_\alpha}{T_{i\alpha}}\right)\frac{(T_{i\alpha}\,s+1)}{s}; \quad F_\alpha(s) = 1 \tag{P4.9}$$

$$G_\beta(s) = \left(\frac{k_\beta}{T_{i\beta}}\right)\frac{(T_{i\beta}\,s+1)}{s}; \quad F_\beta(s) = 1 \tag{P4.10}$$

- *Case 2. Open design.*
  - *Frequency loops*: The blocks $G_{a1}(s)$, $G_{b1}(s)$, $G_{b2}(s)$, $H_{a1}(s)$, $H_{b1}(s)$, and $H_{b2}(s)$ have a non-restricted structure and are designed with the robust QFT control methodology.
  - *Area control error*: The feedback controllers $G_\alpha(s)$, $G_\beta(s)$ and the prefilters $F_\alpha(s)$, $F_\beta(s)$ have a non-restricted structure and are designed with the robust QFT control methodology.

6. *Questions*
   a. Case 1. Classical droop control and PI control. For the frequency control loops of both micro-grids, design the parameters R_{a1}, k_{ia1}, R_{b1}, k_{ib1}, R_{b2}, and k_{ib2} of the droop controllers—see Equations P4.6 through P4.8—to meet the specification defined in Equation P4.1. Use QFT with the restricted droop control structure.

b. Case 1. Classical droop control and PI control. To control the interchange of power with the area control error loops, design the parameters k_α, $T_{i\alpha}$, k_β, and $T_{i\beta}$ of the PI controllers—see Equations P4.9 through P4.10—to meet the specifications defined in Equations P4.2 through P4.5. Use QFT with the restricted PI control structure.

c. Case 1. Classical droop control and PI control. Simulate the frequency control of both micro-grids and the power exchange between them with the following events:

$\omega_{ref} = 60$ Hz. Total time of simulation $= 2500$ s.

$\Delta P_{load\_a}(t = 0) = 0.6$; $\Delta P_{load\_a}(t = 750) = 0.3$;

$\Delta P_{load\_b}(t = 0) = 0.7$; $\Delta P_{load\_b}(t = 1500) = 0.2$;

$\Delta P_{WF} = 0.1 \sin(t\, 2\pi/500) + 0.1$;

$P_{si\_ab}(t = 0) = 0$; $P_{si\_ab}(t = 2000) = -0.4$ (grid b gives 0.4 pu to grid a at $t = 2000$).

d. Case 2. Open design. For the frequency control loops of both micro-grids, design non-restricted structure functions $G_{a1}(s)$, $G_{b1}(s)$, $G_{b2}(s)$, $H_{a1}(s)$, $H_{b1}(s)$, and $H_{b2}(s)$ with the robust QFT control methodology to meet the specification defined in Equation P4.1.

e. Case 2. Open design. To control the interchange of power with the area control error loops, design non-restricted structure feedback controllers $G_\alpha(s)$ and $G_\beta(s)$, and the prefilters $F_\alpha(s)$ and $F_\beta(s)$, with the robust QFT control methodology to meet the specifications defined in Equation P4.2 through P4.5.

f. Case 2. Open design. Simulate the frequency control of both micro-grids and the power exchange between them with the following events:

$\omega_{ref} = 60$ Hz. Total time of simulation $= 2500$ s.

$\Delta P_{load\_a}(t = 0) = 0.6$; $\Delta P_{load\_a}(t = 750) = 0.3$;

$\Delta P_{load\_b}(t = 0) = 0.7$; $\Delta P_{load\_b}(t = 1500) = 0.2$;

$\Delta P_{WF} = 0.1 \sin(t\, 2\pi/500) + 0.1$;

$P_{si\_ab}(t = 0) = 0$; $P_{si\_ab}(t = 2000) = -0.4$ (grid b gives 0.4 pu to grid a at $t = 2000$).

Project P5. *Distillation Column Control*

1. *Chapters and Sections Needed*: Chapters 2, 8; Appendix 2; QFTCT

2. *Project Description*: Consider the 2×2 distillation column shown in Figure P5.1. The inputs are the reflux $u_1(s)$ and the boil up $u_2(s)$, and the controlled outputs are the top and bottom product compositions $y_1(s)$ and $y_2(s)$. The transfer function matrix is[260]

$$\begin{bmatrix} y_1(s) \\ y_2(s) \end{bmatrix} = \frac{1}{\tau s + 1} \begin{bmatrix} k_1 & -86.4 \\ 108.2 & k_2 \end{bmatrix} \begin{bmatrix} u_1(s) \\ u_2(s) \end{bmatrix} \tag{P5.1}$$

with the parameters: $k_1 = 87.8$, $k_2 = -109.6$, and $\tau = 75$.

3. *Control Specifications*: For each loop, the control specifications are

• *Type 1: Stability specification*

$$\left| \frac{p_{ii}(s)\, g_{ii}(s)}{1 + p_{ii}(s)\, g_{ii}(s)} \right| \le \delta_1(\omega) = W_s = 1.305 \tag{P5.2}$$

$\omega \in [0.001\ \ 0.005\ \ 0.01\ \ 0.05\ \ 0.1\ \ 0.5\ \ 1\ \ 5\ \ 10\ \ 50\ \ 100\ \ 500]$ rad/s

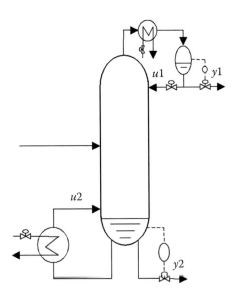

FIGURE P5.1
A 2 × 2 distillation column.

which is equivalent to PM = 45.06°, GM = 4.94 dB.

- *Type 3: Sensitivity or disturbances at plant output specification*

$$\left|\frac{y_i(s)}{d_i(s)}\right| = \left|\frac{1}{1+p_{ii}(s)\,g_{ii}(s)}\right| \le \delta_3(\omega) = \frac{(s/a_d)}{(s/a_d)+1}; a_d = 1 \tag{P5.3}$$

$$\omega \in [0.001\ \ 0.005\ \ 0.01\ \ 0.05\ \ 0.1\ \ 0.5\ \ 1]\ \text{rad/s}$$

4. *Questions. Group 1*

   a. Calculate the Bristol relative gain analysis array.
   b. Find the best input/output pairing and describe the coupling (in %) between the loops.
   c. Design two SISO controllers, $g_{11\text{siso}}(s)$ and $g_{22\text{siso}}(s)$, with QFT. Suggested structure: 1 zero, 1 pole, and 1 integrator.
   d. Simulate the MIMO system with the two independent SISO controllers calculated in the previous question.
   e. Design a MIMO QFT controller according to Method 2 (Sections 8-6.2 and 8.9, with a 2 × 2 G_α, $g_{11}{}^\beta$ and $g_{22}{}^\beta$). Suggested structure for $g_{ii}{}^\beta(s)$: 1 zero, 1 pole, 1 integrator. Hint: for matrix transfer operations, use *minreal(A,0.1)*.
   f. Simulate the MIMO system with the MIMO QFT control solution calculated in the previous question.

NOTE: for the simulations, use the following parameters: $t_{final} = 80$ s, $t_{inc} = 0.01$ s, reference for the first loop: $r_1 = 0$ (from $t = 0$ to $t = 1$ s), and $r_1 = 1$ (from $t = 1$ to $t = 80$ s), and reference for the second loop: $r_2 = 0$ (from $t = 0$ to $t = 40$ s), and $r_2 = 1$ (from $t = 40$ to $t = 80$ s).

5. *Questions. Group 2*

   a. Design a MIMO QFT controller according to Method 1 (Sections 8-5 and 8.8). Suggested structure for $g_{ii}(s)$: 1 zero, 1 pole, and 1 integrator. Hint: for matrix transfer operations, use *minreal*(A,0.1).

   b. Simulate the MIMO system with the MIMO QFT control solution calculated in the previous question.

6. *Questions. Group 3*

   Consider again the model of the distillation column (P5.1), now with parametric uncertainty as $k_1 \in [85, 95]$, $k_2 \in [-115, -105]$, and $\tau \in [70, 80]$.

   a. Design a MIMO QFT controller according to Method 2 (Sections 8-6.2 and 8.9, with a 2×2 $\boldsymbol{G}_\alpha$, $g_{11}{}^\beta$ and $g_{22}{}^\beta$). Hint: for matrix transfer operations, use *minreal*(A,0.1).

   b. Simulate the MIMO system with the MIMO QFT control solution calculated in the previous question.

7. *Questions. Group 4*

   Again with uncertainty $k_1 \in [85, 95]$, $k_2 \in [-115, -105]$, and $\tau \in [70, 80]$.

   a. Design a MIMO QFT controller according to Method 1 (Sections 8-5 and 8.8). Hint: for matrix transfer operations, use *minreal*(A,0.1).

   b. Simulate the MIMO system with the MIMO QFT control solution calculated in the previous question.

   For the simulation parts, we suggest the following MATLAB code:

```
tau = 75;   k1 = 87.8;   k2 = -109.6;
P0 = [k1 -86.4;108.2 k2];   P = tf(1,[tau 1]) * P0;
P.InputName = {'u'};   P.OutputName = {'y'};

G = Ga*[g11b 0; 0 g22b]; % or G = [g11 0; 0 g22]; Group 1 or 3
                         % or G = [g11 g12; g21 g22]; Group 2 or 4

G.InputName = {'e'};   G.OutputName = {'u'};
Sum = sumblk('e = r - y',2);
CLry = connect(P,G,Sum,'r','y'); % Closed loop MIMO system
tinc = 0.01;   t = [0:tinc:80];
U1 = [[0:tinc:1]*0,1+[1+tinc:tinc:80]*0]; % input 1
U2 = [[0:tinc:40]*0,1+[40+tinc:tinc:80]*0]; % input 2
figure,lsim(CLry,[U1;U2],t); % Fig
```

Project P6. *Central Heating System Control*

1. *Chapters and Sections Needed*: Chapters 2, 4, 9; Appendix 2; QFTCT.

2. *Project Description*: Consider the central heating system of a three floor building shown in Figure P6.1. The heating system consists of a number of radiators on every floor connected through a network of pipes to a 3-way mixing valve in the basement, which in turn connects to the boilers that heat up the water

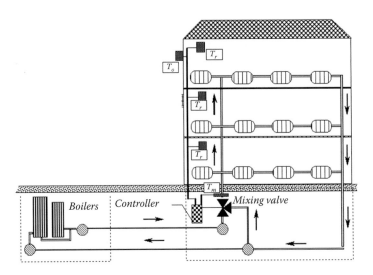

FIGURE P6.1
Central heating system.

that flows through the pipes. The variable to be controlled is the room temperature $T_r(s)$, which is measured by appropriate sensors located inside the building.

The actuator is the 3-way valve. The water temperature at the output of the 3-way valve is the mixed temperature $T_m(s)$, also measured by an appropriate sensor.

3. *Modeling*: The transfer function between the desired mixed temperature $T_{md}(s)$, or controller output, and the actual mixed temperature $T_m(s)$ at the output of the 3-way valve is

$$\frac{T_m(s)}{T_{md}(s)} = P_2(s) = \frac{k_2}{\tau_2 s + 1} \tag{P6.1}$$

Also, the transfer function between the actual mixed temperature $T_m(s)$ at the output of the 3-way valve and the room temperature $T_r(s)$ is

$$\frac{T_r(s)}{T_m(s)} = P_1(s) = \frac{k_1}{\tau_1 s + 1} e^{-sL_1} \tag{P6.2}$$

and the transfer function between the disturbances $d(s)$ and the actual mixed temperature $T_m(s)$ at the output of the 3-way valve is

$$\frac{T_m(s)}{d(s)} = M(s) = 1 \tag{P6.3}$$

Using system identification techniques with experimental data, we found the following parameters: $k_2 \in [5, 8]$; $\tau_2 \in [2, 5]$ s; $k_1 \in [40, 60]$; $\tau_1 \in [800, 1200]$ s; and $L_1 \in [100, 200]$ s.

4. *Control System Topology*: The proposed control system topology consists of a cascade control diagram, with an inner loop (secondary loop) for the valve and water temperature control, and an outer loop (primary loop) with a Smith Predictor for the room temperature control—see Figure P6.2.

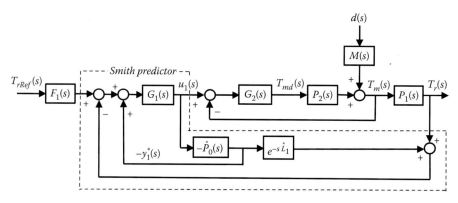

FIGURE P6.2
Control system: Cascade control and Smith predictor.

5. Control Specifications

- *Inner loop*: Taking into account the dynamics of the 3-way valve, we define the following robust control specifications for the inner loop:
  - *Type 1: Stability specification*

$$|T_1(j\omega)| = \left| \frac{P_2(j\omega)G_2(j\omega)}{1 + P_2(j\omega)G_2(j\omega)} \right| \leq \delta_1(\omega) = W_s = 1.01 \tag{P6.4}$$

$$\omega \in [0.001\ 0.005\ 0.01\ 0.05\ 0.1\ 0.5\ 1\ 5\ 10\ 50\ 100\ 500\ 1000]\ \mathrm{rad/s}$$

which is equivalent to $PM = 59.34°$, $GM = 5.97$ dB—see Equations 2.30 and 2.31.

  - *Type 3: Sensitivity or disturbances at plant output specification*

$$|T_3(j\omega)| = \left| \frac{T_m(j\omega)}{d(j\omega)} \right| = \left| \frac{M(j\omega)}{1 + P_2(j\omega)G_2(j\omega)} \right| \leq \delta_3(\omega) = \frac{(s/a_d)}{(s/a_d)+1} \ ; \ a_d = 10 \tag{P6.5}$$

$$\omega \in [0.001\ 0.005\ 0.01\ 0.05\ 0.1\ 0.5\ 1\ 5\ 10]\mathrm{rad/s}$$

- *Outer loop*: Also, taking into account the dynamics of thermal behavior of the room, we define the following robust control specifications for the outer loop:
  - *Type 1: Stability specification*

$$|T_1(j\omega)| = \left| \frac{P_e(j\omega)G_1(j\omega)}{1 + P_e(j\omega)G_1(j\omega)} \right| \leq \delta_1(\omega) = W_s = 1.46 \tag{P6.6}$$

$$\omega \in [0.001\ 0.005\ 0.01\ 0.05\ 0.1\ 0.5\ 1\ 5\ 10\ 50\ 100\ 500\ 1000]\ \mathrm{rad/s}$$

which is equivalent to $PM = 40.05°$, $GM = 4.53$ dB—see Equations 2.30 and 2.31.

  - *Type 6: Reference tracking specification*

$$\delta_{6\_lo}(\omega) < |T_6(j\omega)| = \left| F_1(j\omega) \frac{P_e(j\omega)G_1(j\omega)}{1 + P_e(j\omega)G_1(j\omega)} \right| \leq \delta_{6\_up}(\omega) \tag{P6.7}$$

$$\omega \in [0.01\ 0.05\ 0.1\ 0.5\ 1]\ \mathrm{rad/s}$$

$$\delta_{6\_lo}(s) = \frac{(1-\varepsilon_{lo})}{\left[(s/a_L)+1\right]^2}; \; a_L = 0.08; \; \varepsilon_{lo} = 0 \tag{P6.8}$$

$$\delta_{6\_up}(s) = \frac{\left[(s/a_U)+1\right](1+\varepsilon_{up})}{\left[(s/\omega_n)^2+(2\zeta s/\omega_n)+1\right]}; \; a_U = 0.05; \; \zeta = 0.8; \; \omega_n = \frac{1.25\,a_U}{\zeta}; \; \varepsilon_{up} = 0 \tag{P6.9}$$

6. *Questions*

   a. Design a feedback controller $G_2(s)$ for the inner loop to control the water temperature $T_m(s)$ at the output of the 3-way valve according to the stability and performance specifications defined above—Equations P6.4 and P6.5.

   b. Design the feedback controller $G_1(s)$, the prefilter $F_1(s)$, and the Smith predictor blocks for the outer loop to control the room temperature $T_r(s)$ according to the stability and performance specifications defined above—Equations P6.6 through P6.9. Note that the plant $P_e(s)$ that this controller $G_1(s)$ sees is

$$P_e(s) = P_1(s)\frac{P_2(s)G_2(s)}{1+P_2(s)G_2(s)} \tag{P6.10}$$

and its model

$$\hat{P}_e(s) = \hat{P}_1(s)\frac{\hat{P}_2(s)G_2(s)}{1+\hat{P}_2(s)G_2(s)} = \hat{P}_0(s)\,e^{-s\hat{L}_1} \tag{P6.11}$$

with (see Figure P6.2)

$$\hat{P}_0(s) = \frac{\hat{k}_1}{(\hat{\tau}_1 s+1)}\frac{\left[\hat{P}_2(s)G_2(s)\right]}{\left[1+\hat{P}_2(s)G_2(s)\right]} \tag{P6.12}$$

$$\hat{P}_2(s) = \frac{\hat{k}_2}{\hat{\tau}_2 s+1} \tag{P6.13}$$

   The feedback controller $G_1(s)$ and the prefilter $F_1(s)$ have to be designed for the plant $P_0(s)$—Equation P6.12, and according the control specifications defined in Equations P6.6 through P6.9. Then, the Smith predictor blocks (see Figure P6.2) have to be added following the methodology presented in Chapter 4. Hint: when calculating Equation P6.12 in MATLAB, use the command *minreal*(P_0) to avoid MATLAB round error problems.

   c. Simulate the complete system: inner loop, outer loop, and Smith predictor— see Figure P6.2. Consider a unitary step reference $T_{rRef}(s)$ at $t = 1$ s, and a unitary step disturbance $d(s)$ at $t = 400$ s. Simulation time $t = 800$ s.

Project P7. *Multi-Tank Hydraulic Control System*

1. *Chapters and Sections Needed*: Chapters 2, 9; Appendix 2; QFTCT.

2. *Project Description*: Consider the multi-tank hydraulic system shown in Figure P7.1. It is composed of three tanks of water placed at different levels and connected through a network of pipes, two valves [$v_1(s)$ and $v_2(s)$], four temperature sensors [$T_1(s)$ to $T_4(s)$], a flow sensor $F_1(s)$, and a heat exchanger fed with hot oil in the second tank.

   The water that enters into the system through the first valve has a low temperature—measured by $T_1(s)$. The heat exchanger in the second tank can increase the water temperature in the system. The project's objective is to control the output temperature $T_4(s)$ following the reference $T_{ref}(s)$. The main disturbance of the system is the variation of the water temperature at the input of the first valve.

3. *Modeling*: The transfer functions between the input water temperature $T_1(s)$ and the temperature at the output of the second tank $T_3(s)$, between the second valve $v_2(s)$ and the temperature at the output of the second tank $T_3(s)$, and between the temperature at the output of the second tank $T_3(s)$ and the final output water temperature $T_4(s)$ are, respectively,

$$\frac{T_3(s)}{T_1(s)} = P_1(s) = \frac{1}{\tau_2 s + 1} \frac{1}{\tau_1 s + 1} \tag{P7.1}$$

$$\frac{T_3(s)}{v_2(s)} = P_2(s) = \left(\frac{1}{F_Q \rho C_p}\right) \frac{1}{\tau_2 s + 1} \tag{P7.2}$$

$$\frac{T_4(s)}{T_3(s)} = P_3(s) = \frac{1}{\tau_3 s + 1} \tag{P7.3}$$

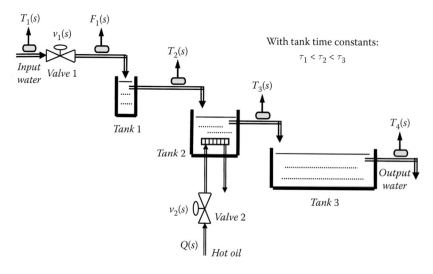

FIGURE P7.1
Multi-tank hydraulic system.

and

$$T_4(s) = P_3(s)\, T_3(s) \tag{P7.4}$$

$$T_3(s) = P_2(s)\, v_2(s) + P_1(s)\, T_1(s) \tag{P7.5}$$

with $\tau_1 = 10$ s, $\tau_2 = 30$ s, $\tau_3 = 90$ s, and $[1/(F_Q\,\rho\,C_p)] \in [2.9, 3.1]$.

4. *Control Specifications*: Taking into account the dynamics of the multi-tank hydraulic system and the operation objectives, we define the following robust control specifications:

- *Type 1: Stability specification*

$$|T_1(j\omega)| = \left| \frac{P(j\omega)G(j\omega)}{1 + P(j\omega)G(j\omega)} \right| \le \delta_1(\omega) = W_s = 1.46 \tag{P7.6}$$

$$\omega \in [0.0001\ 0.0005\ 0.001\ 0.005\ 0.01\ \ 0.05\ \ 0.1\ \ 0.5\ \ 1\ \ 5\ \ 10]\ \mathrm{rad/s}$$

which is equivalent to $PM = 40.05°$, $GM = 4.53$ dB—see Equations 2.30 and 2.31.

- *Type 3: Sensitivity or disturbances at plant output specification*

$$|T_3(j\omega)| = \left| \frac{x(j\omega)}{d_o(j\omega)} \right| = \left| \frac{1}{1 + P(j\omega)G(j\omega)} \right| \le \delta_3(\omega) = \frac{(s/a_d)}{(s/a_d)+1}\,;\ a_d = 0.01 \tag{P7.7}$$

$$\omega \in [0.0001\ 0.0005\ 0.001\ 0.005\ 0.01]\ \mathrm{rad/s}$$

- *Type 6: Reference tracking specification*

$$\delta_{6\_lo}(\omega) < |T_6(j\omega)| = \left| F(j\omega)\, \frac{P(j\omega)G(j\omega)}{1 + P(j\omega)G(j\omega)} \right| \le \delta_{6\_up}(\omega) \tag{P7.8}$$

$$\omega \in [0.0001\ 0.0005\ 0.001\ 0.005\ 0.01\ \ 0.05\ \ 0.1]\ \mathrm{rad/s}$$

$$\delta_{6\_lo}(s) = \frac{(1 - \varepsilon_{lo})}{[(s/a_L)+1]^2}\,;\quad a_L = 0.01;\quad \varepsilon_{lo} = 0 \tag{P7.9}$$

$$\delta_{6\_up}(s) = \frac{[(s/a_U)+1]\,(1+\varepsilon_{up})}{[(s/\omega_n)^2 + (2\zeta s/\omega_n)+1]}\,;\ a_U = 0.01;\ \zeta = 0.8;\ \omega_n = \frac{1.25\,a_U}{\zeta}\,;\varepsilon_{up} = 0 \tag{P7.10}$$

5. *Questions*
   a. Design a single feedback controller $G_{1s}(s)$ and a prefilter $F_{1s}(s)$ for a control loop with the sensor $T_4(s)$ according to Figure P7.2. The project's objective is to control the water temperature $T_4(s)$ at the output of the system following the performance specifications defined in Equations P7.6 through P7.10, and rejecting the $T_1(s)$ disturbances as much as possible.
   b. Design a cascade control system, with an inner controller $G_{2c}(s)$ that uses the sensor $T_3(s)$, and an outer controller $G_{1c}(s)$ and a prefilter $F_{1c}(s)$ with the sensor $T_4(s)$ according to Figure P7.3. The project's objective is to control the

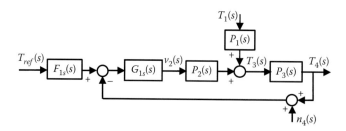

FIGURE P7.2
Single feedback control system for the multi-tank process.

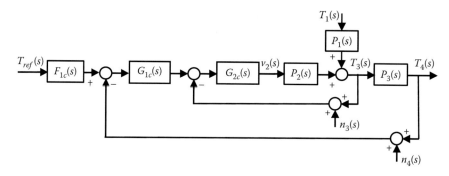

FIGURE P7.3
Cascade control system for the multi-tank process.

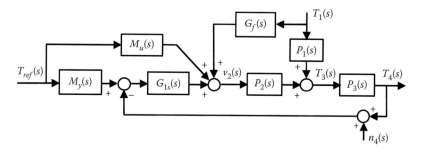

FIGURE P7.4
Feedforward/feedback control system for the multi-tank process.

water temperature $T_4(s)$ at the output of the system following the performance specifications defined in Equations P7.6 through P7.10, and rejecting the $T_1(s)$ disturbances as much as possible.

c. Design a feedforward/feedback control system, with the feedback controller $G_{1s}(s)$ designed in the first question, with a feedforward controller $G_f(s)$ from the sensor $T_1(s)$ for disturbance rejection, and with a model matching (or reference feedforward) controller $M_u(s)$ and $M_y(s)$ from the reference $T_{ref}(s)$ according to Figure P7.4. The project's objective is to control the water temperature $T_4(s)$ at the output of the system following the performance specifications defined in Equations P7.6 through P7.10, and rejecting the $T_1(s)$ disturbances as much as possible.

d. Simulate the system according to the block diagrams of Figures P7.2, P7.3 and P7.4, and with the controllers designed in questions 1, 2, and 3, respectively. The total time of simulation is 3000 s. The reference signal $T_{ref}(s)$ is a unitary step input that changes from 0 to 1 at time $t = 1$ s. The disturbance signal $T_1(s)$ is a step input that changes from 0 to 2 at time $t = 1500$ s.

Project P8. *Attitude Control of a Satellite with Tanks Partially Filled*

1. *Chapters and Sections Needed*: Chapter 2; Appendix 2; QFTCT.

2. *Project Description*: Large satellites need a substantial amount of fuel to place them into orbit or make orbit corrections. In some geostationary and deep space satellites the initial fuel mass is typically about 40% of the total satellite mass, and is contained in tanks in a liquid state. When the tanks are partially filled and the satellite accelerates, the fuel moves uncontrollably inside the tanks, producing the so-called *sloshing effect*.

   There are several approaches to model this sloshing effect. In this project, we apply the approach discussed by M. Sidi—see Reference 363. According to this approach, a mass connected to a spring and a damper models the movement of the liquid inside the tank. Figure P8.1 shows the variables and parameters for attitude dynamics of a satellite in deep space and for the fuel movement inside the tanks. The satellite has several thrusters, of which Th.0 to Th.4 appear in the figure. The main thruster Th.0 moves the satellite in the main direction (y_1), and the additional Th.1 to Th.4 modify the angle $\theta(t)$.

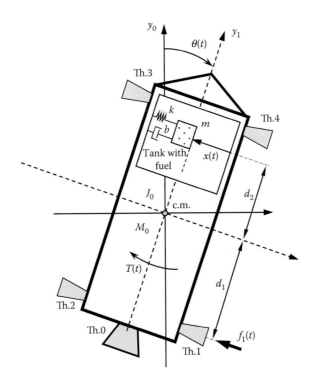

FIGURE P8.1
Satellite description. Attitude dynamics with fuel slosh.

The moment of inertia of the satellite without fuel and with respect to the center of mass (c.m.) is J_0. The total mass of the satellite without fuel is M_0. The mass of the fuel is m, and its relative movement to the satellite frame is $x(t)$, with second-order dynamics with k as the stiffness and b as the damping coefficients. As we fire the thrusters Th.1/3 (or Th.2/4), a force $f_1(t)$ (or $-f_1(t)$) at a distance d_1 of the c.m. is applied. This means that we apply a clockwise (or counter clockwise) torque on the satellite c.m., which is:

$$T(t) = d_1 f_1(t) \tag{P8.1}$$

The relative movement of the mass m of fuel is $x(t)$, and its absolute movement is

$$\theta(t) d_2 - x(t) \tag{P8.2}$$

The parameters of the satellite are the following:
- $J_0 = 575$ (kg m^2) is the moment of inertia of the satellite without fuel and respect to c.m.
- $M_0 = 900$ (kg) is the mass of the satellite without fuel.
- $m \in [100, 200]$ (kg) is the mass of fuel.
- $k = 337.5$ (N/m) is the stiffness coefficient of the fuel model.
- $b \in [0.01, 0.02]$ (N/m s^{-1}) is the damping coefficient of the fuel model.
- $d_1 = 1.2$ (m) is the distance between the thrusters and c.m.
- $d_2 = 0.9$ (m). This is the distance between the mass m that represents the fuel and the c.m. of the satellite.

The variable we want to control is $\theta(t)$, which is the angle of motion of the central body, in rad. The actuators input is the force $f_1(t)$, in N, applied by the thrusters Th.1/3 (or Th.2/4).

3. *Questions*

a. *Modeling I.* Using the Euler–Lagrange approach and the symbolic MATLAB toolbox, calculate the state-space model of the attitude dynamics of the satellite with the fuel slosh. Find the equations in terms of the parameters J_0, m, k, b, d_1, and d_2. As in previous projects, the general form of the Euler–Lagrange equation is

$$\frac{d}{dt}\left(\frac{\partial L}{\partial \dot{q}_i}\right) - \frac{\partial L}{\partial q_i} + \frac{\partial D_d}{\partial \dot{q}_i} = Q_i, \quad i = 1, 2, \dots n \quad \text{and} \quad L = E_k - E_p \tag{P8.3}$$

where q_i are the generalized coordinates or degrees of freedom of the system, Q_i the generalized forces applied to each subsystem "i", E_k and E_p the kinetic and potential energies, respectively, D_d the dissipator co-content, and L the Lagrangian. The generalized coordinates for the satellite studied here are the angle of motion of the central body and the relative movement of the liquid in the tank, which are respectively: $q_1 = \theta(t)$ and $q_2 = x(t)$. The generalized force applied to the satellite by the thrusters Th.1/3 (or by Th.2/4 in the opposite direction) is $f_1(t)$. It produces a torque—see Equation P8.1, about the axis of rotation of the central body, $i = 1$, being: $Q_1 = f_1(t)$ and $Q_2 = 0$. Knowing that the general equation of motion of a mechanical system is

$$\mathcal{M}\,\ddot{q} + \mathcal{C}\,\dot{q} + \mathcal{K}\,q = \mathcal{Q}\,(\dot{q}, q, u, t) \tag{P8.4}$$

where $\mathcal{M}$, $\mathcal{C}$, and $\mathcal{K}$ are the *Mass*, *Damping*, and *Stiffness* matrices, $\mathcal{Q}$ the matrix of inputs, q the vector of generalized coordinates, and u the vector of inputs, find the symbolic state-space representation of the dynamics of the satellite. The terms for the Euler–Lagrange equation are

$$\frac{d}{dt}\left(\frac{\partial L}{\partial \dot{q}_i}\right) = \frac{d}{dt}\begin{bmatrix} \dfrac{\partial E_k}{\partial \dot{\theta}} \\[2mm] \dfrac{\partial E_k}{\partial \dot{x}} \end{bmatrix} = \mathcal{M}\begin{bmatrix} \ddot{\theta} \\ \ddot{x} \end{bmatrix} = \mathcal{M}\,\ddot{q} \tag{P8.5}$$

$$-\frac{\partial L}{\partial q_i} = \frac{\partial E_p}{\partial q_i} = \begin{bmatrix} \dfrac{\partial E_p}{\partial \theta} \\[2mm] \dfrac{\partial E_p}{\partial x} \end{bmatrix} = \mathcal{K}\begin{bmatrix} \theta \\ x \end{bmatrix} = \mathcal{K}\,q \tag{P8.6}$$

$$\frac{\partial D_d}{\partial \dot{q}_i} = \begin{bmatrix} \dfrac{\partial D_d}{\partial \dot{\theta}} \\[2mm] \dfrac{\partial D_d}{\partial \dot{x}} \end{bmatrix} = \mathcal{C}\begin{bmatrix} \dot{\theta} \\ \dot{x} \end{bmatrix} = \mathcal{C}\,\dot{q} \tag{P8.7}$$

$$\mathcal{Q} = \mathcal{R}\begin{bmatrix} f_1 \\ 0 \end{bmatrix} = \mathcal{R}\,u \tag{P8.8}$$

Rearranging the general equation of motion as

$$\ddot{q} = -\mathcal{M}^{-1}\mathcal{C}\,\dot{q} - \mathcal{M}^{-1}\mathcal{K}\,q + \mathcal{M}^{-1}\mathcal{R}\,u, \tag{P8.9}$$

the state-space representation ($\dot{x} = Ax + Bu$; $y = Cx$) is

$$\dot{x} = \left[\begin{array}{c|c} -\mathcal{M}^{-1}\mathcal{C} & -\mathcal{M}^{-1}\mathcal{K} \\ \hline I & 0 \end{array}\right]x + \left[\begin{array}{c} \mathcal{M}^{-1}\mathcal{R} \\ \hline 0 \end{array}\right]u$$

$$y = \begin{bmatrix} 0 & | & I \end{bmatrix}x \tag{P8.10}$$

where the state vector is $x = \begin{bmatrix} \dot{q} \\ q \end{bmatrix} = \begin{bmatrix} \dot{\theta} & \dot{x} & \theta & x \end{bmatrix}^T$, the input vector $u = [f_1\ 0]^T$, and the output vector $y = [\theta\ x]^T$.

Please, provide the following answers:

a.1 The symbolic expressions of the state-space matrices A, B, C, and D, in terms of the parameters J_0, m, k, b, d_1, and d_2.

a.2 The MATLAB m.file that calculates the symbolic expressions.

b. *Modeling II.* From the state-space representation of the dynamics of the satellite calculated in the previous question, and applying the expression for $P(s) = C(sI - A)^{-1}B$ for $y(s) = P(s)u(s)$, find the two transfer functions of the satellite: $p_{11}(s) = \theta(s)/f_1(s)$ and $p_{21}(s) = x(s)/f_1(s)$, in terms of the parameters J_0, m, k, b, d_1, d_2, and s. Use the symbolic MATLAB toolbox.
Please, provide the following answers:

b.1 The symbolic expressions of the two transfer functions $p_{11}(s)$ and $p_{21}(s)$ in terms of the parameters J_0, m, k, b, d_1, d_2, and s.

b.2 The MATLAB m.file that calculates the symbolic expressions.

c. *Controller Design I.* For the transfer functions found in the previous question, the diagram of Figure P8.2, the numerical values of the parameters of the satellite, and $H(s) = 1$, $F(s) = 1$:

c.1 For the nominal case (first values of the parametric uncertainty), find the numerical expressions of $p_{11}(s) = \theta(s)/f_1(s)$ and $p_{21}(s) = x(s)/f_1(s)$.

c.2 Plot the Bode diagrams of $p_{11}(s)$ and $p_{21}(s)$ and analyze the characteristics.

c.3 Design a *lead/lag controller* $G(s)$ to control the angle of the satellite $\theta(s)$ with the thruster force $f_1(s)$. The control specifications are: *Stability*: $PM \geq 35°$; *Steady-state error*: $E_{ss} = 0$ (for a reference step input); *Transient specifications*: Overshoot $M_p \leq 0.10$ per unit, Settling time $t_s \leq 150$ s.

c.4 Using Simulink, simulate the closed-loop control system. Show the evolution of the angle of the satellite $\theta(t)$ and the movement of the liquid $x(t)$ when the reference $r(t)$ changes from 0 to 1 rad at $t = 1$ s, and when an external disturbance $d_0(t)$ changes from 0 to 0.2 rad at $t = 300$ s. Total simulation $= 600$ s.

d. *Controller Design II.* Improve the controller $G(s)$ designed in the previous question with an additional *notch filter* to reduce the vibration produced by the liquid slosh. Consider the same control specifications in c.3. Also, using Simulink,

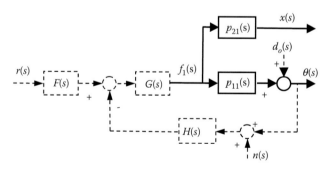

FIGURE P8.2
Control block diagram of the satellite.

simulate the closed-loop control system in the same conditions described in c.4. Analyze the vibration of the angle of the satellite $\theta(t)$ and compare the two cases, with this new controller and with the previous controller.

A1.2 Quick Problems

Problem Q1. Definition of Uncertainty

Determine whether the transfer function $P_1(s)$ with three uncertain parameters can be reduced to the transfer function $P_2(s)$ with only two related uncertain parameters, in order to reduce the number of operations needed to compute the templates.

$$P_1(s) = \frac{c}{as+b}; \quad a \in [1,5]; \ b \in [10,12]; \ c \in [1,2] \tag{Q1.1}$$

$$P_2(s) = \frac{(c/a)}{s+(b/a)}$$

$$(c/a) \in \left[\frac{\min c}{\max a} = 0.2, \frac{\max c}{\min a} = 2 \right]; \quad (b/a) \in \left[\frac{\min b}{\max a} = 2, \frac{\max b}{\min a} = 12 \right] \tag{Q1.2}$$

Obtain, on the Nichols chart, the templates at $\omega = 5$ rad/s for both transfer functions. Explain the effect of the reduction (selection of P_2 instead of P_1), if any, in the controller design.

Problem Q2. Control of First-Order System with Uncertainty

Consider the simple first-order plant with parametric uncertainty shown in Equation Q2.1.

$$\frac{y(s)}{u(s)} = P(s) = \frac{k}{((s/p)+1)}; \quad \text{with } k \in [3,4], \ p \in [0.1,0.2] \tag{Q2.1}$$

Design a $G(s)$ feedback controller and a prefilter $F(s)$ to regulate automatically the plant output $y(s)$ by varying the plant input $u(s)$, and following the next specifications:

- *Type 1: Stability specification*

$$|T_1(j\omega)| = \left| \frac{P(j\omega)G(j\omega)}{1+P(j\omega)G(j\omega)} \right| \leq \delta_1(\omega) = W_s = 1.05 \tag{Q2.2}$$
$$\omega \in [0.001 \ 0.005 \ 0.01 \ 0.05 \ 0.1 \ 0.2 \ 0.5 \ 1 \ 5 \ 10 \ 20 \ 50] \text{ rad/s}$$

which is equivalent to $PM = 56.87°$, $GM = 5.81$ dB—see Equations 2.30 and 2.31.

- *Type 3: Sensitivity or disturbances at plant output specification*

$$|T_3(j\omega)| = \left| \frac{y(j\omega)}{d_o(j\omega)} \right| = \left| \frac{1}{1+P(j\omega)G(j\omega)} \right| \leq \delta_3(\omega) = \frac{(s/a_d)}{(s/a_d)+1}; \ a_d = 1 \tag{Q2.3}$$
$$\omega \in [0.001 \ 0.005 \ 0.01 \ 0.05 \ 0.1 \ 0.2 \ 0.5 \ 1] \text{ rad/s}$$

- *Type 6: Reference tracking specification*

$$\delta_{6\_lo}(\omega) < \left| T_6(j\omega) \right| = \left| F(j\omega) \, \frac{P(j\omega)G(j\omega)}{1+P(j\omega)G(j\omega)} \right| \leq \delta_{6\_up}(\omega)$$

$$\omega \in [0.001 \ 0.005 \ 0.01 \ 0.05 \ 0.1 \ 0.2 \ 0.5 \ 1 \ 5 \ 10] \ \text{rad/s}$$

(Q2.4)

$$\delta_{6\_lo}(s) = \frac{(1-\varepsilon_{lo})}{[(s/a_L)+1]^2}; \quad a_L = 1; \quad \varepsilon_{lo} = 0$$

(Q2.5)

$$\delta_{6\_up}(s) = \frac{[(s/a_U)+1] \, (1+\varepsilon_{up})}{[(s/\omega_n)^2 + (2\zeta s/\omega_n) + 1]}; \quad a_U = 1; \quad \zeta = 0.8; \quad \omega_n = \frac{1.25 \, a_U}{\zeta}; \quad \varepsilon_{up} = 0$$

(Q2.6)

Problem Q3. Control of Third-Order State-Space System with Uncertainty

Consider the third-order plant with parametric uncertainty described according to the state-space representation shown in Equations Q3.1 and Q3.2.

$$A = \begin{bmatrix} a_1 & a_2 & a_3 \\ a_4 & 0 & 0 \\ 0 & 1 & 0 \end{bmatrix}; \quad B = \begin{bmatrix} 2 \\ 0 \\ 0 \end{bmatrix}; \quad C = \begin{bmatrix} 1 & 1 & 0.75 \end{bmatrix}; \quad D = \begin{bmatrix} 0 \end{bmatrix}$$

(Q3.1)

$$\dot{x} = Ax + Bu; \quad y = Cx + Du$$

(Q3.2)

The parameters with uncertainty of the plant are

$$a_1 \in [-9,-7]; \quad a_2 \in [-5,-3]; \quad a_3 \in [-2,-1]; \quad \text{and} \quad a_4 \in [3,5]$$

(Q3.3)

Design a $G(s)$ feedback controller and a prefilter $F(s)$ to regulate automatically the first plant output $y_1(s)$ by varying the first plant input $u_1(s)$, and following the next specifications:

- *Type 1: Stability specification*

$$\left| T_1(j\omega) \right| = \left| \frac{P(j\omega)G(j\omega)}{1+P(j\omega)G(j\omega)} \right| \leq \delta_1(\omega) = W_s = 1.46$$

$$\omega \in [0.001 \ 0.005 \ 0.01 \ 0.05 \ 0.1 \ 0.5 \ 1 \ 5 \ 10 \ 50 \ 100 \ 500] \ \text{rad/s}$$

(Q3.4)

which is equivalent to $PM = 40.05°$, $GM = 4.53$ dB—see Equations 2.30 and 2.31.

- *Type 3: Sensitivity or disturbances at plant output specification*

$$\left| T_3(j\omega) \right| = \left| \frac{y(j\omega)}{d_o(j\omega)} \right| = \left| \frac{1}{1+P(j\omega)G(j\omega)} \right| \leq \delta_3(\omega) = \frac{(s/a_d)}{(s/a_d)+1}; \quad a_d = 1$$

$$\omega \in [0.001 \ 0.005 \ 0.01 \ 0.05 \ 0.1 \ 0.5 \ 1 \ 5] \ \text{rad/s}$$

(Q3.5)

- *Type 6: Reference tracking specification*

$$\delta_{6\_lo}(\omega) < |T_6(j\omega)| = \left| F(j\omega) \frac{P(j\omega)G(j\omega)}{1+P(j\omega)G(j\omega)} \right| \leq \delta_{6\_up}(\omega) \tag{Q3.6}$$

$$\omega \in [0.001 \ 0.005 \ 0.01 \ 0.05 \ 0.1 \ 0.5 \ 1 \ 5 \ 10] \text{ rad/s}$$

$$\delta_{6\_lo}(s) = \frac{(1-\varepsilon_{lo})}{[(s/a_L)+1]^2}; \quad a_L = 1; \quad \varepsilon_{lo} = 0 \tag{Q3.7}$$

$$\delta_{6\_up}(s) = \frac{[(s/a_U)+1] \ (1+\varepsilon_{up})}{[(s/\omega_n)^2+(2\zeta s/\omega_n)+1]}; \quad a_U = 1; \quad \zeta = 0.8; \quad \omega_n = \frac{1.25 \, a_U}{\zeta}; \quad \varepsilon_{up} = 0 \tag{Q3.8}$$

Problem Q4. Field-Controlled DC Motor
Direct current (DC) electrical motors can also vary the voltage applied to the field windings to change the magnetic field affecting the armature, to maintain a constant armature current i_a—see Figure 2.2. In this case, the dynamics of the motor differs from the model derived in Example 2.1—see Equation 2.17, and can be expressed as Equation Q4.1.

$$\frac{\theta_m(s)}{v(s)} = P(s) = \frac{K_T}{s(Js+b)(L_f s+R_f)} = \frac{K_T/(b \ R_f)}{s((s/(b/J))+1)((s/(R_f/L_f))+1)} \tag{Q4.1}$$

The parameters of the motor are: $K_T \in [0.3, 0.5]$ Nm/A; $J \in [5, 9] \times 10^{-4}$ Nm/s²; $b \in [5, 8] \times 10^{-5}$ rad/s; $R_f \in [0.2, 0.5]$ Ω; $L_f = 2 \times 10^{-3}$ H.

Design a $G(s)$ feedback controller and a prefilter $F(s)$ to regulate automatically the angular position $\theta_m(s)$ of the motor by varying the applied voltage $v(s)$, and following the next specifications:

- *Type 1: Stability specification*

$$|T_1(j\omega)| = \left| \frac{P(j\omega)G(j\omega)}{1+P(j\omega)G(j\omega)} \right| \leq \delta_1(\omega) = W_s = 1.33 \tag{Q4.2}$$

$$\omega \in [0.01 \ 0.1 \ 0.5 \ 1 \ 2 \ 5 \ 10 \ 20 \ 50 \ 100 \ 200 \ 500] \text{ rad/s}$$

which is equivalent to $PM = 44.16°$, $GM = 4.87$ dB—see Equations 2.30 and 2.31.

- *Type 3: Sensitivity or disturbances at plant output specification*

$$|T_3(j\omega)| = \left| \frac{\theta_m(j\omega)}{d_o(j\omega)} \right| = \left| \frac{1}{1+P(j\omega)G(j\omega)} \right| \leq \delta_3(\omega) = \frac{(s/a_d)}{(s/a_d)+1}; \quad a_d = 10 \tag{Q4.3}$$

$$\omega \in [0.01 \ 0.1 \ 0.5 \ 1 \ 2 \ 5 \ 10] \text{rad/s}$$

- *Type 6: Reference tracking specification*

$$\delta_{6\_lo}(\omega) < |T_6(j\omega)| = \left| F(j\omega) \frac{P(j\omega)G(j\omega)}{1+P(j\omega)G(j\omega)} \right| \leq \delta_{6\_up}(\omega) \tag{Q4.4}$$

$$\omega \in [0.01 \ 0.1 \ 0.5 \ 1 \ 2 \ 5 \ 10 \ 20 \ 50] \text{ rad/s}$$

$$\delta_{6\_lo}(s) = \frac{(1-\varepsilon_{lo})}{[(s/a_L)+1]^2}; \quad a_L = 10; \quad \varepsilon_{lo} = 0 \tag{Q4.5}$$

$$\delta_{6\_up}(s) = \frac{[(s/a_U)+1]\,(1+\varepsilon_{up})}{[(s/\omega_n)^2+(2\zeta s/\omega_n)+1]}; \quad a_U = 5; \quad \zeta = 0.8; \quad \omega_n = \frac{1.25\,a_U}{\zeta}; \quad \varepsilon_{up} = 0 \tag{Q4.6}$$

Problem Q5. Formation Flying Spacecraft Control. Deep Space
Consider two spacecraft flying in a formation in deep space. The second spacecraft is fixed, while the first spacecraft can activate the thrusters (u_{x1}, u_{y1}, and u_{z1}) to control the relative distances (dx, dy, and dz) between both of them. The distances are measured by RF and laser-based metrology. The satellites present the plant model shown in Equations Q5.1 through Q5.3, so that

$$dx = x_2 - x_1; \quad dy = y_2 - y_1; \quad dz = z_2 - z_1 \tag{Q5.1}$$

$$\begin{bmatrix} dx \\ dy \\ dz \end{bmatrix} = \begin{bmatrix} P_{1x}(s) & 0 & 0 \\ 0 & P_{1y}(s) & 0 \\ 0 & 0 & P_{1z}(s) \end{bmatrix} \begin{bmatrix} u_{x1} \\ u_{y1} \\ u_{z1} \end{bmatrix} \tag{Q5.2}$$

$$P_{1x}(s) = P_{1y}(s) = P_{1z}(s) = P_1(s) = \frac{-1}{m_1 s^2}, \quad m_1 \in [360, \, 460]\,\text{kg} \tag{Q5.3}$$

Note that the formation has three independent SISO systems, one for each axis (x, y, and z), and the plant models have uncertainty due to fuel consumption.

Design a feedback controller $G(s)$ and a prefilter $F(s)$ to regulate automatically and with no ground intervention the relative distance dx between the two spacecraft—see Figure Q5.1, and for the following specifications:

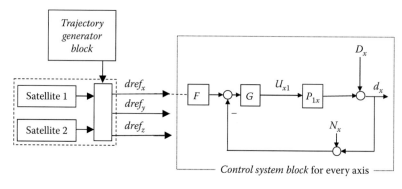

FIGURE Q5.1
Formation flying spacecraft control system.

- *Type 1: Stability specification*

$$\left|T_1(j\omega)\right| = \left|\frac{P_{1x}(j\omega)G(j\omega)}{1+P_{1x}(j\omega)G(j\omega)}\right| \leq \delta_1(\omega) = W_s = 1.1$$

$$\omega \in [0.00001 \ 0.00005 \ 0.0001 \ 0.0005 \ 0.001 \ 0.005 \ 0.01 \ 0.05 \ 0.1 \ 0.5 \ 1.0 \ 2.0 \ 3.0 \ 5.0] \ \text{rad/s} \tag{Q5.4}$$

which is equivalent to $PM = 54.1°$, $GM = 5.6$ dB—see Equations 2.30 and 2.31.

- *Type 6: Reference tracking specification*

$$\delta_{6\_lo}(\omega) < \left|T_6(j\omega)\right| = \left|F(j\omega)\frac{P_{1x}(j\omega)G(j\omega)}{1+P_{1x}(j\omega)G(j\omega)}\right| \leq \delta_{6\_up}(\omega) \tag{Q5.5}$$

$$\omega \in [0.00001 \ 0.00005 \ 0.0001 \ 0.0005 \ 0.001 \ 0.005 \ 0.01 \ 0.05 \ 0.1] \ \text{rad/s}$$

$$\delta_{6\_lo}(s) = \frac{(1-\varepsilon_{lo})}{[(s/a_L)+1]^2}; \quad a_L = 0.015; \quad \varepsilon_{lo} = 0 \tag{Q5.6}$$

$$\delta_{6\_up}(s) = \frac{[(s/a_U)+1](1+\varepsilon_{up})}{[(s/\omega_n)^2+(2\zeta s/\omega_n)+1]}; \ a_U = 0.008; \ \zeta = 0.8; \ \omega_n = \frac{1.25\,a_U}{\zeta}; \varepsilon_{up} = 0 \tag{Q5.7}$$

Problem Q6. Helicopter Control
Consider the helicopter prototype shown in Figure Q6.1. It is a laboratory scale plant with three degrees of freedom: roll angle ϕ, pitch angle θ, and yaw angle ψ. Each angle is measured by an absolute encoder. Two electrical DC motors are attached to the helicopter body, making the two propellers turn, providing an aerodynamic force F that makes the system turn around the angles. A counterweight of mass M balances the body weight.

The dynamics of the pitch angle is obtained by applying Euler–Lagrange's equations, so that

$$Fl_1 - mg\left[(h+d)\sin\theta + l_1\cos\theta\right] + Mg(l_2+l_3\cos\alpha)\cos\theta + Mg(l_3\sin\theta - h)\sin\theta - b_e\frac{d\theta}{dt} = J_e\frac{d^2\theta}{dt^2} \tag{Q6.1}$$

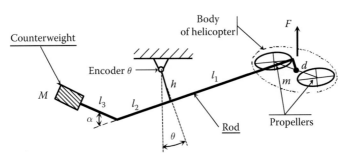

FIGURE Q6.1
Lab-scale helicopter.

where h, d, l_1, l_2, and l_3 are lengths; m the sum of both motors' mass; M the counterweight mass; b_e is the dynamic friction coefficient; g the gravity acceleration; J_e the moment of inertia of the system around the pitch angle θ, and α a fixed construction angle. The nonlinear model in Equation Q6.1 can be simplified by linearizing it around the operational point $\theta_0 = 0$ degrees. It yields a second-order transfer function between the motor control signal $u(s)$ and the pitch angle $\theta(s)$ given by

$$P(s) = \frac{\theta(s)}{u(s)} = \frac{k\,\omega_n^2}{s^2 + 2\zeta\,\omega_n\,s + \omega_n^2}\,e^{-sT} \qquad \text{(Q6.2)}$$

Using system identification techniques from experimental data, the parametric uncertainty found is: $k \in [0.0765, 0.132]$ rad/volt; $\zeta \in [0.025, 0.05]$; $\Omega_n \in [0.96, 1.58]$ rad/s; $T \in [0.09, 0.11]$ s. The motor control signal $u(t)$ presents a saturation limit of ± 10 volt.

Design the feedback controller $G(s)$ and the prefilter $F(s)$ to meet the following control objectives: (a) minimize the reference tracking error, (b) increase the damping, (c) increase the system stability, (c) reduce the overshoot, (d) reject the high-frequency noise introduced at the feedback sensors, and (e) deal with the actuator constraints.

Problem Q7. Two Cart Problem
Consider the classical benchmark problem shown in Figure Q7.1. It is a frictionless train composed of two carts of masses M_1 and M_2, coupled by a link of stiffness γ. The objective is to control the position $x_2(t)$ of the second cart by applying a force $u(t)$ to the first cart.

The model of the system in state space is

$$A = -\begin{bmatrix} 0 & 0 & 1 & 0 \\ 0 & 0 & 0 & 1 \\ \dfrac{\gamma}{M_1} & \dfrac{\gamma}{M_1} & 0 & 0 \\ \dfrac{\gamma}{M_2} & -\dfrac{\gamma}{M_2} & 0 & 0 \end{bmatrix}; \quad B = \begin{bmatrix} 0 \\ 0 \\ \dfrac{1}{M_1} \\ 0 \end{bmatrix}; \quad C = \begin{bmatrix} 0 & 1 & 0 & 0 \end{bmatrix}; \quad D = \begin{bmatrix} 0 \end{bmatrix} \quad \text{(Q7.1)}$$

$$\dot{x} = Ax + Bu; \quad y = Cx + Du \qquad \text{(Q7.2)}$$

being $x_1(t)$ and $x_2(t)$ the positions of the carts 1 and 2, respectively, $x_3(t)$ and $x_4(t)$ the velocities of the carts 1 and 2, respectively, $u(t)$ the control signal or force that acts on the first cart, and $y(t) = x_2(t)$ the sensor output, which is also the variable to be controlled. The parametric uncertainty is given by

$$M_1 \in [0.9,\ 1.1]; \quad M_2 \in [0.9,\ 1.1]; \quad \gamma \in [0.4,\ 0.6] \qquad \text{(Q7.3)}$$

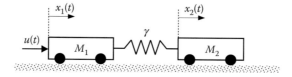

FIGURE Q7.1
Two cart problem.

Design a feedback controller $G(s)$ to regulate automatically the position $x_2(t)$ of the second cart by varying the force $u(t)$ applied to the first cart—see Figure Q7.1, and the following the next specifications:

- *Type 1: Stability specification*

$$|T_1(j\omega)| = \left|\frac{P_{1x}(j\omega)G(j\omega)}{1+P_{1x}(j\omega)G(j\omega)}\right| \le \delta_1(\omega) = W_s = 1.2 \qquad (Q7.4)$$

$$\omega \in [0.001\ 0.005\ 0.01\ 0.05\ 0.1\ 0.5\ 1.0\ 2.0\ 3.0\ 5.0\ 10]\,\text{rad/s}$$

which is equivalent to $PM = 49.24°$, $GM = 5.26$ dB—see Equations 2.30 and 2.31.

- *Type 3: Sensitivity or disturbances at plant output specification*

$$|T_3(j\omega)| = \left|\frac{x_2(j\omega)}{d_o(j\omega)}\right| = \left|\frac{1}{1+P(j\omega)G(j\omega)}\right| \le \delta_3(\omega) = \frac{(s/a_d)}{(s/u_d)+1}\,; \quad a_d = 0.1 \qquad (Q7.5)$$

$$\omega \in [0.001\ 0.005\ 0.01\ 0.05\ 0.1]\,\text{rad/s}$$

- *Type 6: Reference tracking specification*

$$\delta_{6\_lo}(\omega) < |T_6(j\omega)| = \left|F(j\omega)\frac{P(j\omega)G(j\omega)}{1+P(j\omega)G(j\omega)}\right| \le \delta_{6\_up}(\omega) \qquad (Q7.6)$$

$$\omega \in [0.001\ 0.005\ 0.01\ 0.05\ 0.1\ 0.5]\,\text{rad/s}$$

$$\delta_{6\_lo}(s) = \frac{(1-\varepsilon_{lo})}{[(s/a_L)+1]^2}\,; \quad a_L = 0.1\,; \quad \varepsilon_{lo} = 0 \qquad (Q7.7)$$

$$\delta_{6\_up}(s) = \frac{[(s/a_U)+1]\,(1+\varepsilon_{up})}{[(s/\omega_n)^2+(2\zeta s/\omega_n)+1]}\,; \quad a_U = 0.08\,; \quad \zeta = 0.8\,; \quad \omega_n = \frac{1.25\,a_U}{\zeta}\,; \quad \varepsilon_{up} = 0.05 \qquad (Q7.8)$$

Problem Q8. Two Flow Problem

Consider two gaseous elements A and B in the tanks showed in Figure Q8.1. Tank 1 has an 80% of A ($X_1 = 80\%$) and a 20% of B, whereas Tank 2 has a 20% of A ($X_2 = 20\%$) and a 80% of B. The hydraulic system represented in the figure mixes two flows, F_1 and F_2, which

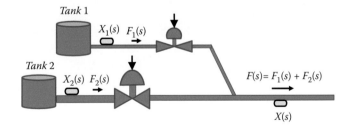

FIGURE Q8.1
Two flow problem.

come from the Tanks 1 and 2, respectively. The product has a composition of $X\%$ of A and a flow F, so that

$$F = F_1 + F_2 \tag{Q8.1}$$

$$F\,X = F_1 X_1 + F_2 X_2 \tag{Q8.2}$$

The plant inputs are the flows F_1 and F_2, and the plant outputs are the flow F and the concentration X of element A. In order to control X around $X_{ref} = 60\%$ and F around $F_{ref} = 200$ mol/h:

1. Calculate Bristol's relative gain analysis array
2. Find the best input/output pairing
3. Analyze the existing coupling between loops
4. Find the coupling worst case according to the set point X_{ref}

Problem Q9. 2 × 2 MIMO System
Consider a 2 × 2 linear multivariable system with uncertainty, whose transfer function matrix is

$$\boldsymbol{P}(s) = \begin{bmatrix} \dfrac{k_{11}}{\tau_{11}\,s+1} & \dfrac{k_{12}}{\tau_{12}\,s+1} \\[2mm] \dfrac{k_{21}}{\tau_{21}\,s+1} & \dfrac{k_{22}}{\tau_{22}\,s+1} \end{bmatrix} \tag{Q9.1}$$

with

$$k_{11} \in [0.5,3],\ k_{12} \in [-2.2,-1.8]\ ,\ k_{21} \in [11,15]\ ,\ k_{22} \in [2,7]$$
$$\tau_{11} \in [0.5,3],\ \tau_{12} \in [8,12],\ \tau_{21} \in [3,8],\ \tau_{22} \in [5,10]$$

1. Calculate the Bristol's relative gain analysis array.
2. Find the best input/output pairing.
3. Analyze the existing coupling between loops.
4. Using *Method 1* (see Sections 8-5 and 8-8), design a MIMO QFT controller that meets the three specifications defined below.
5. Using *Method 2* (see Sections 8-6 and 8-9), design a MIMO QFT controller that meets the three specifications defined below:

   Specifications

   - *Type 1: Stability specification, i = 1,2*
     $|t_{ii,1}(j\omega)| \le \delta_{ii,1}(\omega) = W_s = 1.2$ for $i = 1, 2,\ \forall\omega$, where the terms $t_{ii}(j\omega)$ are the diagonal elements of the matrix $T_{y/r}$. This condition implies at least 50° lower phase margin and at least 1.833 (5.26 dB) lower gain margin.

   - *Type 3: Sensitivity or disturbances at plant output specification, i = 1,2*

     $$|t_{ii,3}(s)| = \left|\dfrac{y(s)}{d_o(s)}\right| = \left|\dfrac{1}{1 + p_{ii}(s)\,g_{ii}(s)}\right| \le \delta_{ii,3}(\omega) = \left|\dfrac{s}{s+10}\right|$$
     $$\omega < 50\,\text{rad/s}, \quad \text{for } i = 1,2$$

- *Minimize the coupling effects c_{ij}* as much as possible and within the range of frequencies: $\omega < 10^{-2}$ rad/s.

Problem Q10. 2 × 2 MIMO System
Consider a 2 × 2 linear multivariable system with uncertainty and the following transfer function matrix:

$$P(s) = \begin{bmatrix} p_{11}(s) & p_{12}(s) \\ p_{21}(s) & p_{22}(s) \end{bmatrix} = \begin{bmatrix} \dfrac{10\alpha\beta}{2s+\beta} & \dfrac{-4\beta}{s+\beta} \\ \dfrac{2\alpha\gamma}{s+\gamma} & \dfrac{3\alpha\gamma}{0.5s+\gamma} \end{bmatrix} \quad \text{(Q10.1)}$$

$$\alpha \in [0.8, 1.2]; \quad \beta \in [8, 10]; \quad \gamma \in [5, 6]$$

1. Calculate the Bristol's relative gain analysis array.
2. Find the best input/output pairing.
3. Analyze the existing coupling between loops.
4. Using *Method 1* (see Sections 8-5 and 8-8), design a MIMO QFT controller that meets the three specifications defined below.
5. Using *Method 2* (see Sections 8-6 and 8-9), design a MIMO QFT controller that meets the three specifications defined below.

   Specifications

   - *Type 1: Stability specification, i = 1,2*

$$\left| t_{ii,1}(j\omega) \right| = \left| \frac{\left(p_{ii}^{*e}\right)^{-1} g_{ii}}{1 + \left(p_{ii}^{*e}\right)^{-1} g_{ii}} \right| \leq \delta_{ii,1}(\omega) = W_s = 1.4, \quad i = 1, 2 \quad \text{(Q10.2)}$$

   - *Type 6: Reference tracking specification, i = 1,2*

$$\delta_{ii,6\_lo}(\omega) < \left| t_{ii,6}^{Y/R}(j\omega) \right| \leq \delta_{ii,6\_up}(\omega), \, i = 1, 2 \quad \text{(Q10.3)}$$

   where

$$\delta_{11,6\_lo}(s) = \frac{16}{(s^2 + 7.6s + 16)\,((s/10)+1)}; \; \delta_{11,6\_up}(s) = \frac{16\,((s/15)+1)}{s^2 + 3.6s + 16}$$

$$\delta_{22,6\_lo}(s) = \frac{100}{(s^2 + 16s + 100)\,((s/20)+1)}; \; \delta_{22,6\_up}(s) = \frac{100\,((s/25)+1)}{s^2 + 14s + 100} \quad \text{(Q10.4)}$$

- *Minimize the coupling effects:*

   Reduce the interaction, $t_{12}^{Y/R}$ and $t_{21}^{Y/R}$ as much as possible. \quad (Q10.5)

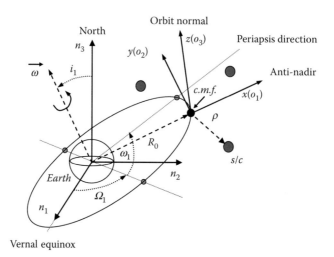

North

Orbit normal

$z(o_3)$

Periapsis direction

n_3

$y(o_2)$

$\vec{\omega}$ i_1

Anti-nadir

c.m.f.

$x(o_1)$

ρ

ω_1

R_0

s/c

Earth

n_2

Ω_1

n_1

Vernal equinox

FIGURE Q11.1
Formation flying spacecraft problem.

Problem Q11. Spacecraft Flying in Formation in Low Earth Orbit
A formation of spacecraft (s/c) in Low Earth Orbit (LEO) is shown in Figure Q11.1. The model of the formation relative to the center of mass of the formation (c.m.f.) is

$$\begin{bmatrix} x \\ y \\ z \end{bmatrix} = \frac{1}{m} \begin{bmatrix} \dfrac{1}{s^2 + \omega_0^2} & \dfrac{2\,\omega_0}{s\,(s^2 + \omega_0^2)} & 0 \\ \dfrac{-2\,\omega_0}{s\,(s^2 + \omega_0^2)} & \dfrac{s^2 - 3\,\omega_0^2}{s^2\,(s^2 + \omega_0^2)} & 0 \\ 0 & 0 & \dfrac{1}{s^2 + \omega_0^2} \end{bmatrix} \begin{bmatrix} Q_x \\ Q_y \\ Q_z \end{bmatrix} \tag{Q11.1}$$

where $\rho = [x, y, z]^T$ are the distances of the s/c to the c.m.f. and $[Q_x, Q_y, Q_z]^T$ the actuator forces (thrusters). Note that this is a 2×2 MIMO system (axes x and y) plus a SISO (single-input single-output) system (axis z). Note also that although the p_{22} element of the system is non-minimum phase, the MIMO plant does not present RHP transmission zeros. Its Smith-McMillan decomposition yields a double integrator and a conjugate pole at $s = \pm j\omega_0$, but no transmission zeros.

For an s/c flying in formation in a circular geostationary Earth orbit, the parameters are: mean orbit rate $\omega_0 \in [7.2637 \times 10^{-5}, 7.3208 \times 10^{-5}]$ rad/s, and mass of the satellite $m \in [1600, 1650]$ kg.

Design a MIMO QFT controller according to *Method 2* (see Sections 8-6 and 8-9) to achieve the following specifications:
Matrix Specifications on $S(s)$ and $T(s)$:

$$\|T(jw)\|_\infty < 2 \quad \text{and} \quad \|S(jw)\|_\infty < 2 \tag{Q11.2}$$

Classical loop-by-loop robust stability and performance specifications:

- *Type 1: Stability specification, $i = 1,2,3$*

$$|t_{ii,1}(s)| = \frac{\left|\left[\left(p_{ii}^x(s)\right)^{*e}\right]^{-1} g_{ii}^\beta(s)\right|}{\left|1 + \left[\left(p_{ii}^x(s)\right)^{*e}\right]^{-1} g_{ii}^\beta(s)\right|} \leq \delta_{ii,1}(\omega) = W_s = 1.1 , \quad \forall \omega , i = 1,2,3 \qquad \text{(Q11.3)}$$

- *Type 3: Sensitivity or disturbances at plant output specification, $i = 3$*

$$|t_{33,3}(s)| = \left|\frac{z(s)}{d_o(s)}\right| = \left|\frac{1}{1 + p_{33}(s) g_{33}(s)}\right| \leq \delta_{33,3}(\omega) = 2 \quad \forall \omega \qquad \text{(Q11.4)}$$

- *Type 4: Disturbances at plant input specification, $i = 1,2,3$*

$$|t_{ii,4}(s)| = \frac{\left|\left[\left(p_{ii}^x(s)\right)^{*e}\right]^{-1}\right|}{\left|1 + \left[\left(p_{ii}^x(s)\right)^{*e}\right]^{-1} g_{ii}^\beta(s)\right|} \leq \delta_{ii,4}(\omega) = \left|\frac{20\,s}{\left((s/0.008) + 1\right)\left((s/0.06) + 1\right)\left((s/0.6) + 1\right)}\right| \qquad \text{(Q11.5)}$$

$\forall \omega < 1 \text{ rad/s}$

Problem Q12. 3×3 MIMO System
Consider a 3×3 MIMO system whose transfer function matrix is

$$\begin{bmatrix} y_1(s) \\ y_2(s) \\ y_3(s) \end{bmatrix} = \begin{bmatrix} \dfrac{0.1\,e^{-0.4\,s}}{0.92s + 1} & \dfrac{2(3s+1)}{4s+1} & \dfrac{-1}{2s+1} \\[2mm] \dfrac{1\,e^{-0.1\,s}}{7s+1} & \dfrac{1}{3s+1} & \dfrac{-0.1\,e^{-0.2\,s}}{0.87s+1} \\[2mm] \dfrac{-2(s+1)}{0.92s+1} & \dfrac{-3\,e^{-0.4\,s}}{0.54s+1} & \dfrac{1e^{-0.3\,s}}{6s+1} \end{bmatrix} \begin{bmatrix} m_1(s) \\ m_2(s) \\ m_3(s) \end{bmatrix} \qquad \text{(Q12.1)}$$

1. Calculate Bristol's relative gain analysis array.
2. Find the best input/output pairing.
3. Analyze the existing coupling between loops.
4. Suppose that the three control loops are designed and working. In that case, study the effect of an instantaneous opening of the loop $y_2(s)$.
5. Using *Method 1* (see Sections 8-5 and 8-8), design a MIMO QFT compensator so that the system reaches a good level of stability, disturbance rejection, and decoupling (no uncertainty).
6. Using *Method 2* (see Sections 8-6 and 8-9), design a MIMO QFT compensator so that the system reaches a good level of stability, disturbance rejection, and decoupling (no uncertainty).

Appendix 2: QFT Control Toolbox—User's Guide[175]

A2.1 Introduction

This appendix presents the *QFT Control Toolbox*, or *QFTCT*, for MATLAB. It is the interactive and object-oriented computer tool used through the book to design QFT controllers. The toolbox includes the latest quantitative robust control techniques within a user-friendly and interactive environment. The toolbox, developed by Prof. Mario García Sanz, has been tested in numerous courses, universities, companies, and centers over the years.

A demo version of the toolbox can be found at http://cesc.case.edu. The student and standard versions of the QFTCT is at http://codypower.com. For additional information, see http://crcpress.com.

The QFT control toolbox runs under MATLAB and shows an architecture based on seven principal windows, which are: *W1—Plant Definition*; *W2—Templates*; *W3—Control Specifications*; *W4—Bounds*; *W5—Controller Design*; *W6—Prefilter Design*; and *W7—Analysis*—see Figure A2.2 and also Figure 2.1.

The toolbox also includes a library of advanced functions in Windows, offers a user-friendly and interactive environment, and allows the user to easily re-scale the problem from SISO to MIMO problems.

Following the QFT robust control methodology, the objective of the toolbox is to design 2DOF (two degree of freedom) robust control systems, including plants with model uncertainty and a multi-objective set of performance specifications—see Figure A2.1. The compensator $G(s)$ and prefilter $F(s)$ of the 2DOF control system are to be designed to meet the robust stability and performance specifications, including reference tracking $r(s)$, disturbance rejection $d(s)$, $d_e(s)$, $d_i(s)$, $d_o(s)$, signal noise attenuation $n(s)$, and control effort minimization $u(s)$ specifications, while reducing the complexity of the controllers (order) and quantifying the cost of feedback (gain at each frequency and bandwidth).

Figure A2.2 lists the steps involved in the QFT design procedure. Figure A2.3 represents the toolbox flowchart of the QFT design procedure and Figure A2.4 presents an overview of the QFT design process.

Required products:	Matlab 2015b (version 8.5) or later version.
	Control System Toolbox 9.9 or later version.
Platform:	PC, Windows 10, 8, or 7.
Installation:	To install the QFTCT, copy the files of the toolbox in a folder. Then add the folder to the path: file → set path → add folder → save → close.
Start:	To start a new session with the QFT Control Toolbox. type "QFTCT" in the MATLAB Command Window.

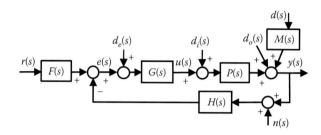

FIGURE A2.1
2-Degree-of-freedom (2DOF) feedback control system.

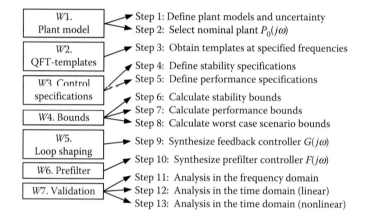

FIGURE A2.2
QFT design procedure and QFTCT windows.

A2.2 QFT Control Toolbox Windows

This section describes the seven main windows of *QFTCT* for MATLAB—see Figure A2.2 and also Figure 2.1.

A2.2.1 General Description

The toolbox contains windows of the form shown in Figure A2.5. As mentioned, they fall into seven categories: W1: *Plant Definition*; W2: *Templates*; W3: *Control Specifications*; W4: *Bounds*; W5: *Controller Design*; W6: *Prefilter Design*; and W7: *Analysis*. All the windows have a navigator panel at the top—see Figure A2.6. This allows the user to change the active window among the seven options. Not all the windows are available from the beginning, as some windows have prerequisites that have to be met in order to activate them. The seven categories are listed next, following the order that they have to be executed in the design process:

1. *Plant definition*: No pre-requisites
2. *Templates*: Needs plant definition
3. *Specifications*: Needs plant definition

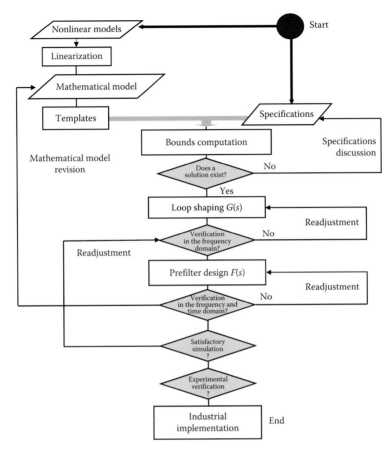

FIGURE A2.3
QFT flowchart for a multi-input single-output (MISO) control system. (Adapted from Garcia-Sanz, M. and Houpis, C.H. 2012. *Wind Energy Systems: Control Engineering Design (Part I: QFT Control, Part II: Wind Turbine Design and Control).* A CRC Press book, Taylor & Francis Group, Boca Raton, FL.)

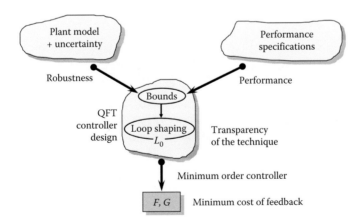

FIGURE A2.4
An overview of the QFT design process. (Adapted from Garcia-Sanz, M. and Houpis, C.H. 2012. *Wind Energy Systems: Control Engineering Design (Part I: QFT Control, Part II: Wind Turbine Design and Control).* A CRC Press book, Taylor & Francis Group, Boca Raton, FL.)

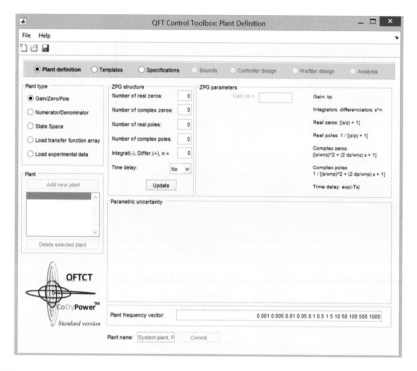

FIGURE A2.5
Plant definition window.

FIGURE A2.6
Window toolbar with the seven window categories. (Adapted from Garcia-Sanz, M. and Houpis, C.H. 2012. *Wind Energy Systems: Control Engineering Design (Part I: QFT Control, Part II: Wind Turbine Design and Control).* A CRC Press book, Taylor & Francis Group, Boca Raton, FL.)

FIGURE A2.7
Common toolbars. (Adapted from Garcia-Sanz, M. and Houpis, C.H. 2012. *Wind Energy Systems: Control Engineering Design (Part I: QFT Control, Part II: Wind Turbine Design and Control).* A CRC Press book, Taylor & Francis Group, Boca Raton, FL.)

4. *Bounds*: Needs plant definition, templates, and specifications

5. *Controller Design*: Needs plant definition, templates, specifications, and bounds

6. *Prefilter*: Needs plant definition, templates, specifications, bounds, and controller design. It is only for reference tracking problems

7. *Analysis*: Needs plant definition, templates, specifications, bounds, and controller design

- *Common Toolbars*: All the windows of the toolbox have a toolbar—see Figure A2.7a, which allows the user to create a new project, open an existing one, or save the current project. The toolbar of the windows which have figures (Templates, Bounds, Controller Design, Prefilter Design, and Analysis) also has controls to zoom in and out on the plots—see Figure A2.7b.
- *Menu* Additionally, all the windows have the following two menus:
  - *File*: Allows the user to create new projects, open existing projects, save the current project, and save the current project in a new file.
  - *Help*: Provides information about the toolbox version, etc.

  In addition,
  - The windows that have figures include a menu item in the *File menu* that allows the user to save the plot in a file with the format .emf, .bmp, or .fig.
  - Also, some windows have the following specific menu items:
    - **Templates**: In the *File menu*, there is an option that allows the user to export the nominal plant as a transfer function.
    - **Controller Design**: In the *File menu*, there are menu items that allow the user to—see Figure A2.8a:
      - *Load a controller* from the hard disk (*.contr).
      - *Load a list of controllers* from the hard disk (*.contrList).
      - *Save selected controller* to the hard disk (*.contr).
      - *Save controller list* to the hard disk (*.contrList).
      - *Export* the selected controller as a transfer function to workspace.
      - *Save Figure* to the hard disk (*.emf, *.bmp, or *.fig).
      - *Check nominal plant*. It shows the selected nominal plant.

(a)

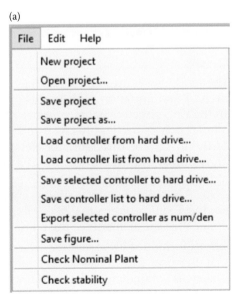

(b)

FIGURE A2.8
(a) File menu, controller design window. (b) Check stability.

- *Check stability.* It checks the stability of the closed-loop system with the nominal plant and the selected controller. It applies the method presented in Chapter 3, Section 3.4, and the algorithms included in Appendix 3: *The Nyquist Stability Criterion in the Nichols Chart*—see example in Figure A2.8b.

The controller design window has also an Edit menu that allows the user to:

- Undo changes
- Redo changes
- Set the frequency vector for $L(s)$: minimum frequency and maximum frequency (rad/s), and number of points.

- **Prefilter Design**: Similarly, this window category has the following menu items in the File menu:
  - Load prefilter from the hard disk (*.prefltr)
  - Load prefilter list from the hard disk (*.prefltrList)
  - Save selected prefilter to the hard disk as (*.prefltr)
  - Save prefilter list to the hard disk (*.prefltrList)
  - Export the selected prefilter as a transfer function to the workspace
  - Save Figure to the hard disk (*.emf, *.bmp, or *.fig)

A2.2.2 Plant Definition Window

In the Plant Definition window, see Figure A2.9, the user can define the plant type, model structure, parameters, uncertainty, and frequencies of interest, either for the System plant $P(s)$, or for additional plants like $M(s)$ in Figure A2.1 (*add new plant* button). In the *Plant type* panel, the user can select the way to describe the plant model, which can be: (1) Gain/Zero/Pole or ZPG transfer functions, (2) Numerator/Denominator transfer functions, (3) State Space representation, (4) a general array of transfer functions, and (5) experimental data.

- *Gain/Zero/Pole Transfer Function:* The model structure and its elements are defined using the syntax listed in Table A2.1 and Equation A2.1. In this option, the user has to enter the information in the *ZPG structure* panel: number of real zeros, complex zeros, real poles, and complex poles—see Figures A2.9 and A2.10. The user also has to enter the value of the integrator/differentiator element: 0 if it is not used, a positive integer for the number of differentiators and a negative integer for the number of integrators. In addition, the user has to specify whether the plant has time delay (Yes/No). After all this information is entered, the user clicks the *Update* button. Then, the Toolbox opens the *ZPG parameters* panel and the user can write the expressions for the elements of the plant.

   Note that, at this point, it is possible to introduce *alpha-numeric information* with numbers, letters, or expressions. If the user introduces expressions, the Toolbox identifies the letters in the expressions as parameters with uncertainty, and automatically adds their names to the *Parametric uncertainty* panel. In Figures A2.9 and A2.10, a gain is introduced as "$k/(a*b)$", a real pole as "a", and another real pole as "b"—see Equation A2.2.

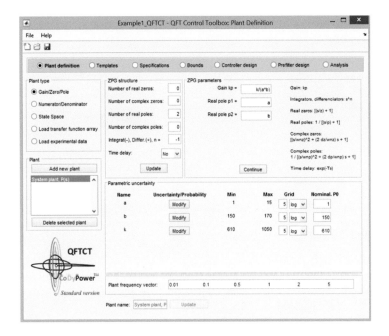

FIGURE A2.9
Plant definition window.

TABLE A2.1

Gain/Zero/Pole/Delay Element Syntax

Gain	k_p	Complex zero	$\dfrac{s^2}{\omega_n^2} + \dfrac{2\zeta s}{\omega_n} + 1 \; (\zeta < 1)$
Real pole	$\dfrac{1}{\dfrac{s}{p} + 1}$	Integrator	$\dfrac{1}{s^n}$
Real zero	$\dfrac{s}{z} + 1$	Differentiator	s^n
Complex pole	$\dfrac{1}{\dfrac{s^2}{\omega_n^2} + \dfrac{2\zeta s}{\omega_n} + 1} \; (\zeta < 1)$	Time delay	e^{-Ts}

ZPG structure

Number of real zeros:	0
Number of complex zeros:	0
Number of real poles:	2
Number of complex poles:	0
Integrat(-), Differ.(+), n =	-1
Time delay:	No ∨

Update

ZPG parameters

Gain kp =	k/(a*b)
Real pole p1 =	a
Real pole p2 =	b

Continue

Gain: kp

Integrators, differenciators: s^n

Real zeros: [(s/z) + 1]

Real poles: 1 / [(s/p) + 1]

Complex zeros:
[(s/wnz)^2 + (2 dz/wnz) s + 1]

Complex poles:
1 / [(s/wnp)^2 + (2 dp/wnp) s + 1]

Time delay: exp(-Ts)

FIGURE A2.10
Gain/Zero/Pole or ZPG structure and ZPG parameters panels.

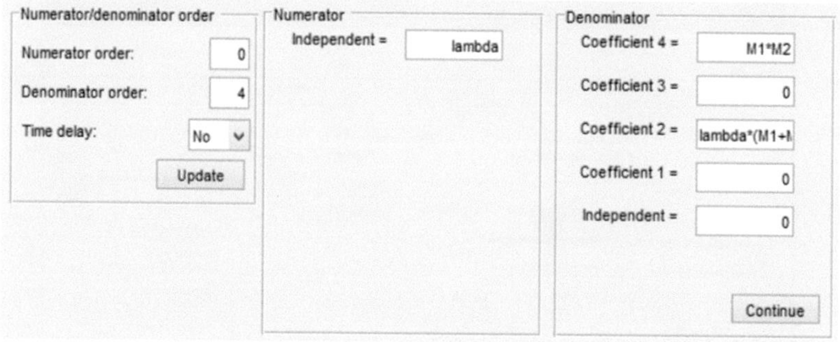

FIGURE A2.11
Numerator/denominator panels. (Adapted from Garcia-Sanz, M. and Houpis, C.H. 2012. *Wind Energy Systems: Control Engineering Design (Part I: QFT Control, Part II: Wind Turbine Design and Control)*. A CRC Press book, Taylor & Francis Group, Boca Raton, FL.)

$$P(s) = \frac{k_p \; zeros(s)}{poles(s)} = \frac{k_p \left(\dfrac{s}{z_1}+1\right)\left(\dfrac{s}{z_2}+1\right)\cdots\left(\dfrac{s^2}{\omega_{ni}^2}+\dfrac{2\zeta_i s}{\omega_{ni}}+1\right)\cdots}{s^n \left(\dfrac{s}{p_1}+1\right)\left(\dfrac{s}{p_2}+1\right)\cdots\left(\dfrac{s^2}{\omega_{nj}^2}+\dfrac{2\zeta_j s}{\omega_{nj}}+1\right)\cdots} e^{-Ts} \tag{A2.1}$$

$$P(s) = \frac{k}{(s+a)(s+b)} = \frac{\left(\dfrac{k}{a\,b}\right)}{\left(\dfrac{s}{a}+1\right)\left(\dfrac{s}{b}+1\right)} \tag{A2.2}$$

- *Numerator/Denominator Transfer Function*: The plants can also be defined as a transfer function, with a numerator and a denominator as Laplace polynomials—see Equation A2.3. The first step consists in entering the model structure: numerator and denominator polynomial orders— see Figure A2.11 and Equation A2.3. At this point the user can also specify whether there is or there is not time delay in the plant. Then, after pressing the *Update* button, the Toolbox updates the *Numerator* and *Denominator* panels, and the user can enter the expressions for the coefficients of the polynomials and time delay. Note that, at this point, it is possible to introduce numbers or letters. If the user introduces letters, the Toolbox identifies them as parameters with uncertainty, and automatically adds their names to the *parametric uncertainty* panel, which will be defined afterward. As an example, the expression defined in Equation A2.4 is introduced in Figure A2.11.

$$P(s) = \frac{n(s)}{d(s)} = \frac{Ncoef_n s^n + Ncoef_{n-1} s^{n-1} + \cdots + Ncoef_1 s + Ncoef_{indep}}{Dcoef_m s^m + Dcoef_{m-1} s^{m-1} + \cdots + Dcoef_1 s + Dcoef_{indep}} e^{-Ts} \tag{A2.3}$$

$$P(s) = \frac{Lambda}{(M_1 M_2)s^4 + 0 s^3 + Lambda(M_1 + M_2)s^2 + 0s + 0} \tag{A2.4}$$

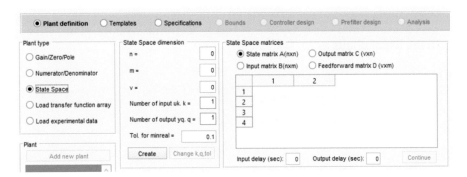

FIGURE A2.12
State-space panel.

- *State Space*: The plants can also be described with a state space representation. In this case, we enter the dimension (n, m, and v) of the matrixes A ($n \times n$), B ($n \times m$), C ($v \times n$), and D ($v \times m$)—see Equation A2.5 and Figure A2.12. Also, we need to choose the input k ($k \leq m$) and output q ($q \leq v$) channels for the $y_q(s)/u_k(s) = P_{qk}(s)$, with $P(s)_{v \times m} = [C(sI-A)^{-1}B + D]_{v \times m}$. As the conversion from state space to transfer function typically involves some round errors, we can introduce a tolerance error for the *minreral*(.) function to remove the extra zeros and poles that could be added by MATLAB in the conversion. After clicking *Create* button, the Toolbox opens the *State Space matrices* panel, and we can enter the expressions for the elements of the four matrices and the input and output time delays. Note that, at this point, we can introduce numbers or letters. If we introduce letters, the Toolbox identifies them as parameters with uncertainty, and automatically adds their names to the *parametric uncertainty* panel, which is defined later.

$$\underset{n \times 1}{\dot{x}} = \underset{n \times n}{A} \underset{n \times 1}{x} + \underset{n \times m}{B} \underset{}{u}$$
$$\underset{v \times 1}{y} = \underset{v \times n}{C} \underset{}{x} + \underset{v \times m}{D} \underset{}{u}$$
(A2.5)

- *Load Transfer Function Array*: If the plant cannot be defined using the three previous options, yet we can load an array of transfer functions from a file. After uploading the file, we enter the row of the array that represents the nominal plant.

Notice that this is a very powerful tool that can define any kind of plant, with different structures and parametric and nonparametric uncertainty.

For the *SystemPlant*, the technique is as follows: (1) Run first in MATLAB an m.file like the one described in Example A2.1; (2) then go to the Workspace; (3) click on P with the right bottom and save as PP.mat (or other name) in the hard disk; (4) then go to the Plant definition window; (5) click "Load transfer function array"; (6) select Nominal plant (usually number 1); (7) click "Import. mat"; (8) select in the hard disk PP.mat and click "Open"; (9) click "Commit"; (10) the plant will appear in the list of plants as *System Plant, P(s)*.

For an additional plant: steps (1–6) are the same—now save as MM.mat, for instance. Then (7) click "Add new plant"; (8) put a name in the "Plant name" cell at the bottom (for example M); (9) click "Import.mat"; (10) select in the hard disk MM.mat and click "Open"; (11) click "Commit"; (12) the plant will appear in the list of plants as *M*.

EXAMPLE A2.1: M.FILE

```
c = 0;
for k = linspace(610,1050,3)
    for a = linspace(1,15,15)
        for b = linspace(150,170,2)
            c = c+1;
            P(1,1,c) = tf(k,[1 (a+b) a*b 0]);
            M(1,1,c) = tf(k,[1/a 1]);
        end
    end
end
```

Notice also that we have three nested "for" loops in the above example because there are three parameters with uncertainty: $k \in [610,1050]$, $a \in [1,15]$, $b \in [150,170]$. We can also put another type of grid, different from *linespace*, like *logspace* or others, or a mix. The definition of the plants in this example is

```
P(1,1,c)=tf(k,[1 (a+b) a*b 0]);
M(1,1,c)=tf(k,[1/a 1]);
```

which are: $P(s) = k/[s^3 + (a + b)s^2 + ab\ s]$, and $M(s) = k/[(s/a) + 1]$.

As we see, this option allows the user to include in these lines of the algorithm any kind of structure or expression, even with different structures (uncertainty in the structure) and with any parametric or nonparametric uncertainty, interdependence in the uncertain parameters, and many other special features.

Finally, note that as this option is so open, it is important to make sure that the numbers of the plants are consistent, that is, the index "*c*" in $P(1,1,c)$ and $M(1,1,c)$ is related to the same values of the parameters that $P(1,1,c)$ and $M(1,1,c)$ share. This is easily done by defining $P(1,1,c)$ and $M(1,1,c)$ within the same for-loop in the m.file, as shown in Example A2.1.

In case of a combination of *"load transfer function array"* plants with other plants, like *Gain/Zero/Pole* or *Numerator/Denominator* or *State-Space* plants, make sure that the definition of the parametric uncertainty (parameters and gridding) is consistent in all of them. Also select the same gridding in the parametric uncertainty of the *Analysis* window.

- **Load Experimental Data**: The user can also define the system by uploading experimental data in a frequency response data vector. The technique is as follows: (1) prepare an frd (Frequency Response Data model) system in MATLAB: *freq* = vector of frequencies in rad/s; *resp* = vector of complex numbers with the response of the system at each frequency $(a + jb)$, *experimentalData* = frd(*resp,freq*). Note that the name of the frd structure can be *experimentalData* or any other.

  For example, in MATLAB we can introduce the following arrays: *freq* = logspace(–1,2,50); *resp* = 0.05*(*freq*).*exp(i*2**freq*); *experimentalData* = frd(*resp,freq*)—or a collection of real data in *resp* and *freq*; (2) Then go to the Workspace; (3) Click on

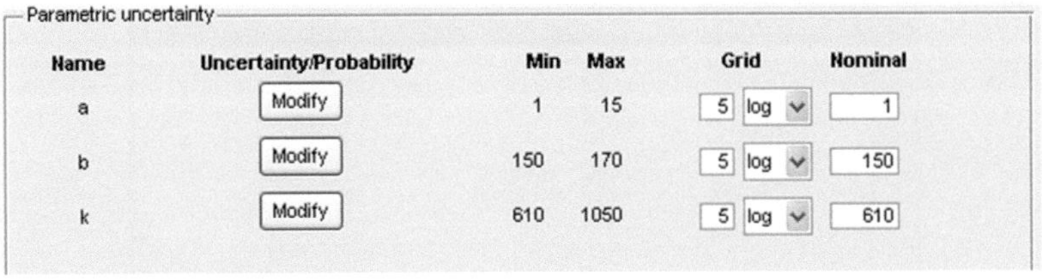

FIGURE A2.13
Parametric uncertainty panel. (Adapted from Garcia-Sanz, M. and Houpis, C.H. 2012. *Wind Energy Systems: Control Engineering Design (Part I: QFT Control, Part II: Wind Turbine Design and Control)*. A CRC Press book, Taylor & Francis Group, Boca Raton, FL.)

experimentalData with right bottom and save as sys.mat (or other name) in the hard disk; (4) Then go to the Plant definition window; (5) Click "Load experimental data"; (6) Select Nominal plant (usually num.1); (7) Click "Import experimental…"; (8) Select in the hard disk sys.mat and click "Open"; (9) Click "Commit"; (10) The plant will appear in the list of plants as *System Plant, P(s)*.

- *Expressions and Parametric Uncertainty*: As seen, it is possible to introduce alpha-numeric expressions. The toolbox automatically recognizes letters as parameters with uncertainty. After entering the expressions of the plant, either as parameters in the zpk model, or in the numerator/denominator model, or in the state space matrices, and after pressing the *Continue* button, a new *parametric uncertainty* panel appears to define the parametric uncertainty of the plant—see Figure A2.13. This panel displays the following information of each parameter:
  - The name of the parameter
  - A button to open a window to modify the probability distribution of the parameter
  - The minimum and the maximum value of the parameter (given by its probability distribution)
  - Number of points in the grid for each parameter (two is the minimum number, which means that there are only two points in the grid, the minimum and the maximum), and its distribution (logarithmic or lineal)
  - The nominal value of the parameter (it must be included in the minimum–maximum range of the parameter)
- *Probability Distribution of the Parameters*: In order to define the probability distribution for the uncertainty of the parameters, the user needs to press the *Modify* button in the *Parametric uncertainty* panel to open a new sub-window—see Figure A2.14. This sub-window has three probability distributions available: Uniform, Normal, and Weibull. After selecting the distribution and entering its parameters, a graphical representation of the distribution is plotted. Also, the user can define the percentage to be reached. This percentage is applied to the distribution, and the resulting values will be the minimum and the maximum values of the parameter.
- *System Plant Frequency Vector*: The user has also to enter a vector with the frequencies of interest of the plant—see for example Figures A2.5 and A2.9. The textbox accepts an array of numbers or a MATLAB command that produces the array of numbers, e.g., logspace(–2,3,100). Note that it is very

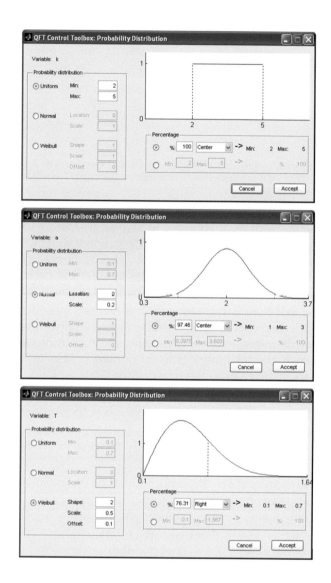

FIGURE A2.14
Probability distribution: uniform, normal, and the Weibull cases. (Adapted from Garcia-Sanz, M. and Houpis, C.H. 2012. *Wind Energy Systems: Control Engineering Design (Part I: QFT Control, Part II: Wind Turbine Design and Control)*. A CRC Press book, Taylor & Francis Group, Boca Raton, FL.)

important to define this vector correctly, populating the vector with an enough number of points in the frequency regions where there are resonances or quick changes in magnitude or phase, as well as selecting properly the lowest and the highest value of frequency.

A2.2.3 Templates Window

A template is the representation of the frequency response of the plants, including the uncertainty, in the Nichols chart at a particular frequency. There is a specific window to define the templates—see Figure A2.15.

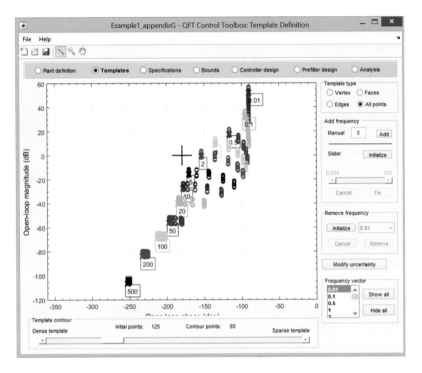

FIGURE A2.15
Template definition window.

- *Number of Template Points*: The number of points of the templates depends on three parameters: template type, parametric uncertainty, and template contour. If the template is too sparse, then it might be not accurate enough to represent the plant. On the contrary, if the template has too many points, the number of calculations to generate the bounds in the next step might be too high.

- *Template Type*: The user can select four different template types, ordered from sparse to dense: "Vertex", "Edges", "Faces," and "All points". Figure A2.16 shows different template types for a plant which has three uncertain parameters. The filled circle in each of the templates denotes the nominal plant.

- *Grid of the Parametric Uncertainty Variables*: The more points are considered—see Figure A2.13, the denser the templates are (unless the template type is "Vertex").

- *Template Contour*: In the lower part of the window, there is a slider which can be used to adjust the contour of the templates. When the slider is moved to the right, the *inner points* of the template become sparser. Figure A2.17 shows the template of a plant that has initially 1000 points (a), and the template of the same plant after its contour has been adjusted (b).

- *Plant Frequencies*: There is a template for each frequency of interest. The user can add and remove frequencies using the Templates window (or the Plant definition window as well).

  - *Add Frequency*: Frequencies can be added using two different methods:
    - By entering the value of the new frequency in the textbox beside "Manual" and pressing the *Add* button—see Figure A2.15.

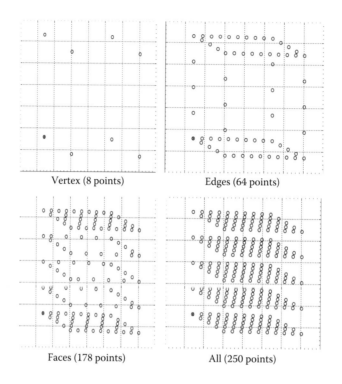

Vertex (8 points)   Edges (64 points)

Faces (178 points)   All (250 points)

FIGURE A2.16
Template types. (Adapted from Garcia-Sanz, M. and Houpis, C.H. 2012. *Wind Energy Systems: Control Engineering Design (Part I: QFT Control, Part II: Wind Turbine Design and Control).* A CRC Press book, Taylor & Francis Group, Boca Raton, FL.)

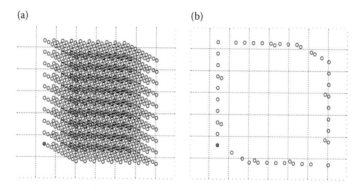

(a)   (b)

FIGURE A2.17
Template without contour adjustment [1000 points, (a)], and with contour adjustment with the slider [51 points, (b)]. (Adapted from Garcia-Sanz, M. and Houpis, C.H. 2012. *Wind Energy Systems: Control Engineering Design (Part I: QFT Control, Part II: Wind Turbine Design and Control).* A CRC Press book, Taylor & Francis Group, Boca Raton, FL.)

- By using the slider—see the *Add frequency* panel in Figure A2.15. When the user clicks the *Initialize* button in the *Add frequency* panel, the slider is enabled and the user can pre-visualize the position and shape of the new template. The user can then use the slider to enter the value of the new frequency. Then, by clicking the *Fix* button, the new frequency is added to the list.

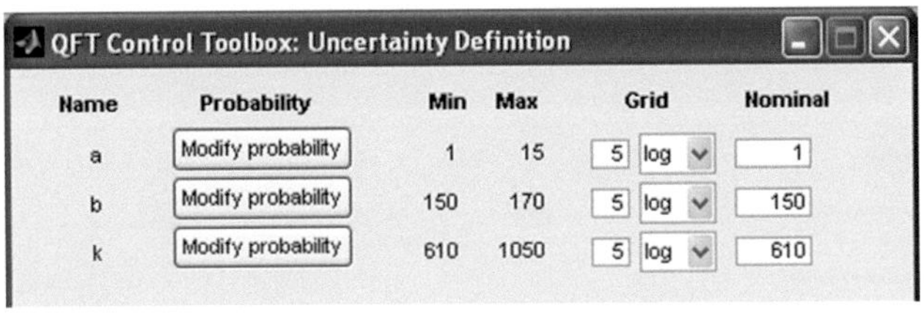

FIGURE A2.18
Parameter uncertainty window. (Adapted from Garcia-Sanz, M. and Houpis, C.H. 2012. *Wind Energy Systems: Control Engineering Design (Part I: QFT Control, Part II: Wind Turbine Design and Control)*. A CRC Press book, Taylor & Francis Group, Boca Raton, FL.)

- *Remove Frequency*: To remove a frequency, the user has to click the *Initialize* button in the *Remove frequency* panel. A list box appears, and enables the user to select the frequency to be removed. The points markers of the template associated with the frequency are indicated by 'x' instead of 'o'. To remove the selected frequency, click "Remove".

- *Frequency Vector*: The user can use the *Frequency vector* panel (see Figure A2.15) to change the visibility of the templates associated with the frequencies. If the user double clicks on a number of the frequency list that represents a template, the visibility of the template is switched (on/off). The user can also use the *Show all* and *Hide all* buttons to show all the templates or to hide them all, respectively.

- *Modify Parametric Uncertainty*: The user can modify the parametric uncertainty by pressing the *Modify uncertainty* button—see Figure A2.15. Then, a new sub-window allows the user to modify the parametric uncertainty in the same way as in the plant definition window—see Figure A2.18. Also, using this sub-window, the user can see how the changes in the parametric uncertainty are applied in real time to the templates window.

- *Export Nominal Plant*: The templates window has a menu item that allows the user to export the nominal plant as a transfer function. The nominal plant is saved in the current directory as a *.mat file. Afterward, the user can load that file to the MATLAB's Workspace (load *.mat) to use it.

A2.2.4 Specifications Window

This window allows the user to introduce robust stability and performance control specifications—see Figure A2.19. The *Choose specification type* panel includes seven groups of specifications. The first six options are the classical specifications that correspond to Equations A2.9 through A2.14, and the last one is a general specification that corresponds to Equation A2.15. This last option is able to generate the first five specifications and many other possibilities, including additional transfer functions like $M(s)$ in Figure A2.1 and others. See Table A2.2 for more details.

- *Predefined Specifications*: (see Equations A2.9 through A2.14). The user can add specifications for any transfer function extracted from Figure A2.1, as defined in Equations A2.6 through A2.8. The most common control specifications, including

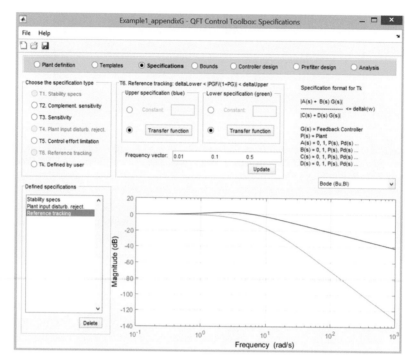

FIGURE A2.19
Specifications window.

stability, disturbance rejection, reference tracking, and control effort attenuation, are defined as $T_1, T_2, \ldots, T_6$ in the *Choose specification type* panel. They correspond to Equations A2.9 through A2.14, or Equations 2.35 through 2.40 in Chapter 2—see also Figure A2.19.

$$y = \frac{PG}{1+PGH} Fr + \frac{1}{1+PGH} (PG\, d_e + P\, d_i + d_o + M\, d) - \frac{PGH}{1+PGH} n \qquad \text{(A2.6)}$$

$$u = \frac{G}{1+PGH} Fr + \frac{G}{1+PGH} (d_e - HP\, d_i - H\, d_o - H\, M\, d) - \frac{GH}{1+PGH} n \qquad \text{(A2.7)}$$

$$e = \frac{1}{1+PGH} Fr - \frac{H}{1+PGH} (PG\, d_e - P\, d_i - d_o - M\, d) - \frac{H}{1+PGH} n \qquad \text{(A2.8)}$$

- **User-Defined Specifications**: (see Equation A2.15). All the predefined specifications, except the reference tracking, can be expressed using the T_k *Defined by user* specifications. Moreover, the option T_k opens the door to a wide variety of control specifications. It corresponds to Equations A2.15 and 2.41. By selecting the functions *A*, *B*, *C*, and *D*, we can define any one of the previous specifications (T_1–T_5) and also other cases that include additional plants introduced in the Plant definition window. In this way, we can define general disturbance rejection specifications, cascade control loops, asymmetric topologies, specifications involving Smith predictor blocks, MIMO equivalent plants, DPS plants, etc. For instance, see

TABLE A2.2

Control System Specifications

T1: Stability specs

$$|T_1(j\omega)| = \left|\frac{P(j\omega)G(j\omega)}{1 + P(j\omega)G(j\omega)}\right| \leq \delta_1(\omega) = W_s, \quad \omega \in \Omega_1$$

$$\text{where } |T_1(j\omega)| = \left|\frac{y(j\omega)}{F(j\omega)r(j\omega)}\right|$$

(A2.9)

T2: Complementary sensitivity specs

$$|T_2(j\omega)| = \left|\frac{P(j\omega)G(j\omega)}{1 + P(j\omega)G(j\omega)}\right| \leq \delta_2(\omega), \quad \omega \in \Omega_2$$

$$\text{where } |T_2(j\omega)| = \left|\frac{y(j\omega)}{F(j\omega)r(j\omega)}\right| = \left|\frac{y(j\omega)}{n(j\omega)}\right| = \left|\frac{u(j\omega)}{d_i(j\omega)}\right| = \left|\frac{y(j\omega)}{d_e(j\omega)}\right| = \left|\frac{e(j\omega)}{d_e(j\omega)}\right|$$

(A2.10)

T3: Sensitivity or Disturbances at plant output specs

$$|T_3(j\omega)| = \left|\frac{1}{1 + P(j\omega)G(j\omega)}\right| \leq \delta_3(\omega), \quad \omega \in \Omega_3$$

$$\text{where } |T_3(j\omega)| = \left|\frac{y(j\omega)}{d_o(j\omega)}\right| = \left|\frac{e(j\omega)}{d_o(j\omega)}\right| = \left|\frac{e(j\omega)}{F(j\omega)r(j\omega)}\right| = \left|\frac{e(j\omega)}{n(j\omega)}\right|$$

(A2.11)

T4: Disturbances at plant input specs

$$|T_4(j\omega)| = \left|\frac{P(j\omega)}{1 + P(j\omega)G(j\omega)}\right| \leq \delta_4(\omega), \quad \omega \in \Omega_4$$

$$\text{where } |T_4(j\omega)| = \left|\frac{y(j\omega)}{d_i(j\omega)}\right| = \left|\frac{e(j\omega)}{d_i(j\omega)}\right|$$

(A2.12)

T5: Control effort reduction specs

$$|T_5(j\omega)| = \left|\frac{G(j\omega)}{1 + P(j\omega)G(j\omega)}\right| \leq \delta_5(\omega), \quad \omega \in \Omega_5$$

$$\text{where } |T_5(j\omega)| = \left|\frac{u(j\omega)}{d_e(j\omega)}\right| = \left|\frac{u(j\omega)}{d_o(j\omega)}\right| = \left|\frac{u(j\omega)}{n(j\omega)}\right| = \left|\frac{u(j\omega)}{F(j\omega)r(j\omega)}\right|$$

(A2.13)

T6: Reference tracking specs

$$\delta_{6\_lo}(\omega) < |T_6(j\omega)| = \left|\frac{y(j\omega)}{r(j\omega)}\right| = \left|F(j\omega)\frac{P(j\omega)G(j\omega)}{1 + P(j\omega)G(j\omega)}\right| \leq \delta_{6\_up}(\omega),$$

$$\omega \in \Omega_6$$

$$\frac{|P_d(j\omega)G(j\omega)| \, |1 + P_e(j\omega)G(j\omega)|}{|P_e(j\omega)G(j\omega)| \, |1 + P_d(j\omega)G(j\omega)|} \leq \delta_6(\omega) = \frac{\delta_{6\_up}(\omega)}{\delta_{6\_lo}(\omega)}, \quad \omega \in \Omega_6$$

(A2.14)

Tk: General specs

$$\left|\frac{A(j\omega) + B(j\omega)G(j\omega)}{C(j\omega) + D(j\omega)G(j\omega)}\right| \leq \delta_k(\omega), \, \omega \in \Omega_k$$

(A2.15)

the code for Example 4.1 in Appendix 4, or for Example 5.1 in Appendix 5, or for Example 8.1 (Sections 8.8 and 8.9) in Appendix 8.

- Equation A2.9 [$T_1(j\omega)$] defines the robust closed-loop stability specification.
- Equation A2.10 [$T_2(j\omega)$] defines complementary sensitivity, robust sensor noise attenuation, robust control effort limitation from the plant input disturbance, and robust rejection of disturbances at the input of the controller.
- Equation A2.11 [$T_3(j\omega)$] defines the sensitivity, robust rejection of disturbances at the output of the plant, and robust sensor noise attenuation.
- Equation A2.12 [$T_4(j\omega)$] defines robust rejection of disturbances at the input of the plant.
- Equation A2.13 [$T_5(j\omega)$] defines robust control effort limitation from disturbances at the controller input, plant output, sensor noise, and filtered reference signal.
- Equation A2.14 [$T_6(j\omega)$] defines the robust reference tracking specification.
- Equation A2.15 [$T_k(j\omega)$] defines any specification from type $T_2(j\omega)$ to $T_5(j\omega)$, and many other options in general, where $A(j\omega)$, $B(j\omega)$, $C(j\omega)$, and $D(j\omega)$ can be defined by the user, from several options like 0, 1, $P(j\omega)$, or any other plant introduced in the Plant definition window, as for example $M(j\omega)$ to define the specification $|M(j\omega)/[1+P(j\omega)G(j\omega)]|$.

The value of $\delta_i(\omega)$ denotes the upper limit of the magnitude of the objective (the specification) at every frequency of interest. Each specification can be defined for a different set of frequencies ω_i, always a sub-set of the original set of frequencies of interest—see also Figure 2.14.

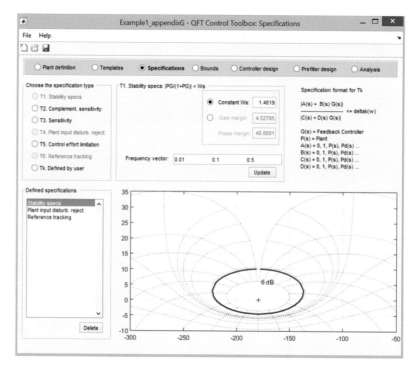

FIGURE A2.20
Defining stability specifications, either as W_s or as *GM* and *PM*.

- **Defining a Specification**: The user has to enter the value of the performance specification $\delta_i(\omega)$ [$\delta_{6\_up}(\omega)$ and $\delta_{6\_lo}(\omega)$ in the reference tracking case]. The method to enter the value of $\delta_i(\omega)$ depends on the type of specification being defined, so that:

  - **Robust Stability Specs**: (see Table A2.2, Equation A2.9, and Figure A2.20). $\delta_1(\omega)$ is a constant (W_s). The user can enter directly the value of $\delta_1(\omega) = W_s$ for the stability specification. W_s is the closed-loop constant magnitude circle (in magnitude) in the Nichols chart. In this case, the gain and phase margins are calculated automatically. If the user introduces the phase margin PM, then W_s and GM are calculated automatically. The following equations are used in the calculations:

  Gain margin:

  $$GM = 20\log_{10}\left(1+\frac{1}{W_s}\right), \text{ in dB} \tag{A2.16}$$

  Phase margin:

  $$PM = 180 - 2\left(\frac{180}{\pi}\right)\text{acos}\left(\frac{0.5}{W_s}\right), \text{ in deg} \tag{A2.17}$$

  Closed-loop M circle:

  $$W_s = \frac{0.5}{\cos\left(\left(\frac{\pi}{180}\right)\frac{180-PM}{2}\right)}, \text{ in magnitude} \tag{A2.18}$$

- **Robust Performance Specs**: (see Table A2.2, Equations A2.10 through A2.13, and Figures A2.21 and A2.22). $\delta_i(\omega)$, $i = 2,3,4,5$ can be defined as a constant (see Figure A2.21), a vector of constants of the same length as the specification frequency vector, or as transfer function (zero/pole/gain or num/den, see Figure A2.22). In this case, if $\delta_i(\omega)$ is a transfer function, its Bode diagram is also displayed in the Specification window (similar to Figure A2.19).

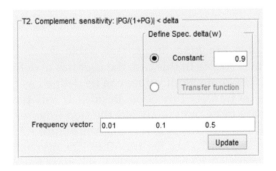

FIGURE A2.21
Defining $\delta_2(\omega) = 0.9$, as a constant. Similar for $\delta_i(\omega)$, $i = 2,3,4$, and 5.

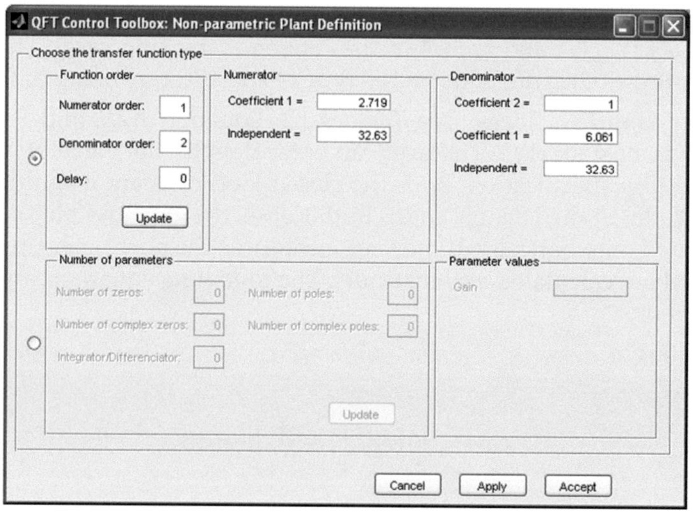

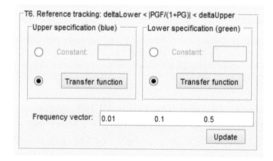

FIGURE A2.22
Defining $\delta_i(\omega)$, $i = 2,3,4$, and 5 as a transfer function. Click: Transfer function in Figure A2.21. Then: $\delta_2(s) =$ $(2.719\ s + 32.63)/(s^2 + 6.061\ s + 32.63)$. (Adapted from Garcia-Sanz, M. and Houpis, C.H. 2012. *Wind Energy Systems: Control Engineering Design (Part I: QFT Control, Part II: Wind Turbine Design and Control)*. A CRC Press book, Taylor & Francis Group, Boca Raton, FL.)

FIGURE A2.23
Defining $\delta_{6\_up}(\omega)$ and $\delta_{6\_lo}(\omega)$ specifications. (Adapted from Garcia-Sanz, M. and Houpis, C.H. 2012. *Wind Energy Systems: Control Engineering Design (Part I: QFT Control, Part II: Wind Turbine Design and Control)*. A CRC Press book, Taylor & Francis Group, Boca Raton, FL.)

- *Robust Reference Tracking Specs*: (see Table A2.2, Equation A2.14, and Figures A2.23 and A2.24). In this case the user has to define $\delta_{6\_up}(\omega)$ and $\delta_{6\_lo}(\omega)$ (see Figure A2.23). Again, both values can be defined as constants, as a vector of constants or as transfer functions. Besides, if this type of specification is selected, there are more plots available: the step time-response of $\delta_{6\_up}(\omega)$ and $\delta_{6\_lo}(\omega)$, the Bode diagram of $\delta_{6\_up}(\omega) - \delta_{6\_lo}(\omega)$, see Figure A2.24, and the Bode diagram of $\delta_{6\_up}(\omega)$ and $\delta_{6\_lo}(\omega)$, see Figure A2.19. The *reference tracking specification* type button (Figure A2.20) is active only if the plant has model uncertainty.

- *Defined by User Specs*: (see Table A2.2, Equation A2.15 and Figure A2.25). In this case the user not only has to enter $\delta_k(\omega)$, but also $A(j\omega)$, $B(j\omega)$, $C(j\omega)$, and

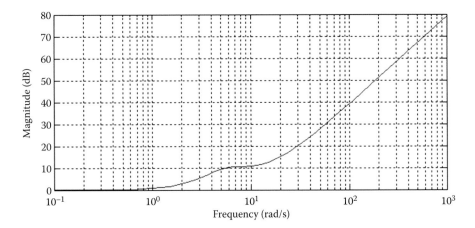

FIGURE A2.24
Bode diagram of the difference $\delta_{6\_up}(\omega)-\delta_{6\_lo}(\omega)$. (Adapted from Garcia-Sanz, M. and Houpis, C.H. 2012. *Wind Energy Systems: Control Engineering Design (Part I: QFT Control, Part II: Wind Turbine Design and Control)*. A CRC Press book, Taylor & Francis Group, Boca Raton, FL.)

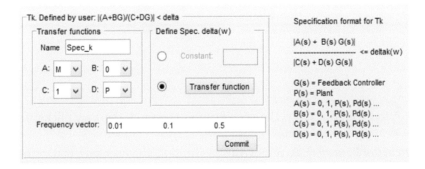

FIGURE A2.25
Defining *Defined by user* specs. Equation A2.15, having for example $M(j\omega)$ from the *Plant definition* window, and doing $|M(j\omega)/[1+P(j\omega)G(j\omega)]|$.

$D(j\omega)$, for a specification $|[A(j\omega)+B(j\omega)G(j\omega)]/[C(j\omega)+D(j\omega)G(j\omega)]|\leq\delta_k(\omega)$. They can be constants (0 or 1), the system plant $P(j\omega)$, or other auxiliary plants like $M(j\omega)$, defined in the *Plant Definition* window. Notice that $G(j\omega)$ is the feedback controller.

- *Frequency Vector*: The user can define a different frequency vector for each specification. These vectors must be a subset of the initial frequency vector of the system plant.

- *Specification Addition*: The user can define new specifications by selecting them in the *Choose specification type* panel—see Figure A2.19 or A2.20. Then, once the parameters of the specification are entered, the user clicks the *Commit* button to finally add the specification.

- *Specification Edition*: To edit a specification, the user has to select it in the *Defined specifications* panel. Then, the data of the specification are displayed in the center

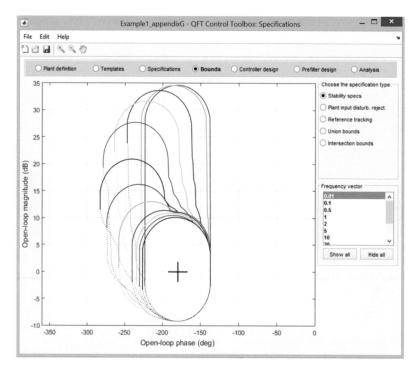

FIGURE A2.26
Bounds window.

panel of the window. Once the specification is edited, the user clicks the *Update* button to apply the changes.

- *Specification Removal*: To remove a specification, the user selects it in the *Defined specifications* panel and clicks the *Delete* button.

A2.2.5 Bounds Window

Given the plant templates and the control specifications, QFT converts the closed-loop magnitude specifications $[T_1(j\omega)$ to $T_k(j\omega)]$ into magnitude and phase constraints for a nominal open-loop function $L_0(j\omega) = P_0(j\omega)G(j\omega)$. These constraints are called QFT bounds—see Figure A2.26. After the design of the controller $G(s)$ (next section), the nominal open-loop function $L_0(j\omega)$ must remain above the solid-line bounds and below the dashed-line bounds at each specific frequency of interest, to meet the specifications.

The Bounds window (see Figure A2.26) shows the QFT bounds of each specification defined in the Specification window, as well as the Union of all the bounds and the Intersection (worst case scenario) of all the bounds.

By clicking "Show all" or "Hide all", the toolbox plots all the bounds or hides all of them. By double clicking on its corresponding frequency, the bound of that frequency is shown or hidden.

The user can define the minimum, maximum, and step value of the phase vector of the bounds with the *Edit phase vector* submenu in the *Edit* menu.

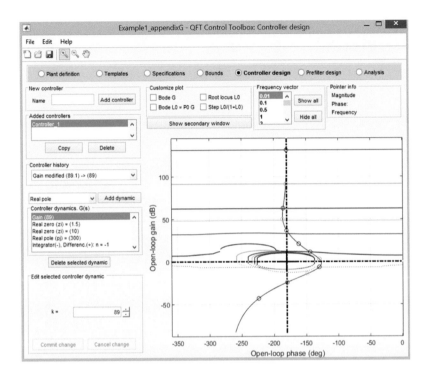

FIGURE A2.27
Controller design window.

A2.2.6 Controller Design Window

Once the user has introduced the information of the plant and the control specifications, and once the templates and bounds have been calculated, the next step involves the design (loop shaping) of the feedback controller $G(s)$, so that the nominal open-loop transfer function $L_0(s) = P_0(s)G(s)$ meets the bounds—see Figure A2.27. Generally speaking, the loop shaping, or $G(s)$ design, requires to change the gain and add poles and zeros, either real or complex, until the nominal loop $L_0(s)$ lies above the solid-line bounds and below the dashed-line bounds at each frequency of interest.

- Controller management
  - Controller addition. There are two ways of adding a new controller:
    - To create the new controller from scratch, the user has to enter the name of the controller and press the *Add controller* button. If there is not any other controller defined with that name, the new controller appears in the list of added controllers.
    - If the user wants to add a new controller based of the dynamics of an existing controller, the user has to select it from the "Added controllers" list and press the *Copy* button. The user has then to enter the name of the new controller in the emerging sub-window.
  - Controller removal. To remove a controller, the user has to select it from the list of "Added controllers" and press the *Delete* button.

TABLE A2.3

Controller Elements, $G_i(s)$

Gain	$\pm k$	Integrators, Differentiators	$\dfrac{1}{s^n}, s^n$
Real zero	$\dfrac{s}{z_i}+1$	Real pole	$\dfrac{1}{\dfrac{s}{p_j}+1}$
Complex zero	$\dfrac{s^2}{\omega_n^2}+\dfrac{2\zeta s}{\omega_n}+1,(\zeta<1)$	Complex pole	$\dfrac{1}{\dfrac{s^2}{\omega_n^2}+\dfrac{2\zeta s}{\omega_n}+1},(\zeta<1)$
Lead/Lag network	$\dfrac{\left(\dfrac{s}{z_i}+1\right)}{\left(\dfrac{s}{p_j}+1\right)}$	Notch filter	$\dfrac{\left(\dfrac{s^2}{\omega_n^2}+\dfrac{2\zeta_1 s}{\omega_n}+1\right)}{\left(\dfrac{s^2}{\omega_n^2}+\dfrac{2\zeta_2 s}{\omega_n}+1\right)}$
P.I. controller	$K_p\left(1+\dfrac{1}{T_i s}\right)$	P.D. controller	$K_p\left(1+\dfrac{T_d s}{\dfrac{T_d}{N}s+1}\right)$
P.I.D. controller		$K_p\left(1+\dfrac{1}{T_i s}+\dfrac{T_d s}{\dfrac{T_d}{N}s+1}\right)$	

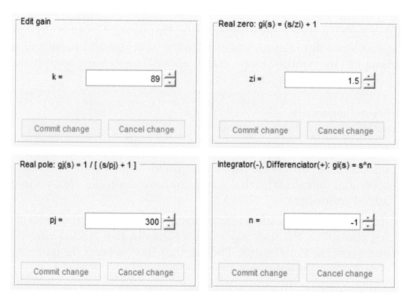

FIGURE A2.28
Windows for gain, real zero, real pole, and integrator/differentiator.

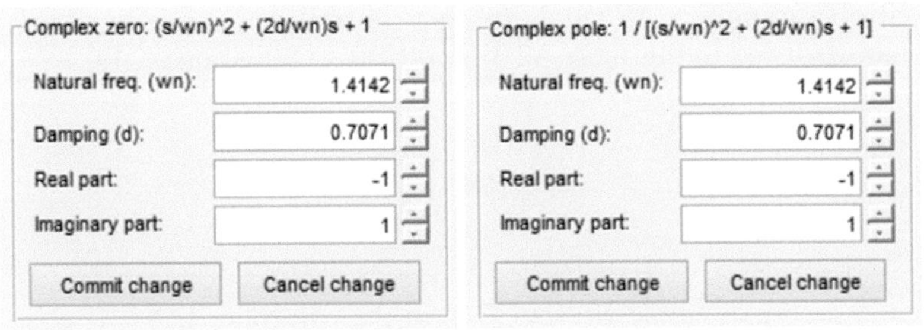

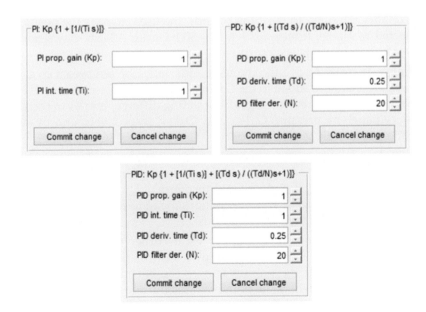

FIGURE A2.29
Window for complex zero and complex pole.

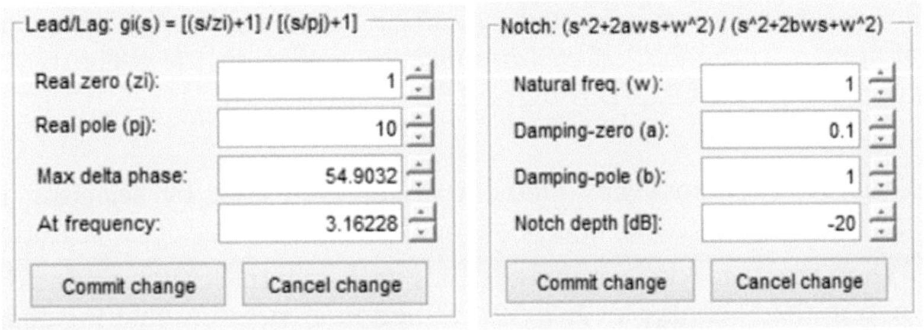

FIGURE A2.30
Window for Lead/Lag element and Notch filter.

FIGURE A2.31
Window for PI, PD, and PID controllers.

- Controller dynamics. When defining a controller, the user can work with the dynamics listed in the Table A2.3, according to Equation A2.19.

$$L_0(s) = P_0(s) \prod_{i=0}^{n} [G_i(s)] \qquad (A2.19)$$

To add a new dynamic element $G_i(s)$ in the controller $G(s) = \Pi\; G_i(s)$, the user has to press the *Add dynamic* button. The dynamic is added with its predefined values. In the lower left corner of the window, there is a panel which shows information about the selected dynamic. The information displayed depends on the type of the element, as shown in Figures A2.28 through A2.31.

- Dynamic edition.

  All the dynamics can be edited by entering their new values in the textboxes that appear in the *Edit selected controller dynamic* panel (see Figure A2.27), as shown in Figures A2.28 through A2.31.

- Dynamic removal.

  The user can remove any element of the controller by selecting it in the *Controller dynamics* panel and pressing the *Delete selected dynamic* button.

  When the user edits some dynamics of the controller, the updated $L_0(j\omega)$ is shown in the Nichols plot as a red-dashed line, meaning that the changes are provisional. To commit the changes, the user has to press the *Commit* button. Then the red-dotted line is drawn as a solid black line. To discard the changes, the user has to press the *Cancel* button.

  Figures A2.32 illustrate an example of the graphic dynamic edition. Figure A2.32a shows the Controller design window and two details of the secondary window. To meet the bounds, the user changes the position of the selected real pole, from $p = 0.1$ to $p = 300$. Figure A2.32b shows how all the plots have been updated accordingly.

- *Change History.* Each time the user adds, edits, or removes a dynamic element, the change is added to the *Controller history* panel (see Figure A2.33). The user can undo and redo changes by selecting different entries in the *Controller history* panel, by pressing Ctrl+x and Ctrl+y, respectively, and by using the Edit menu as well.

- *Secondary Window.* The user can open a secondary window with the following plots (see Figures A2.32 and A2.34): Bode diagram of $G(s)$, Bode diagram of $L_0(s)$, root locus of $L_0(s)$, and unit step response of $L_0(s)/[1+L_0(s)]$. The secondary window can be resized, and the changes made to the selected controller are applied in real time to all the diagrams.

- *Pointer Information*: In the upper right corner of the Controller design window, there is a panel that shows information about the pointer—see Figure A2.35. If the pointer is over the Nichols plot, the panel shows information about where the pointer is (magnitude and phase). If the pointer is placed over the $L_0(j\omega)$ line, the panel also includes the frequency ω associated with that point.

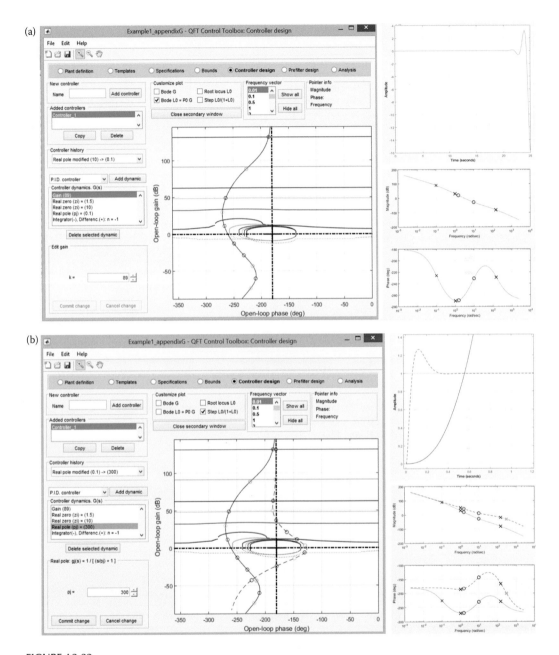

FIGURE A2.32
An example of the graphic dynamic edition. From (a) pole $pj = 0.1$ to (b) pole $pj = 300$. NC: Original $L_0(s)$ in black solid line. New $L_0(s)$ in red dashed line. Secondary window: plant poles (x) and zeros (o) in black, controller poles (x) and zeros (o): blue = original, or red, green = new.

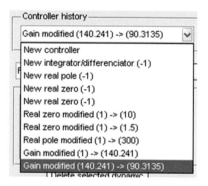

FIGURE A2.33
Controller history panel. (Adapted from Garcia-Sanz, M. and Houpis, C.H. 2012. *Wind Energy Systems: Control Engineering Design (Part I: QFT Control, Part II: Wind Turbine Design and Control).* A CRC Press book, Taylor & Francis Group, Boca Raton, FL.)

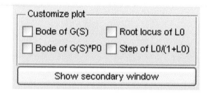

FIGURE A2.34
Panel to open the secondary window. (Adapted from Garcia-Sanz, M. and Houpis, C.H. 2012. *Wind Energy Systems: Control Engineering Design (Part I: QFT Control, Part II: Wind Turbine Design and Control).* A CRC Press book, Taylor & Francis Group, Boca Raton, FL.)

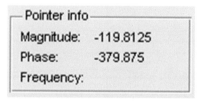

FIGURE A2.35
Pointer info panel. (Adapted from Garcia-Sanz, M. and Houpis, C.H. 2012. *Wind Energy Systems: Control Engineering Design (Part I: QFT Control, Part II: Wind Turbine Design and Control).* A CRC Press book, Taylor & Francis Group, Boca Raton, FL.)

A2.2.7 Prefilter Design Window

If the plant has model uncertainty and the control problem requires reference tracking specifications, then the Prefilter design window is active after the design of the $G(s)$ controller.

Figure A2.36 shows the Prefilter design window, with the following plots: the upper and lower reference tracking specifications [$\delta_{6\_up}(\omega)$, $\delta_{6\_lo}(\omega)$, dashed blue lines], see Equation A2.14 and Figure A2.23, and the maximum and minimum cases of $L_0(s)F(s)/[1+L_0(s)]$ over the plant uncertainty (dashed black lines).

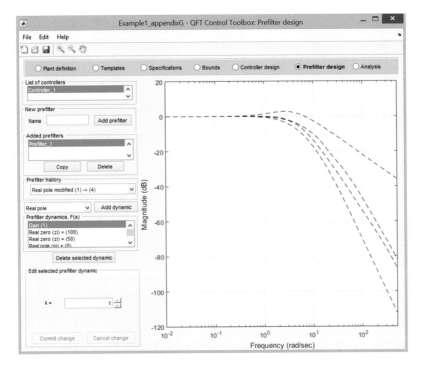

FIGURE A2.36
Prefilter design window.

The design of the prefilter calculates automatically the worst upper and lower closed-loop response cases of $L_0(s)F(s)/[1+L_0(s)]$ over the plant uncertainty. These cases should be between the upper and lower reference tracking functions to meet the specifications. The Prefilter design window is similar to the Controller Design window. The way in which the prefilters are added, edited, and removed is the same.

The Prefilter design window has an additional *List of controllers* panel, which allows the user to select among the feedback controllers $G(s)$ designed in the previous window.

A2.2.8 Analysis Window

Once the user finishes the controller (and prefilter) design, the Analysis window is active. The analysis is performed in both the frequency domain and time domain. The window analyzes the controller $G(s)$ and prefilter $F(s)$ performance in the worst case scenario over the plant uncertainty.

- The window allows the user to perform two types of analysis:

  Frequency-Domain Analysis: Select the specification of interest in the *Specification* panel, and the Frequency-domain option in the *Analysis type* panel. The Toolbox shows the Bode diagram for the specification and the control system with the selected controller and prefilter (if any). The dashed line represents the desired specification $\delta_i(\omega)$ and the solid line the worst case of the control system over the

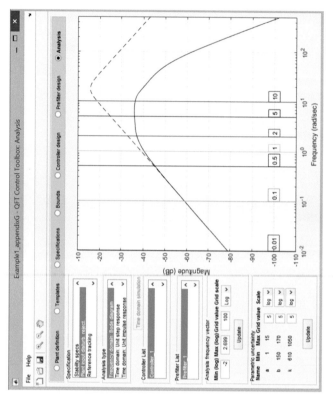

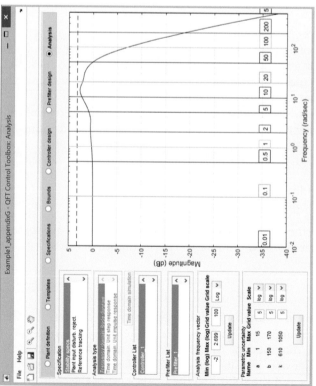

FIGURE A2.37

Analysis window: Frequency-domain specifications.

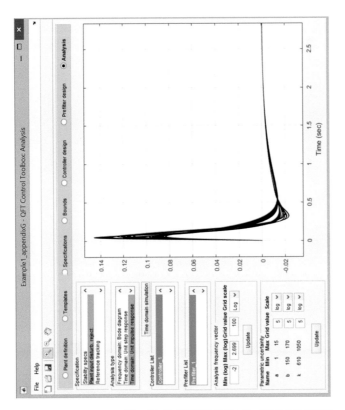

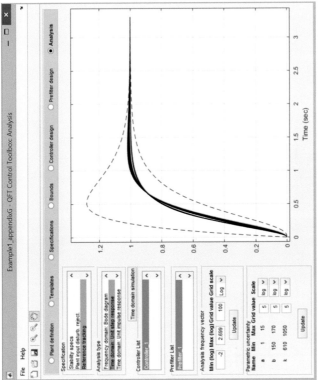

FIGURE A2.38
Analysis window: Time-domain specifications.

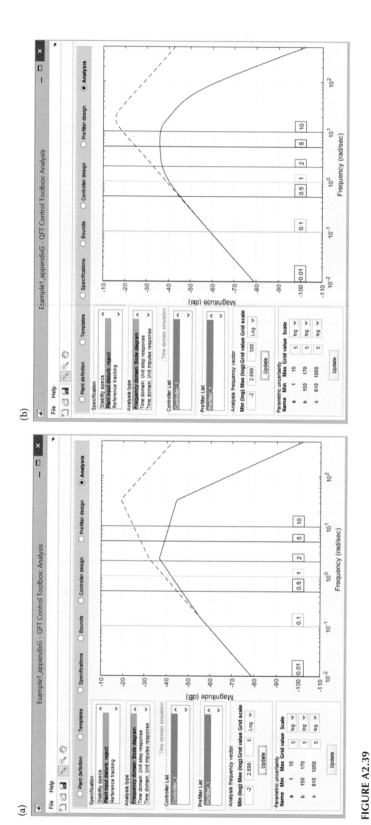

FIGURE A2.39

Stability analysis: (a) with not enough points in the frequency vector and (b) with a more populated grid.

plant uncertainty at each frequency—see Figure A2.37 for the analysis of the stability and plant input disturbance rejection specifications in the frequency domain. Notice that the solid line is not a transfer function, but the worst case among all the transfer functions over the plant uncertainty at every frequency.

Time-Domain Analysis: After selecting the specification, the input (step or impulse), the controller $G(s)$, and prefilter $F(s)$, and clicking the *time domain simulation* button, the toolbox analyzes the time response of the control system with every plant defined in the *Parametric uncertainty* panel. Figure A2.38 shows two cases: (a) $y(t)$ response for $L_0(s)F(s)/[1+L_0(s)]$ with $r(s)$ = unitary step; and (b) $y(t)$ response for $P_0(s)/[1+L_0(s)]$ with $d(s)$ = unitary impulse. The number of plants analyzed (number of lines plotted) depends on the values introduced in the *Parametric uncertainty* panel.

- *Controller/Prefilter Combinations*: The user can select any combination of controllers (from the *Controller List* panel) and prefilters (from the *Prefilter List* panel) previously defined (if any).

- *Frequency Vector Panel for the Analysis*: The *Analysis frequency vector* panel allows the user to enter the frequency vector to be used in the frequency-domain analysis. If there are not enough points in the frequency vector, the resulting analysis may not be accurate enough—see Figure A2.39.

- *Parametric Uncertainty Panel*: As was explained, this panel allows the user to modify the grid of the parametric uncertainty variables. Again, if too few points are selected, the analysis may not be accurate enough. On the other hand, if too many points are selected, the analysis may be slow. This panel allows the user to analyze the system in both, (1) the points previously defined in the plant definition window (with the *parametric uncertainty* panel), and (2) in new points of uncertainty, defined now in the analysis window with the *parametric uncertainty* panel.

If the responses for all the *selected plants* satisfy the desired control performance specifications at both frequency and time domains, then the design is completed. If the design fails at any frequency or time, you may decide to re-design the controller or prefilter, or to check the plant model, uncertainty, or specification definitions.

Appendix 3: Algorithm—Nyquist Stability Criterion in Nichols Chart

This appendix includes the MATLAB code for the algorithm that calculates the stability criterion presented in Chapter 3. This is the practical method developed by the author to study the Nyquist stability criterion for feedback control systems in the Nichols chart.[125]

The algorithm is included in the QFTCT, *Controller design* window, *File* menu, *Check stability* option. Also, it can be free-downloaded from the website:

http://cesc.case.edu/Stability_Nyquist_GarciaSanz.htm

The code is presented next and includes references to the main equations, figures, cases, and rules of the method as developed in Chapter 3.

```
function [zc,N,num_p_RHP,Na,Nb,Nc,Nd,zpCancel,k,sigm,alpha,gamma]=c
alc_zc(L)

% Function to calculate closed-loop system stability
% - Input: L(s) = P(s) C(s)
% - Outputs: zc,N,num_p_RHP,Na,Nb,Nc,Nd,zpCancel,k,sigm,alpha,gamma
%            according to method introduced by Mario Garcia-Sanz (2016)
% - Code free-download at:
%   http://cesc.case.edu/Stability_Nyquist_GarciaSanz.htm
%   And also included in the QFTCT, Controller design window
% ------------------------------------------------------------

% 1. Initial values
% ----------------
tolPh = 1e-3; % tolerance phase
tolPh1 = 0.002; % tolerance phase
tolPh2 = 5; % tolerance phase
tolM = 0.05; % tolerance magnitude
tolA = 0.05; % tolerance for change around 0.
wmin = 1e-16; % lowest freq.
zpCancel = 0; % RHP zero-pole cancelations. "0"=no, "1"=yes
z_LHP = []; z_RHP = []; z_0 = []; z_i = [];
p_LHP = []; p_RHP = []; p_0 = []; p_i = [];

% 2. gain, zeroes, poles, delay
% ----------------------------
dc_gain = abs(dcgain(L));
[zzz,ppp,kkk]=tf2zpk(L.num{1},L.den{1});
zzz_roots = roots([L.num{1}]);
ppp_roots = roots([L.den{1}]);

orderNum = length(zzz_roots);
orderDen = length(ppp_roots);
```

```
timeDelay = L.iodelay;

zzzRe = real(zzz_roots);
pppRe = real(ppp_roots);

zzzMag = abs(zzz_roots);
pppMag = abs(ppp_roots);

[row_z_LHP,column_z_LHP] = find(zzzRe<0);
num_z_LHP = length(row_z_LHP);
for ii=1:num_z_LHP
    z_LHP(ii,1) = zzz_roots(row_z_LHP(ii),column_z_LHP(ii));
end

[row_z_RHP,column_z_RHP] = find(zzzRe>0);
num_z_RHP = length(row_z_RHP);
for ii=1:num_z_RHP
    z_RHP(11,1) = zzz_roots(row_z_RHP(ii),column_z_RHP(ii));
end

[row_z_0,column_z_0] = find(zzzMag==0);
num_z_0 = length(row_z_0);
for ii=1:num_z_0
    z_0(ii,1) = zzz_roots(row_z_0(ii),column_z_0(ii));
end

[row_z_i,column_z_i] = find(zzzRe==0 & zzzMag~=0);
num_z_i = length(row_z_i);
for ii=1:num_z_i
    z_i(ii,1) = zzz_roots(row_z_i(ii),column_z_i(ii));
end

[row_p_LHP,column_p_LHP] = find(pppRe<0);
num_p_LHP = length(row_p_LHP);
for ii=1:num_p_LHP
    p_LHP(ii,1) = ppp_roots(row_p_LHP(ii),column_p_LHP(ii));
end

[row_p_RHP,column_p_RHP] = find(pppRe>0);
num_p_RHP = length(row_p_RHP);
for ii=1:num_p_RHP
    p_RHP(ii,1) = ppp_roots(row_p_RHP(ii),column_p_RHP(ii));
end

[row_p_0,column_p_0] = find(pppMag==0);
num_p_0 = length(row_p_0);
for ii=1:num_p_0
    p_0(ii,1) = ppp_roots(row_p_0(ii),column_p_0(ii));
end

[row_p_i,column_p_i] = find(pppRe==0 & pppMag~=0);
num_p_i = length(row_p_i);
```

```
for ii=1:num_p_i
    p_i(ii,1) = ppp_roots(row_p_i(ii),column_p_i(ii));
end

% 3. RHP zero-pole cancelation
% --------------------------   (Sec.3-4. Rule 1)
cc = 0;
dd = [];
if ~isempty(z_RHP) & ~isempty(p_RHP)
    for pp=1:num_p_RHP
        dd = find(p_RHP(pp)==z_RHP);
        if ~isempty(dd)
            cc=1;
            break;
        end
    end
end
if (num_z_0>0 & num_p_0>0) || cc==1
  zpCancel = 1;
end

% 4. mag, pha, ww
% ---------------
% LN
[LNnum,LNden] = zp2tf([z_LHP;z_RHP],[p_LHP;p_RHP],kkk);
LN = tf(LNnum,LNden);
[magLN0,phaseLN0] = bode(LN,0);

% L
[magL0,phaseL0] = bode(L,0); % phase at w=0
phaseL0 = round(phaseL0*100)/100; % Protection numerical accuracy
[mag2,pha2,ww2] = nichols(L);
ww1 = logspace(log10(ww2(1)),log10(ww2(end)),5000);
[mag1,pha1] = nichols(L,ww1);
nn = length(pha1);
mag = [];
pha = [];
ww = [];
for jj=1:nn
    mag(jj) = mag1(1,1,jj);
    pha(jj) = pha1(1,1,jj);
    ww(jj) = ww1(jj);
end
indPh2 = find(phaseL0~=pha2,1,'first'); % the first that is "~="
phaseL1 = pha2(indPh2); % phase at w=0+
leftRightAt0 = sign(phaseL0-phaseL1);

% 5. Find crosses at -900, -540, -180, +180 etc and mag>1
% -------------------------------------------------------
pha_neg360_0 = pha;
for jj=1:nn
    if pha_neg360_0(jj)>tolPh
```

```
            d1 = ceil(pha_neg360_0(jj)/360);
            pha_neg360_0(jj) = pha_neg360_0(jj) - 360*d1;
        elseif pha_neg360_0(jj)<=-360-tolPh
            d1 = floor(-pha_neg360_0(jj)/360);
            pha_neg360_0(jj) = pha_neg360_0(jj) + 360*d1;
        end
end
changeAround0 = pha_neg360_0 + 180;
indSignChange = [];
for jj=1:nn-1
    if (sign(changeAround0(jj+1))~=sign(changeAround0(jj))) &
            abs(pha_neg360_0(jj+1)-pha_neg360_0(jj))<(180-tolPh1)…
            & abs(pha_neg360_0(jj+1)-pha_neg360_0(jj))>tolPh1
        indSignChange = [indSignChange jj];
    end
end
mm = length(indSignChange);
ind_kk_100 — [];
for jj=1:mm
    if mag(indSignChange(jj)+1)>1 & mag(indSignChange(jj))>1
        ind_kk_180 = [ind_kk_180 indSignChange(jj)];
    end
end
if ~isempty(ind_kk_180)
    n_ind_kk_180 = length(ind_kk_180);
end

% 6. Na. Fig.3.7(a)
% ----------------
Na = 0;
Na_1 = 0;
nMaxPha = length(pha);
if ~isempty(ind_kk_180)
    for jj=1:n_ind_kk_180
        Na_1(jj) = 0;
        if mag(ind_kk_180(jj))>(1+tolM) % if greater than 0 dB
            if (ind_kk_180(jj)+1)<nMaxPha % Protection
                pha21 = pha(ind_kk_180(jj)+1)-pha(ind_kk_180(jj));
                if abs(pha21)<tolPh
                    pha21 = 0;
                end
                ss = 1;
                while pha21==0
                    ss = ss+1;
                    if (nMaxPha-ind_kk_180(jj))>ss
                        pha21 = pha(ind_kk_180(jj)+ss)…
                                -pha(ind_kk_180(jj));
                        if abs(pha21)<tolPh
                            pha21 = 0;
                        end
                    else
                        pha21 = 0;
```

```
                                   return;
                             end
                       end
                 if pha21>0
                       Na_1(jj) = -2; % Fig.3.7(a) to the right
                 elseif pha21<0
                       Na_1(jj) = +2; % Fig.3.7(a) to the left
                 else
                       Na_1(jj) = 0; % Fig.3.7(a) in axis
                 end
           end
     end
  end
end
Na = sum(Na_1);
```

% 7. **Nb**. **Fig.3.7(b)**
```
% ------------------
Nb = 0;
if isfinite(dc_gain)==1  % dc_gain is finite
    k_at_w0 = ((phaseL0/(-180))-1)/2; % at -900,-540,-180,+180,etc
    if abs(k_at_w0-round(k_at_w0))<tolPh
        if dc_gain>1 % dc_gain is >1
            pha21 = pha(2)-pha(1);
            ss = 2;
            while pha21==0
                ss = ss+1;
                if nMaxPha>ss
                    pha21 = pha(ss)-pha(1);
                else
                    pha21 = 0;
                    return;
                end
            end
            if pha21>0
                Nb = -1; % Fig.3.7(b) to the right
            elseif pha21<0
                Nb = +1; % Fig.3.7(b) to the left
            else
                Nb = 0; % Fig.3.7(b) in axis
            end
        else % dc_gain is <1
            Nb = 0; % Fig.3.7(b)
        end
    else
        Nb = 0; % Fig.3.7(b)
    end
end
```

% 8. **Nc**. **Fig.3.7(c)**
```
% ------------------
Nc = 0;
```

```
if mag(end)>1  % mag(w=inf)>1
    k_at_wInf = ((pha(end)/(-180))-1)/2;
    if abs(k_at_wInf-floor(k_at_wInf))<0.001
        % at -900,-540,-180,+180,etc
        pha21 = pha(end)-pha(end-1);
        ss = 1;
        while pha21==0
            ss = ss+1;
            if nMaxPha>ss
                pha21 = pha(end)-pha(end-ss);
            else
                pha21 = 0;
                return;
            end
        end
        if pha21>0
            Nc = -1; % Fig.3.7(c) to the right
        elseif pha21<0
            Nc = +1; % Fig.3.7(c) to the left
        else
            Nc = 0; % Fig.3.7(c) in axis
        end
    else
        Nc = 0; % Fig.3.7(c)
    end
end

% 9. Nd. Fig.3.7(d)
% ----------------
Nd = 0; k = 0; sigm = 0; alpha = 0; gamma = 0; % Initialization

if num_p_0>0;
    if isfinite(dc_gain)==1  % dc_gain is finite
        Nd = 0; % Fig.3.7(d)
    else % dc_gain is infinite
        % -- k --
            if phaseL1>-180 & phaseL1<90 % Case [a]
                k = -1; % Eq.(3.8)
            elseif phaseL1>90 & phaseL1<180
                if leftRightAt0==1 % to the left. Case [c]
                    k = 0; % Eq.(3.10)
                elseif leftRightAt0==-1 % to the right. Case [b]
                    k = -1; % Eq.(3.9)
                end
            elseif phaseL1<=-180 % to the left and right. Case [d]
                k = -ceil((phaseL1+180)/360); % Eq.(3.11)
            elseif phaseL1>=180 % to the left and right. Case [e]
                k = -ceil((phaseL1-180)/360); % Eq.(3.12)
            end

        % -- sigm --
            dcgainLN = dcgain(LN);
```

```
                    if dcgainLN>=0 % Case [a]
                        sigm = 0; % Eq.(3.13)
                    else % Case [b]
                        sigm = 1; % Eq.(3.14)
                    end


            % -- gamma --
                    numZP = num_z_RHP + num_p_LHP - num_z_LHP - num_p_RHP;
                    gamma = 2 * numZP/max(abs(numZP),1); % Eq.(3.20)


            % -- alpha --
                    if phaseL0<=0
                        if leftRightAt0==1 % to the left. Case [a]
                            alpha = -ceil((phaseL0+90*(num_p_0))/360);
                            % Eq.(3.15)
                        elseif leftRightAt0==-1 % to the right. Case [b]
                            alpha = -ceil((phaseL0+90*(num_p_0-2))/360);
                            % Eq.(3.16)
                        end
                    elseif phaseL0>0
                        if leftRightAt0==1 % to the left. Case [c]
                            alpha = floor((phaseL0+90*(num_p_0-2))/360);
                            % Eq.(3.17)
                        elseif leftRightAt0==-1 % to the right. Case [d]
                            alpha = -floor((phaseL0-90*(num_p_0))/360);
                            % Eq.(3.18)
                        end
                    end
                    if orderNum==orderDen & …
                       ( (pha(end)<180+tolPh2 & pha(end)>180-tolPh2)…
                       | (pha(end)<tolPh2 & pha(end)>-tolPh2) ) % Case [e]
                        alpha = 0; % Eq.(3.19)
                    end


            Nd = 2*(k+1) + sigm + alpha*gamma; % Eq.(3.6)
        end
end


% 10. N sum
% ---------
N = Na + Nb + Nc + Nd; % Eq.(3.5)

% 11. Zc sum
% ----------   (Sec.3-4. Rule 2)
zc = N + num_p_RHP;

% The closed-loop system is stable if:
% - Rule 1: "zpCancel=0". RHP zero-pole cancelations "0"=no,"1"=yes
% - Rule 2: "zc = 0". Being zc = N + num_p_RHP
% ---------------------------------------------
```

Appendix 4: Algorithms—Smith Predictor Robust Control

This appendix includes the MATLAB code of the algorithms presented in Chapter 4, Example 4.1, for the QFT design of a Smith predictor.

```
% Example 4.1
% ===========

% TABLE 4.1 (The First algorithm). Section 4-2.1
% ===============================================

% Table 4.1. Step 1 ----
n1 = 5;
n2 = 5;
n3 = 5;
k = linspace(8.1,9.9,n1);
tau = linspace(2.25,2.75,n2);
td = linspace(0.63,0.77,n3);

% Table 4.1. Step 2 ----
omegaBW = 2.25; % rad/sec
magSelecUp = 1.4125; % 3 dB = 1.4125 in magnitude
magSelecLo = 1/1.4125; % 3 dB = 1.4125 in magnitude
rejectedModels = [];
r3 = 0;

% Table 4.1. Step 3 ----
r1 = 0;
for i1=1:1:n1
  km0 = k(i1);
  for j1=1:1:n2
    taum0 = tau(j1);
    for k1=1:1:n3
      tdm0 = td(k1);
      r1 = r1 + 1;
      Prm(r1) = tf(km0,[taum0 1]); % SP plant model, no delay
      Pm(r1) = tf(km0,[taum0 1],'iodelay',tdm0); % SP plant model
      models(r1).km0 = km0;
      models(r1).taum0 = taum0;
      models(r1).tdm0 = tdm0;

      % Table 4.1. Step 4 ----
      r2 = 0;
      for i2=1:1:n1
        k0 = k(i2);
```

```matlab
        for j2=1:1:n2
          tau0 = tau(j2);
          for k2=1:1:n3
            td0 = td(k2);
            r2 = r2 + 1;
            P(r2) = tf(k0,[tau0 1],'iodelay',td0); % Real plant
            P0(r2) = tf(k0,[tau0 1]); % Real plant, no delay
            exp_mod_cx = exp(-tdm0*omegaBW*j);
            exp_plant_cx = exp(-td0*omegaBW*j);
            Prm_cx = freqresp(Prm(r1),omegaBW);
            P0_cx = freqresp(P0(r2),omegaBW);
            H = (1 - exp_mod_cx) * (Prm_cx/P0_cx) + exp_plant_cx;
            Q = exp_plant_cx / H;
            magQ = abs(Q);

            % Table 4.1. Step 5 ----
            if magQ > magSelecUp || magQ < magSelecLo
              r3 - r3 + 1;
              rejectedModels(r3) = r1;
            end
          end
        end
      end
    end
  end
end

nModels = [1 : 1 : n1*n2*n3];
rejectedModels = unique(rejectedModels);
acceptedModels = setxor(nModels,rejectedModels);

r1 = 0;
rr = 0;
for i1=1:1:n1
  km0 = k(i1);
  for j1=1:1:n2
    taum0 = tau(j1);
    for k1=1:1:n3
      tdm0 = td(k1);
      r1 = r1 + 1;
      if ~isempty (find(acceptedModels==r1))
        rr = rr + 1;
        modelsStep1(rr).k = km0;
        modelsStep1(rr).tau = taum0;
        modelsStep1(rr).td = tdm0;
        modelsStep1(rr).wBW = omegaBW;
        modelsStep1(rr).magUp = magSelecUp;
        modelsStep1(rr).magLo = magSelecLo;
      end
    end
  end
end
```

```
vv(:,1) = [modelsStep1(1:rr).tau]';
vv(:,2) = [modelsStep1(1:rr).td]';
vv(:,3) = [modelsStep1(1:rr).k]';
figure;
plot3(vv(:,1),vv(:,2),vv(:,3),'.','MarkerSize',10);
grid on;
xlabel('{\tau}','FontSize',16,'FontWeight','bold','Color','k');
ylabel('td','FontSize',12,'FontWeight','bold','Color','k');
zlabel('K','FontSize',12,'FontWeight','bold','Color','k');
title(['wBW = ',num2str(omegaBW),' rad/sec']);
bb = boundary(vv);
hold on;
trisurf(bb,vv(:,1),vv(:,2),vv(:,3),'Facecolor','red','FaceAl
pha',0.1);
axis([tau(1)*0.95,tau(end)*1.05,td(1)*0.95,td(end)*1.05,k(1)*0.95,
k(end)*1.05]);
view(-61.5,20);

% -----------------------------------------------------------------

% TABLE 4.2 (The Second algorithm). Section 4-2.2
% ================================================

% Model selection for SP. From modelsStep1(1:rr)

ww = [0.5 1 1.5 2];
numFreq = length(ww);

for ff=1:numFreq
  atW(ff).w0 = ww(ff);
  % Table 4.2. Step 1 ----
  for numModel = 1:1:rr % select from 1 to end
    taum0 = modelsStep1(numModel).tau; % model selected for SP
    tdm0 = modelsStep1(numModel).td; % model selected for SP
    km0 = modelsStep1(numModel).k; % model selected for SP
    % Table 4.2. Step 2 ----
    c = 0;
    for i1=1:1:n1
      for j1=1:1:n2
        for k1=1:1:n3
          c = c + 1;
          Prm = tf(km0,[taum0 1]); % model without delay
          Pm = tf(km0,[taum0 1],'iodelay',tdm0); % model with delay
          P = tf(k(i1),[tau(j1) 1],'iodelay',td(k1)); % real plant
          Prm_w0 = freqresp(Prm,ww(ff)); % at ww(ff)
          Pm_w0 = freqresp(Pm,ww(ff)); % at ww(ff)
          P_w0 = freqresp(P,ww(ff)); % at ww(ff)
          % Eq.(4.19)
          atW(ff).Peq(numModel).complex(c) = Prm_w0 - Pm_w0 + P_w0;
          atW(ff).Peq(numModel).dB(c) = ...
            20*log10(abs(atW(ff).Peq(numModel).complex(c)));
          atW(ff).Peq(numModel).ph(c) = ...
```

```
                angle(atW(ff).Peq(numModel).complex(c))*180/pi;
              if numModel==1 % only once
                atW(ff).Preal.complex(c) = P_w0; % P real a ww(ff)
                atW(ff).Preal.dB(c) = …
                  20*log10(abs(atW(ff).Preal.complex(c)));
                atW(ff).Preal.ph(c) = …
                  angle(atW(ff).Preal.complex(c))*180/pi;
              end
            end
          end
        end
      end
end

colors = ['r' 'g' 'b' 'c' 'm' 'k'];
figure;
hold on;

% Table 4.2. Step 3 ----
for ff=1:numFreq
  xr = atW(ff).Preal.ph;
  yr = atW(ff).Preal.dB;
  hr = convhull(xr,yr);
  plot(xr(hr),yr(hr),'-k',xr(hr),yr(hr),'.k','MarkerSize',10)
  atW(ff).areaTemplateReal = polyarea(xr(hr),yr(hr));
  for numModel=1:1:rr
    xx = atW(ff).Peq(numModel).ph;
    yy = atW(ff).Peq(numModel).dB;
    hh = convhull(xx,yy);
    plot(xx(hh),yy(hh),['-',colors(mod…
      (numModel-1,6)+1)],xx(hh),yy(hh),…
      ['.',colors(mod(numModel-1,6)+1)],'MarkerSize',10)
    atW(ff).areaTemplate(numModel) = polyarea(xx(hh),yy(hh));
    atW(ff).areaRatio(numModel) = …
      atW(ff).areaTemplate(numModel) / …
      atW(ff).areaTemplateReal; % Eq.(4.22)
  end
end
grid;
xlabel('Phase (deg)');
ylabel('Magnitude (dB)');
WeightF = 1 + 0 * [1:numFreq];
for numModel=1:1:rr
  sumAreaRatio = 0;
  for ff=1:numFreq
    sumAreaRatio = sumAreaRatio + WeightF(ff) * …
      atW(ff).areaRatio(numModel);
  end
  Icost(numModel) = (1/numFreq) * sumAreaRatio;
end

figure;
```

```
bar(Icost,0.5);
axis([0,numModel+1,0.93,1.02]);
grid;
xlabel('model number');
ylabel('Icost');

% ------------------------------------------------------------------

% Peq for the QFT Control Toolbox. Section 4-2.2
% ================================================

% The model selected for the SP is model 6
km0 = 8.55; % model 6. Eq.(4.26)
taum0 = 2.75; % model 6. Eq.(4.26)
tdm0 = 0.77; % model 6. Eq.(4.26)
c = 0;
for i1=1:1:n1
  for j1=1:1:n2
    for k1=1:1:n3
      c = c + 1;
      Prm = tf(km0,[taum0 1]); % model without delay
      Pm = tf(km0,[taum0 1],'iodelay',tdm0); % model with delay
      P = tf(k(i1),[tau(j1) 1],'iodelay',td(k1)); % real plant
      Peq(1,1,c) = Prm - Pm + P; % Peq for the QFTCT. Eq.(4.19)
      % Note that Peq includes internal delays. For this reason
      % Matlab automatically calculates it in State Space
    end
  end
end
save Peq Peq; % Plant to import, from disk into QFTCT

% ----------------------------------------------
```

Appendix 5: Algorithms—DPS Robust Control

This appendix includes the MATLAB code of the algorithms presented in Chapter 5, Example 5.1, for the design of a QFT robust control solution for a heat transmission distributed parameter system.

```matlab
% Example 5.1
% ============

% This m.file calculates the transfer functions with parametric
% uncertainty Pxoxd, Pxsxd, Pxoxa, Pxsxa and B1 for the QFTCT.
% =========================================================
====

% Symbolic representation ---

syms k1 k2 k3 R c s
q11 = (1/(k1*R)) + (k1*c*s/2) + (1/(k2*R)) + (k2*c*s/2);
q12 = -(1/(k2*R));
q21 = -(1/(k2*R));
q22 = (1/(k2*R)) + (k2*c*s/2) + (1/(k3*R)) + (k3*c*s/2);
Q = [q11 q12 ; q21 q22];
P = inv(Q); % P contents Eqs. (5.39) to (5.47)

% Transfer functions ---
% The following sentences introduce the parameters (both
% fixed parameters and parameters with uncertainty) in P

k1 = pi/4; % fixed parameter
k2 = pi/4; % fixed parameter

P1 = subs(P);
[P0,sigma] = subexpr(P1); % sigma is the common subexpression
commonDenP0 = 1/sigma;
commonDenP0Subs = subs(commonDenP0);
denP = coeffs(commonDenP0Subs,s);
P0withoutSigma = subs(P0,'sigma',1);
PxoxdNum = coeffs(P0withoutSigma(1,1),s);
PxoxaNum = coeffs(P0withoutSigma(1,2),s);
PxsxdNum = coeffs(P0withoutSigma(2,1),s);
PxsxaNum = coeffs(P0withoutSigma(2,2),s);

k3v0 = pi/2; % parameters with uncertainty
Rv0 = 1;
cv0 = 1;
k3v = linspace(k3v0*0.9,k3v0*1.1,5);
Rv = linspace(Rv0*0.9,Rv0*1.1,5);
cv = linspace(cv0*0.9,cv0*1.1,5);
```

```
n1Max = length(k3v);
n2Max = length(Rv);
n3Max = length(cv);
jj = 0;
for n1=1:n1Max
  k3 = k3v(n1);
  for n2=1:n2Max
    R = Rv(n2);
    for n3=1:n3Max
      c = cv(n3);
      PxoxdNumSubs = double(vpa(subs(PxoxdNum),5));
      PxoxaNumSubs = double(vpa(subs(PxoxaNum),5));
      PxsxdNumSubs = double(vpa(subs(PxsxdNum),5));
      PxsxaNumSubs = double(vpa(subs(PxsxaNum),5));
      denPSubs = double(vpa(subs(denP),5));
      jj = jj + 1;
      Pxoxd(1,1,jj) = tf(PxoxdNumSubs,denPSubs);
      Pxoxa(1,1,jj) = tf(PxoxaNumSubs,denPSubs);
      Pxsxd(1,1,jj) = tf(PxsxdNumSubs,denPSubs);
      Pxsxa(1,1,jj) = tf(PxsxaNumSubs,denPSubs);
      B = Pxsxa(1,1,jj)*Pxoxd(1,1,jj)-Pxoxa(1,1,jj)*Pxsxd(1,1,jj);
      B1(1,1,jj) = minreal(B);
    end
  end
end
save Pxoxd Pxoxd; % Pxoxd plant. To import, from disk into QFTCT
save Pxsxd Pxsxd; % Pxsxd plant. To import, from disk into QFTCT
save Pxoxa Pxoxa; % Pxoxa plant. To import, from disk into QFTCT
save Pxsxa Pxsxa; % System plant. To import, from disk into QFTCT
save B1 B1; % For spec yxo/uxd. To import, from disk into QFTCT
% ----------------------------------------------
```

Appendix 6: Algorithms—Gain Scheduling/Switching Control

This appendix includes the MATLAB code of the algorithms presented in Chapter 6, Examples 6.1 and 6.2, for the design of gain scheduling/QFT switching robust control solutions.

```matlab
% Example 6.1, Case 1
% ===================

A1 = [0 1 0;0 0 1;-1 -2 -3]; % Eq.(6.20)
B1 = [-1 0 0]';
C1 = [0 1 1];
D1 = [0];
[numT1,denT1] = ss2tf(A1,B1,C1,D1);
numL1 = numT1;
denL1 = denT1-numT1;
L1 = tf(numL1,denL1); % Eq.(6.21)

A2 = [0 1 0;0 0 1;-2 -3 -1]; % Eq.(6.20)
B2 = [-1 0 0]';
C2 = [0 1 1];
D2 = [0];
[numT2,denT2] = ss2tf(A2,B2,C2,D2);
numL2 = numT2;
denL2 = denT2-numT2;
L2 = tf(numL2,denL2); % Eq.(6.22)

w = [0.01:0.01:10]; %  frequency vector

FR_1L1 = freqresp((1+L1),w);
An_FR_1L1 = angle(FR_1L1(1,:));

FR_1L2 = freqresp((1+L2),w);
An_FR_1L2 = angle(FR_1L2(1,:));

Phi12 = abs((An_FR_1L2 - An_FR_1L1))*180/pi; %  Phi12. Eq.(6.16)

FR_alpha = freqresp(tf(denL2,denL1),w);
alpha = abs(angle(FR_alpha(1,:)))*180/pi; %  alpha. Eq.(6.17)

figure;
plot(w,Phi12+alpha,w,90+w*0); % Fig.6.5. To check Eq.(6.18)
xlabel ('w (rad/s)');
ylabel('Phi12+alpha & 90 degrees');
```

```
% Example 6.1, Case 2
% ====================

A1 = [-3 -2 -1; 1 0 0; 0 1 0]; % Eq.(6.23)
B1 = [1 0 0]';
C1 = [3 2 0];
D1 = [0];
[numT1,denT1] = ss2tf(A1,B1,C1,D1);
numL1 = numT1;
denL1 = denT1-numT1;
L1 = tf(numL1,denL1); % Eq.(6.24)

A2 = [-1 -3 -2; 1 0 0; 0 1 0]; % Eq.(6.23)
B2 = [1 0 0]';
C2 = [1 3 1];
D2 = [0];
[numT2,denT2] = ss2tf(A2,B2,C2,D2);
numL2 - numT2,
denL2 = denT2-numT2;
L2 = tf(numL2,denL2); % Eq.(6.25)

w = [0.01:0.01:10]; %  frequency vector

FR_1L1 = freqresp((1+L1),w);
An_FR_1L1 = angle(FR_1L1(1,:));

FR_1L2 = freqresp((1+L2),w);
An_FR_1L2 = angle(FR_1L2(1,:));

Phi12 = abs((An_FR_1L2 - An_FR_1L1))*180/pi; %  Phi12. Eq.(6.16)

figure;
plot(w,Phi12,w,90+w*0); % Fig.6.6. To check Eq.(6.18)
xlabel ('w (rad/s)');
ylabel('Phi12 & 90 degrees,(alpha=0)');

% Example 6.2
% ===========

A1 = [-3 -2 -1;1 0 0;0 1 0]; % Eq.(6.26)
B1 = [1 0 0]';
C1 = [0 0 1];
D1 = [0];
[numT1,denT1] = ss2tf(A1,B1,C1,D1);
numL1 = numT1;
denL1 = denT1-numT1;
L1 = tf(numL1,denL1); % Eq.(6.27)

A2 = [-3 -1 -1;1 0 0;0 1 0]; % Eq.(6.26)
B2 = [1 0 0]';
C2 = [0 0 1];
D2 = [0];
```

```
[numT2,denT2] = ss2tf(A2,B2,C2,D2);
numL2 = numT2;
denL2 = denT2-numT2;
L2 = tf(numL2,denL2); % Eq.(6.28)

w = [0.01:0.01:10]; %  frequency vector

FR_1L1 = freqresp((1+L1),w);
An_FR_1L1 = angle(FR_1L1(1,:));

FR_1L2 = freqresp((1+L2),w);
An_FR_1L2 = angle(FR_1L2(1,:));

Phi12 = abs((An_FR_1L2 - An_FR_1L1))*180/pi; %  Phi12. Eq.(6.16)

FR_alpha = freqresp(tf(denL2,denL1),w);
alpha = abs(angle(FR_alpha(1,:)))*180/pi; %  alpha. Eq.(6.17)

figure;
plot(w,Phi12+alpha,w,90+w*0); % Fig.6.7. To check Eq.(6.18)
xlabel('w (rad/s)');
ylabel('Phi12+alpha & 90 degrees');
% -----------------------------------------------
```

Appendix 7: Algorithms—Nonlinear Dynamic Control

This appendix includes the MATLAB code of the cases presented in Chapter 7, Sections 7.4 and 7.5, for the design of nonlinear dynamic controllers.

```
% Section 7-4.
% Example 7.1. PID with anti wind-up
% ====================================

% (1) Model Definition. Nominal Plant. Eq.(7.7)
P0 = tf(0.5,conv(conv([1/0.83 1],[1/0.83 1]),[1/0.83 1])); %nominal

% (2) Controller G1(s). Designed with QFTCT. Eq.(7.11)
G1 = tf(0.7*[1/0.5^2 2*0.9/0.5 1],[1/300 1 0]); % QFTCT pid1.mat

% (3) Function L1(s)
L1 = minreal(P0 * G1);

% (4) Function A(s).
A = tf(0.9,[1 0]); % Eq.(7.13)

% (5) Function LE(s).
Le = minreal((L1-A)/(1+A)); % Eq.(7.14)

% (6) Figures
figure;step(L1/(1+L1));
ww1 = logspace(-2,4,10000);
figure; bode(G1,'g-',A,'r--',ww1);grid; % Fig.7.15a
ww2 = logspace(-3,2,10000);
figure; bode(L1,'b-',Le,'--r',ww2);grid; % Fig.7.15b

% (7) Circle criterion
n1 = 10; n2 = 10;
k1v = linspace(0.5,1.5,n1);
m1v = linspace(0.83,1.25,n2);
cc = 0;
figure; hold on;
for ii=1:n1
    k1 = k1v(ii);
    for jj=1:n2
        cc = cc + 1;
        m1 = m1v(jj);
        p = tf(k1,[1/m1 1]);
        L1 = p*G1;
```

```
        Le(cc) = minreal((L1-A)/(1+A));
        nyquist(Le(cc)); % Fig.7.16
    end
end

% Section 7-5. NDC
% Example 7.2. NDC with several nonlinearities
% ============================================

% (1) Model Definition. Nominal Plant P0(s).
k = 610;
a = 1;
b = 150;
numP = k/(a*b);
denP = conv(conv([1/a 1],[1/b 1]),[1 0]);
P0 = tf(numP,denP); % Nominal plant P0(s). Eq.(7.33)

% (2) Controller G1(s). Designed with QFTCT. NDC_Control_G1.mat
numG1 = 89 * conv([1/1.2 1],[1/5 1]);
denG1 = [1/500 1 0];
G1 = tf(numG1,denG1); % Eq.(7.37)

% (3) Controller G2(s). Designed with QFTCT. NDC_Control_G2.mat
numG2 = 0.0035 * conv([1/0.002 1],[1/3 1]);
denG2 = [1/10 1 0];
G2 = tf(numG2,denG2); % Eq.(7.42)

% (4) Function L01(s) and L02(s)
L10 = P0 * G1;
L20 = P0 * G2;
ww = logspace(-3,3,10000);
figure; bode(L10,L20,ww); grid; % Fig.7.38

% (5) Function A(s)
A = tf(4.5,[1 0]); % Eq.(7.44)
ww0 = logspace(-6,4,10000);
figure;
bode(G1,'-.b',G2,'--g',A,'-r',ww0); % Fig.7.39a
grid;

% (6) Functions Le10(s), Le20(s)
Le10 = minreal((L10-A)/(1+A));
Le20 = minreal((L20-A)/(1+A));
ww1 = logspace(-4,5,10000);
figure;
bode(L10,'-b',Le10,'-.b',L20,'-g',Le20,'-.g',ww1); % Fig.7.39b
grid;

% (7) Circle criterion for Le1 and Le2
n1 = 4; n2 = 4; n3 = 4;
kv = linspace(610,1050,n1);
```

```
av = linspace(1,15,n2);
bv = linspace(150,170,n2);
cc = 0;
figure(21); hold on;
figure(22); hold on;
for i1=1:n1
    k = kv(i1);
    for i2=1:n2
        a = av(i2);
        for i3=1:n3
            b = bv(i3);
            cc = cc + 1;
            numP = k/(a*b);
            denP = conv(conv([1/a 1],[1/b 1]),[1 0]);
            p = tf(numP,denP); % plants
            L1 = p*G1;
            L2 = p*G2;
            Le1(cc) = minreal((L1-A)/(1+A));
            Le2(cc) = minreal((L2-A)/(1+A));
            figure(21); nyquist(Le1(cc)); % Fig.7.40a
            figure(22); nyquist(Le2(cc)); % Fig.7.40b
        end
    end
end

% (8) Iso-lines
UpperLimitSatAct = 1.1;
LowerLimitSatAct = -1.1;
ModelUpperLimitSatAct = 1.1;
ModelLowerLimitSatAct =  -1.1;
delta_high = 0.15;
delta_low = 0.075;
fact_delta = delta_high^2 - delta_low^2;
clear phLc; clear dBLc;
e_st = -5;
e_end = 2;
e_num = 60;
error = logspace(e_st,e_end,e_num); % inputs: E = error
w_st = -4;
w_end = 5;
w_num = 100;
www = logspace(w_st,w_end,w_num); % frequencies

for jj=1:1:w_num
    wc = www(jj);
    [mod,deg] = bode(P0,wc); ph = deg*pi/180;
    pp = mod*(cos(ph)+j*sin(ph)); % plant at wc
    [mod,deg] = bode(G1,wc); ph = deg*pi/180;
    c1 = mod*(cos(ph)+j*sin(ph)); % controller G1(s) at wc
    [mod,deg] = bode(G2,wc); ph = deg*pi/180;
    c2 = mod*(cos(ph)+j*sin(ph)); % controller G2(s) at wc
    [mod,deg] = bode(A,wc); ph = deg*pi/180;
```

```
    aa = mod*(cos(ph)+j*sin(ph)); % function A(s) at wc

    for ii=1:1:e_num

        x0 = error(ii); % error (input to L)
        if (abs(x0)<=delta_low)
            N1 = 1; % contr.nonlinearity N1
            N2 = 1 - N1; % contr.nonlinearity N2
        else
            N1 = j * (1/pi/x0^2)*fact_delta; % contr.nonlinearity N1
            N2 = 1 - N1; % contr.nonlinearity N2
        end;
        x1 = c1 * N1 * x0; % output of G1
        x2 = c2 * N2 * x0; % output of G2
        x3 = x1 + x2; % output of G1 + G2

        x4 = x3; % block inner-lopp with A(s) and Nsaturation
        mod_x4 - abs(x4);
        if mod_x4>ModelUpperLimitSatAct % Dead zone = 1-Saturation
            dE = ModelUpperLimitSatAct/mod_x4;
            Ndz = (1 - (2/pi)*(asin(dE)+dE*sqrt(1-dE^2)));
            x4 = (1/(1+real(aa)*Ndz)) * x4;
        end

        if (abs(x4)<=UpperLimitSatAct) % Actuator Saturation
            Nsat = 1;
        else
            mod_x4 = abs(x4);
            dE = UpperLimitSatAct/mod_x4;
            Nsat = abs((2/pi)*(asin(dE)+dE*sqrt(1-dE^2)));
        end
        x7 = Nsat * x4; % output of actuator saturation (= v)

        x8 = pp * x7; % plant output (= y)

        Lc(ii) = x8/x0;
        dBLc(ii,jj)=20*log10(abs(Lc(ii)));

        angLc=unwrap(angle(Lc(ii)));
        nn=angLc/2/pi; nn=ceil(nn); angLc=-(2*pi)*nn+angLc;
        phLc(ii,jj)=angLc*180/pi;
    end
end
figure; hold on;
nichols(L10,'b.',L20,'g.');
for kk=1:1:e_num,
    plot(phLc(kk,:),dBLc(kk,:),'k:'); % plot iso-lines. Fig.7.41
end;
title('NDC. Iso-lines.');
axis([-270 0 -120 120]);
% ----------------------------------
```

Appendix 8: Algorithms—MIMO Robust Control

This appendix includes the MATLAB code for the examples and cases presented in Chapter 8: MIMO QFT robust control.

```matlab
% Example 8.3. Section 8-3.3
% ===========================

p11 = tf(4*[1 -1],[1 1]);
p12 = tf([1 0],[1 2]);
p21 = tf(-6,[1 1]);
p22 = tf([1 -2],[1 1]);
P = [p11 p12 ; p21 p22]; % Eq.(8.30)
Pinv = inv(P);
Pinv = minreal(Pinv); % Eq.(8.32)
zeros = roots(Pinv.den{1});

% Example 8.4. Section 8-3.3
% ===========================

p11 = tf([1 1],[5 1]);
p12 = tf([1 4],[5 1]);
p21 = tf([1 1],[5 1]);
p22 = tf([2 2],[5 1]);
P = [p11 p12 ; p21 p22]; % Eq.(8.33)
Pinv = inv(P);
Pinv = minreal(Pinv); % Eq.(8.34)
zeros = roots(Pinv.den{1});

P0 = dcgain(P); % Eq.(8.35)
rga = P0 .* (inv(P0))'; % Eq.(8.36)

Pinf = (1/5)*[1 1;1 2]; % Eq.(8.37)
rga = Pinf .* (inv(Pinf))'; % Eq.(8.38)

% Example 8.1. Section 8-3.5
% ===========================

p11 = tf(7,[1/0.25 1]);
p12 = tf(9,[1/0.2 1]);
p21 = tf(3,[1/0.125 1]);
p22 = tf(5,[1/0.5 1]);
P = [p11, p12 ; p21 , p22]; % Eq.(8.4)
P0 = dcgain(P); % Eq.(8.39)
[U,sig,V] = svd(P0); % Eq.(8.40)

% Example 8.1. Section 8-1
% =========================
```

```
% Heat exchanger example.

% (1) Model Definition. Nominal Plant. Eq.(8.4)
p11 = tf(7,[1/0.25 1]);
p12 = tf(9,[1/0.2 1]);
p21 = tf(3,[1/0.125 1]);
p22 = tf(5,[1/0.5 1]);
Pn = [p11, p12 ; p21 , p22];
Pn.InputName = {'u'};
Pn.OutputName = {'y'};

% (2) Conversions
Pss = ss(Pn); % from transfer function matrix to state space model
Ptf = tf(Pss); % from state space to transfer function matrix

% (3) Independent design. g11 and g22 (PIDs with filter). Fig.8.2
% --- for p11. QFTCT project: main_indep_g11.mat
g11i = tf(0.18*[1/0.2 1],[1/390 1 0]); % Eq.(8.9)

% --- for p22. QFTCT project: main_indep_g22.mat
g22i = tf(0.25*[1/0.35 1],[1/190 1 0]);  % Eq.(8.10)

% --- Diagonal controller matrix
Gi = [g11i, 0; 0, g22i];
Gi.InputName = {'e'};
Gi.OutputName = {'u'};

% (4) Closed loop MIMO system
Sum = sumblk('e = r - y',2);
CLry = connect(Pn,Gi,Sum,'r','y'); % Closed loop MIMO system

% (5) Simulation closed loop MIMO system
tinc = 0.01;
tt = [0:tinc:300];
nn = length(tt);
U1 = [[0:tinc:20]*0,1+[20+tinc:tinc:300]*0]; % input 1
U2 = [[0:tinc:170]*0,1+[170+tinc:tinc:300]*0]; % input 2
figure,lsim(CLry,[U1;U2],tt); % Fig.8.9

% Example 8.1. Section 8-8 (MIMO QFT Method 1)
% ============================================
% Heat exchanger example.

% (1) Step A
% ----------

% --- RGA. According to nominal plant.
p11 = tf(7,[1/0.25 1]);
p12 = tf(9,[1/0.2 1]);
p21 = tf(3,[1/0.125 1]);
p22 = tf(5,[1/0.5 1]);
Pn = [p11, p12 ; p21 , p22];
```

```
P0 = dcgain(Pn);
rga = P0 .* inv(P0)';  % Eq.(8.5) and Eq.(8.163)

% --- Order. According to nominal plant.
Pninv = inv(Pn);
Pninv = minreal(Pninv);
pinv11_inv = 1/Pninv(1,1);
pinv22_inv = 1/Pninv(2,2);
figure; bode(pinv11_inv,pinv22_inv);grid; % Fig.8.21.

% (2) Step B
% ----------

% --- Inverse plant "P(s)*" with uncertainty
% -- grid for uncertainty
n1 = 2; n2 = 2; n3 = 2; n4 = 2;
n5 = 2; n6 = 2; n7 = 2; n8 = 2;
% -- parameters
k11v = linspace(6.3,7.7,n1); % parameters, see Eq. (8.4)
k12v = linspace(8.9,9.1,n2);
k21v = linspace(2.9,3.1,n3);
k22v = linspace(4.5,5.5,n4);
tau11v = linspace(3.6,4.4,n5);
tau12v = linspace(4.9,5.1,n6);
tau21v = linspace(7.9,8.1,n7);
tau22v = linspace(1.8,2.2,n8);
cc = 0;

% --- for QFTCT, design of g22_med1(s) and f22_med1(s)
for i1=1:n1
  k11 = k11v(i1);
  for i2=1:n2
    k12 = k12v(i2);
    for i3=1:n3
      k21 = k21v(i3);
      for i4=1:n4
        k22 = k22v(i4);
        for i5=1:n5
          tau11 = tau11v(i5);
          for i6=1:n6
            tau12 = tau12v(i6);
            for i7=1:n7
              tau21 = tau21v(i7);
              for i8=1:n8
                cc = cc + 1;
                tau22 = tau22v(i8);
                p11 = tf(k11,[tau11 1]); % from Eq.(8.4)
                p12 = tf(k12,[tau12 1]); % from Eq.(8.4)
                p21 = tf(k21,[tau21 1]); % from Eq.(8.4)
                p22 = tf(k22,[tau22 1]); % from Eq.(8.4)
                Plant(cc).tfm = [p11, p12 ; p21 , p22];
                Pinverse = inv(Plant(cc).tfm); % Eq.(8.164)
```

```
                       Pinv(cc).tfm(1,1) = minreal(Pinverse(1,1));
                       Pinv(cc).tfm(1,2) = minreal(Pinverse(1,2));
                       Pinv(cc).tfm(2,1) = minreal(Pinverse(2,1));
                       Pinv(cc).tfm(2,2) = minreal(Pinverse(2,2));
                     end
                   end
                 end
               end
             end
           end
         end
       end
numPlants = cc;
for ii=1:numPlants
  q11(1,1,ii) = 1/Pinv(ii).tfm(1,1); % Eq.(8.166)
end
save q11 q11; % for QFTCT, design of g11_med1(s) and f11_med1(s)
g11_med1 = tf(20*conv([1/0.12 1],[1/0.24 1]),…
             [1/0.014 1 0]); % Eq.(8.167)
f11_med1 = tf(1,[1/1.5 1]);    % Eq.(8.168)

% --- for g21_med1(s)
g21_med1 = tf(0); % Eq.(8.169)

% --- for QFTCT, design of g22_med1(s) and f22_med1(s)
cc = 0;
for i1=1:n1
  for i2=1:n2
    for i3=1:n3
      for i4=1:n4
        for i5=1:n5
          for i6=1:n6
            for i7=1:n7
              for i8=1:n8
                cc = cc + 1;
                gg = Pinv(cc).tfm(2,2) - ((Pinv(cc).tfm(2,1) + …
                    g21_med1) * Pinv(cc).tfm(1,2) / …
                    (Pinv(cc).tfm(1,1) + g11_med1));
                p22_ast_eq2(cc) = minreal(gg,0.01); % Eq.(8.170)
              end
            end
          end
        end
      end
    end
  end
end
numPlants = cc;
for ii=1:numPlants
  q22(1,1,ii) = 1/p22_ast_eq2(ii); % Eq.(8.171)
end
save q22 q22; % for QFTCT, design of g22_med1(s) and f22_med1(s)
```

```
g22_med1 = tf(30*[1/0.6 1],[1/300 1 0]);; % Eq.(8.172)
f22_med1 = tf(1,[1/0.75 1]); % Eq.(8.173)

% --- for g12_med1(s)
figure; hold on;
for jj=1:10:numPlants
  kkk2(jj) = g22_med1 * Pinv(jj).tfm(1,2) / Pinv(jj).tfm(2,2);
  bode(kkk2(jj)); % Fig.8.28
end
grid;
g12_med1 = g22_med1 * Pninv(1,2) / Pninv(2,2);
g12_med1 = minreal(g12_med1,0.01);
g12_med1 = tf(-40*[1/0.833 1],[1/300 1 0]); Eq.(8.174)
bode(g12_med1); % Fig.8.28

% (3) Step C
% ----------

% --- Closed loop MIMO system
Gmed1 = [g11_med1, g12_med1; g21_med1, g22_med1];
Gmed1.InputName = {'e'};
Gmed1.OutputName = {'u'};
% Fmed1 = [1, 0; 0, 1]; % for Fig.8.29
Fmed1 = [f11_med1, 0; 0, f22_med1]; % for Fig.8.30
Sum = sumblk('e = r - y',2);
CLry = connect(Pn,Gmed1,Sum,'r','y')*Fmed1; % Closed loop MIMO syst

% --- Simulation closed loop MIMO system
tinc = 0.01;
tt = [0:tinc:300];
nn = length(tt);
U1 = [[0:tinc:20]*0,1+[20+tinc:tinc:300]*0]; % input 1
U2 = [[0:tinc:170]*0,1+[170+tinc:tinc:300]*0]; % input 2
figure,lsim(CLry,[U1;U2],tt); % Fig.8.29 and Fig.8.30

% Example 8.1. Section 8-9 (MIMO QFT Method 2)
% ============================================
% Heat exchanger example.

% (1) Step A
% ----------

% --- RGA. According to nominal plant.
p11 = tf(7,[1/0.25 1]);
p12 = tf(9,[1/0.2 1]);
p21 = tf(3,[1/0.125 1]);
p22 = tf(5,[1/0.5 1]);
Pn = [p11, p12 ; p21 , p22];
P0 = dcgain(Pn);
rga = P0 .* inv(P0)'; % Eq.(8.5) and Eq.(8.163)

% (2) Step B. --- Ga
```

```
% ------------------

% --- Inverse plant "P(s)*" with uncertainty
% grid for uncertainty
n1 = 2; n2 = 2; n3 = 2; n4 = 2;
n5 = 2; n6 = 2; n7 = 2; n8 = 2;
% parameters
k11v = linspace(6.3,7.7,n1); % parameters, see Eq. (8.4)
k12v = linspace(8.9,9.1,n2);
k21v = linspace(2.9,3.1,n3);
k22v = linspace(4.5,5.5,n4);
tau11v = linspace(3.6,4.4,n5);
tau12v = linspace(4.9,5.1,n6);
tau21v = linspace(7.9,8.1,n7);
tau22v = linspace(1.8,2.2,n8);
cc = 0;
% --- for g11a, g12a(s), g21a(s), g22a(s)
for i1=1:n1
  k11 = k11v(i1);
  for i2=1:n2
    k12 = k12v(i2);
    for i3=1:n3
      k21 = k21v(i3);
      for i4=1:n4
        k22 = k22v(i4);
        for i5=1:n5
          tau11 = tau11v(i5);
          for i6=1:n6
            tau12 = tau12v(i6);
            for i7=1:n7
              tau21 = tau21v(i7);
              for i8=1:n8
                cc = cc + 1;
                tau22 = tau22v(i8);
                p11 = tf(k11,[tau11 1]); % from Eq.(8.4)
                p12 = tf(k12,[tau12 1]); % from Eq.(8.4)
                p21 = tf(k21,[tau21 1]); % from Eq.(8.4)
                p22 = tf(k22,[tau22 1]); % from Eq.(8.4)
                Plant(cc).tfm = [p11, p12 ; p21 , p22];
                Pdiag(cc).tfm = [p11, 0 ; 0 , p22];
                Pinverse = inv(Plant(cc).tfm);
                Pinv(cc).tfm(1,1) = minreal(Pinverse(1,1));
                Pinv(cc).tfm(1,2) = minreal(Pinverse(1,2));
                Pinv(cc).tfm(2,1) = minreal(Pinverse(2,1));
                Pinv(cc).tfm(2,2) = minreal(Pinverse(2,2));
                pp = Pinv(cc).tfm * Pdiag(cc).tfm; % Eq.(8.176)
                PinvPdiag(cc).tfm(1,1) = minreal(pp(1,1),0.01);
                k_PinvPdiag(cc).tfm(1,1) = …
                  dcgain(PinvPdiag(cc).tfm(1,1));
                PinvPdiag(cc).tfm(1,2) = minreal(pp(1,2),0.01);
                k_PinvPdiag(cc).tfm(1,2) = …
                  dcgain(PinvPdiag(cc).tfm(1,2));
```

```
                    PinvPdiag(cc).tfm(2,1) = minreal(pp(2,1),0.01);
                    k_PinvPdiag(cc).tfm(2,1) = …
                      dcgain(PinvPdiag(cc).tfm(2,1));
                    PinvPdiag(cc).tfm(2,2) = minreal(pp(2,2),0.01);
                    k_PinvPdiag(cc).tfm(2,2) = …
                      dcgain(PinvPdiag(cc).tfm(2,2));
                  end
                end
              end
            end
          end
        end
      end
    end
  end
end
numPlants = cc;
kvector11 = [];kvector12 = [];kvector21 = [];kvector22 = [];
for jj=1:numPlants
  kvector11 = [kvector11 k_PinvPdiag(jj).tfm(1,1)];
  kvector12 = [kvector12 k_PinvPdiag(jj).tfm(1,2)];
  kvector21 = [kvector21 k_PinvPdiag(jj).tfm(2,1)];
  kvector22 = [kvector22 k_PinvPdiag(jj).tfm(2,2)];
end
kmean11 = mean(kvector11);
kmean12 = mean(kvector12);
kmean21 = mean(kvector21);
kmean22 = mean(kvector22);
e11 = inf; e12 = inf; e21 = inf; e22 = inf;

for jj=1:numPlants
  ne11 = abs((kmean11 - k_PinvPdiag(jj).tfm(1,1)));
  if ne11<=e11;
    nPinvPdiag11 = jj;
    e11 = ne11;
  end
end
g11a = minreal(PinvPdiag(nPinvPdiag11).tfm(1,1),0.1);
g11a = tf([1.24 0.16],[1 0.01]); % Eq.(8.177)
figure; hold on;
for jj=1:10:numPlants
  bode(PinvPdiag(jj).tfm(1,1)); % Fig.8.33
end
bode(g11a,'b--',g11a,'b*'); % Fig.8.33
title('PinvPdiag(1:end).tfm(1,1) and g11-alpha');

for jj=1:numPlants
  ne12 = abs((kmean12 - k_PinvPdiag(jj).tfm(1,2)));
  if ne12<=e12;
    nPinvPdiag12 = jj;
    e12 = ne12;
  end
end
g12a = minreal(PinvPdiag(nPinvPdiag12).tfm(1,2),0.1);
```

```
g12a = tf(-[1.53 0.19],[1 0.01]); % Eq.(8.178)
figure; hold on;
for jj=1:10:numPlants
  bode(PinvPdiag(jj).tfm(1,2)); % Fig.8.34
end
bode(g12a,'b--',g12a,'b*'); % Fig.8.34
title('PinvPdiag(1:end).tfm(1,2) and g12-alpha');

for jj=1:numPlants
  ne21 = abs((kmean21 - k_PinvPdiag(jj).tfm(2,1)));
  if ne21<=e21;
    nPinvPdiag21 = jj;
    e21 = ne21;
  end
end
g21a = minreal(PinvPdiag(nPinvPdiag21).tfm(2,1),0.1);
g21a = tf(-[0.18 0.1],[1 0.01]); % Eq.(8.179)
figure; hold on;
for jj=1:10:numPlants
  bode(PinvPdiag(jj).tfm(2,1)); % Fig.8.35
end
bode(g21a,'b--',g21a,'b*'); % Fig.8.35
title('PinvPdiag(1:end).tfm(2,1) and g21-alpha');

for jj=1:numPlants
  ne22 = abs((kmean22 - k_PinvPdiag(jj).tfm(2,2)));
  if ne22<=e22;
    nPinvPdiag22 = jj;
    e22 = ne22;
  end
end
g22a = minreal(PinvPdiag(nPinvPdiag22).tfm(2,2),0.1);
g22a = tf([1.24 0.16],[1 0.01]); % Eq.(8.180)
figure; hold on;
for jj=1:10:numPlants
  bode(PinvPdiag(jj).tfm(2,2)); % Fig.8.36
end
bode(g22a,'b--',g22a,'b*'); % Fig.8.36
title('PinvPdiag(1:end).tfm(2,2) and g22-alpha');

% (3) Step C
% ----------

% for QFTCT, design of g11b(s) and f11(s)
Ga = [g11a g12a; g21a g22a];
for jj=1:numPlants
  Px(jj).tfm = minreal(Plant(jj).tfm * Ga);
  Pxinv(jj).tfm = inv(Px(jj).tfm); % Eq.(8.181)
end
```

```
for ii=1:numPlants
  qx11(1,1,ii) = 1/Pxinv(ii).tfm(1,1); % Eq.(8.182)
end
save qx11 qx11; % for QFTCT, design of g11_med2(s) and f11_med2(s)
g11b = tf(0.9*[1/0.24 1],[1/40 1 0]); % Eq.(8.183)
f11 = tf([1/20 1],[1/2 1]); % Eq.(8.184)

% for QFTCT, design of g22b(s) and f22(s)
cc = 0;
for i1=1:n1
  for i2=1:n2
    for i3=1:n3
      for i4=1:n4
        for i5=1:n5
          for i6=1:n6
            for i7=1:n7
              for i8=1:n8
                cc = cc + 1;
                gg = Pxinv(cc).tfm(2,2) - …
                     ((Pxinv(cc).tfm(2,1) * Pxinv(cc).tfm(1,2)) / …
                     (Pxinv(cc).tfm(1,1) + g11b));
                px22_ast_eq2(cc) = minreal(gg,0.01); % Eq.(8.185)
              end
            end
          end
        end
      end
    end
  end
end
numPlants = cc;
for ii=1:numPlants
  qx22(1,1,ii) = 1/px22_ast_eq2(ii); % Eq.(8.186)
end
save qx22 qx22; % for QFTCT, design of g22a(s) and f11(s)
g22b = tf(0.4*[1/0.1 1],[1/40 1 0]); % Eq.(8.187)
f22 = tf([1/20 1],[1/2 1]); % Eq.(8.188)

% (4) Step D
% ----------

% --- Closed loop MIMO system
Gb = [g11b 0; 0 g22b];
Gmed2 = Ga * Gb;
Gmed2.InputName = {'e'};
Gmed2.OutputName = {'u'};
%Fmed21 = [1, 0; 0, 1]; % for Fig.8.43
Fmed2 = [f11, 0; 0, f22]; % for Fig.8.44
Sum = sumblk('e = r - y',2);
```

```
CLry = connect(Pn,Gmed2,Sum,'r','y')*Fmed2; % Closed loop MIMO syst

% --- Simulation closed loop MIMO system
tinc = 0.01;
tt = [0:tinc:300];
nn = length(tt);
U1 = [[0:tinc:20]*0,1+[20+tinc:tinc:300]*0]; % input 1
U2 = [[0:tinc:170]*0,1+[170+tinc:tinc:300]*0]; % input 2
figure,lsim(CLry,[U1;U2],tt); % Fig.8.43 and Fig.8.44

% ---------------------------------------------
```

Appendix 9: Conversion of Units

	Unit	Symbol	Conversion
International System of units (Metric system)	Meter	m	1 m = 3.2808 ft
	Meter	m	1 m = 39.37 in
	Meter	m	1 m = 1.0936 yd
	Square meter	m²	1 m² = 10.7639 ft²
	Square meter	m²	1 m² = 2.4710 × 10⁻⁴ ac
	Square kilometer	km²	1 km² = 0.3861 mi²
	Cubic meter	m³	1 m³ = 35.3145 ft³
	Cubic meter	m³	1 m³ = 264.17 gal
	Meter/second	m s⁻¹	1 m s⁻¹ = 3.2808 ft s⁻¹
	Meter/second	m s⁻¹	1 m s⁻¹ = 2.2369 mi h⁻¹
	Kilometer/hour	km h⁻¹	1 km h⁻¹ = 0.6214 mi h⁻¹
	Kilogram	kg	1 kg = 2.2046 lb
	Celsius degree	°C	$T(°C) = 5/9[T(°F) - 32]$
	Watt	W	1 W = 1.3405 × 10⁻³ HP
American System of units (US)	Foot	ft	1 ft = 0.3048 m
	Inch	in	1 in = 2.54 × 10⁻² m
	Yard	yd	1 yd = 0.9144 m
	Square foot	sq. ft (ft²)	1 ft² = 9.2903 × 10⁻² m²
	Acre	ac	1 ac = 4.0469 × 10³ m²
	Square mile	sq. mi (mi²)	1 mi² = 2.59 km²
	Cubic foot	cu. ft (ft³)	1 ft³ = 2.8317 × 10⁻² m³
	Gallon	gal	1 gal = 3.7854 × 10⁻³ m³
	Foot/second	ft s⁻¹	1 ft s⁻¹ = 0.3048 m s⁻¹
	Mile/hour	mph (mi h⁻¹)	1 mi h⁻¹ = 1.6093 km h⁻¹
	Mile/hour	mph (mi h⁻¹)	1 mi h⁻¹ = 0.4470 m s⁻¹
	Pound	lb	1 lb = 0.4536 kg
	Fahrenheit degree	°F	$T(°F) = (9/5)T(°C) + 32$
	Horsepower	HP	1 HP = 746 W
Energy	Joule	J	1 J = 2.7778 × 10⁻⁷ kWh
	Kilowatt-hour	kWh	1 kWh = 3.6 × 10⁶ J
Angle	Radian	rad	1 rad = (180/π)°
	Degree	°	1° = (π/180) rad
Frequency	Hertz	Hz	1 Hz = 60 rpm
	Revolution/minute	rpm	1 rpm = (1/60) Hz
	Radian/second	rad/s	1 rad/s = 9.5493 rpm
	Revolution/minute	rpm	1 rpm = 0.1047 rad/s

References

This section compiles the main references used along the book as well as fundamental bibliography. It is arranged according to subject, and chronologically within each subject.

Books Related to QFT and Frequency-Domain Methods

1. Bode, H.W. 1945. *Network Analysis and Feedback Amplifier Design*. Van Nostrand Company, Princeton, NY.
2. Horowitz, I. 1963. *Synthesis of Feedback Systems*. Academic Press, New York, NY.
3. Horowitz, I. 1993. *Quantitative Feedback Design Theory (QFT)*. QFT Publishers, Denver, CO.
4. Yaniv, O. 1999. *Quantitative Feedback Design of Linear and Non-linear Control Systems*. Kluwer Academic Publisher, Norwell, MA.
5. Sidi, M. 2002. *Design of Robust Control Systems: From Classical to Modern Practical Approaches*. Krieger Publishing, Malabar, FL.
6. Houpis, C.H., Rasmussen S.J., and Garcia-Sanz, M. 2006. *Quantitative Feedback Theory: Fundamentals and Applications*. 2nd Edition, A CRC Press book, Taylor & Francis Group, Boca Raton, FL.
7. Garcia-Sanz, M. and Houpis, C.H. 2012. *Wind Energy Systems: Control Engineering Design (Part I: QFT Control, Part II: Wind Turbine Design and Control)*. A CRC Press book, Taylor & Francis Group, Boca Raton, FL.

Special Issues about QFT

8. Nwokah, O.D.I. (Guest Editor), 1994. Horowitz and QFT design methods special issue. *Int. J. Robust Nonlinear Control*, 4(1), 1–230.
9. Houpis, C.H. (Guest Editor), 1997. Quantitative feedback theory special issue. *Int. J. Robust Nonlinear Control*, 7(6), 513–674.
10. Eitelberg, E. (Guest Editor), 2001–2002. Isaac Horowitz special issue. *Int. J. Robust Nonlinear Control*, Part 1, 11(10), 883–999; Part 2, 12(4), 287–402.
11. Garcia-Sanz, M. (Guest Editor), 2003. Robust frequency domain special issue. *Int. J. Robust Nonlinear Control*, 13(7), 595–688.
12. Garcia-Sanz, M. and Houpis, C.H. (Guest Editors), 2007. Quantitative feedback theory: In memoriam of Isaac Horowitz, special issue. *Int. J. Robust Nonlinear Control*, 17(2–3), 91–264.

International QFT Symposia

13. Houpis, C.H. and Chander, P. (Editors), 1992. *1st International Symposium on QFT and Robust Frequency Domain Methods*. Writght Patterson Airforce Base, Dayton, OH.
14. Nwokah, O.D.I. and Chander, P. (Editors), 1995. *2nd International Symposium on QFT and Robust Frequency Domain Methods*. Purdue University, West Lafayette, IN.

15. Petropoulakis, L. and Leithead, W.E. (Editors), 1997. *3rd International Symposium on QFT and Robust Frequency Domain Methods*. University of Strathclyde, Glasgow, Scotland, UK.
16. Boje, E. and Eitelberg, E. (Editors), 1999. *4th International Symposium on QFT and Robust Frequency Domain Methods*. University of Natal, Durban, South Africa.
17. Garcia-Sanz, M. (Editor), 2001. *5th International Symposium on QFT and Robust Frequency Domain Methods*. Public University of Navarra, Pamplona, Spain.
18. Boje, E. and Eitelberg, E. (Editors), 2003. *6th International Symposium on QFT and Robust Frequency Domain Methods*. University of Cape Town, Cape Town, South Africa.
19. Colgren, R. (Editor), 2005. *7th International Symposium on Quantitative Feedback Theory and Robust Frequency Domain Methods*. University Kansas, Lawrence, KS.
20. Gutman, P-O. (Editor), 2007. *8th International Symposium on Quantitative Feedback Theory and Robust Frequency Domain Methods*. Weizmann Institute of Science, Rehovot, Israel.
21. From 2009 on, the *International Symposium on Quantitative Feedback Theory and Robust Frequency Domain Methods* merged the *IFAC Symposium on Robust Control Design*. The *9th QFT International Symposium* merged the *6th ROCOND*, June 2009, Haifa, Israel.

Tutorials about QFT

22. Horowitz, I. 1982. Quantitative feedback theory. *IEE Control Theory Appl.*, 129(6), 215–226.
23. Horowitz, I. 1991. Survey of quantitative feedback theory. *Int. J. Control*, 53(2), 255–291.
24. Houpis, C.H. 1996. Quantitative feedback theory (QFT) technique. In *The Control Handbook*. (Editor W.S. Levine), Chapter 44, CRC Press, Boca Raton, FL, pp. 701–717.
25. Garcia-Sanz, M. 2005. Control Robusto Cuantitativo: Historia de una Idea, (in Spanish). *Rev. Iberoam Autom.*, 2(3), 25–38.
26. Garcia-Sanz, M. 2015. Quantitative feedback theory. In *Encyclopedia of Systems and Control*. (Editors T. Samad, J. Baillieul). Springer Verlag, Berlin, Germany. Article ID: 366609, Chapter ID: 238.

About the QFT History

27. Horowitz, I.M. 1992. QFT—Past, present and future. Plenary address. *1st International Symposium on QFT & Robust Frequency Domain Methods*, Dayton, OH, pp. 9–14.
28. Horowitz, I.M. 1999. Frequency response in control. Plenary. *4th International Symposium on QFT and Robust Frequency Domain Methods*, Durban, South Africa, pp. 233–239.
29. Garcia-Sanz, M. 2001. QFT international symposia: Past, present and future. *Editorial of 5th International Symposium on QFT and Robust Frequency Domain Methods*, Pamplona, Spain, pp. ix–xii.
30. Horowitz, I.M. 2002. It was not easy: A personal view. *Int. J. Robust Nonlinear Control*, 12(4), 289–293.
31. Houpis, C.H. 2002. Horowitz: Bridging the gap. *Int. J. Robust Nonlinear Control*, 12(4), 293–302.

First QFT Papers

32. Horowitz, I.M. 1959. Fundamental theory of automatic linear feedback control systems. *I.R.E. Trans. Autom. Control*, 4, 5–19.

33. Horowitz, I.M. and Sidi, M. 1972. Synthesis of feedback systems with large plant ignorance for prescribed time-domain tolerances. *Int. J. Control*, 16(2), 287–309.

34. Horowitz, I.M. 1973. Optimum loop transfer function in single-loop minimum-phase feedback systems. *Int. J. Control*, 18(1), 97–113.

QFT Templates

35. Bailey, F.N. and Hui, C.H. 1989. A fast algorithm for computing parametric rational functions. *IEEE Trans Autom. Control*, 34(11), 1209–1212.

36. Gutman, P.O., Baril, C., and Neumann, L. 1990. An image processing approach for computing value sets of uncertain transfer functions. *Proceedings of the 29th IEEE Conference on Decision and Control*, Honolulu, Hawaii, CA, pp. 1209–1212.

37. Bartlett, A.C. 1993. Computation of the frequency response of systems with uncertain parameters: A simplification. *Int. J. Control*, 57(6), 1293–1309.

38. Bartlett, A.C., Tesi, A., and Vicino, A. 1993. Frequency response of uncertain systems with interval plants. *IEEE Trans. Autom. Control*, 38(6), 929–933.

39. Gutman, P.O., Baril, C., and Neuman, L. 1994. An algorithm for computing value sets of uncertain transfer functions in factored real form. *IEEE Trans. Autom. Control*, 39(6), 1268–1273.

40. Cohen, B., Nordin, M., and Gutman, P. 1995. Recursive grid methods to compute value sets for transfer functions with parametric uncertainty. *Proceedings of the American Control Conference*, Seattle, pp. 3861–3865.

41. Ballance, D.J. and Hughes, G. 1996. A survey of template generation methods for quantitative feedback theory. *UKACC International Conference on Control '96*, Exeter, UK, pp. 172–174.

42. Sardar, G. and Nataraj, P.S.V. 1997. A template generation algorithm for non-rational transfer functions in QFT designs. *Proceedings of the 36th Conference on Decision and Control*, San Diego, USA, pp. 2684–2689.

43. Lasky, T.A. and Ravani, B. 1997. Use of convex hulls for plant template approximation in QFT design. *ASME J. Dyn. Syst. Meas. Control*, 119(3), 598–600.

44. Ballance, D.J. and Chen, W. 1998. Symbolic computation in value sets of plants with uncertain parameters. *UKACC International Conference on Control '98*, Exeter, UK, pp. 1322–1327.

45. Chen, W. and Balance, D.J. 1999. Plant template generation for uncertain plants in QFT. *ASME J. Dyn. Syst. Meas. Control*, 121(3), 358–364.

46. Garcia-Sanz, M. and Vital, P. 1999. Efficient computation of the frequency representation of uncertain systems. *4th International Symposium on QFT and Robust Frequency Domain Methods*, Durban, South Africa, pp. 117–126.

47. Nataraj, P.S.V. and Sardar, G. 2000. Template generation for continuous transfer functions using interval analysis. *Automatica*, 36, 111–119.

48. Boje, E. 2000, Finding non-convex hulls of QFT templates. *Trans. ASME*, 122, 230–232.

49. Nataraj, P.S.V. and Sheela, S. 2002. Template generation algorithm using vectorized function evaluations and adaptive subdivisions. *ASME J. Dyn. Syst. Meas. Control*, 124(4), 585–588.

50. Hwang, C. and Yang, S. 2002. QFT template generation for time-delay plants based on zero-inclusion test. *Syst. Control Lett.*, 45, 179–191.

51. Martin, J.J., Gil-Martinez, M., and Garcia-Sanz, M. 2007. Analytical formulation to compute QFT templates for plants with a high number of uncertain parameters. *15th Mediterranean Conference on Control and Automation, MED'07*, Athens, Greece, doi: 10.1109/MED.2007.4433934.

52. Diaz, J.M., Dormido, S., and Aranda, J. 2008. An interactive approach to template generation in QFT methodology. *Asian J. Control*, 10(3), 361–367.

53. Yang, S. 2009. An improvement of QFT plant template generation for systems with affinely dependent parametric uncertainties. *J. Franklin Inst.*, 346(7), 663–675.

QFT Bounds

54. Longdon, L. and East, D.J. 1978. A simple geometrical technique for determining loop frequency bounds which achieve prescribed sensitivity specifications. *Int. J. Control*, 30(1), 153–158.

55. Bailey, F.N., Panzer, D., and Gu, G. 1988. Two algorithms for frequency domain design of robust control systems. *Int. J. Control*, 48(5), 1787–1806.

56. Brown, M. and Petersen, I.R. 1991. Exact computation of the Horowitz bound for interval plants. *Proceedings of the 30th IEEE Conference on Decision and Control*, Brighton, England, pp. 2268–2273.

57. Nwokah, O.D.I., Jayasuriya, S., and Chait, Y. 1991. Parametric robust control by quantitative feedback theory. *American Control Conference*, Boston, MA, pp. 1975–1980.

58. Wang, G.G., Chen, C.W., and Wang, S.H. 1991. Equations for loop bound in quantitative feedback theory. *Proceedings of the 30th Conference on Decision and Control*, Brighton, pp. 2968–2969.

59. Fialho, I.J., Pande, V., and Nataraj, P.S.V. 1992. Design of feedback systems using Kharitonov's segments in quantitative feedback theory. *Proceedings of the 1st QFT Symposium*, Dayton, OH, pp. 457–470.

60. Chait, Y. and Yaniv, O. 1993. Multi-input/single-output computer-aided control design using the quantitative feedback theory. *Int. J. Robust Nonlinear Control*, 3(3), 47–54.

61. Zhao, Y. and Jayasuriya, S. 1994. On the generation of QFT bounds for general interval plants. *Trans. ASME*, 116, 618–627.

62. Chait, Y., Borghesani, C., and Zheng, Y. 1995. Single-loop QFT design for robust performance in the presence of non-parametric uncertainties. *ASME J. Dyn. Syst. Meas. Control*, 117(3), 420–425.

63. Moreno, J.C., Baños, A., and Montoya, J.F. 1997. An algorithm for computing QFT múltiple-valued performance bounds. *International Symposium on QFT and Robust Frequency Domain Methods*, Scotland, pp. 29–34.

64. Rodrigues, J.M., Chait, Y., and Hollot, C.V. 1997. An efficient algorithm for computing QFT bounds. *Trans. ASME*, 119, 548–552.

65. Eitelberg, E. 2000. Quantitative feedback design for tracking error tolerance. *Automatica*, 36, 319–326.

66. Nataraj, P. and Sardar, G. 2000. Template generation for continuous transfer functions using interval analysis. *Automatica*, 36(1), 111–119.

67. Nataraj, P. and Sardar, G. 2000. Computation of QFT bounds for robust sensitivity and gain-phase margin specifications. *Trans. ASME*, 122, 528–534.

68. Nataraj, P. 2002. Computation of QFT bounds for robust tracking specifications. *Automatica*, 38(2), 327–334.

69. Nataraj, P. 2002. Interval QFT: A mathematical and computational enhancement of QFT. *Int. J. Robust Nonlinear Control*, 12(4), 385–402.

70. Boje, E. 2003. Pre-filter design for tracking error specifications in QFT. *Int. J. Robust Nonlinear Control*, 13, 637–642.

71. Moreno, J.C., Baños, A., and Berenguel, M. 2006. Improvements on the computation of boundaries in QFT. *Int. J. Robust Nonlinear Control*, 16, 575–597.

72. Gutman, P.-O., Nordin, M., and Cohen, B. 2007. Recursive grid methods to compute value sets and Horowitz-Sidi bounds. *Int. J. Robust Nonlinear Control*, 17, 155–171.

73. Martin-Romero, J.J., Gil-Martinez, M., and Garcia-Sanz, M. 2009. Analytical formulation to compute QFT bounds: The envelope method. *Int. J. Robust Nonlinear Control*, 19(17), 1959–1971.

74. Yang, S.F. 2011. Generation of QFT bounds for robust tracking specifications for plants with affinely dependent uncertainties. *Int. J. Robust Nonlinear Control*, 21(3), 237–247.

75. Elso, J., Gil-Martinez, M., and Garcia-Sanz, M. 2012. Non-conservative QFT bounds for tracking error specifications. *Int. J. Robust Nonlinear Control*, 22, 2014–2025.

QFT Loop-Shaping. Controller Synthesis

76. Gera, A. and Horowitz, I.M. 1980. Optimization of the loop transfer function. *Int. J. Control*, 31(2), 389–398.

77. Thompson, D.F. and Nwokah, O.D.I. 1989. Stability and optimal design in quantitative feedback theory. *Proceedings of the ASME Winter Annual Meeting Conference*, San Francisco, CA, ASME Paper No. 89-WA/DSC-39.

78. Thompson, D.F. and Nwokah, O.D.I. 1994. Analytic loop shaping methods in quantitative feedback theory. *ASME J. Dyn. Syst. Meas. Control*, 116(2), 169–177.

79. Chait, Y., Chen, Q., and Hollot, C.V. 1997. Automatic loop-shaping of QFT controllers via linear programming. *3rd International Symposium on QFT and Other Robust Frequency Domain Methods*, Glasgow, UK, pp. 13–28.

80. Chait, Y., Chen, Q., and Hollot, C.V. 1999. Automatic loop-shaping of QFT controllers via linear programming. *ASME J. Dyn. Syst. Meas. Control*, 121(3), 351–357.

81. Garcia-Sanz, M. and Guillen, J.C. 2000. Automatic loop-shaping of QFT robust controllers via genetic algorithms. *3rd IFAC Symposium on Robust Control Design*, Praha.

82. Khaki-Sedigh, A. and Lucas, C. 2000. Optimal design of robust quantitative feedback controllers using random optimization techniques. *Int. J. Syst. Sci.*, 31(8), 1043–1052.

83. Lee, J.W., Chait, Y., and Steinbuch, M. 2000. On QFT tuning of multivariable μ controllers. *Automatica*, 36(11), 1701–1708.

84. Sidi, M.J. 2002. A combined QFT/Hinfinity design technique for TDOF uncertain feedback systems. *Int. J. Control*, 75(7), 475–489.

85. Garcia-Sanz, M., Brugarolas, M.J., and Eguinoa, I. 2004. Quantitative analysis of controller fragility in the frequency domain. *23rd IASTED International Symposium. Modelling, Identification and Control*, Grindelwald, Switzerland.

86. Garcia-Sanz, M. and Oses, J.A. 2004. Evolutionary algorithms for automatic tuning of QFT controllers. *23th IASTED International Symposium on Modelling, Identification and Control*, Grindelwald, Switzerland.

87. Molins, C. and Garcia-Sanz, M. 2009. Automatic loop-shaping of QFT robust controllers. *61st National Aerospace & Electronics Conference, NAECON*, July 2009, Dayton, OH, doi: 10.1109/NAECON.2009.5426643.

88. Garcia-Sanz, M. and Molins, C. 2010. Automatic loop-shaping of QFT robust controllers with multi-objective specifications via nonlinear quadratic inequalities. *62nd National Aerospace & Electronics Conference, NAECON*, July 2010, Dayton, OH, doi: 10.1109/NAECON.2010.5712975.

Existence Conditions for QFT Controllers

89. Nwokah, O.D.I., Thompson, D.F., and Perez, R.A. 1990. On some existence conditions for QFT controllers. *DSC*, 24, 1–10.

90. Jayasuriya, S. and Zhao, Y. 1994. Stability of quantitative feedback designs and the existence of robust QFT controllers. *Int. J. Robust Nonlinear Control*, 4(1), 21–46.

91. Gil-Martinez, M. and Garcia-Sanz, M. 2003. Simultaneous meeting of robust control specifications in QFT. *Int. J. Robust Nonlinear Control*, 13(7), 643–656.

MIMO QFT

92. Horowitz, I.M. 1979. Quantitative synthesis of uncertain multiple input-output feedback systems. *Int. J. Control*, 30(1), 81–106.

93. Horowitz, I.M. and Sidi, M. 1980. Practical design of feedback systems with uncertain multi-variable plants. *Int. J. Control*, 11(7), 851–875.

94. Horowitz, I.M. and Loecher, C. 1981. Design 3 × 3 multivariable feedback system with large plant uncertainty. *Int. J. Control*, 33, 677–699.

95. Horowitz, I.M., Neumann, L., and Yaniv, O. 1981. A synthesis technique for highly uncertain interacting multivariable flight control system (TYF16CCV). *Proceedings on NAECON Conference*, Dayton, OH, pp. 1276–1283.

96. Horowitz, I.M. 1982. Improved design technique for uncertain multiple input-output feedback systems. *Int. J. Control*, 36, 977–988.

97. Nwokah, O.D.I. 1984. Synthesis of controllers for uncertain multivariable plants for described time domain tolerances. *Int. J. Control*, 40, 1189–1206.

98. Yaniv, O. and Horowitz, I.M. 1986. A quantitative design method for MIMO linear feedback systems having uncertain plants. *Int. J. Control*, 43(2), 401–421.

99. Nwokah, O.D.I. 1988. Strong robustness in uncertain multivariable systems. *IEEE Conference on Decision and Control*, Austin, TX, doi: 10.1109/CDC.1988.194715.

100. Park, M.S., Chait, Y., and Steinbuch, M. 1994. A new approach to multivariable quantitative feedback theory: Theoretical and experimental results. *1994 American Control Conference: ACC '94*, Baltimore, MD, doi: 10.1109/ACC.1994.751755.

101. Franchek, M.A. and Nwokah, O.D.I. 1995. Robust multivariable control of distillation columns using non-diagonal controller matrix. *DSC-Vol. 57-1, IMECE, ASME Dynamics Systems and Control Division*, San Francisco, CA, pp. 257–264.

102. Yaniv, O. 1995. MIMO QFT using non-diagonal controllers. *Int. J. Control*, 61(1), 245–253.

103. Zhao, Y. and Jayasuriya, S. 1996. A new formulation of multiple-input multiple-output quantitative feedback theory. *ASME J. Dyn. Syst. Meas. Control*, 118(4), 748–752.

104. Franchek, M.A., Herman, P., and Nwokah, O.D.I. 1997. Robust nondiagonal controller design for uncertain multivariable regulating systems. *ASME J. Dyn. Syst. Meas. Control*, 119, 80–85.

105. Piedmonte, M.D., Meckl, P.H., Nwokah, O.D.I., and Franchek, M.A. 1998. Multivariable vibration control of a coupled flexible structure using QFT. *Int. J.Control*, 69(4), 475–498.

106. Boje, E. and Nwokah, O.D.I. 2001. Quantitative feedback design using forward path decoupling. *ASME J. Dyn. Syst. Meas. Control*, 123(1), 129–132.

107. Boje, E. 2002. Non-diagonal controllers in MIMO quantitative feedback design. *Int. J. Robust Nonlinear Control*, 12(4), 303–320.

108. Boje, E. 2002. Multivariable quantitative feedback design for tracking error specifications. *Automatica*, 38, 131–138.

109. De Bedout, J.M. and Franchek, M.A. 2002. Stability conditions for the sequential design of non-diagonal multivariable feedback controllers. *Int. J. Control*, 75(12), 910–922.

110. Garcia-Sanz, M. and Egaña, I. 2002. Quantitative non-diagonal controller design for multivariable systems with uncertainty. *Int. J. Robust Nonlinear Control*, 12(4), 321–333.

111. Eitelberg, E. 2003. On multivariable tracking. *6th International Symposium on Quantitative Feedback Theory*, Cape Town, South Africa, Vol. 2, pp. 514–519.

112. Kerr, M. and Jayasuriya, S. 2003. Sufficient conditions for robust stability in non-sequential MIMO QFT. *42nd IEEE International Conference on Decision and Control*, Maui, HI, doi: 10.1109/CDC.2003.1272876.

113. Lan, C.Y., Kerr, M.L., and Jayasuriya, S. 2004. Synthesis of controllers for non-minimum phase and unstable systems using non-sequential MIMO quantitative feedback theory. *2004 American Control Conf. (ACC)*, Boston, MA.

114. Garcia-Sanz, M., Egaña, I., and Barreras, M. 2005. Design of quantitative feedback theory non-diagonal controllers for use in uncertain MIMO systems. *IEE Control Theory Appl.*, 152(2), 177–187.

115. Garcia-Sanz, M. and Eguinoa, I. 2005. Improved non-diagonal MIMO QFT design technique considering non-minimum phase aspects. *7th International Symposium on QFT and Robust Frequency Domain Methods*, Lawrence, KS.

116. Kerr, M.L., Jayasuriya, S., and Asokanthan, S.F. 2005. On stability in non-sequential MIMO QFT designs. *ASME J. Dyn. Syst. Meas. Control*, 127(1), 98–104.
117. Kerr, M.L. and Jayasuriya, S. 2006. An improved non-sequential multi-input multi-output quantitative feedback theory design methodology. *Int. J. Robust Nonlinear Control*, 16(8), 379–95.
118. Mahdi Alavi, S.M., Khaki-Sedigh, A., Labibi, B., and Hayes, M.J. 2007. Improved multivariable feedback design for tracking error specifications. *IET Control Theory Appl.*, 1, 1046–1053.
119. Garcia-Sanz, M., Eguinoa, I., and Bennani, S. 2009. Non-diagonal MIMO QFT controller design reformulation. *Int. J. Robust Nonlinear Control*, 19(9), 1036–1064.
120. Elso, J., Gil-Martinez, M., and Garcia-Sanz, M. 2014. A quantitative feedback solution to the multivariable tracking error problem. *Int. J. Robust Nonlinear Control*, 24(16), 2331–2346.
121. Elso, J., Gil-Martinez, M., and Garcia-Sanz, M. 2017. Quantitative feedback control for multivariable model matching and disturbance rejection. *Int. J. Robust Nonlinear Control*, 27(1), 121–134.

Time-Delay Systems. QFT Controller Design

122. Garcia-Sanz, M. and Guillen, J.C. 1999. Smith Predictor for Uncertain Systems in the QFT Framework. *Lecture Notes in Control and Information Sciences*, Ed. Springer Verlag. vol. 243. *Progress in System and Robot Analysis and Control Design*. Chapter 20, pp. 243–250.

Unstable Systems and Non-Minimum Phase Systems. QFT Controller Design

123. Chen, W. and Balance, D. 1998. QFT design for uncertain non-minimum phase and unstable plants. *American Control Conference*, vol. 4, Philadelphia, PA, pp. 2486–2490.
124. Chen, W. and Balance, D. 2001. QFT design for uncertain non-minimum phase and unstable plants revisited. *Int. J. Control*, 74(9), 957–965.
125. Garcia-Sanz, M. 2016. The Nyquist stability criterion in the Nichols chart. *Int. J. Robust Nonlinear Control*, 26(12), 2643–2651.

Digital QFT

126. Horowitz, I.M. and Liao, Y. 1986. Quantitative feedback design for sampled-data systems. *Int. J. Control*, 44, 665–675.
127. Yaniv, O. and Chait, Y. 1991. Direct robust control of uncertain sampled-data systems using the quantitative feedback theory. *Proceedings of the ACC Conference*, Boston, MA, pp. 1987–1988.
128. Houpis, C.H. and Lamont, B.G. 1992. *Digital Control Systems: Theory, Hardware, Software*. 2nd Edition, McGraw Hill, New York, NY.
129. Yaniv, O. and Chait, Y. 1993. Direct control design in sampled-data uncertain systems. *Automatica*, 29(2), 365–372.

Distributed Parameter Systems. QFT Controller Design

130. Horowitz, I.M. and Azor, R. 1983. Quantitative synthesis of feedback systems with distributed uncertain plants. *Int. J. Control*, 38(2), 381–400.
131. Horowitz, I.M. and Azor, R. 1984. Uncertain partially non-casual distributed feedback systems. *Int. J. Control*, 40(5), 989–1002.
132. Chait, Y., Maccluer, C.R., and Radcliffe, C.J. 1989. A Nyquist stability criterion for distributed parameter systems. *IEEE Trans. Autom. Control*, 34(1), 90–92.
133. Horowitz, I.M., Kannai, Y., and Kelemen, M. 1989. QFT approach to distributed systems control and applications. *1989 IEEE International Conference on Control and Applications*, Jerusalem, Israel, pp. 516–519.
134. Kelemen, M., Kanai, Y., and Horowitz, I.M. 1989. One-point feedback approach to distributed linear systems. *Int. J. Control*, 49(3), 969–980.
135. Kelemen, M., Kanai, Y., and Horowitz, I.M. 1990. Improved method for designing linear distributed feedback systems. *Int. J. Adapt. Control Signal Process.*, 4, 249–257.
136. Hedge, M.D. and Nataraj, P.S.V. 1995. The two-point feedback approach to linear distributed systems. *International Conference on Automatic Control*, Indore, India, pp. 281–284.
137. Garcia-Sanz, M., Huarte, A., and Asenjo, A. 2007. A quantitative robust control approach for distributed parameter systems. *Int. J. Robust Nonlinear Control*, 17(2–3), 135–153.

Feedforward and QFT Controller Design

138. Elso, J., Gil-Martinez, M., and Garcia-Sanz, M. 2013. Quantitative feedback-feedforward control for model matching and disturbance rejection. *IET Control Theory Appl.*, 7(6), 894–900.

Non-Minimum Phase Systems. QFT Controller Design

139. Horowitz, I.M. and Sidi, M. 1978. Optimum synthesis of non-minimum phase systems with plant uncertainty. *Int. J. Control*, 27(3), 361–386.
140. Horowitz, I.M. 1979. Design of feedback systems with non-minimum phase unstable plants. *Int. J. Systems Sci.*, 10, 1025–1040.
141. Horowitz, I.M. and Liao, Y. 1984. Limitations on non-minimum phase feedback systems. *Int. J. Control*, 40(5), 1003–1015.
142. Horowitz, I.M. 1986. The singular-G method for unstable non-minimum phase plants. *Int. J. Control*, 44(2), 533–541.
143. Horowitz, I.M., Oldak, S., and Yaniv, O. 1986. An important property of non-minimum phase multi-inputs multi-outputs feedback systems. *Int. J. Control*, 44(3), 677–688.
144. Chen, W. and Ballance, D. 1998. QFT design for uncertain non-minimum phase and unstable plants. *American Control Conference*, Philadelphia, PA, pp. 2486–2490.

Multi-Loop Systems. QFT Controller Design

145. Horowitz, I.M., Neumann, L., and Yaniv, O. 1985. Quantitative synthesis of uncertain cascade multi-input multi-output feedback systems. *Int. J. Control*, 42(2), 273–303.

146. Horowitz, I.M. and Yaniv, O. 1985. Quantitative cascade MIMO synthesis by an improved method. *Int. J. Control*, 42(2), 305–331.
147. Eitelberg, E. 1999. *Load Sharing Control*. NOYB Press, Durban, South Africa.
148. Baños, A. and Horowitz, I.M. 2000. QFT design of multi-loop nonlinear control systems. *Int. J. Robust Nonlinear Control*, 10(15), 1263–1277.

Nonlinear Systems. QFT Controller Design

149. Horowitz, I.M. 1976. Synthesis of feedback systems with non-linear time-varying uncertain plants to satisfy quantitative performance specifications. *IEEE Proc.*, 64, 123–130.
150. Horowitz, I.M. 1981. Quantitative synthesis of uncertain non-linear feedback systems with non-minimum phase inputs. *Int. J. Syst. Sci.*, 1(12), 55–76.
151. Horowitz, I.M. 1981. Improvements in quantitative non-linear feedback design by cancellation. *Int. J. Control*, 34(3), 547–560.
152. Breiner, M. and Horowitz, I.M. 1981. Quantitative synthesis of feedback systems with uncertain nonlinear multivariable plants. *Int. J. Syst. Sci.*, 12, 539–563.
153. Horowitz, I.M. 1982. Feedback systems with non-linear uncertain plants. *Int. J. Control*, 36, 155–171.
154. Horowitz, I.M. 1983. A synthesis theory for a class of saturating systems. *Int. J. Control*, 38(1), 169–187.
155. Horowitz, I.M. and Liao, Y. 1986. Quantitative non-linear compensation design for saturating unstable uncertain plants. *Int. J. Control*, 44, 1137–1146.
156. Oldak, S., Baril, C., and Gutman, P.O. 1994. Quantitative design of a class of nonlinear systems with parameter uncertainty. *Int. J. Robust Nonlinear Control*, 4(1), 101–117.
157. Baños, A. and Bailey, F.N. 1998. Design and validation of linear robust controllers for nonlinear plants. *Int. J. Robust Nonlinear Control*, 8(9), 803–816.
158. Baños, A. and Barreiro, A. 2000. Stability of non-linear QFT designs based on robust absolute stability criteria. *Int. Journal of Control*, 73(1), 74–88.
159. Baños, A., Barreiro, A., Gordillo, F., and Aracil, J. 2002. A QFT framework for nonlinear robust stability. *Int. J. Robust Nonlinear Control*, 12(4), 357–372.

Linear-Time-Variant (LTV) Systems. QFT Controller Design

160. Horowitz, I.M. 1975. A synthesis theory for linear time-varying feedback systems with plant uncertainty. *IEEE Trans. Autom. Control*, AC-20, 454–463.
161. Yaniv, O. and Boneh, R. 1997. Robust LTV feedback synthesis for SISO nonlinear plants. *Int. J. Robust Nonlinear Control*, 7, 11–28.
162. Yaniv, O. 1999. Robust LTV feedback synthesis for nonlinear MIMO plants. *Trans. ASME*, 121, 226–232.
163. Garcia-Sanz, M. and Elso, J. 2009. Beyond the linear limitations by combining switching & QFT. Application to wind turbines pitch control systems. *Int. J. Robust Nonlinear Control*, 19(1), 40–58.

Stability Analysis and Controller Design in the Nichols Chart

164. Cohen, N., Chait, Y. Yaniv, O., and Borghesani, C. 1994. Stability analysis using Nichols Charts. *Int. J. Robust Nonlinear Control*, 4(1), 3–20.

165. Garcia-Sanz, M. 2016. The Nyquist stability criterion in the Nichols chart. *Int. J. Robust Nonlinear Control*, 26(12), 2643–2651.

CAD Tools for QFT Controller Design

166. Bailey, F.N. and Hul, C.H. 1989. CACSD tools for loop gain-phase shaping design of SISO robust controllers. *Proceedings of the IEEE Control System Society Workshop on Computer Aided Control System*, Berkeley, CA, pp. 151–157.

167. Sating, R.R. 1992. Development of an Analog MIMO Quantitative Feedback Theory (QFT) CAD Package, MS Thesis, AFIT/GE/ENG/92J-04, Air Force Institute of Technology, Wright Patterson AFB, OH.

168. Houpis, C.H. and Sating, R.R. 1997. MIMO QFT CAD package (Ver.3). *Int. J. Control*, 7(6), 533–549.

169. Borghesani, C., Chait, Y., and Yaniv, O. 1994, 2002. *Quantitative Feedback Theory Toolbox—For Use with MATLAB*. Terasoft, San Diego, CA.

170. Gutman, P.-O. 1996, 2001. *Qsyn—the Toolbox for Robust Control Systems Design for Use with Matlab, User's Manual*, NovoSyn AB, Jonstorp, Sweden; Electro-Optics Industries Ltd, Rehovot, Israel.

171. Houpis, C.H., Rasmussen, S.J., and Garcia-Sanz, M. 2006. CAD Tool for controller design. In *Quantitative Feedback Theory: Fundamentals and Applications*. 2nd Edition, CRC, Taylor & Francis, Boca Raton, FL.

172. Garcia-Sanz, M., Vital, P., Barreras, M., and Huarte, A. 2001. InterQFT. Public University of Navarra. Also as Interactive Tool for Easy Robust Control Design. In *IFAC International Workshop, Internet Based Control Education*, Madrid, Spain, pp. 83–88.

173. Diaz, J.M., Dormido, S., and Aranda, J. 2004. SISO-QFTIT, una herramienta software interactiva para diseño de controladores robustos usando QFT. *UNED*, Madrid, Spain.

174. Garcia-Sanz, M., Mauch, A., and Philippe, C. 2009. QFT control toolbox: An interactive object-oriented Matlab CAD tool for quantitative feedback theory. *6th IFAC Symposium on Robust Control Design, ROCOND'09*, Haifa, Israel.

175. Garcia-Sanz, M. 2008–2017. The QFT Control Toolbox for Matlab—QFTCT. Standard and Student versions at http://codypower.com.

Real-World Applications with QFT

176. Horowitz, I., Neumann, L., and Yaniv, O. 1981. A synthesis technique for highly uncertain interacting multivariable flight control system (TYF16CCV). *Proceedings of the IEEE Naecon Conference*, Dayton, OH.

177. Horowitz, I.M. et al. 1982. *Multivariable Flight Control Design with Uncertain Parameters (YF16CCV)*. AFWAL-TR-83-3036, Air Force Wright Aeronautical Laboratories, Wright-Patterson AFB, OH.

178. Walke, J., Horowitz, I., and Houpis, C. 1984. Quantitative synthesis of highly uncertain MIMO flight control system for the forward swept wing X-29 aircraft. *Proceedings of the IEEE Naecon Conference*, Dayton, OH, pp. 576–583.

179. Bossert, D.E. 1989. *Design of Pseudo-Continuous-Time Quantitative Feedback Theory Robot Controllers*. AFIT/GE/ENG/89D-2, Air Force Institute of Technology, Wright-Patterson AFB, OH.

180. Trosen, D.W. 1993. *Development of an Prototype Refueling Automatic Flight Control System Using Quantitative Feedback Theory.* AFIT/GE/ENG/93-J-03, Air Force Institute of Technology, Wright-Patterson AFB, OH.

181. Kelemen, M. and Bagchi, A. 1993. Modeling and feedback control of a flexible arm of a robot for prescribed frequency domain tolerances. *Automatica*, 29, 899–909.

182. Reynolds, O.R., Pachter, M., and Houpis, C.H. 1994. Design of a subsonic flight control system for the Vista F-16 using quantitative feedback theory. *Proceedings of the American Control Conference*, Baltimore, MD, pp. 350–354.

183. Rasmussen, S.J. and Houpis, C.H. 1994. Development implementation and flight of a MIMO digital flight control system for an unmanned research vehicle using quantitative feedback theory. *Proceedings of the ASME Dynamic Systems and Control, Winter Annual Meeting of ASME*, Chicago, IL.

184. Bentley, A.E. 1994. Quantitative feedback theory with applications in welding. *Int. J. Robust Nonlinear Control*, 4(1), 119–160.

185. Yau, C.H., Gallagher, J.E., and Nwokah, O.D.I. 1994. A model reference quantitative feedback design theory with application to turbomachinery. *Int. J. Robust Nonlinear Control*, 4(1), 181–210.

186. Miller, R.B., Horowitz, I.M., Houpis, C.H., and Finley, B. 1994. Multi-input, multi-output flight control system design for the YF-16 using nonlinear QFT and pilot compensation. *Int. J. Robust Nonlinear Control*, 4(1), 211–230.

187. Miller, R.B., Horowitz, I.M., Houpis, C.H., and Barfield, A.F. 1994. Multi-input, multi-output flight control system design for the YF-16 using nonlinear QFT and pilot compensation. *Int. J. Robust Nonlinear Control*, 4(1), 211–230.

188. Osmon, C., Pachter, M., and Houpis, C.H. 1996. Active flexible wing control using QFT. *IFAC 13th World Congress*, vol. H, San Francisco, CA, pp. 315–320.

189. Franchek, M. and Hamilton, G.K. 1997. Robust controller design and experimental verification of I.C. engine speed control. *Int. J. Robust Nonlinear Control*, 7, 609–628.

190. Pachter, M., Houpis, C.H., and Kang, K. 1997. Modelling and control of an electro-hydrostatic actuator. *Int. J. Robust Nonlinear Control*, 7, 591–608.

191. Boje, E. and Nwokah, O.D.I. 1999. Quantitative multivariable feedback design for a turbofan engine with forward path decoupling. *Int. J. Robust Nonlinear Control*, 9(12), 857–882.

192. Garcia-Sanz, M. and Ostolaza, J.X. 2000. QFT-control of a biological reactor for simultaneous ammonia and nitrates removal. *Int. J. Syst. Anal. Model. Simul. SAMS*, 36, 353–370.

193. Egaña, I., Villanueva, J., and Garcia-Sanz, M. 2001. Quantitative multivariable feedback design for a SCARA robot arm. *5th International Symposium on QFT and Robust Frequency Domain Methods*, Pamplona, Spain, pp. 67–72.

194. Rueda, T.M. and Velasco, F.J. 2001. Robust QFT controller for marine course-changing control. *5th International Symposium on QFT and Robust Frequency Domain Methods*, Pamplona, Spain, pp. 79–84.

195. Kelemen, M. and Akhrif, O. 2001. Linear QFT control of a highly nonlinear multi-machine power system. *Int. J. Robust Nonlinear Control*, 11(10), 961–976.

196. Bentley, A.E. 2001. Pointing control design for high precision flight telescope using quantitative feedback theory. *Int. J. Robust Nonlinear Control*, 11(10), 923–960.

197. Liberzon, A., Rubinstein, D., and Gutman, P.O. 2001. Active suspension for single wheel satin of on-road track vehicle. *Int. J. Robust Nonlinear Control*, 11(10), 977–999.

198. Garcia-Sanz, M., Guillen, J.C., and Ibarrola, J.J. 2001. Robust controller design for time delay systems with application to a pasteurisation process. *Control Eng. Pract.* 9, 961–972.

199. Yaniv, O., Fried, O., and Furst-Yust, M. 2002. QFT application for headphone's active noise cancellation. *Int. J. Robust Nonlinear Control*, 12(4), 373–383.

200. Niksefat, N. and Sepehri, N. 2002. A QFT fault-tolerant control for electrohydraulic positioning systems. *IEEE Trans. Control Syst. Technol.*, 10(4), 626–632.

201. Gutman, P.O., Horesh, E., Guetta, R., and Borshchevsky, M. 2003. Control of the aero-electric power station—An exciting QFT application for the 21st century. *Int. J. Robust Nonlinear Control*, 13(7), 619–636.

202. Torres, E. and Garcia-Sanz, M. 2004. Experimental results of the variable speed, direct drive multipole synchronous wind turbine: TWT1650. *Wind Energy*, 7(2), 109–118.

203. Garcia-Sanz, M. and Hadaegh, F.Y. 2004. *Coordinated Load Sharing QFT Control of Formation Flying Spacecrafts. 3D Deep Space and Low Earth Keplerian Orbit Problems with Model Uncertainty.* NASA-JPL, JPL Document, D-30052, Pasadena, CA.

204. Kerr, M. 2004. Robust Control of an Articulating Flexible Structure Using MIMO QFT, PhD Dissertation, University of Queensland, Australia.

205. Barreras, M. and Garcia-Sanz, M. 2004. Multivariable QFT controllers design for heat exchangers of solar systems. *International Conference on Renewable Energy and Power Quality*, Barcelona, Spain.

206. Garcia-Sanz, M. and Barreras, M. 2006. Non-diagonal QFT controller design for a 3-input 3-output industrial Furnace. *Int. J. Dyn. Syst. Meas. Control ASME*, 128(2), 319–329.

207. Barreras, M., Villegas, C., Garcia-Sanz, M., and Kalkkuhl, J. 2006. Robust QFT tracking controller design for a Car equipped with 4-wheel steer-by-wire. *IEEE International Conference on Control Applications, CCA*, Munich, Germany, doi: 10.1109/CACSD-CCA-ISIC.2006.4776832.

208. Garcia-Sanz, M., Eguinoa, I., Ayesa, E., and Martin, C. 2006. Non-diagonal multivariable robust QFT control of a wastewater treatment plant for simultaneous nitrogen and phosphorus removal. *Robust Control Design Conference, ROCOND'06, IFAC*, Toulouse, France.

209. Garcia-Sanz, M. and Hadaegh, F.Y. 2007. Load-sharing robust control of spacecraft formations: Deep space and low earth elliptic orbits. *IET Control Theory Appl. (former IEE)*, 1(2), 475–484, UK.

210. Kerr, M.L., Lan, C.Y., and Jayasuriya, S. 2007. Non-sequential MIMO QFT control of the X-29 aircraft using a generalized formulation. *Int. J. Robust Nonlinear Control*, 17(2–3), 107–134.

211. Garcia-Sanz, M., Eguinoa, I., Barreras, M., and Bennani, S. 2008. Non-diagonal MIMO QFT controller design for Darwin-type spacecraft with large flimsy appendages. *J. ASME J. Dyn. Syst. Meas. Control*, 130, 011006-1–011006-15.

212. Garcia-Sanz, M., Eguinoa, I., Gil-Martinez, M., Irizar, I., and Ayesa, E. 2008. MIMO quantitative robust control of a wastewater treatment plant for biological removal of nitrogen and phosphorus. *16th Mediterranean Conference on Control and Automation, MED'08*, Ajaccio, France.

213. Garcia-Sanz, M. and Molins, C. 2008. QFT robust control of a Vega-type space launcher. *16th Mediterranean Conference on Control and Automation*, Ajaccio, France.

214. Garcia-Sanz, M. 2009. QFT: New developments and advanced real-world applications. Plenary session. *6th IFAC Symposium on Robust Control Design*, Haifa, Israel.

215. Karpenko, M. and Sepehri, N. 2010. On quantitative feedback design for robust position control of hydraulic actuators. *Control Engineering Practice*, 18(3), 289–299.

216. Garcia-Sanz, M., Eguinoa, I., and Barreras, M. 2011. Advanced attitude and position MIMO robust control strategies for telescope-type spacecraft with large flexible appendages. In *Advances in Spacecraft Technologies*. INTECH. (Editor Jason Hall), ISBN: 978-953-307-551-8, doi: 10.5772/14506

217. Garcia-Sanz, M., Ranka, T., and Joshi, B.C. 2011. Advanced nonlinear robust controller design for high-performance servo-systems in large radar antennas. *63th National Aerospace & Electronics Conference, IEEE-NAECON*, Dayton, OH.

218. Tierno, N., White, N., and Garcia-Sanz, M. 2011. Longitudinal flight control for a novel airborne wind energy system: Robust MIMO control design techniques. *ASME International Mechanical Engineering Congress & Exposition, IMECE-2011*, November 11–17, Denver, CO.

219. Karpenko, M. and Sepehri, N. 2012. Electrohydraulic force control design of a hardware-in-the-loop load emulator using a nonlinear QFT technique. *Control Eng. Pract.*, 20(6), 598–609.

220. Garcia-Sanz, M., Ranka, T., and Joshi, B.C. 2012. High-performance switching QFT control for large radio telescopes with saturation constraints. *64th National Aerospace & Electronics Conference, IEEE-NAECON*, Dayton, OH.

221. Lounsbury, W. and Garcia-Sanz, M. 2014. High-performance quantitative robust switching control for optical telescopes. *2014 SPIE Astronomical Telescopes and Instrumentation Conference, SPIE, Software and Cyberinfrastructure for Astronomy III*, Montreal, QC, doi: 10.1117/12.2056910.

222. Garcia-Sanz, M., Labrie, H., and Cavalcanti, J. 2014. Wind farm lab test-bench for research/education on optimum design and cooperative control of wind turbines. In *Wind Turbine Control and Monitoring*. (Editors Luo, Vidal, and Acho), Springer Verlag, Berlin, Germany, Chapter 14, Series of Green Energy and Technology, ISBN: 978-3-319-08412-1.

NATO/RTO Lecture Series about QFT

223. Garcia-Sanz, M. 2003. Quantitative Feedback Theory (QFT): Bridging the Gap. NATO/RTO Lecture Series SCI-236. Systems Concepts and Integration Panel. Robust Integrated Control System Design Methods for the 21st Century Military Applications. Setubal, Portugal; Forli, Italy; and Los Angeles, CA.
224. Garcia-Sanz, M. 2005. Quantitative Robust Control Engineering. Theory and Applications. NATO/RTO Lecture Series SCI-166. Partnership for Peace (PFP). Systems Concepts and Integration Panel. Achieving Successful Robust Integrated Control System Designs for the 21st Century Military Applications. Stockholm, Sweden; Zurich, Switzerland; Bucharest, Romania.
225. Garcia-Sanz, M. 2007. Quantitative Robust Control of Spacecraft Formations. NATO/RTO Lecture Series SCI-175. Systems Concepts and Integration Panel: System Control Technologies, Design Considerations & Integrated Optimization Factors For Distributed Nano Unmanned Air Vehicle (UAV) Applications. Davis, CA; Rostock, Germany; Florence, Italy.
226. Garcia-Sanz, M., Eguinoa, I., and Elso, J. 2008. Beyond the Classical Performance Limitations Controlling Uncertain MIMO Systems: UAV Applications. NATO/RTO Lecture Series SCI-195. Systems Concepts and Integration Panel: Advanced autonomous formation control and trajectory management techniques for multiple micro UAV applications. Glasgow, UK; Pamplona, Spain; Cleveland, OH.
227. Garcia-Sanz, M. 2009. MIMO QFT Controller Design Reformulation. Application to spacecraft with flexible appendages and to spacecraft flying in formation in a low Earth orbit. NATO/RTO Lecture Series SCI-209. Systems Concepts and Integration Panel: Small Satellite Formations for Distributed Surveillance: System Design and Optimal Control Considerations. Stanford, CA; Wurzburg, Germany; Rome, Italy.

Miscellaneous. QFT Control

228. Nordgren, R.E., Nwokah, O.D.I., and Franchek, M.A. 1994. New formulations for quantitative feedback theory. *Int. J. Robust Nonlinear Control*, 4(1), 47–64.

Books Related to Control Engineering

229. Rosenbrock, H.H. 1970. *State-Space and Multivariable Theory*. Thomas Nelson, London, UK.
230. Takahaschi, Y., Rabims, M., and Auslander, D. 1970. *Control and Dynamic Systems*. Addison Wesley, Reading, MA.
231. Rosenbrock, H.H. 1974. *Computer-Aided Control System Design*. Academic Press, New York, NY.
232. Wolovich, W.A. 1974. *Linear Multivariable Systems*, 11, Springer-Verlag, Berlin, Germany.

233. Desoer, C.A. and Vidyasagar, M. 1975. *Feedback Systems: Input–Output Properties*. Academic Press, New York, NY.

234. Frank, P.M. 1978. *Introduction to System Sensitivity Theory*. Academic Press, New York, NY.

235. Postlethwaite, I. and MacFarlane, A.G.J. 1979. *A Complex Variable Approach to the Analysis of Linear Multivariable Feedback Systems*, 12. Springer-Verlag, Berlin, Germany.

236. Wellstead, P.E. 1979. *Introduction to Physical System Modelling*. Academic Press, New York, NY.

237. Kailath, T. 1980. *Linear Systems*. Prentice-Hall, Englewood Cliffs, NJ.

238. Hung, Y.S. and MacFarlane, A.G.J. 1982. *Multivariable Feedback: A Quasi-Classical Approach*, 40. Springer-Verlag, Berlin.

239. McAvoy, T.J. 1983. *Interaction Analysis—Principles and Applications*. Instrument Society of America, Research Triangle Park, NC.

240. Ljung, L. 1987. *System Identification: Theory for the User*. Prentice Hall, Englewood Cliffs, NJ. Information and System Science Series. And Ljung, L., (1997), System Identification Toolbox User's Guide. The Mathworks, Inc.

241. O'Reilly, J. 1987. *Multivariable Control for Industrial Applications*. Peter Peregrinus Ltd, London.

242. Franklin, G.F. and Powell, J.D. 1988. *Digital Control of Dynamic Systems*. 2nd Edition, Addison-Wesley, Reading, MA.

243. Freudenberg, J.S. and Looze, D.P. 1988. *Frequency Domain Properties of Scalar and Multivariable Feedback Systems*. Springer-Verlag, Berlin.

244. Deshpande, P.B. 1989. *Multivariable Process Control*. Instrument Society of America, Research Triangle Park, NC.

245. Maciejowski, J.M. 1989. *Multivariable Feedback Design*. Addison-Wesley Publishing Company, Reading, MA.

246. Morari, M. and Zafiriou, E. 1989. *Robust Process Control*. Prentice-Hall International, Englewood Cliffs, NJ.

247. Houpis, C.H. and Lamont, G. 2016. *Digital Control Systems: Theory, Hardware, Software*. 3rd Edition, McGraw-Hill, NY.

248. Franklin, G.F., Powell, J.D., and Emani-Naeini, A. 1994. *Feedback Control of Dynamic Systems*. Addison Wesley, New York, NY.

249. Astrom, K.J. and Hagglund, T. 1995. *PID Controllers: Theory, Design, and Tuning*. 2nd Edition, ISA, North Carolina. Also 2nd ed., (2006) as Advanced PID Control.

250. Bhattacharyya, S.P., Chapellat, H., and Keel, L.H. 1995. *Robust Control: The Parametric Approach*. Prentice Hall, UK.

251. Zhou, K., Doyle, J., and Glover, K. 1996. *Robust and Optimal Control*. Prentice Hall, Englewood Cliffs, NJ.

252. Dutton, K., Thompson, S., and Barraclough, B. 1997. *The Art of Control Engineering*. Pearson/Prentice-Hall, UK.

253. Datta, A., Ho, M.T., and Bhattacharyya, S.P. 2000. *Structure and Synthesis of PID Controllers*. Springer-Verlag, Berlin, Germany.

254. Dorato, P. 2000. *Analytic Feedback System Design. An Interpolation Approach*. Brooks-Cole, Pacific Grove, CA.

255. Kailath, T., Sayed, A.H., and Hassibi, B. 2000. *Linear Estimation*. Prentice Hall, Information and System Sciences Series, New Jersey.

256. Lurie, B.J. and Enright, P.J. 2000. *Classical Feedback Control with MATLAB*. Marcel Dekker, New York, NY.

257. Rosenwasser, E. and Yusupov, R. 2000. *Sensitivity of Automatic Control Systems*. CRC Press, Taylor & Francis Group, Boca Raton, FL.

258. Pintelon, R. and Shoukens, J. 2012. *System Identification: A Frequency Domain Approach*. Wiley-IEEE Press, Hoboken, NJ.

259. Houpis, C.H. and Sheldon, S.N. 2013. *Linear Control System Analysis and Design with Matlab*. 6th Edition, CRC Press, Taylor & Francis Group, Boca Raton, FL.

260. Skogestad, S. and Postlethwaite, I. 2005. *Multivariable Feedback Control. Analysis and Design*. 2nd Edition, John Wiley & Sons Ltd, UK.

261. Verhaegen, M. and Verdult, V. 2007. *Filtering and System Identification: An Introduction.* Cambridge University Press, Cambridge, UK.

262. Ogata, K. 2009. *Modern Control Engineering.* 5th edition, Prentice-Hall, New Jersey.

263. Dorf, R.C. and Bishop, R.H. 2016. *Modern Control Systems.* 13th Edition, Pearson, New Jersey.

General MIMO Systems

264. McMillan, B. 1952. Introduction to formal realizability theory—I, II. *Bell Syst. Tech. J.*, 31(2, 3), 217–279, 541–600.

265. Bristol, E.H. 1966. On a new measure of interaction for multi-variable process control. *IEEE Trans. Autom. Control*, AC-11(1), 133–134.

266. Hsu, C.-H. and Chen, C.-T. 1968. A proof of the stability of multivariable feedback systems. *Proc. IEEE*, 56(11), 2061–2062.

267. Rosenbrock, H.H. 1969. Design of multivariable control systems using the inverse Nyquist array. *IEE-Proc.*, 116(11), 1929–1936.

268. Luyben, W.L. 1970. Distillation decoupling. *AIChE J.*, 16, 198–203.

269. Mayne, D.Q. 1973. The design of linear multivariable systems. *Automatica*, 9(2), 201–207.

270. Rosenbrock, H.H. 1973. The zeros of a system. *Int. J. Control*, 18(2), 297–299.

271. Barman, J.F. and Katzenelson, J. 1974. A generalized Nyquist-type stability criterion for multivariable feedback systems. *Int. J. Control*, 20(4), 593–622.

272. Davison, E. and Wang, S.H. 1974. Properties and calculation of transmission zeros of linear multivariable systems. *Automatica*, 10(6), 643–658.

273. Desoer, C.A. and Schulman, J.D. 1974. Zeros and poles of matrix transfer functions and their dynamical interpretation. *IEEE Trans. Circuits Syst.*, CAS-21(1), 3–8.

274. Rosenbrock, H.H. 1974. Corrections to "the zeros of a system". *Int. J. Control*, 20(3), 525–527.

275. MacFarlane, A.G.J. and Karcanias, N. 1976. Poles and zeros of linear multivariable systems: A survey of the algebraic, geometric and complex-variable theory. *Int. J. Control*, 24(1), 33–74.

276. Shaked, U., Horowitz, I., and Golde, S. 1976. Synthesis of multivariable, basically non-interacting systems with significant plant uncertainty. *Automatica*, 12(1), 61–71.

277. MacFarlane, A.G.J. and Postlethwaite, I. 1977. The generalized Nyquist stability criterion and multivariable root loci. *Int. J. Control*, 25(1), 81–127.

278. Postlethwaite, I. 1977. A generalized inverse Nyquist stability criterion. *Int. J. Control*, 26(3), 325–40.

279. Witcher, M.F. and McAvoy, T.J. 1977. Interacting control systems: Steady state and dynamic measurement of interaction. *ISA Trans.*, 16(3), 35–41.

280. Doyle, J.C. 1978. Robustness of multi-loop linear feedback systems. *IEEE Conference on Decision and Control, 17th Symposium. Adaptive Processes*, Fort Lauderdale, FL.

281. MacFarlane, A.G.J. and Karcanias, N. 1978. Relationships between state space and frequency response concepts. *Proceedings of the 7th IFAC Congress*, Lisbon, Portugal, pp. 1771–1779.

282. MacFarlane, A.G.J. and Scott-Jones, D.F.A. 1979. Vector gain. *Int. J. Control*, 29(1), 65–91.

283. Mayne, D.Q. 1979. Sequential design of linear multivariable systems. *Proc. IEE*, 126(6), 568–572.

284. Desoer, C. and Wang, Y.-T. 1980. On the generalized Nyquist stability criterion. *IEEE Trans. Autom. Control*, 25(2), 187–196.

285. Wall, J.E., Doyle, J.C., and Harvey, C.A. 1980. Tradeoffs in the design of multivariable feedback systems. *18th Allerton Conference on Communication, Control and Computing*, Monticello, IL, pp. 715–725.

286. Weischedel, K. and McAvoy, T.J. 1980. Feasibility of decoupling in conventionally controlled distillation columns. *Ind. Eng. Chem. Fund.*, 19, 379–384.

287. MacFarlane, A.G.J. and Hung, Y.S. 1981. Use of parameter groups in the analysis and design of multivariable feedback systems. *20th IEEE Conference on Decision and Control, Symposium on Adaptive Processes*, San Diego, CA, doi: 10.1109/CDC.1981.269507.

288. Mees, A.I. 1981. Achieving diagonal dominance. *Syst. Control Lett.*, 1(3), 155–158.

289. Postlethwaite, I., Edmunds, J., and MacFarlane, A. 1981. Principal gains and principal phases in the analysis of linear multivariable feedback systems. *IEEE Trans. Autom. Control*, 26(1), 32–46.

290. Doyle, J. 1982. Analysis of feedback systems with structured uncertainties. *IEE Proc. D (Control Theory Appl.)*, 129(6), 242–50.

291. Doyle, J.C., Wall, J.E., and Stein, G. 1982. Performance and robustness analysis for structured uncertainty. *21st IEEE Conference on Decision and Control*, Orlando, FL, pp. 629–636.

292. Arkun, Y., Manousiouthakis, B., and Palazoglu, A. 1984. Robustness analysis of process control systems. A case study of decoupling control in distillation. *Ind. Eng. Chem. Process Design Develop.*, 23, 93–101.

293. Bryant, G.F. and Yeung, L.F. 1994. New sequential design procedures for multivariable systems based on Gauss-Jordan factorization. *IEE Control Theory Appl.*, 141(6), 427–436.

294. Grosdidier, P., Morari, M., and Holt, B.R. 1985. Closed-loop properties from steady-state gain information. *Ind. Eng. Chem. Fund.*, 24(2), 221–235.

295. Marino-Galarraga, M., Marlin, T.E., and McAvoy, T.J. 1985. Using the relative disturbance gain to analyse process operability. *1985 American Control Conference (Catalogue No. 85CH2119-6)*, Boston, MA, pp. 1078–1083.

296. Stanley, G., Marino-Galarraga, M., and McAvoy, T.J. 1985. Shortcut operability analysis. I. The relative disturbance gain. *Ind. Eng. Chem. Proc. Des. Dev.*, 24(4), 1181–1188.

297. Grosdidier, P. and Morari, M. 1986. Interaction measures for systems under decentralized control. *Automatica*, 22(3), 309–319.

298. Manousiouthakis, V., Savage, R., and Arkun, Y. 1986. Synthesis of decentralized process control structures using the concept of block relative gain. *AIChE J.*, 32(6), 991–1003.

299. Mijares, G., Cole, J.D., Naugle, N.W., Preisig, H.A., and Holland, C.D. 1986. New criterion for the pairing of control and manipulated variables. *AIChE J.*, 32(9), 1439–1449.

300. Slaby, J. and Rinard, I.H. 1986. Complete interpretation of the dynamic relative gain array. *American Institute of Chemical Engineers, Annual Meeting*, Miami, FL.

301. Grosdidier, P. and Morari, M. 1987. A computer aided methodology for the design of decentralized controllers. *Comp. Amp; Chem. Eng.*, 11(4), 423–433.

302. Nett, C.N. and Spang, H.A. 1987. Control structure design: A missing link in the evolution of modem control theories. *American Control Conference*, Minneapolis, MN.

303. Skogestad, S. and Morari, M. 1987. Implications of large RGA elements on control performance. *Ind. Eng. Chem. Res.*, 26(11), 2323–2330.

304. Skogestad, S. and Morari, M. 1987. Effect of disturbance directions on closed-loop performance. *Ind. Eng. Chem. Res.*, 26(10), 2029–2035.

305. Chiu, M.S. and Arkun, Y. 1990. Decentralized control structure selection based on integrity considerations. *Ind. Eng. Chem. Res.*, 29(3), 369–373.

306. Yu, C.-C. and Fan, M.K.H. 1990. Decentralized integral controllability and D-stability. *Chem. Eng. Sci.*, 45(11), 3299–3309.

307. Chang, J.-W. and Yu, C.-C. 1992. Relative disturbance gain array. *AIChE J.*, 38(4), 521–534.

308. Hovd, M. and Skogestad, S. 1992. Simple frequency-dependent tools for control system analysis, structure selection and design. *Automatica*, 28(5), 989–996.

309. Campo, P.J. and Morari, M. 1994. Achievable closed-loop properties of systems under decentralized control: Conditions involving the steady-state gain. *IEEE Trans. Autom. Control*, 39(5), 932–943.

310. Franchek, M.A., Herman, P.A., and Nwokah, O.D.I. 1995. Robust multivariable control of distillation columns using a non-diagonal controller matrix. *1995 ASME International Mechanical Engineering Congress and Exposition. Part 1 (of 2)*, November 12–17, 1995, San Francisco, CA.

311. Van de Wal, M. and de Jager, B. 1995. Control structure design: A survey. *American Control Conference*, Seattle, WA, pp. 225–229.

312. Skogestad, S. and Havre, K. 1996. The use of RGA and condition number as robustness measures. *Comput. Chem. Eng.*, 20, S1005–S1010.
313. Franchek, M.A., Herman, P., and Nwokah, O.D.I. 1997. Robust non-diagonal controller design for uncertain multivariable regulating systems. *Transactions of the ASME. J. ASME J. Dyn. Syst. Meas. Control*, 119(1), 80–85.
314. Wade, H.L. 1997. Inverted decoupling: A neglected technique. *ISA Trans.*, 36, 3–10.

Papers Related to Fragility

315. Keel, L.H. and Bhattacharyya, S.P. 1997. Robust, fragile, or optimal? *IEEE Trans. Automat. Control*, 42, 1098–1105.
316. Dorato, P. 1998. Non-fragile controller design, an overview. *Proceedings of the American Control Conference*, Philadelphia, PA, pp. 2829–2931.
317. Famularo, P.D.D., Abdallah, C.T., Jadbabaie, A., and Haddad, W. 1998. Robust non-fragile LQ controllers: The static state feedback case. *Proceedings of the American Control Conference*, Philadelphia, PA, pp. 1109–1113.
318. Haddad, W.M. and Corrado, J.R. 1998. Robust resilient dynamic controllers for systems with parametric uncertainty and controller gain variations. *Proceedings of the American Control Conference*, Philadelphia, PA, pp. 2837–2841.
319. Jadbabaie, A., Chaouki, T., Famularo, D., and Dorato, P. 1998. Robust, non-fragile and optimal controller design via linear matrix inequalities. *Proceedings of the American Control Conference*, Philadelphia, PA, pp. 2842–2846.
320. Keel, L.H. and Bhattacharyya, S.P. 1998. Authors' reply. *IEEE Trans. Automat. Contr.*, 43, 1268.
321. Keel, L.H. and Bhattacharyya, S.P. 1998. Stability margins and digital implementation of controllers. *Proceedings of the American Control Conference*, Philadelphia, PA, pp. 2852–2856.
322. Mäkilä, P.M. 1998. Comments on "Robust, Fragile, or Optimal?". *IEEE Trans. Automat. Control*, 43, 1265–1267.
323. Corrado, J.R. and Haddad, W.M. 1999. Static output feedback controllers for Systems with parametric uncertainty and controller gain variation. *Proceedings of the 1999 American Control Conference*, San Diego, CA, pp. 915–919.
324. Yang, G.H., Wang, J.L., and Lin, C. 1999. H∞ control for linear systems with controller uncertainty. *Proceedings of the 1999 American Control Conference*, San Diego, CA, pp. 3377–3381.
325. Paattilammi, J. and Mäkilä, P.M. 2000. Fragility and robustness: A case study on paper machine headbox control. *IEEE Control Syst. Mag.*, 20, 13–22.

Papers Related to Hybrid and Switching Control Systems

326. Narendra, K. and Goldwyn, R. 1964. A geometrical criterion for the stability of certain nonlinear nonautonomous systems. *IEEE Trans. Circuits Syst.*, 11, 406–408.
327. Willems, J. 1973. The circle criterion and quadratic Lyapunov functions for stability analysis. *IEEE Trans. Autom. Control*, 18, 184.
328. Molchanov, A.P. and Pyatnitskii, E.S. 1989. Criteria of asymptotic stability of differential and difference inclusions encountered in control theory. *Syst. Control Lett.*, 13, 59–64.
329. Feuer, A., Goodwin, G.V., and Salgado, M. 1997. Potential benefits of hybrid control for linear time invariant plants. *Proceedings of the American Control Conference*, Albuquerque, NM.

330. Seron, M.M., Braslavsky, J.H., and Goodwin, G.C. 1997. *Fundamental Limitations in Filtering and Control*. Springer, London.
331. Dayawansa, W.P. and Martin, C.F. 1999. A converse Lyapunov theorem for a class of dynamical systems which undergo switching. *IEEE Trans. Autom. Control*, 44, 751–760.
332. Liberzon, D. and Morse, A.S. 1999. Basic problems in stability and design of switched systems. *IEEE Control Syst. Mag.*, 19, 59–70.
333. Decarlo, R.A., Branicky, M.S., Pettersson, S., and Lennartson, B. 2000. Perspectives and results on the stability and stabilizability of hybrid systems. *Proc. IEEE*, 88, 1069–1082.
334. Mcclamroch, N.H. and Kolmanovsky, I. 2000. Performance benefits of hybrid control design for linear and nonlinear systems. *Proc. IEEE*, 88, 1083–1096.
335. Shorten, R.N., Mason, O., O'Cairbre, F., and Curran, P. 2004. A unifying framework for the SISO circle criterion and other quadratic stability criteria. *Int. J. Control*, 77, 1–8.
336. Shorten, R., Wirth, F., Mason, O., Wulff, K., and King, C. 2007. Stability criteria for switched and hybrid systems. *SIAM Rev.*, 49, 545–592.

Miscellaneous. Control

337. Tustin, A. 1947. A method of analyzing the behavior of linear systems in terms of time series. *JIEE, London*, 94(Pt IIA), 152–160.
338. Smith, O.J.M. 1957. Closer control of loops with dead time. *Chem. Eng. Progr.*, 53, 217–219.
339. Lepschy, A. and Ruberty, A. 1960. A rule for direct verification of the Nyquist criterion in non-polar diagrams. *1st IFAC World Congress, Part I, vol. I*, Moscow, URSS, pp. 13–17.
340. Horowitz, I.M. and Shaked, U. 1975. Superiority of transfer function over state-variable methods in linear time-invariant feedback system design. *IEEE Trans. Automat. Control*, AC-20, 84–97.
341. Powell, J.D., Parsons, E., and Tashka, G. 1976. A comparison of flight control design methods. *Guidance and Control Conference*, San Diego, CA.
342. Ioannides, A.C., Rogers, G.J., and Latham, V. 1979. Stability limits of a Smith controller in simple systems containing a dead-time. *Int. J. Control*, 29, 557–563.
343. Palmor, Z. 1980. Stability properties of Smith dead-time compensator controllers. *Int. J. Control*, 32, 937–949.
344. Kannai, Y. 1982. Causality and stability of linear systems described by partial differential operators. *Siam J. Control Opt.*, 10(5), 669–674.
345. Horowitz, I. 1983. Some properties of delayed controls (Smith regulator). *Int. J. Control*, 38, 977–990.
346. Anderson, B.D.O. and Parks, P.C. 1985. Lumped approximation of distributed systems and controllability questions. *IEE Proc.*, 132(3), 89–94.
347. Yamanaka, K. and Shimemura, E. 1987. Effects of mismatched Smith controller on stability in systems with time-delay. *Automatica*, 23, 787–791.
348. Vidyasagar, M., Bertschmann, R.K., and Sallaberger, C.S. 1988. Some simplifications of the graphical Nyquist criterion. *IEEE Trans. Automat. Control*, 33(3), 301–305.
349. Collins, Jr., E.G., King, J.A., and Bernstein, D.S. 1991. Robust control design for a benchmark problem using the maximum entropy approach. *Proceedings of the American Control Conference*, Boston, MA.
350. Wie, B. and Bernstein, D. 1992. Benchmark problems for robust control design. *J. Guid. Control Dyn.*, 15, 1057–1059.
351. Astrom, K.J., Hang, C.C., and Lim, B.C. 1994. A new Smith predictor for controlling a process with an integrator and long dead-time. *IEEE Trans. Automat. Control*, 39, 343–345.

352. Matausek, M.R. and Micic, A.D. 1996. A modified Smith predictor for controlling a process with an integrator and long dead-time. *IEEE Trans. Automat. Contr.*, 41, 1199–1203.
353. Normey-Rico, J.E. and Camacho, E.F. 1999. Smith predictor and modifications: A comparative study. *Proceedings of the European Control Conference on ECC'99*, Karlsruhe, Germany.
354. Glad, T. and Ljung, L. 2000. *Control Theory. Multivariable and Nonlinear Methods.* Taylor & Francis Group, London.
355. Sein, G 2003. Respect the unstable. *IEEE Control Syst. Mag.*, 23(4), 12–25.
356. Belanger, P.R. 2005. *Control Engineering: A Modern Approach.* Oxford University Press, Orlando, FL.
357. McCullough, D. 2015. *The Wright Brothers.* Simon & Schuster, New York, NY.

Miscellaneous

358. Farlow, S.J. 1937. *Partial Differential Equations for Scientists and Engineers.* Dover Publications. Reprint edition 1993, New York, NY.
359. Blakelock, J.H. 1965. *Automatic Control of Aircraft and Missiles.* Wiley, New York, NY.
360. Isermann, R. 2005. *Mechatronic systems.* Springer, London, UK.
361. Palm, III, W.J. 1999. *Modeling, Analysis and Control of Dynamic Systems.* 2nd Edition, Wiley, New York, NY.
362. Egeland, O. and Gravdahl, T. 2002. *Modeling and Simulation for Automatic Control.* Marine Cybernetics, Norway.
363. Sidi, M. 1997. *Spacecraft Dynamics and Control. A Practical Engineering Approach.* Cambridge University Press, Cambridge, UK.
364. Jonkman, J., Butterfield, S., Musial, W., and Scott, G. 2009. *Definition of a 5-MW Reference Wind Turbine for Offshore System Development.* Technical Report NREL/TP-500-38060, Colorado, CO.
365. Ranka, T., Garcia-Sanz, M., Symmes, A., Ford, J., and Weadon, T. 2016. Dynamic analysis of the Green Bank Telescope structure and servo system. *J. Astron. Telesc. Instrum. Syst.*, 2(1), 014001.1–014001.11, doi: 10.1117/1.JATIS.2.1.014001.
366. Verschuur, G.L. 2007. *The Invisible Universe: The Story of Radio Astronomy.* 2nd Edition, Springer, New York, NY.
367. Marlin, T.E. 2000. *Process Control. Designing Processes and Control Systems for Dynamic Performance.* 2nd Edition, McGraw-Hill, Singapore.

Index